WOOD AND CELLULOSICS:
Industrial Utilisation, Biotechnology, Structure and Properties

WOOD AND CELLULOSICS:
Industrial Utilisation, Biotechnology, Structure and Properties

Editors:
J. F. KENNEDY
B.Sc., Ph.D., D.Sc., C.Biol., C.Chem, F.B.I.M., F.I.Biol., F.I.F.S.T., F.R.S.C.
Professor of Applied Chemistry
North East Wales Institute of Higher Education, and
Department of Chemistry, University of Birmingham

G. O. PHILLIPS, B.Sc., Ph.D., D.Sc., F.R.S.C.
First Principal, North East Wales
Institute of Higher Education

P. A. WILLIAMS, B.Sc., Ph.D., C.Chem., M.R.S.C.
Research Scientist, North East Wales
Institute of Higher Education

ELLIS HORWOOD LIMITED
Publishers · Chichester

Halsted Press: a division of
JOHN WILEY & SONS
New York · Chichester · Brisbane · Toronto

First published in 1987 by
ELLIS HORWOOD LIMITED
Market Cross House, Cooper Street,
Chichester, West Sussex, PO19 1EB, England
The publisher's colophon is reproduced from James Gillison's drawing of the ancient Market Cross, Chichester.

Distributors:
Australia and New Zealand:
JACARANDA WILEY LIMITED
GPO Box 859, Brisbane, Queensland 4001, Australia
Canada:
JOHN WILEY & SONS CANADA LIMITED
22 Worcester Road, Rexdale, Ontario, Canada
Europe and Africa:
JOHN WILEY & SONS LIMITED
Baffins Lane, Chichester, West Sussex, England
North and South America and the rest of the world:
Halsted Press: a division of
JOHN WILEY & SONS
605 Third Avenue, New York, NY 10158, USA

British Library Cataloguing in Publication Data

Wood and cellulosics: industrial utilisation, biotechnology, structure and properties. —
(Biological chemistry and biotechnology).
1. Cellulose
I. Kennedy, John F. (John Frederick)
II. Philips, Glyn O. III. Williams, Peter A.
IV. Series
661'.802 TP1180.C6

Library of Congress CIP data available

ISBN 0–7458–0113–7 (Ellis Horwood Limited)
ISBN 0–470–20884–8 (Halsted Press)

Printed in Great Britain by Unwin Bros. of Woking

Table of Contents

Acknowledgements

MEMBERS OF THE ORGANISING COMMITTEE

This meeting owed its success to the invaluable assistance of the Organising Committee

Professor G O Phillips (Chairman)	The North East Wales Institute
Professor J F Kennedy (Vice Chairman and Treasurer)	The North East Wales Institute and University of Birmingham
Dr M B Huglin (Hon. Secretary)	University of Salford
Dr P A Williams (Secretariat)	The North East Wales Institute
Mr H Hughes (Secretariat)	The North East Wales Institute
Mr J C Arthur Jr.	Cellulose Consultant
Mr P Bale	Hercules Ltd
Dr W B Banks	University College of North Wales
Dr C Bucke	Polytechnic of Central London
Mr K Geddes	Crown Decorative Products Ltd
Mr A Grayson	Forestry Commission
Mr T Greenway	Berol Kemi (UK) Ltd
Dr T J Lewis	RARDE, Ministry of Defence
Professor Y Nakamura	University of Gunma, Japan
Mr W B Painting	Hoechst UK Ltd
Mr A Reveley	Courtaulds Chemicals and Plastics
Mr A M Sellar	Hercules Ltd
Dr D J Wedlock	Shell Research Ltd
Dr C A White	Biotechnics

The Organising Committee would like to express their appreciation to those Companies and Institutions who gave their financial assistance and support to the Conference.

Sponsored by

The Biochemical Society – Biotechnics – Hoechst UK Ltd – The North East Wales Institute – USAF European Office of Aerospace Research and Development – US Army Research, Development and Standardisation Group, UK – Welsh Development Agency.

Supported by

The American Chemical Society (Cellulose, Paper and Textile Division) – Berol Kemi (UK) Ltd – Courtaulds Chemicals and Plastics – Ministry of Defence – Crown Decorative Products – Hercules Ltd – The Forestry Commission.

Introduction

GLYN O. PHILLIPS
Chairman, Organising Committee

Cellucon-86 was an even greater success than Cellucon-84 as this book testifies. This second volume on cellulose chemistry and applications places greater emphasis on the value of cellulose as a renewable resource and its potential in the biotechnology area.

The contributions were truly international: **AUSTRALIA, CANADA, CZECHOSLOVAKIA, ENGLAND, EGYPT, FINLAND, FRANCE, IRELAND, JAPAN, KENYA, KUWAIT, MALAYSIA, NETHERLANDS, POLAND, ROMANIA, SWEDEN, USA, USSR, WEST GERMANY.....** and of course, **WALES**, the host country. Within one volume, therefore, it is now possible to obtain a comprehensive evaluation of current developments, in a field which is becoming truly interdisciplinary.

There was an extremely strong representation from Japan, and this volume is testimony to the faith placed by that country in the industrial exploitation of research findings. For example, wood as a rich source of chemical materials has all but been neglected except in Japan.

As we were hoping in initiating the Cellucon series of international meetings, new and better communications have been forged between cellulose chemists. New research has been stimulated, and despite parallel sessions, very many papers could not be accepted. It augers well for the future in the partnership between basic research and industrial scientists in this field. This book will be invaluable for all working in this subject, and the educationalists who strive to keep up with an expanding, yet scattered literature.

Again, my thanks to the sponsors. Additionally to the members represented on the Organising Committee, the Welsh Development Agency, the US Air Force European Office of Aerospace Research Development and the US Army Research Development and Standardisation Group-UK, have provided invaluable assistance.

My debt is also substantial to the members of the Organising Committee, who have regularly met to plan all aspects of the meeting. Professor John Kennedy (Deputy Chairman and Treasurer) has supported me in every way, particularly keeping a tight grip on all matters when I could not attend. The secretariat, Dr.Peter Williams and Mr.Haydn Hughes, again have borne the brunt of the arrangements.

GLYN O. PHILLIPS
Chairman

OVERVIEW

The Forestry Commission-Wood Production and Processsing

G. J. Francis – The Forestry Commission, Alice Holt Lodge, Wrecclesham, Farnham, Surrey GU10 4LH.

INTRODUCTION

It gives me great pleasure to attend the opening session of this the Second International Cellucon Conference and to have the privilege of addressing such a distinguished group of scientists. You have a full and interesting programme before you, with an impressive array of scientific papers and eminent speakers, which has attracted delegates to attend from many parts of the world. During the course of the Conference you will have the opportunity to visit Cynwyd Forest, one of the oldest forests managed by the Forestry Commission, and also the newsprint mill owned by the Shotton Paper Company which is the newest of our forest industries and the single largest consumer of our domestic wood supply. You will therefore see at first hand some of the results of a forest policy which dates back to the passing of the first Forestry Act in 1919. This Act established the Forestry Commission and initiated a programme of afforestation, public and private, which has continued ever since. Clearly wood is of interest to the cellulose industry and my purpose in this opening address is to describe some of the main features of forestry and the wood supply in this country because it may have some long term relevance to the British cellulose industry.

THE EXISTING FOREST AREA

In the past 60 years the area of productive forest in Britain has increased from 4% of the total land area to 10% (which is some 2 million hectares) but despite this increase the area of land devoted to forestry is among the lowest in Europe. Only

Ireland at 5% and the Netherlands at 9% have a smaller proportion. France has 27% and West Germany 30% of the land area occupied by forests. The increase in the forest area has occurred in the main as a result of the planting of coniferous trees on bare land chiefly in the uplands of North and West Britain. In general it has occurred on poor land where agriculture has proved to be marginal. Few opportunities have occurred for forestry on better land at lower elevations. Despite this the productivity of these forests is relatively high by comparison with the other major forest areas of Northern Europe and North America and although the proportion of land under forest in Britain is small the volume of timber from these forests is increasing very rapidly. Not only is it increasing rapidly but it is doing so in a consistent and predictable way because of the reliable area, growth and yield information which is available and because of the intensive management practised in the majority of woodlands.

Currently the total production of (softwoods) from British forests is about 4 million cubic metres annually. By the end of this century we expect this to reach a total of almost 9 million cubic metres annually. On the demand side the consumption of wood and wood products in Britain represents the equivalent of 38 million cubic metres of which domestic supply including hardwoods and softwoods represents 4 million cubic metres or some 10%. Among the major sectors of the wood market the degree of self sufficiency varies considerably, for example, it is about 30% ir wood based panels and 18% in sawn softwood to the complete absence of a domestic supply of chemical pulp products.

THE RAW MATERIAL

The expansion of the forest area which has taken place in this country over the last 65 years or so has involved the use of a number of coniferous species all of which apart from Scots pine, are non-indigenous. Among the most important of these are Sitka and Norway spruce. The result is that of the current wood production some 45% is composed of spruce, some 30% of pine while the larches and Douglas fir account for the remainder. The size of wood, in terms of diameter, determines to a large extent the nature of the industrial use to which it is put. The larger diameter wood is normally marketed for sawmilling which accounts for about 55% of the total output. The smaller diameter wood is used by a variety of industries but principally the wood based panel industry and the pulp and paper industry. As wood production increases the proportions of larger and smaller diameter wood seem unlikely to change significantly although the proportion of spruce is expected to increase from the present 45% to 55% while the proportion of larch and Douglas fir declines marginally. The general position therefore is one where the composition of the wood output will not change significantly, apart from the increase in the importance of spruce, in terms of the species or the sizes which are available.

Finally, on the subject of the raw material, reference should be made to the characteristic properties of the main species groups in terms of their suitability for the principal markets for wood and wood products. The spruces, alone among the commonly planted softwoods, have a uniformly pale wood. They are the lightest in weight of the British-grown softwoods and are marginally less dense than Scandinavian whitewood, though this reflects more the differences arising from the younger age and greater vigour of these species grown under British conditions. In general, when machine graded, spruce sawnwood is capable of meeting the strength characteristics required for structural timber. Because of their whiteness and fibre length spruces provide an excellent raw material for papers based on mechanical pulp. It is a non-durable wood and with present technology it is difficult to treat with preservatives and even the sapwood is resistant to impregnation.

Pine is probably the most versatile British-grown softwood. As sawnwood, pine is well suited for many structural purposes and, in the better qualities, for joinery. Although acceptable for pulping it is not used on any significant scale for mechanical pulping in Britain where its main use has been in a variety of wood based panels. Sawnwood of both larch and Douglas Fir is well suited for structural use and, although not suitable for mechanical pulping, they have been increasingly used in wood based panel production.

THE DEVELOPMENT OF THE WOOD-USING INDUSTRY IN BRITAIN

National policies on industrial development based on Government financial assistance are, in general, aimed at improving employment opportunities in defined regions. In the main such policies are not specific to any one sector of industry. The wood-using industries, in common with all other sectors of industry, are eligible for grant aid toward the capital cost of new investment, in specified localities, in the form of Regional Development Grant and Selective Financial Assistance. The availability of this assistance can have an important effect on the growth and development of the forest products industry by attracting new investment which has a choice of location internationally. However, there are limitations on the usefulness of these measures in the forest products sector. These stem from the fact that the main objective of such policies is to promote employment and most forest products industries are capital intensive relative to employment creation. Thus although the Regional Development Grant is a specified proportion of fixed capital investment there is a defined limit placed on the sum payable per job, similar constraints are placed on Selective Financial Assistance. The effect is that a new sawmill investment, with a relatively favourable ratio of capital cost to employment is likely to secure a higher proportion of assistance than a wood based panel or pulp and paper project where the ratio is much higher.

The usefulness of these forms of assistance is also limited because the regions selected are those where unemployment has reached high levels and thus typically they consist of the heavily industrialised and more densely populated areas remote from the forests. The significance of this lies in the conflict between the attractions of assistance in areas distant from forests and the cost penalties incurred by transporting large volumes of timber to locations eligible for financial assistance.

Publicly owned forests are frequently a major component of the forest area in many countries. Britain is no exception with about 60% of the conifer forest area in public ownership. This large scale ownership has created opportunities to promote and encourage the development of the wood using industry. This has been achieved by the marketing policies adopted by the Forestry Commission which include the opportunity for long-term contracts of up to 10 years for the supply of wood where major new investments are being considered. These contracts have proved to be an important inducement in attracting investment in large-scale capital intensive industries. In general, they have not been necessary for the more modest investments characteristic of the sawmill industry. However, in the pulp and paper sector, where investment is on a very large scale, they have proved decisive in attracting internationally mobile investment. Careful consideration is always given to the proportion of supplies provided under contracts of this kind so as to ensure a flexibility of market opportunities for the industry concerned.

The wood processing industry in Britain has an impressive record of expansion. The sawmill industry has expanded steadily since the mid 1970's both by the expansion of existing mills and by a limited number of investments in greenfield projects with the trend being toward larger scale mills. The opportunities for new investment afforded by an increasing supply of raw material have been fully exploited. Such has been the expansion, often in areas which do not attract Government assistance, that in some areas sawmilling capacity significantly exceeds the supply of sawlogs.

The years from 1983 onwards have seen major new investment in wood based panels and in pulp and paper. In total the investment has amounted to almost £300 million at plants located throughout the country, (Inverness, Stirling, Hexham, Workington, Shotton and Chirk). The opportunities for further investment in the near future are now limited to Scotland where a new pulp and paper project is now under active consideration. It is quite a remarkable success story.

COMPETING USES FOR WOOD FIBRE

Imports dominate the British market for wood and wood products and they are likely to continue to do so the the foreseeable future. The future world supply and demand for wood has implications for British wood production in relation to wood quality and price. The main factors likely to have an influence on our competitors appear to be

- a gradual decline in the quality and size of logs available as they are forced to switch their supply from natural forests to those based on second growth or more intensively managed forests

- a gradual increase in the cost of wood in real terms as wood has to be harvested in more remote areas.

While there may be a prospect for imported wood product prices to rise in real terms as the balance between supply and demand becomes tighter this effect is unlikely to be significant until after the turn of the century. Thus the competitiveness of the British forest products industry is more likely to be influenced by economic factors other than the cost of wood to industries exporting wood products to this country.

There are perhaps four principal economic factors which are particularly important and they are interrelated. The consumption of most wood products is generally regarded as being correlated with economic growth rate. With an expanding world economy the demand for wood products is likely to rise and industries may have the opportunity of achieving higher prices. However, the overall trend in economic growth rates in the OECD for example, has been downwards.

High inflation and high interest rates can have a major influence on the competitiveness of industries but perhaps the most significant factor for the wood products industry is the exchange rate of sterling against the currencies of our principal competitors in Scandinavia and North America. If sterling should devalue relative to other currencies over the long term the competitiveness of wood products manufactured in Britain should in general increase.

LONGER TERM ISSUES

The characteristics of the wood available to the market to the end of the century and perhaps beyond are determined by the composition of the existing forest area. Nevertheless, forest managers and researchers need to give attention to what should be produced for the future. The key to this is the material needed and this cannot be predicted with certainty but there are some commonly accepted pointers. Currently around 60% of the sawn softwood used is for construction and there appear to be good reasons - low energy input in processing, ease of fabrication and even traditional attitudes on the part of designers and uses - to expect this demand to be maintained. It is therefore usually accepted that the forest manager and researcher should aim to satisfy the construction market because of its size and the scope it offers for the use of wood.

In terms of the technology of growing wood there is some evidence that the characteristics of wood for structural use can be influenced by the management techniques applied. Spacing at planting and thinning are the traditional techniques used to manipulate the quality of timber produced. For example, it appears that wide spacing at planting can reduce the strength of sawnwood in Sitka spruce due to a higher proportion of juvenile wood and larger branches. However, manipulation by spacing and thinning are very imprecise techniques for control of product quality.

Genetic improvement programmes in many countries have reached the point where breeders have a reasonable understanding of the way in which genetic factors influence the growth characteristics of trees. For the most part the main aim of tree breeders has been to improve productivity by selection based on vigour of growth and tree form based on stem straightness and branching. However, the point of concern is the negative relationship between growth rate and density and the implications for the quality of wood produced by fast grown trees.

In contrast relatively little is known about the genetic control of wood quality. The present understanding of the genetic control of wood quality is derived largely from wood on Loblolly pine (Pinus tae da) in the southern United States and Radiata pine (Pinus radiata) in Australia. A number of wood properties which affect the quality of sawnwood have been shown to be under genetic control including a range of tracheid characteristics. There are a number of difficulties in breeding for improved wood quality - one of the most important is the low correlation between the characteristics of juvenile and mature wood which means that the testing period is very extended.

Forest policy in this country is based on a continuing expansion both to reduce our dependence on imported wood in the long term and to provide continued employment in forestry and associated industries. It is widely thought that there should be scope for new planting to continue at broadly the rate of the past 25 years - some 20,000 to 30,000 hectares annually.

The primary objective of forest policy in this country is wood production to meet the needs of the market. The expansion of forestry which has taken place and the prospect of future expansion has probably been the cause of the most widespread change of the rural landscape this century. Change which has been brought about in two main ways - through new planting, through the felling and replacing of the forests created in the first half of this century. In forestry as in any other form of productive land management a balance has to be struck between the requirements of efficient (wood) production and of environmental quality. The first tends towards large scale, simple and uniform systems of work, the latter towards small scale complex and diverse systems of work. It is a familiar dilemma. In recent years we have seen a popular upsurge of interest in nature conservation and in the environment generally. We are fortunate in having a rich and varied countryside in which to live and work. In forestry we have a responsibility to maintain the quality of the countryside and to add to the pleasure of those who visit it, notably in terms of landscape, nature conservation and access for public enjoyment. The standards of environmental management have advanced greatly over the last decade or so. Although forestry is often still judged by the less sensitive standards of the past which where tree planting is involved can persist over 50 years or more, foresters are committed to achieving a reasonable balance between environmental needs and those of wood production.

Long standing priorities for food production have confined forestry to the land that was of least value to agriculture. Today we are perhaps at the onset of a period of change confronting agriculture. The European Community produces surpluses of most temperate foodstuffs at considerable cost to the taxpayer and the UK taxpayer in particular. The desire to bring Community production into a better balance will continue to occupy the minds of policy makers in agriculture and wood production is a crop which could provide an alternative land use and an opportunity for forestry on better quality land with improved productivity.

Although forestry is a long term activity with a production cycle which can extend over 100 years, forest policy can never be static. The relative value of the various benefits change with time. For example, as in Britain, a forest planted to produce timber for wooden ships can become a prized place of recreation - ideas change faster than trees grow. Today the environmental importance of forestry is increasing rapidly but we should not be blind to the fact that the major function still is wood production. The use to which that wood is put may not be what we foresee as we plant trees today and perhaps your deliberations at this Conference will influence how they will be used.

PART 1

STRUCTURE AND PROPERTIES

Chapter 1

Modern analytical methods - application to cellulose and its derivatives

Dr. P. J. Stephenson – Technical Specialist, 3M U.K. PLC, European Technical Centre, Gorseinon, Swansea, SA4 1GD.

INTRODUCTION

Chemical analytical methods applied to the characterisation of polymers have traditionally been destructive in nature with their roots in pyrolysis and related techniques. Fundamental advances, particularly over the last decade, have armed the the analyst with an array of non-destructive methods for the analysis of solids allowing a much more detailed elucidation of structure.

Inspection of the literature reveals that scientists will often select cellulose and its derivatives to demonstrate the potential of a new technique. This is undoubtedly due to the accepted difficulty in examining a material, (i.e. cellulose itself), which is largely insoluble in common solvents and whose morphology and crystalline structure is dependent on source and pretreatments. The derivatised forms also present a challenge due to the suspected dependence of degree of substitution on factors such as amorphous/crystalline ratio or degree of diffusion of reagents through the fibres. Also, surprisingly little was known, until recently, about the primary vs. secondary hydroxyl reactivity; (and the effect of substitution on the reactivity of adjacent cellulose rings), or indeed about the surface vs. bulk characteristics of the finished product.

Table 1 shows the introduction of a series of analytical instrumental methods which have been applied to the problems discussed above with appropriate references.

It is the intention of this paper to summarise the impact of such techniques and to present possible applications of a new generation of methods to cellulose and its derivatives.

Table 1.

^{13}C Solution Nuclear Magnetic Resonance (NMR) [1,2]

^{13}C CP/MAS Solid State NMR [5-9]

Electron Spectroscopy for Chemical Applications (ESCA) [10,12]

Photoacoustic Fourier Transform Infra-Red (PAS-FTIR) [13]

Secondary Ion Mass Spectrometry (SIMS) [15,4,14]

Fast Atom Bombardment Mass Spectrometry (FABMS)

INVESTIGATION OF BULK STRUCTURE

Within the confines of traditional analytical methodology it has been very difficult to follow the progress of solid state transformations in polymers by chemical methods. As examples consider the cure of solid phenolic resins or indeed mercerisation or nitration of cellulose.

The advent of CP/MAS ^{13}C NMR has allowed polymers to be examined directly in the solid state without resorting to dissolution. The principles of this technique have been reviewed elsewhere, notably by Fyfe, [16] and the applications to cellulose were presented in the proceedings of Cellucon 1984. [17]

Essentially it has been found that not only can the technique provide conventional nmr data of solid cellulose derivatives but in the case of cellulose itself can provide additional information relating to the crystalline structure of cellulose polymorphs. [8,9] CP/MAS ^{13}C n.m.r. spectra of cellulose I and II have been refined, (Fig. 1) to show multiplet splittings particularly obvious in the C1 and C4 resonances. These additional spectral features are due to carbons in the solid which are chemically equivalent but magnetically inequivalent due to short range interactions in the lattice. It is probable that such information is complimentary to X-ray diffraction data.

In the case of cellulose II the source of the splittings has been investigated by a thorough analysis of the cellulose oligomers[6] whereas in the case of cellulose I there is still dispute as to the nature of the multiple splittings in the C1 and C4 spectral regions. [7,9,18]

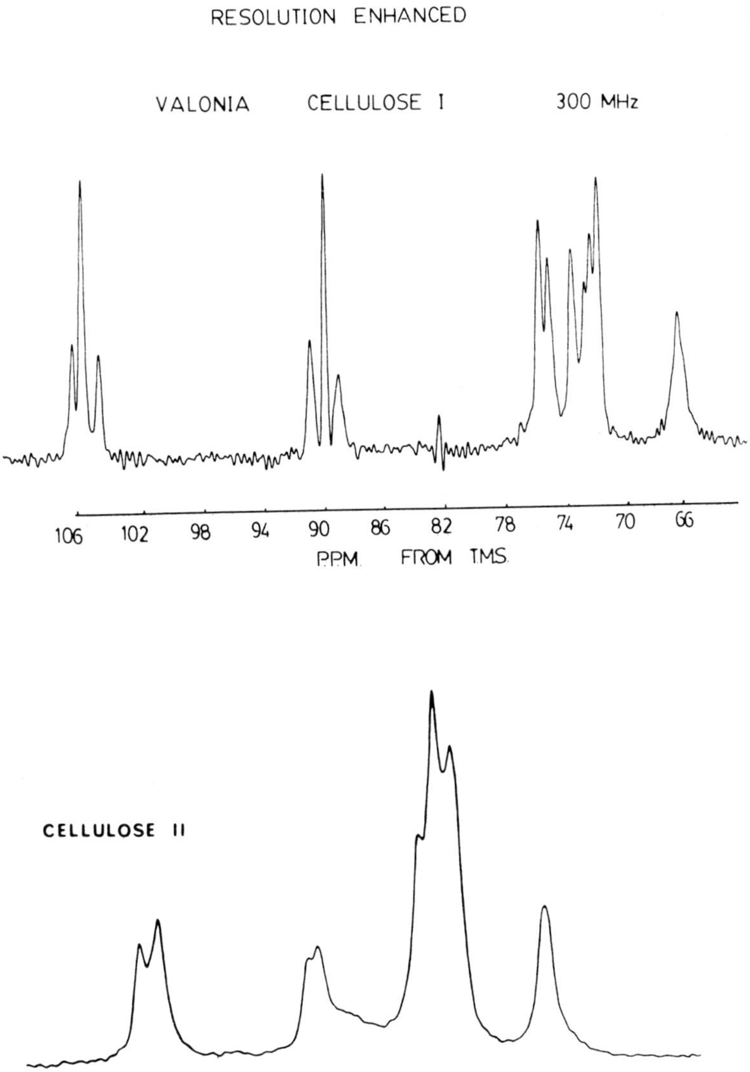
RESOLUTION ENHANCED
VALONIA
CELLULOSE I
300 MHz
106
102
98
94
90
86
82
78
74
70
66
PPM FROM TMS
CELLULOSE II

Fig. 1.

Vanderhart[9] has studied the crystalline/amorphous regions of celluloses of different sources and more recently Horii[19,30] and co-workers have examined the spin relaxation phenomena of crystalline and amorphous components in the CP/MAS ^{13}C nmr spectra of Cellulose I.

In the case of cellulose derivatives Koenig[20] has published a study of the nitration of cellulose in the solid state using CP/MAS methods. Mercerisation has been similarly studied by Fyfe et al[6].

The Fellgett and Jaquinot advantage in Fourier Transform Infra-Red spectroscopy[28] has allowed the use of a series of low throughput accessories for the analysis of polymers and polymer surfaces. Diffuse and specular reflectance have supplemented M.A.T.R. methods and more recently photoacoustic infra-red has been introduced. In this technique the sample without preparation is placed in a low volume sampling chamber in the presence of an inert gas[31]. On irradiation by infra-red, absorbtion occurs which is converted to heat. By thermal diffusion this is transferred to the gas. The gas expands and contracts at the modulation frequency causing a pressure change. This is detected by a sensitive microphone.

The advantage of PAS methods is that no sample preparation is required and by varying the velocity of the moving mirror in the Michelson interferometer the frequency is modulated causing variation in sampling depth. Some applications have been reported on cellulose materials[30]. Again this method is likely to become important in the cellulose industries.

INVESTIGATION OF SURFACE CHARACTERISTICS

Degradation and weathering studies on polymers where chemical changes can be expected in the outermost monolayers require a technique capable of delineating surface from bulk. Such methods are also a pre-requisite for studying the reactions of reagents with cellulose fibres where it may be expected, in the early stages of reaction, that the surface and bulk chemistry may be quite different. E.S.C.A. has been applied to the study of polymer surfaces for many years, notably by Clark[21] but essentially it is possible to obtain data e.g. D.O.S. from the first 50A° of cellulose fibres after treatments and by the use of dual anode or other methods to produce depth profile information. By such methods the X-ray[12,4] and U.V.[12,22] degradation of cellulose nitrates have been studied and a degradation route has been proposed[22,12].

E.S.C.A. as a surface technique is extremely powerful but suffers from a lack of chemical specificity which limits its application in certain cases. More recently, Secondary Ion Mass Spectrometry has been utilised to produce positive ion mass spectra of polymer surfaces[24-25]. This data is obtained by bombarding the surface with a beam of Argon ions producing clusters of organic fragments at a point distant from the point of impact. These can then be studied by a quadropole or time of flight mass analyser. Again this technique has limitations in the study of polymers largely due to the charging of the surface in electrical insulators. In addition the positive ion spectra were dominated in low mass regions by hydrocarbon fragments.

In 1984, working with U.M.I.S.T., we isolated conditions which allowed negative ion spectra of a series of fatty acids located on polymer surfaces to be recorded[26] (Fig. 2). These conditions have now been used to record such spectra for a range of industrial polymers[27].

Combined with the positive ion data a fuller picture of the surface composition can be obtained.

Figs. 3-4 show positive and negative ion SIMS spectra obtained (using an Argon atom beam), of cellulose nitrate (2.1 D.O.S.). Similar spectra have been recorded of methyl, hydroxy propyl derivatives, cellulose acetate and others. In the study of degradation products of cellulose derivatives it seems likely that these techniques will be of vital importance since many degradation products may be expected to be stabilised by a negative charge.

SUMMARY

The application of modern analytical methods in a multi-disciplinary way can resolve many industrial problems which were previously considered intractable.

In the case of cellulose and its derivatives it seems likely that the impact of such techniques will continue to grow over the next few years.

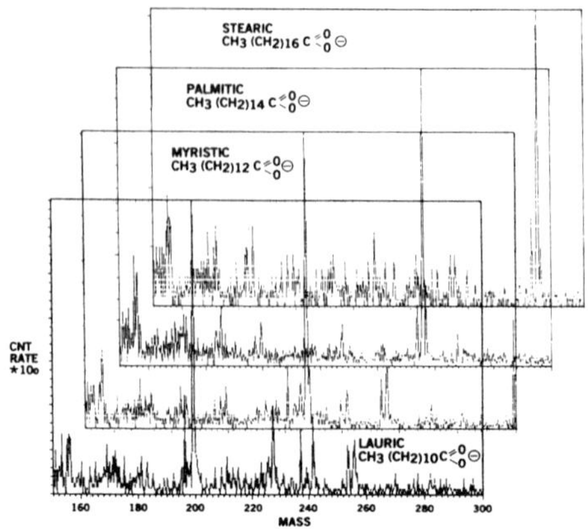
STEARIC
CH3 (CH2)16 C
PALMITIC
CH3 (CH2)14 C
MYRISTIC
CH3 (CH2)12 C
LAURIC
CH3 (CH2)10C
CNT
RATE
*10o
160 180 200 220 240 260 280 300
MASS

Fig. 2.

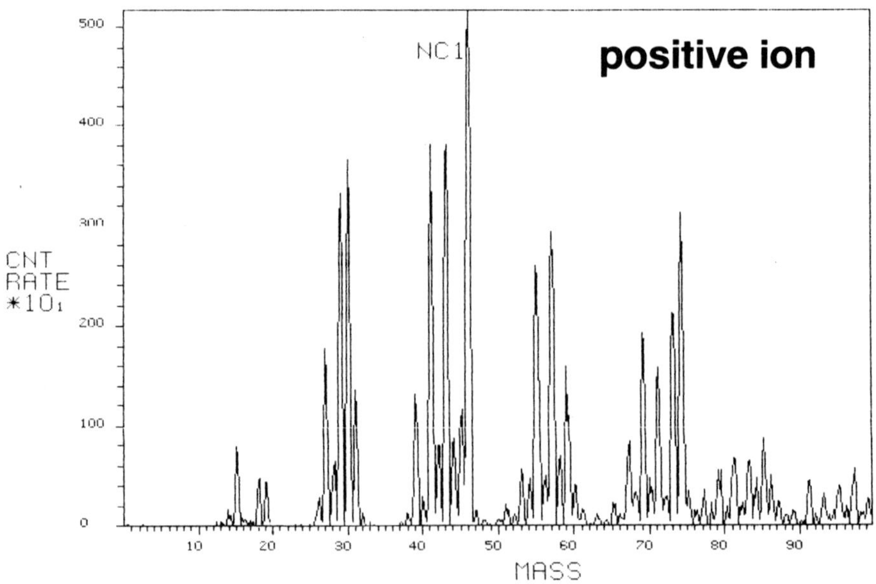
positive ion
NC1
CNT
RATE
*10₁
500
400
300
200
100
0
10 20 30 40 50 60 70 80 90
MASS

Fig. 3.

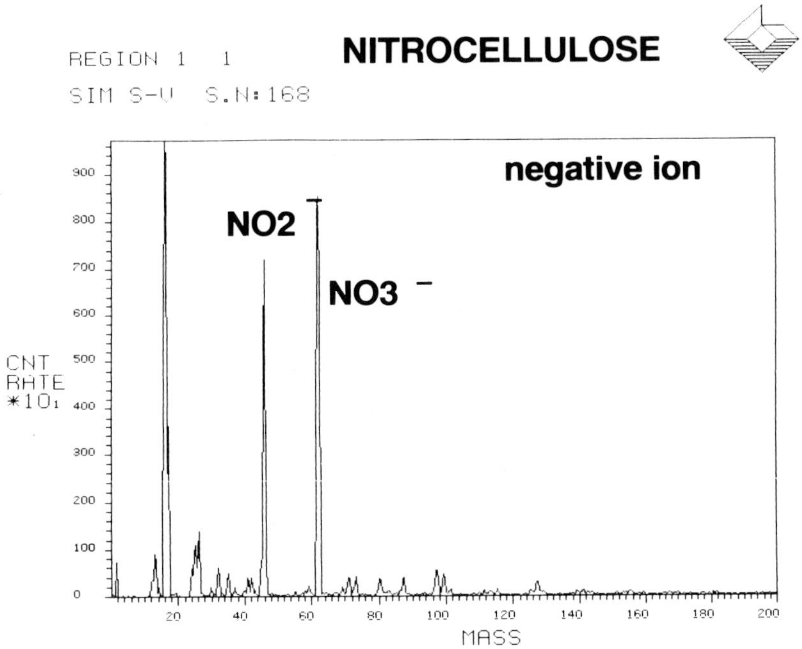
REGION 1 1
NITROCELLULOSE
SIM S-V S.N:168
negative ion
NO2
NO3 −
CNT RATE *10₁
MASS

Fig. 4.

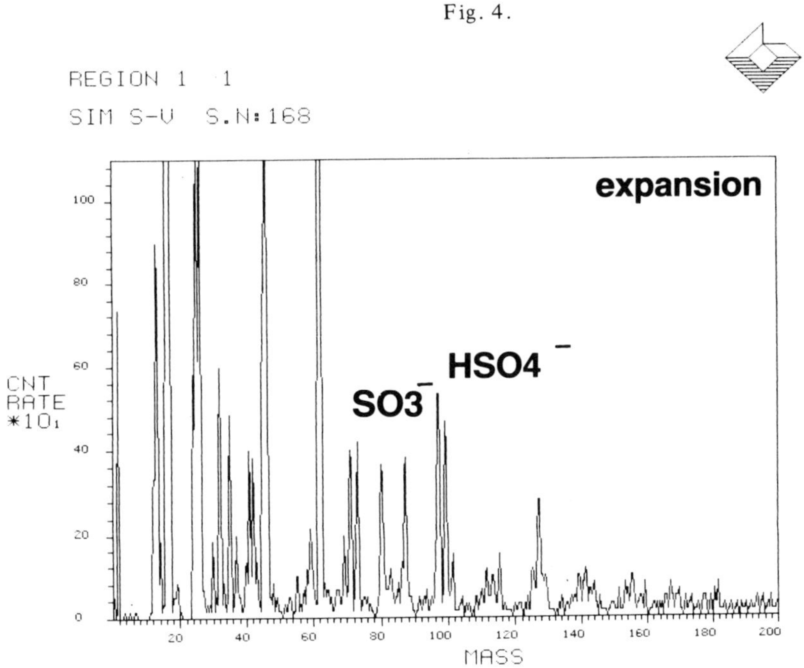
REGION 1 1
SIM S-V S.N:168
expansion
HSO4 −
SO3 −
CNT RATE *10₁
MASS

REFERENCES

1. Clark D.T., Stephenson P.J., and Heatley F. Polymer 1981, 22, 1112.

2. Clark D.T., Stephenson P.J. Polymer, 1982, 23, 1295.

3. Fowler A.H.K., Munro H.S. and Clark D.T. 'Cellulose and its Derivitives' edited by Kennedy J.F., Phillips G.O., Wedlock D.J., and Williams P.A. Published by Ellis Horwood Ltd., Chichester, 1985.

4. Fowler A.H.K., Munro H.S. Polymer Degrad. and Stab. 1985, 13(1), 21-29.

5. Fyfe C.A., Stephenson P.J., Marchessault R.H., Deslandes Y. J.Am. Chem Soc. 105, 2469 1983.

6. Fyfe C.A., Stephenson P.J. Marchessault R.H. Dudley R.L. Hamer G.K., J.M.S. - Rev. Macromol. Chem. Phys. C23 (2) 187 1983.

7. Atalla R.H., Gast J.C., Sindorf D.W., Bartasko V.J., Maciel G.E.: J Am Chem. Soc., 1980, 102, 3249.

8. Hortling B., Lindberg J.J. in Advances in Polymer Science No. 66 Springer-Verlag, Berlin, 1985.

9. Earl W.L., Van Der Hart D.L., J. Am Chem., Soc., 1980, 102 3251.

10. Clark D.T., Stephenson P.J. Polymer 1982, 23, 1034.

11. Clark D.T., Stephenson P.J., Polymer Degradation and Stability 1982. 4, 185.

12. Fowler A.H.K., Munro H.S., Clark D.T. in 'Cellulose and its Derivitives', Ellis Horwood Ltd. 1985, 246-253.

13. Teramae N, Shigeyuki T. Spectroscopy Letters 1981, 14 (10), 687-694.

14. Briggs D. Polymer, 1984 25 1379.

15. Stephenson P.J., Brown A. Unpublished results.

16. Fyfe C.A., 'Solid State N.M.R. for Chemists' C.F.C. Press P.O.Box 1780 Guelph Ontario, Canada, 1985.

17. Stephenson P.J., in Cellulose and its Derivitives Ellis Horwood Ltd., Chichester, 1985.

18. Earl W.L. Vanderhart D.L., Macromolecules 1981, 14, 570.

19. Horri F., Hirai A, Kitamura R., Polymer Bulletin 1983, 10, 357,

20. Koenig J., Patterson P., Blackwell J, J. Polym. Sci., Polym. Phys. Ed., 1985, 23 483-492.

21. Clark D.T. "Advances in E.S.C.A. applied to Polymer Characterisation. Pure Appl. Chem. 1982, 54 415-438.

22. Stephenson P.J., PhD. Thesis University of Durham, England 1981.

23. Brown A., Vickerman J.C., Surf. and Interface Analysis, 1984, 6, 1-14.

24. Briggs D., Surf and Interface Analysis 1982, 4 (4), 151-155.

25. Brown A., Vickerman J.C. Analyst 1984, 109 851-857.

26. Stephenson P.J., Brown A, in preparation.

27. Brown A., Surface + Interface Analysis 1986, 8 75.

28. Griffiths P.R., 'Transform Techniques in Chemistry' Heyden 1978.

29. Teramae N., Hirogughi M, and Tanaka M. Bull. Chem. Soc., Jpn. 1982 55 2097-2100.

30. Horii F., Hirai A., Kitamura R. J. Carb. Chem. 1984, 3(4), 641-662.

31. e.g. 'F.T.I.R. Spectroscopy'. Ed. T. Theophanides. D. Riedel, Publishing Co. 1984.

Chapter 2

Structural change of the amorphous region of cellulose in the presence of water

T. Hatakeyama – Research Institute for Polymers and Textiles, 1-1-4, Yatabe-Higashi, Tsukuba, Ibaraki 305.

Y. Ikeda – Tokyo Metropolitan Textile Institute, 3-15-5, Honjo, Sumida-ku, Tokyo 130.

H. Hatakeyama – Industrial Products Research Institute, 1-1-4, Yatabe-Higashi, Tsukuba, Ibaraki 305, Japan.

INTRODUCTION

The higher-order structure of cellulose has been explained by the two-state model for more than half a century. The concept that cellulose consists of crystalline and amorphous regions has elucidated the results obtained by X-ray diffractometry and also by mechanical measurements. Although this model is still used, it has not been considered that the amorphous region has a homogeneous structure [1,2,3]. In our former reports [1,3], it was indicated that the amorphous region of cellulose was difficult to consider as a homogeneous state, especially when the cellulose coexisted with water molecules. From heat capacity measurements [4], it was suggested that the amorphous region of cellulose I reversibly transformed to the crystalline region when the non-freezing water became bound to the hydroxyl group [4,5,6].

In this study, a transient state between the amorphous and crystalline region of cellulose in the presence of water was followed by differential scanning calorimetry (DSC). Attention was paid to the difference of transient state between cellulose I and II.

EXPERIMENTAL

Sample preparation

A cotton linter, Toyo Tokushi Co, was used as a sample of cellulose I. The crystallinity estimated by -ray diffractometry was 52%. A rayon fiber, Asahi Chemical Co. was used as a sample of cellulose II. The crystallinity was 38%. Both samples were washed in an ethyl-alcohol-benzene solution and dried in vacuo (1×10^{-1} mmHg) for 5 hours at 20 °C. In this study, the sample which showed no heat of vaporization in the thermogravimetric curve was used as a dry sample. Water content (Wc) is defined in this study as follows:

$$Wc(g/g)=(water,g)/(dry\ sample,g) \qquad (1)$$

Differential scanning calorimetry

A Perkin Elmer differential scanning calorimeter DSC-II, equipped with a cooling cell, was used. Sample weight was about 3 to 5 mg and the scanning rate was varied from 1 to 10 K/min. An aluminium sample pan was used, the surface of which had previously been treated with water vapour in order to avoid the reaction of aluminium and water during heating. Temperature and enthalpy were calibrated using pure water. The bound water content was calculated as reported previously [4,6].

X-ray diffractometry

The cellulose samples ground to a fine powder were packed in a hole in an acrylic plate with 2mm thickness and 3mm diameter. After a determined amount of water was added to the cellulose sample, the plate was covered with 7 μm aluminium foil so as not to allow water evaporation. A Rigaku Denki Co. X-ray diffractometer A-3 type was used. X-ray diffraction patterns were recorded with a scintillation counter and a rate meter. CuKα radiation with a Ni filter was used.

Stress-Strain Measurement

A Toyo Baldwin Co, Tensilon UTM-III was used. The dry sample was placed in an unsealed polyethylene envelope and kept in a moisture controlled desiccator for 30 hrs. at 22 °C. The envelope was closed immediately after the conditioning. When the sample in the envelope was set in clamps in the Tensilon, the envelope was loosened so that it would not interfere with the measurements.

RESULTS AND DISCUSSION

Figs. 1 and 2 show the stacked DSC cooling curves of water sorbed on cellulose I (Fig. 1) and II (Fig. 2). The DSC pattern varies as a function of Wc. The broken line in Fig.2 shows the crystallization curve of pure water. Initially Wc values of 0 to 0.15 for cellulose I and 0 to 0.25 for cellulose II were observed. There was no crystallization peak. The water showing no phase transition was classified as non-freezing water. The critical amount of non-freezing water of each sample was found to correspond to the number of hydroxyl groups existing in the amorphous region, where water molecules were allowed to diffuse [4]. The crystallization peak at the low temperature side (Peak II) was observed firstly when Wc exceeded the critical amount in both kinds of cellulose. Peak II was considered to reflect the irregular structure of ice formed under the effect of the presence of cellulose matrix, since the peak temperature was observed at 230K and the peak width was broad. Because of this, water observed as Peak II can be categorized as a bound water.

In the case of cellulose II, the crystallization pattern varies in a complex way. When the cellulose II contained Wc from 0.3 to 0.6, an intermediate peak (Peak II') appeared at the high temperature side of Peak II. Peak II' was found to shift to the high temperature side with increasing water content and when the sample had Wc=0.50, the peak became a shoulder of Peak I. This suggested that Peak II' was also attributed to bound water.

Although the pattern of the main peak (Peak I) of both samples varied under the influence of the low temperature side peak, the main peak almost corresponded to the peak of pure water. Therefore, Peak I was attributed to free water.

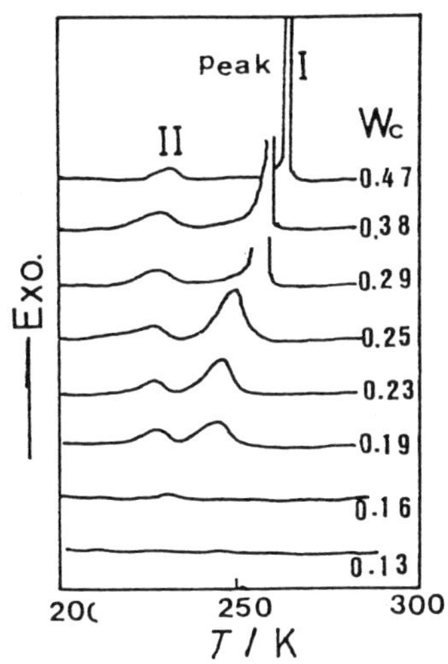

Fig. 1. Stacked DSC cooling curves of water sorbed on cellulose I.

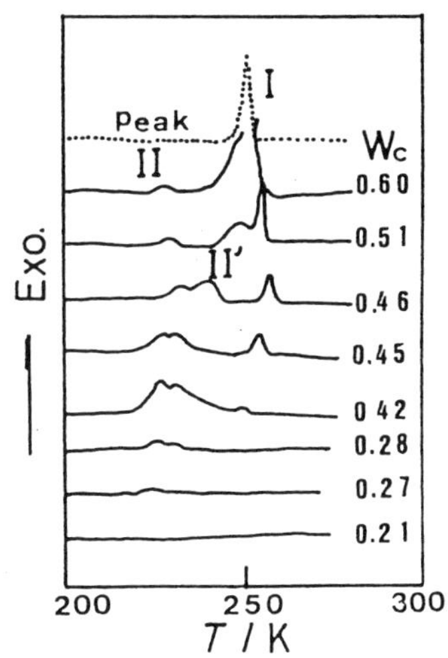

Fig. 2. Stacked DSC cooling curves of water sorbed on cellulose II.
.... pure water

A similar tendency was observed in the melting curves of celluloses I and II. Although melting peaks were not clearly separated as crystallization peaks, it was clear that the melting started from a temperature far lower than 273K.

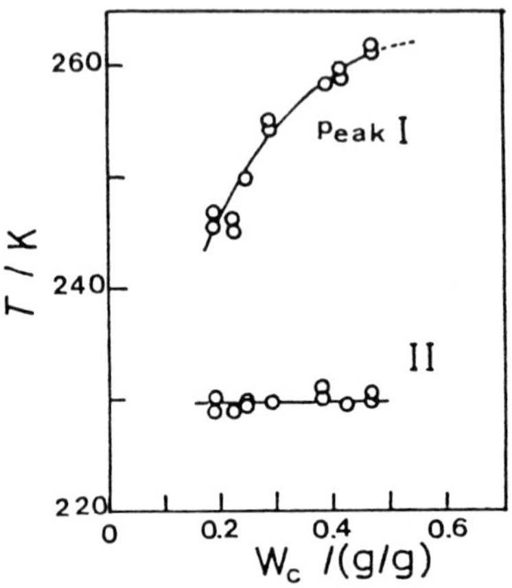

Fig. 3. Relationship between peak temperature of crystallization and water content (Wc) of cellulose I.

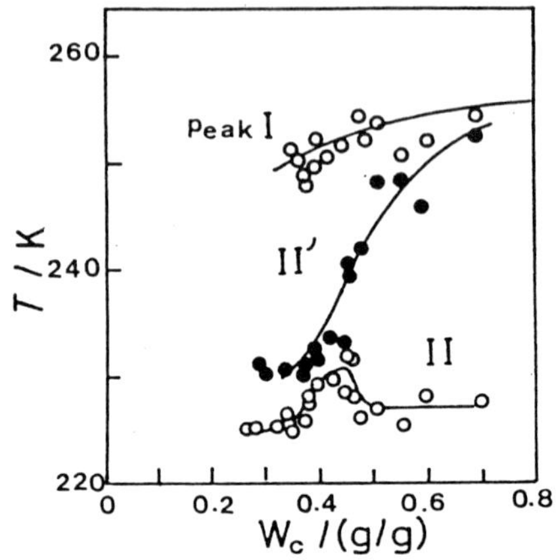

Fig. 4. Relationship between peak temperature of crystallization and water content (Wc) of cellulose II.

Figs. 3 and 4 show the relationship between the peak temperature of crystallization and Wc of celluloses I and II. In the case of cellulose I as shown in Fig. 3, the temperature of Peak II was observed at 228-230K regardless of Wc. The temperature of Peak I increased at the low Wc region and then remained almost stable at about 260K. In contrast, the temperature of Peak II of cellulose II shows a maximum at Wc=0.45. The intermediate peak II' separated from II at Wc=ca.0.3 and merged into Peak I at Wc=0.6. In Wc's ranging from 0.3 to 0.6, water sorbed on cellulose II crystallizes in a complex manner, reflecting the transient structure of the amorphous region of celullose II.

Enthalpies of melting and crystallization of water sorbed on both celluloses were evaluated using pure water as the standard. As indicated in Figs. 1 and 2, from the crystallization curve of water sorbed on cellulose, three kinds of water can be evaluated from the following simple equation:

$$Wc=Wf+Wfb+Wnf \qquad (2)$$

where Wc is water content evaluated from weight, defined as shown in the equation (1); Wf is the amount of free water calculated from enthalpy of Peak I; Wfb is freezing bound water from Peak II for cellulose I and from Peaks II and II' for cellulose II; Wnf is non-freezing water the phase transition of which can not be detected by DSC. Both Wfb and Wnf are categorized as the bound water. The nuclear magnetic relaxation times of bound water of cellulose have been evaluated using the NMR relaxation technique as reported previously [5,7,8] . By DSC, it is possible to obtain the exact amount of Wnf, although the value contains a certain error coming from the fact that the enthalpy of pure water is used for the calibration. Detailed discussion of this error can be found eleswhere [5].

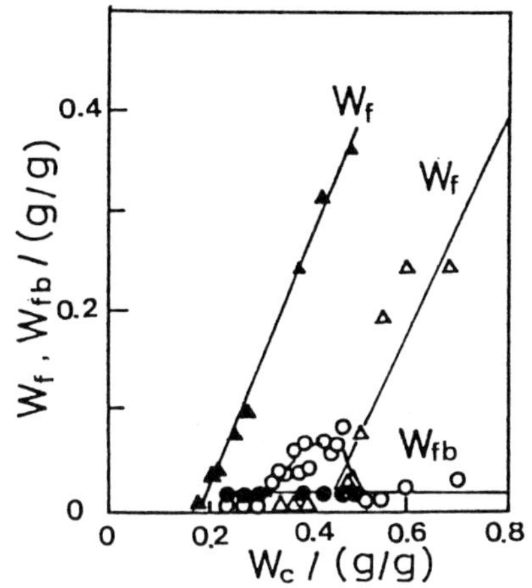

Fig. 5. Relationship between freezing water (Wf), and freezing bound water (Wfb) and water content (Wc).
▲●; cellulose I, △○; cellulose II
▲△; Wf, ●○; Wfb

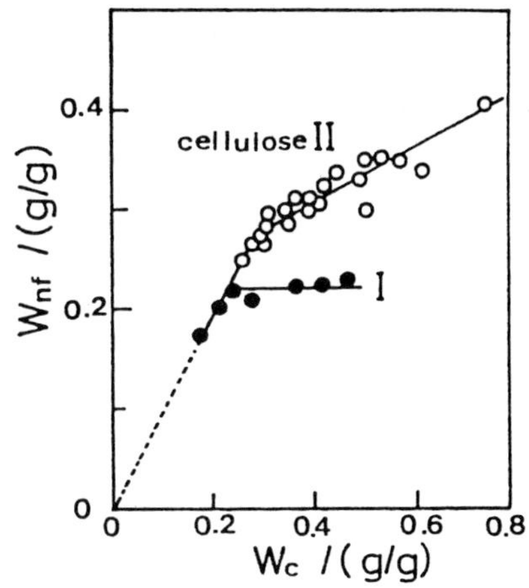

Fig. 6. Relationship between non-freezing water (Wf) and water content (Wc).
●; cellulose I
○; cellulose II

Fig. 5 shows Wf and Wfb of celluloses I and II as a function of Wc. Wf of cellulose I firstly appeared at about Wc=0.16 and linearly increased with increasing Wc. In the case of cellulose II, Wf started at Wc=0.40 and increased. This directly indicates that the amorphous region of cellulose II is larger than that of cellulose I. Moreover, not only the amount of the amorphous region but the molecular arrangement in that region seem to differ between cellulose I and II. The difference can be clearly seen for Wfb. Wfb of cellulose II (summation of Peaks II and II') started to increase at Wc=0.25 and attained a maximum at Wc=0.40, then decreased and levelled off when Wc exceeded 0.50, where Peak II' merged into Peak I. The maximum point coincided with the Wc at the point where free water appeared. This suggested that the size and regularity of the ice, which was considered to reflect the structure of the amorphous region of the sample, changed as a function of water content. It is reasonable to think that the number of water molecules which diffuse into the intermediate region increase with increasing the inter-molecular distance of cellulose II. This was indicated by the fact that the temperature of freezing bound water shifts to the high temperature region. However, when the crystal size of ice exceeded a critical point, which corresponded to the water content 0,60, this crystal merged into ordinal ice and became difficult to distinguish.

On the other hand, Wbf of cellulose I was almost constant throughout the Wc that was tested. The above facts suggest that the molecular arrangement of cellulose II changes dramaticaly with successive increase of water, in contrast the molecular chain of cellulose I scarcely changes.

Fig. 6 shows the relationship between Wnf and Wc of cellulose I and II. The amount of non-freezing water is constant for cellulose I, however, that of cellulose II gradually increases. Cellulose is not water soluble and the Wnf value should reach a constant at a certain Wc. However, in the Wc from 0 to 1.0, the Wnf value of cellulose II increases. This indicates that the amorphous region of cellulose increases in the presence of water.

X-ray diffractograms of cellulose I in the presence of water showed that the peak reflecting (002) plane became more pronounced with increasing Wc. This suggested the increase of molecular regularity of cellulose I. On the other hand, the peaks of cellulose II decreased and became flatter with increasing of water content, suggesting the decrease of the crystallinity of cellulose II.

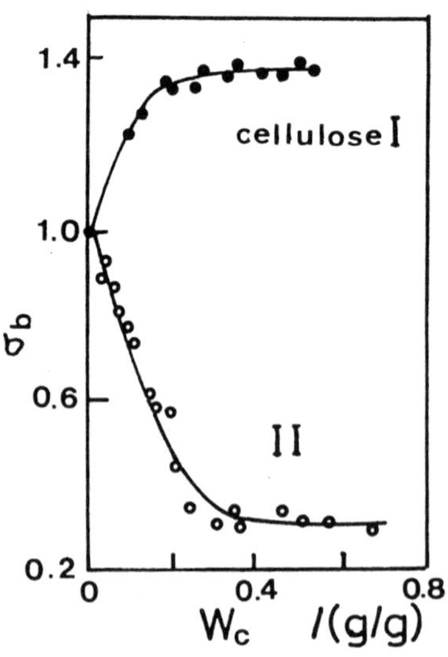

Fig. 7. Relative strength at break (σ_b) as a function of water content (Wc).
●; cellulose I, ○; cellulose II

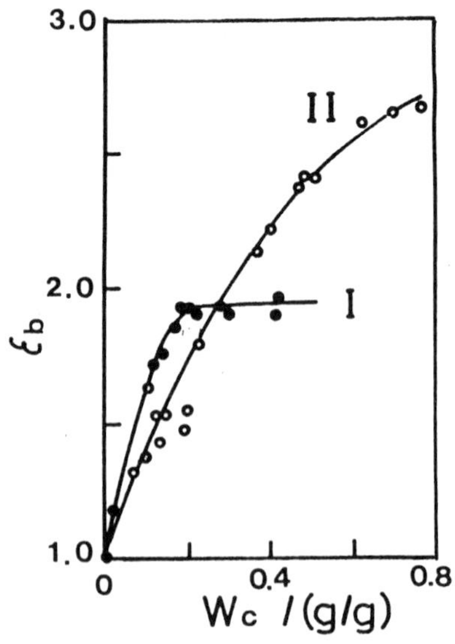

Fig. 8. Relative elongation at break (ϵ_b) as a function of water content (Wc).
●; cellulose I, ○; cellulose II

The change in the molecular arrangement in the amorphous region is thought to affect the mechanical properties of cellulose, since the higher-order structure sensitively corresponds to the stress-strain behaviour. Fig. 7 shows the strength at break of celluloses I and II as a function of Wc. The strength in this figure is described as a relative value as shown in the following equation.

$$\sigma_b = \sigma(\text{wet state}) / \sigma(\text{dry state}) \quad (3)$$

The σ_b of cellulose I increases up to around Wc=0.20 , but that of cellulose II decreases up to Wc=0.40. The σ_b showing opposite variation suggests that the molecular chain of celluloses I and II rearranges in a different way with the presence of water.

Fig. 8 shows the elongation at break of celluloses I and II as a function of Wc. The elongation in this figure is shown as a relative value similar to that in equation 3. ε_b of both samples increases with increasing Wc. However, the ε_b value of cellulose I levels off at around Wc=0.2, while that of cellulose II increases without showing levelling off in the Wc's from 0 to 0.8. This result also indicates the structural difference between cellulose I and II.

Based on the results obtained in this experiment and previously obtained [1,3,5], it is considered that the higher order structures of the amorphous regions of celluloses I and II are different from each other i.e. the molecular chain of cellulose I is extended and tilted through both amorphous and crystalline regions. On the other hand, the chain end and other defects might be concentrated in the amorphous region of cellulose II. Horii and his co-workers [2] suggested from their results obtained by cross-polarization/magic angle spinning 13C-NMR that the conformation of the non-crystalline component was rather hindered for cellulose I, whereas in cellulose II it was relatively relaxed [2]. The assumption was made because of the fact that line width of the C1 carbon of cellulllose I was about half that of the cellulose II and it was believed that the line width was primarily dependent on the distribution of torsion angle by X-ray analysis [9].

When the cellulose I is formed in the plant tissue, water molecules unavoidably coexist with cellulose molecules. Moreover, if we consider that cellulose is synthesized in living plants, it is reasonable to consider that the molecular chain takes an extended form. During the isolation process of natural cellulose, water molecules take a catalytic role, accelerate the formation of hydrogen bonding of cellulose, and once the crystalline region has formed, water itself is excluded from the crystalline part [5]. However, a part of the cellulose molecule, which is included in the crystalline region is distorted and strained. This process takes place reversibly by absorbing and desorbing water. The internal strain of cellulose I is relaxed when water molecules break the hydrogen bonding and the amorphous chains rearrange to a more ordered structure. The results obtained in this study i.e. the small amount of Wfb, the constant Wnf, X-ray patterns, and high σb in the presence of water, support the idea mentioned earlier. This also elucidates the decrease of heat capacity of cellulose in the presence of non-freezing water as reported previously [3] and the results of broadline NMR [10].

In contrast, the higher order structure of cellulose II is artificially formed from a concentrated solution. In this process, chain ends and other defects are concentrated in the amorphous region. A part of the molecular chain is considered to form the entanglement taking a kind of folded chain structure; the surface of these chain molecules can also be counted as a kind of defect. When the water molecule diffuses into such areas, inter-molecular distance might increase. The DSC curves measured by varying water content clearly indicate the presence of the transient state of the amorphous region of cellulose II, judging from the variation of intermediate peak. The increase of Wnf, and marked decrease of breaking strength and elongation also suggest the increase of the amorphous region.

It is concluded that the structural difference between the amorphous region of cellulose I and that of II can be emphasized in the presence of water. The molecular chain in the amorphous region of cellulose I transforms into a more ordered state and is partly included in the crystalline region. In contrast, that of cellulose II changes into a more random state and the amount of the amorphous region increases. These facts strongly suggest that the physico-chemical properties of cellulose are difficult to explain using a simple two-state model.

REFERENCES

1 H. Hatakeyama and T. Hatakeyama, Makromol. Chem., 182, 1655 (1981)

2 F. Horii, A. Hirai and R. Kitamaru, IN Polymers for Fibers and Elastomers, ACS Symposium series No. 260, J. C. Arthur, Ed., American Chemical Soc., Washington D.C., p.27 (1984)

3 T. Hatakeyama and H. Hatakeyama, in Cellulose and Its Derivatives, J. F. Kennedy, G. O. Phillips, D. J. Wedlock and P. A. Williams Eds., Ellis Horwood, Chichester p.87 (1985)

4 K. Nakamura, T. Hatakeyama and H. Hatakeyama, Tex. Res. J., 51, 607 (1981)

5 H. Hatakeyama, T. Hatakeyama and K. Nakamura, J. Appl. Polym. Sci. Appl. Polym. Symp., 37, 979 (1983)

6 K. Nakamura, T. Hatakeyama and H. Hatakeyama, Polymer, 24, 871 (1983)

7 M. F. Froix and A. O. Goedde, Polymer, 9, 428 (1976)

8 T. F. Child, Polymer, 13, 259 (1972)

9 for example, K. J. Kolpak, J. Blackwell, Macromolecules, 9, 273 (1980)

10 A. Hirai, R. Kitamaru, F. Horii and I. Sakurada, Cellulose Chem. Technol., 14, 611 (1980)

Chapter 3

Changes in the ultrastructure of cellulose during pulping

Tuulikki Hattula – The Finnish Pulp and Paper Research Institute, BOX 136, 00101 Helsinki, Finland.

INTRODUCTION

Pulping means changing wood fibers to pulp fibers by chemical or mechanical means. Chemical pulping can be carried out under either alkaline (kraft cooking) or acid (bisulfite cooking) conditions. Both methods involve cooking with aqueous solutions of chemicals at elevated temperatures. Water and elevated temperature are also used in mechanical pulping processes, in which hydrothermal treatment causes softening of chemical components present in the wood. Neither chemical nor mechanical pulping should affect the cellulose component, or at least it should affect it less than the hemicelluloses and lignin.

The purpose of this study was to establish how the ultrastructure of cellulose changes during different pulping processes. The study was carried out by following the state of the ultrastructure of the cellulose during cooking of wood and almost pure wood cellulose dissolving pulp under the conditions of bisulphite and kraft cooking. The evaluation of changes in the ultrastructure of cellulose during mechanical pulping has been limited to the effect of water and elevated temperature. Mechanical pulp and dissolving pulp were treated with saturated steam at different temperatures. In addition to monitoring the state of the ultrastructure of cellulose during cooking the purpose was also to establish which phenomena cause the changes in the ultrastructure of cellulose.

CHARACTERIZATION OF THE ULTRASTRUCTURE OF CELLULOSE

Cellulose consists of crystalline and noncrystalline cellulose. It is not known how the state of order changes from perfect crystalline to disordered [1]. According to some authors [2,3] paracrystalline cellulose coexists with crystalline and disordered cellulose. This uncertainty makes the quantitative determination of crystalline cellulose in fibres difficult. In this work cellulose is regarded as a two phase system. The distinction between crystalline and disordered is made by means of the method used to detect the state of order, namely wide angle X-ray diffraction. The definition of what constitutes crystalline cellulose is that used by Ruland [4], who says that the designation crystalline is restricted to crystalline regions larger than 2-3 nm containing lattice imperfections of the second kind not exceeding root mean square deviations of about 10 % in the nearest neighbour distances. The methods used in the present work were normalized on the basis of Ruland's and Vonk's [4,5] methods. It is not possible here to go into details of the method, which are published elsewhere [6,7]. The most important thing in the methods used is that they give the contents of crystalline and disordered cellulose as mass fractions. This allows the crystallinity of pulp to be corrected by yield, which is necessary when the crystallinities of pulps are compared with each other. Crystallization is defined here as an increase in yield-corrected crystallinity. It is also important to take special care in preparing powder samples from wood or pulp for X-ray diffractometry [7]. For the characterization of the ultrastructure of cellulose the size of crystallites is an important quantity. Using X-ray diffractometry the average width of crystallites was determined from the 002 reflection and the average length from the 040 reflection [8].

In addition to the content of crystalline and disordered cellulose the changes in the ultrastructure of cellulose were also characterized quantitatively by the loss in total cellulose content. The cellulose contents of pulp and wood have been obtained by calculating the polysaccharide composition [9] based on the gas-chromatographically determined monosaccharide composition [10], lignin [11] and extractive contents [12] of the sample.

CRYSTALLIZATION AND DISORDERING DURING PULPING

It is known that higher chain mobility allows rearrangement of lateral bonds between cellulose chains to occur. There are several reasons for higher chain mobility. For example, elevated temperature, especially with some swelling agent, causes rearrangement. Intercrystalline swelling only increases the chain's mobility, but intracrystalline swelling may even destroy the crystalline lattice. The swelling power of water is increased by the addition of bases, salts or organic liquids at elevated temperatures. In the following some examples are presented of the treatment of cellulose in acid milieu and at elevated temperature.

Ca-bisulphite cooking

In the first example Ca-bisulphite cooking was carried out for Norway spruce to different yields from 88 % to 51 %. The temperature was raised to 130 °C over 4 hours and kept at 130 °C for various lengths of time. The cooking liquor was Ca-bisulphite liquor containing 1.2 % CaO and 7 % total SO_2. The liquor to wood ratio was 5:1. The same cooking liquor was used for dissolving pulp. In this case the ratio of liquor to pulp was 10:1. Dissolving pulp was manufactured by the Rauma process [13]. The intrinsic viscosity determined by the SCAN method [14] was 450 dm^3/kg.

The results show that when spruce chips were cooked under conditions of Ca-bisulphite cooking the crystallinity of pulp increased distinctly, but after correction by yield the increase was only slight, i.e. from 18 to 20 % (Fig. 1). On the other hand, the content of disordered cellulose corrected by yield decreased because of crystallization and the small loss of cellulose (1 %).

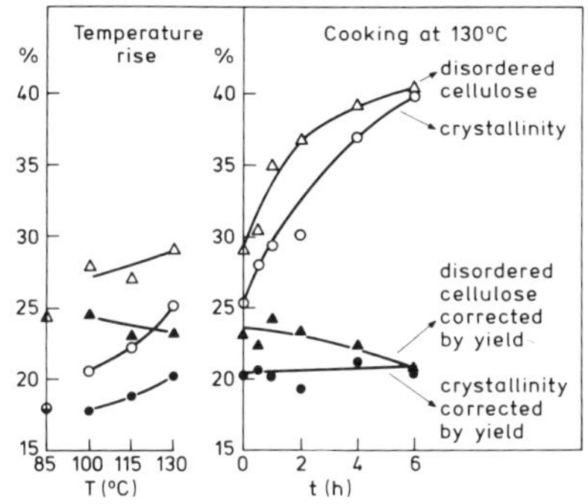

Fig. 1. Crystallinity and content of disordered cellulose as well as corresponding yield-corrected values from Ca-bisulphite cooking of spruce chips. The figures of untreated spruce chips are given on the ordinate.

A similar result was obtained when dissolving pulp was treated under similar conditions. The crystallinity of dissolving pulp increased as the treatment advanced, as did the crystallinity of pulp corrected by yield (from 46 to 53 %) (Fig. 2.). The content of disordered cellulose as well as that corrected by the yield decreased, partly because of crystallization and partly because of the loss of cellulose (from 49 to 39 %).

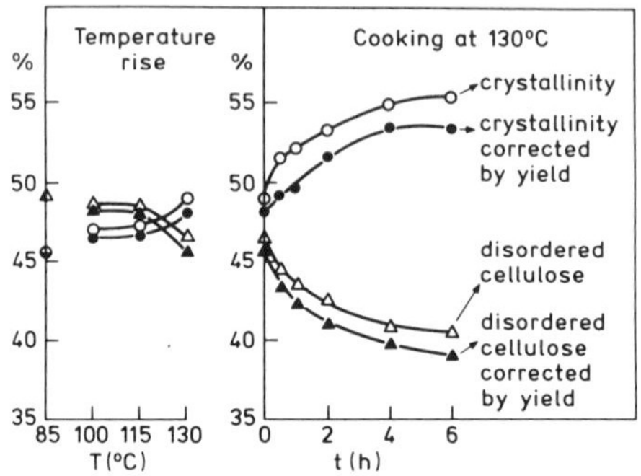

Fig. 2. Crystallinity and content of disordered cellulose as well as corresponding yield-corrected values of pulp during Ca-bisulphite cooking of dissolving pulp. The figures of untreated dissolving pulp are given on the ordinate.

Both examples show that the content of crystalline cellulose corrected by yield increased, i.e. cellulose crystallized. Cellulose underwent some hydrolysis in spruce but more in dissolving pulp.

Hydrothermal treatment

The other example concerns hydrothermal treatment with saturated steam at different temperatures. The question arises, can water alone cause sufficient hydrolysis to affect the ultrastructure of cellulose? To study the effect of hydrothermal treatment of wood or high-yield pulp, air-dry thermomechanical pulp was treated with saturated steam at temperatures up to 200 °C for 60 min. The same treatment was carried out with air-dry dissolving pulp as a model substance for pure cellulose. The yield of pulp dropped to 80 % for TMP and to 98 % for dissolving pulp in the treatment at 200 °C. The pH of the water decreased with increasing treatment temperature and was 3.5 in the treatment at 200 °C.

According to the results the crystallinity of TMP increased up to a temperature of 175 °C and decreased at higher temperatures. The correction by yield made the curve form more distinct (Fig. 3.). The yield-corrected crystallinity increased from 23 to 32 % at 175 °C and dropped to 27 % at 200 °C. This means that cellulose crystallizes when the temperature is 130-175 °C and disorders at higher temperatures. Similar results were obtained from the experiments where dissolving pulp was steamed under the same conditions. Cellulose crystallized and disordered, but to a lesser extent than in the case of thermomechanical pulp (Fig. 3.). Naturally, the content of disordered cellulose changed oppositely because of the loss of cellulose, which was only 1 %.

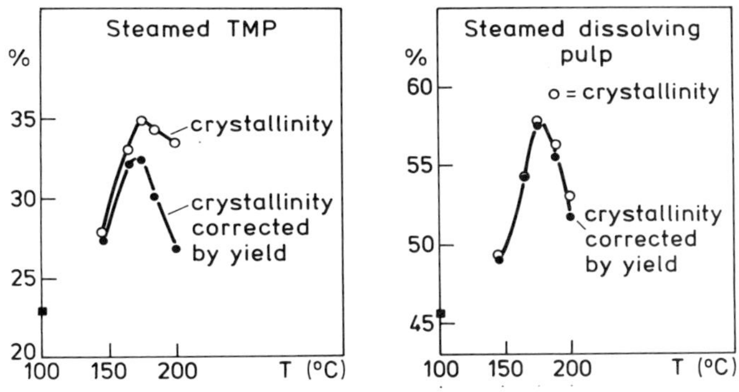

Fig. 3. Crystallinity and yield-corrected crystallinity of TMP and dissolving pulp as a function of steaming temperature. Corresponding figures for untreated samples are given on the ordinate.

The results show that the amorphous cell wall components undergo hydrolysis and that acids are formed in the vicinity of the cellulose chains. At elevated temperatures the acid medium is strong enough to cause crystallization, which becomes disordering when the temperature is increased over 175 °C and the conditions become more drastic.

Kraft cooking

Kraft cooking of pine and wood cellulose in dissolving pulp is presented as an example of alkali treatment. Kraft cooking of pine was carried out to yields varying from 86 % to 42 %. The temperature was raised to 170 °C in 120 min and kept at 170 °C for various lengths of time. The cooking liquor consisted of NaOH and Na_2S. The content of effective alkali was 4.4 mol/kg wood and the sulphidity 30 %. The liquor to wood ratio was 4:1. For kraft cooking of dissolving pulp the effective alkali charge was 2.2 mol/kg, pulp sulphidity 30 % and the liquor to wood ratio 10:1.

Fig. 4 shows that during kraft cooking of pine chips the crystallinity of pulp increased but the yield-corrected crystallinity remained constant (at 17 %). The content of disordered cellulose corrected by yield decreased a little because of the loss in cellulose, the content of which fell from 40 % to 37 % during cooking.

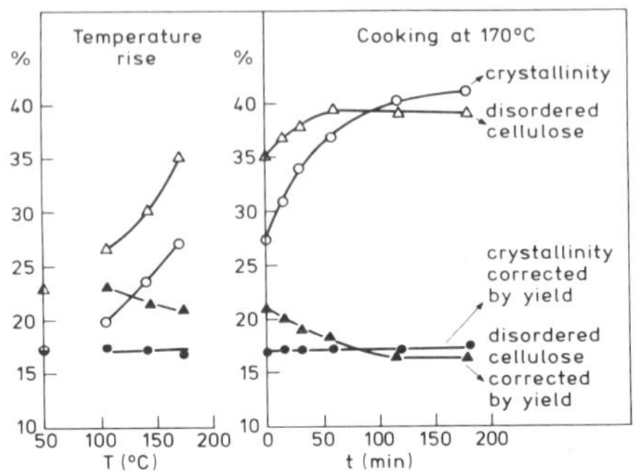

Fig. 4. Crystallinity and content of disordered cellulose as well as corresponding yield-corrected values of pulp during kraft cooking of pine chips. The figures of untreated pine chips are given on the ordinate.

The treatment of dissolving pulp under conditions of kraft cooking showed that a considerable amount of cellulose (about 18 %) was lost during pulping and that pulp crystallinity remained constant but decreased from 46 to 37 % when the correction by yield was made. This means that cellulose disorders. It is probable that the greater effect of kraft cooking liquors on dissolving pulp is caused by the lower degree of polymerization (DP) and the higher carboxylic acid group content in dissolving pulp than in pine cellulose. To confirm the results obtained, dissolving pulp was also treated in water under conditions of kraft cooking. In this case cellulose showed an increase in yield-corrected crystallinity of pulp as the treatment advanced. This result agrees with those presented earlier in this work. When these results are compared with those from the kraft cooking of dissolving pulp it can be confirmed that the kraft cooking liquor changes the phenomenon from crystallization to disordering.

SIZE OF CRYSTALLITES

It was found that the average width of crystallites increased in all the treatments discussed here. The width of cellulose crystallites from wood increased slightly in Ca-bisulphite cooking from 3.0 to 3.2 nm and in kraft cooking from 3.1 to 3.4 nm. The corresponding change during the treatments with dissolving pulp was from 4.4 to 4.7 nm in both types of cook.

The average length of crystallites (13 nm) remained constant during chemical cooking. It was considerably lower than that of the wood cellulose (35 nm). The only case where the average length of crystallites increased (from 13 to 20 nm) was steaming of thermomechanical pulp at temperatures higher than 175 ^{0}C. This can probably be explained by the relaxation of microstrain caused by mechanical refining.

SUMMARY

During chemical cooking the pulp crystallinity increases, mainly due to the decrease in yield. Only in the case of Ca-bisulphite cooking did the yield-corrected crystallinity increase and slight crystallization occur. The results obtained from treatments with dissolving pulp confirm these results. During kraft cooking of pine the cellulose neither crystallized nor disordered. The results from the treatment of dissolving pulp under kraft cooking conditions show that kraft cooking can actually cause disordering. This is probably because the cellulose of dissolving pulp has a lower DP and a higher carboxylic acid content than wood cellulose.

Steaming of both pulp types caused crystallization up to 175 ^{0}C and disordering at higher temperatures. The effect of steaming was more pronounced in the case of high-yield pulp because of the greater hydrolysis of amorphous cell wall components. Based on these results it seems probable that cellulose crystallizes during mechanical pulping, the extent of crystallization being dependent on the process conditions.

It has often been proposed that crystallization of cellulose means an increase in the size of crystallites. X-ray diffractometry permits only average sizes to be determined. The increase in the average width or length of crystallites may be a real increase in crystallite width, though it is more likely to be the result of the change in the distribution of crystallite width. This occurs when small, badly crystallized crystallites are damaged and degraded. The average length of crystallites seemed to drop considerably at the very beginning of the cook, but then remained constant during the cook.

It has been shown that during alkali and acid cooking the ultrastructure of cellulose undergoes some small changes. Steaming seems to change the ultrastructure most. The effect of these changes on the chemical and physical properties of wood and pulp needs to be studied.

ACKNOWLEDGEMENT

The author is grateful to Prof. Timo Paakkari and his working group for carrying out the X-ray diffraction measurements and the analysis of the X-ray diffraction data.

REFERENCES

1. J.A.Howsmon and W.A.Sisson, in "Cellulose and Cellulose Derivatives", (Eds. E.Ott, H.M.Spurlin and M.W.Graffin), Interscience Publishers, New York, 2nd Edition, Part I, 1954,23
2. G.B.Mitra and P.S.Mukerhjee, Polymer 21,(1980),1403
3. R.Hosemann, Progr Colloid and Polymer Sci., 60,(1970),213
4. W.Ruland, Acta Cryst.,14,(1961),1180
5. C.G.Vonk, J.Appl. Cryst.,6,(1976),148
6. M.Ahtee, T.Hattula, J.Mangs and T.Paakkari, Paperi ja Puu, 65,(1983),475
7. T.Hattula, Effect of heat and water on the ultrastructure of wood cellulose, Doctoral dissertation, University of Helsinki, Helsinki, Finland,1985.
8. P.Scherrer, Göttinger Nachr.Math.Phys.,(1918),98
9. J.Janson, Faserforschung und Textiltechnik, 25,(1974),375
10. Standard method KL 27-76
11. Standard method KL 115b:76
12. Standard method SCAN-C7:62, Paperi ja Puu, 44,(1962),275
13. I.Hassinen and R.Räsänen, Pulp Paper Mag.Can., 75,(1974),137
14. Standard method SCAN-C15:62

^1H and ^{23}Na NMR studies of the interaction between water and sodium cellulose sulfate

Hyoe Hatakeyama – Industrial Products Research Institute, Tsukuba, Ibaraki 305.
Hiro Iwata – Japan Pulp and Paper Research Institute, Tokodai, Ibaraki.
Taksuko Hatakeyama – Research Institute for Polymers and Textiles, Tsukuba, Ibaraki.

INTRODUCTION

It is generally known that polyelectrolytes are highly soluble in water through hydration of ionic groups. However, attention has scarcely been paid to the physico-chemical properties of highly concentrated solutions of polyelectrolytes. We have already reported that there are two kinds of water, i.e. freezing water and non-freezing water, around the ionic groups of sodium poly(styrenesulfonate) (NaPSS) in the H_2O-NaPSS system and around those of sodium cellulose sulfate (NaCS) in the H_2O-NaCS system [1, 2]. Judging from the experimental results obtained by differential scanning calorimetry (DSC) and by nuclear magnetic relaxation (NMR) measurements, water molecules are strongly associated with the ionic group and form the hydration shell.

In this study, we have chosen a H_2O-NaCS system with a small amount of water, and ^1H and ^{23}Na nuclear magnetic relaxation times of the system have been investigated. The longitudinal and transverse relaxation times (T_1 and T_2) of the system have been measured as functions of water content and temperature. This paper places special emphasis on (a) evidences of phase transition phenomena of NaCS with small amounts of water, (b) the evaluation of nuclear correlation time, and (c) the evaluation of activation energies of water and counter ion in the system.

EXPERIMENTAL

Sample

The sample used in this experiment was sodium cellulose sulfate (NaCS) obtained from Scientific Polymer Products, Inc. The degree of substitution (D.S.) of NaCS was determined according to the procedure reported by Whistler and Spencer [3]. The obtained D.S. was 2.26.

NUCLEAR MAGNETIC RELAXATION (NMR)

Longitudinal and transverse relaxation times (T_1 and T_2) of ^{1}H and ^{23}Na in the H_2O-NaCS system were measured using a Nicolet FT-NMR, model NT-200WB. The T_1 was measured by 180-τ-90 degree pulse method. The obtained data were calculated according to the method reported by Levy et al. [4]. T_2 was measured by the Meiboom-Gill variant of the Carr-Purcel method [5]. However, in the case of very rapid relaxation, the free induction decay (FID) method was applied [6]. The sample temperature was changed from 30 °C to -70 °C with the assistance of the 1180 NIC system. The accuracy of the temperature control was ±0.5 °C.

The sample for the NMR measurement was prepared by adding a certain amount of degassed water to a dry NaCS sample in a NMR glass tube, then the usual freeze-pump-thaw technique was applied for removing air in the H_2O-NaCS system. Water content (Wc) of each samle was defined according to the following equation:

$$Wc = Ww/Ws \tag{1}$$

where Ww is the weight of added water, and Ws is the absolutely dry weight of each sample which was determined with a Shimazu thermogravimeter TGC-30 by heating the sample to 160 °C at the heating rate of 10 °C/min. The weight when no volatization of water from the sample was observed, was adopted as the weight of a dry sample.

RESULTS AND DISCUSSION

^{1}H Nuclear Magnetic Relaxation

The ^{1}H longitudinal relaxation time (T_1) was studied as a function of temperatures at various Wc's. The T1 data vs. the inverse absolute temperature are shown in Figure 1. Since the observed longitudinal relaxation is only one-component over the entire temperature and Wc range covered in this experiment, the weight-average relaxation is observed. A minimum value for T_1 is observed at a temperature ranging from -25 to -40 °C. The temperature where the T_1 minimum value is observed depends on Wc of the sample, corresponding to the crystallization of water in the system. The T_1 minimum value decreases with decreasing Wc in the region higher than ca. 1.2. However, the T_1 minimum values observed at Wc=1.03 and 0.83 are larger than those observed at Wc's higher than 1.2. The appearance of the T_1 minimum is ordinarily explained in connection with the correlation time (τc). This will be discussed later.

Figure 2 shows the change in T_2 values vs. the inverse absolute temperature at various Wc's. A sudden decrease in T_2 values is seen at a temperature lower than −15 °C. The T_2 values give an average indication of the motion of the sorbed water in NaCS. As we have reported previously [2], two groups of water molecules are considered to exist in the H_2O-NaCS system : i.e. the group corresponding to the shorter transverse relaxation time is considered as being more restricted in the molecular motion than the other group corresponding to the longer transverse relaxation time. In a quantitative sense, the degrees of freedom of the more tightly bound group of water

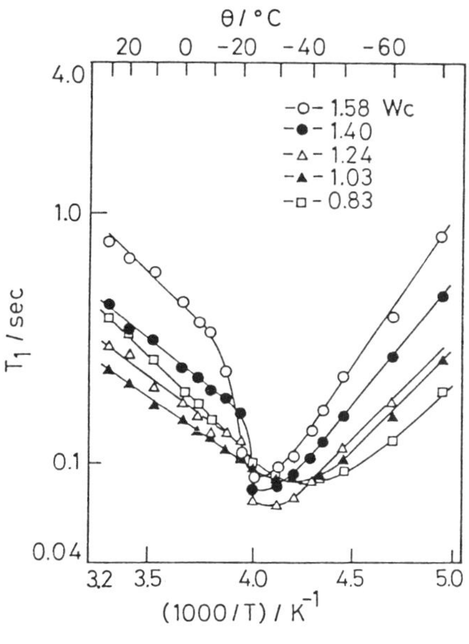

Fig. 1. Temperature dependence of the ^1H longitudinal relaxation time of water in the H_2O–NaCS system.

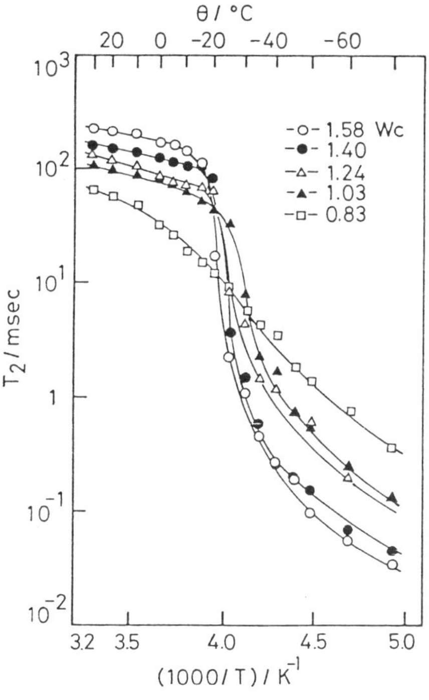

Fig. 2. Temperature dependence of the ^1H transverse relaxation time of water in the H_2O–NaCS system.

molecules would be more restricted than the degrees of freedom of the less tightly bound group. In other words, if we consider the molecular motion of water in the H_2O-NaCS system of a certain Wc, the T_2 values at a higher temperature range are characteristic of more mobile water, while the T_2 values at a lower temperature range are characteristic of more restricted water. At a temperature range lower than where the T_1 minimum is observed, water molecules are in a frozen state. However, if we assume that there are two states of water at the temperature range, the observed T_2 value represents the average T_2 value, which corresponds to the weight-average relaxation of two different states; one short and one long, representing two different states of restriction in the motion of water molecules.

If we assume that the relaxation of water in the frozen H_2O-NaCS system is determined by one average correlation time (τc), thiscorrelation time can be determined by the use of following equation.

$$\frac{T_{2b}}{T_{1b}} = \frac{2\left(\dfrac{1}{1+\omega_0^2\tau_c^2} + \dfrac{4}{1+4\omega_0^2\tau_c^2}\right)}{\left(3+\dfrac{5}{1+\omega_0^2\tau_c^2} + \dfrac{2}{1+4\omega_0^2\tau_c^2}\right)} \qquad (2)$$

where T_{1b} and T_{2b} are average longitudinal and transverse relaxation times of bound water in the frozen H_2O-NaCS system [2]. It might be an oversimplification to only consider one type of average bound water. However, in the practical application of equation (2), it has not been considered worthwhile to separately calculate the two states of bound water.

Figure 3 shows the obtained τc values vs. the inverse absolute temperature. From the τc values, it is possible to estimate the average relaxation behaviour of the bound water in the frozen H_2O-NaCS system. The plots are apparently rectilinear, showing a straight line corresponding to each sample. This temperature dependence of τc may be expressed with good accuracy in the following Arrhenius form:

$$\tau_c = \tau_0 \exp(E_a/RT) \qquad (3)$$

where Ea is the average activation energy for the relaxation process of the bound water which is considered to consist of freezing and non-freezing water in the H_2O-NaCS system in the frozen state.

Figure 4 shows the change of Ea with Wc. The obtained Ea value decreases rectilinearly with decreasing Wc, and the extrapolation of the the straight line shows that Ea at Wc 0.38, which corresponds to a critical water content less than which the water sorbed in NaCS cannot be crystallized [2], is ca. 3

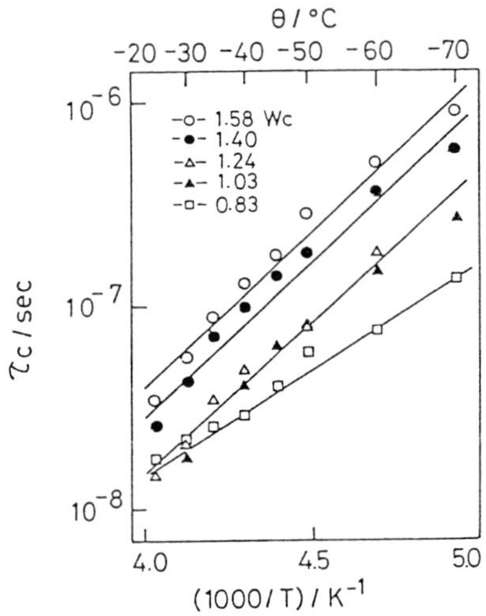

Fig. 3. Temperature dependence of the average correlation time of water in the H_2O–NaCS system.

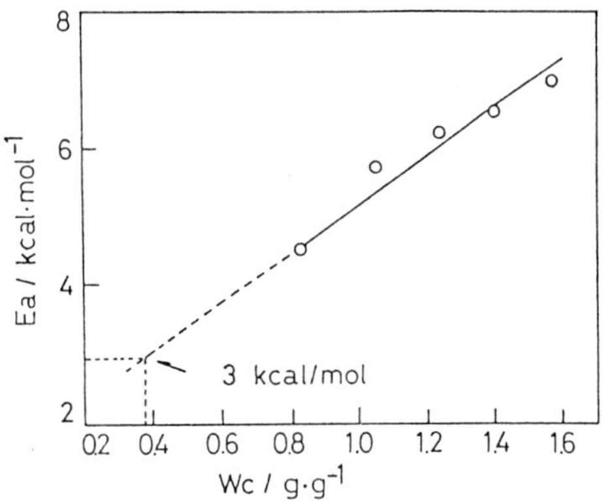

Fig. 4. Change of the activation energy with Wc in the H_2O–NaCS system.

kcal/mol. This value agrees well with the previously reported Ea for the relaxation of non-freezing water [2]. This suggests that the non-freezing water around sodium ions is more mobile than the crystallized water in the H_2O-NaCS system in the temperature range lower than that where the T1 minimum is observed. This indicates that the temperature dependence of the molecular motion of non-freezing water is small in the temperature range covered by this experiment. The glass transiton of the H_2O-NaCS system, which was observed in the temperature range from -70 to -100 ^{o}C [7], may be attributed to the change of the non-freezing water from the mobile state to the glassy state.

^{23}Na Nuclear Magnetic Relaxation

Figure 5 shows the change of T1 values of ^{23}Na in the H_2O-NaCS system with the inverse absolute temperature at various Wc's. T_1 plots are rectilinear in the temperature range where T_1 values were observed in this experiment. In the temperature range lower than -20 ^{o}C, it was difficult to measure the T_1 value of ^{23}Na by the 180-τ-90 degree pulse method because of the extreme broadening of the line-width of NMR peaks.

Figure 6 shows the change of T_2 values of ^{23}Na in the H_2O-NaCS system with the inverse absolute temperature. As seen from Figure 6, two types of transverse relaxations are observed. One is 'slow' and the other is 'fast' relaxation. This means that the transverse relaxation decays in a non-exponential manner.

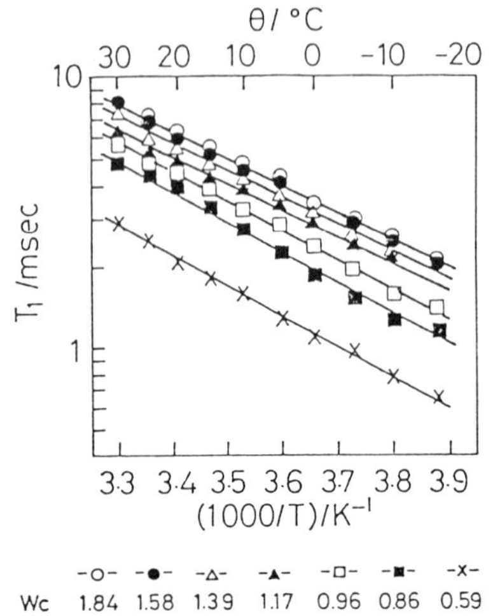

Fig. 5. Temperature dependence of the ^{23}Na longitudinal relaxation time in the H_2O–NaCS system.

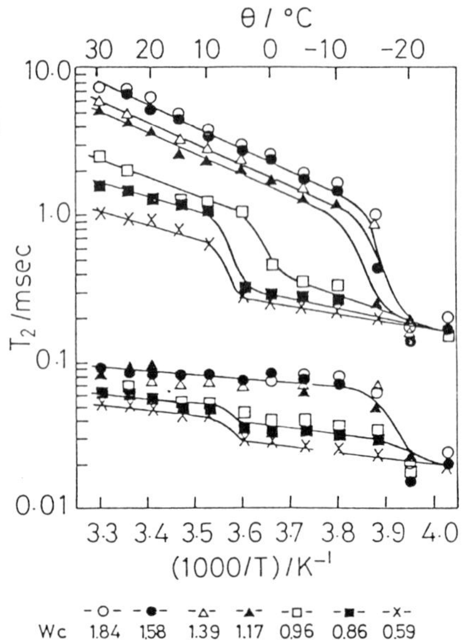

Fig. 6. Temperature dependence of the ^{23}Na transverse relaxation time in the H_2O–NaCS system.

The non-exponential relaxation due to quadrapole relaxation has been calculated by Hubbard [8], who considered the case where the electric field gradients fluctuate less rapidly than the Larmor period of the nucleus. In this case, the transverse relaxations produced by a quadrapole interaction are the sum of two or more decaying exponentials. As shown in Figure 6, the longer transverse relaxation time (T_{2s}) decreases with decreasing temperature, although the shorter transverse relaxation time (T_{2f}) does not change much with temperature. When Wc is higher than ca. 1.2, T_{2s} values decrease suddenly at around −10 °C. This change probably reflects the crystallization of free water. However, when Wc is lower than 1.2, T_{2s} values start to decrease at around 10 °C. This change probably shows the transition from the isotropic phase to the mesophase [7]. However, it is needless to say that the simple comparison of the data obtained by NMR with that obtained by DSC is difficult and that a further investigation is very necessary to explain the transition occurred.

From the slopes of the straight lines shown in Figure 5, the apparent activation energy (Ea) of the relaxation process was calculated. The obtained Ea values are shown in Figure 7. In contrast with the activation energies of ^{1}H of sorbed water in

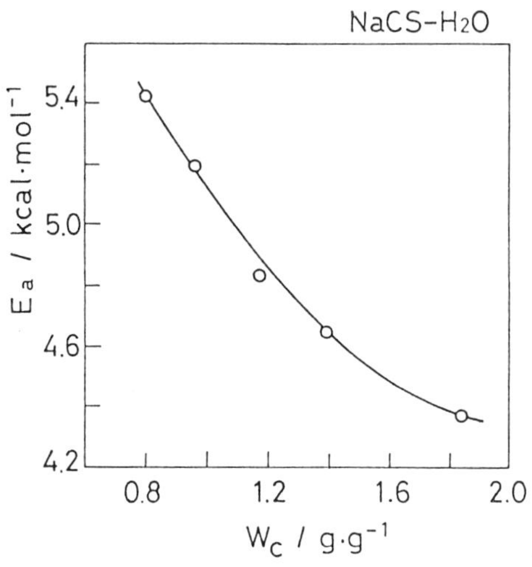

Fig. 7. Change of the activation energy of ^{23}Na in the H_2O–NaCS system.

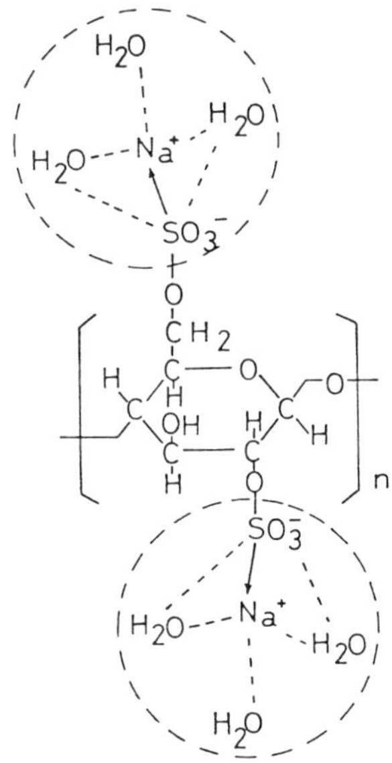

Fig. 8. Schematic of the H_2O–NaCS system.

the temperature region lower than -25 °C, the Ea values shown in Figure 7 increase with decreasing Wc. This suggest that the molecular motion of ^{23}Na is strongly restricted in the low Wc region. However, this result is quite reasonably explained if we assume that the sodium ion is surrounded by the non-freezing water and that this non-freezing water is surrounded by the free water. In the temperature region higher than -20 °C, with decrease of Wc, water molecules around the non-freezing water decrease and form hydrogen bonds between each other. This restricts the motion of the non-freezing water and indirectly restricts that of the sodium ion. With further decrease of Wc, the restriction increases, resulting in the prominent increase of the Ea of ^{23}Na.

Figure 8 shows a schematic model for the hydrated structure for NaCS as assumed from experimental results obtained by DSC and NMR. Each sodium ion in the H_2O-NaCS system is surrounded by three water molecules which are strongly associated with the sodium ion and with each other. Therefore, the non-freezing water does not crystallize even at a low temperature where the free water in the system crystallizes. However, this non-freezing water turns to a glassy state in the temperature range lower than -80 °C, although it is dependent on the value of Wc.

REFERENCES

[1] T. Hatakeyama, K. Nakamura, H. Yoshida and H. Hatakeyama, Thermochimica Acta, **88**, 223 (1985).

[2] H. Hatakeyama, H. Yoshida and T. Hatakeyama, in "Cellulose and Its Derivatives" (Eds. J. F. Kennedy, G. O. Phillips, D. J. Wedlock and P. A. Williams), Ellis Horwood Ltd., Chichester, 1985, Ch. 21.

[3] R. L. Whistler and W. W. Spencer, in "Method in Carbohydrate Chemistry, Vol. III" (Eds. R. L. Whistler, J. W. Green and J.N. BeMiller), Academic Press Inc., New York, 1963, pp.265.

[4] G. C. Levy and I. R. Peat, J. Magn. Resonance, **18**, 500 (1975).

[5] S. Meiboom and D. Gill, Rev. Sci. Inst., **29**, 688 (1958).

[6] K. Fujimoto, T. Nishi and R. Kado, Polymer J. **3**, 448 (1972).

[7] T. Hatakeyama, K. Nakamura, H. Yoshida and H. Hatakeyama, Polymer Preprints, Japan, **34**, 2393 (1985).

[8] P. S. Hubbard, J. Chem. Phys., **53**, 985 (1970).

Chapter 5

A mechanism of transformation from Cellulose I to II. Mercerization of bacterial and valonia cellulose

J. Hayashi and Y. Yaginuma – Applied Chemistry Department, Faculty of Engineering, Hokkaido University.

INTRODUCTION

Allomorphs and derivatives prepared from cellulose 1 (cell 1) and 11 were transformed into cell 1 and 11, respectively, and were classified into cell 1 and 11 families [1]. The results suggested that the memory phenomenon of the original crystal structure is due to conformations of individual chains that are common within each family because the crystal structures are quite different within the families. From the experimental observation that the addition of AcOK to cell 11 enhanced the intensity of (001) of cellulose but that to cell 1 did not, we assumed that the 1 and 11 families have "bent" and "bent & twisted" chain conformations, respectively [2]. Based on the X-ray work of the crystal structure of cell 1 and 11, Sarko proposed that the allomorphs in the 1 and 11 families are based on parallel and anti-parallel chains, respectively [3]. Kolpak et al [4] and Sarko et al [5] have proposed mechanisms of the change from "parallel chain" to "antiparallel chain" for the mercerization process in solid state.

In the present work, mercerization of bacterial and valonia cellulose membranes which have the uniplanar orientation of $(1\bar{1}0)$ was examined. The changes of their structures during mercerization gives us useful information concerning this problem.

EXPERIMENTAL

Purified Valonia macrophysa and Acetobacter Xylinum membranes and commercial cellophane were used. Pellicles of bacterial cellulose were grown under the conditions described by Colvin [6] for 48 hr. To obtain the membrane without the uniplanar orientation of ($1\bar{1}0$), bacterial cellulose was acetylated as follows. Acetic anhydride containing sulfonic acid 4.3 g per 100 ml was heated at 100° for 10 min to form sulfonic acetic acid. The bacterial membrane pre-treated with glacial acetic acid was acetylated with the sulfonic acetic acid solution (25 vol %) and benzene (75 vol %) at 20°C for 48 hr. The acetylated membrane was recrystallized by heat treatment at 210°C for 30 min and followed by saponification with 0.25 N NaOH at 20° for 24 hr after swelling treatment with 75 % aq. solution at 50°C for 30 min.

Mercerization was carried out with 18 or 21 % NaOH aq. solution at 20°C for 2 min to 1150 hr and followed by washing with water of 20°C and air-drying. All the treatments were carried out under slack state. X-Ray diffractograms were measured by the reflection method laying the membrane surface parallel to the reflection plane with Rigaku's diffractometer Cu K_{α}).

The wet membranes after mercerization and washing with water were also analyzed.

RESULTS AND DISCUSSION

Fig. 1 shows X-ray diffractograms of bacterial cellulose membrane mercerized with 18 % NaOH, taken in the wet state. Following reaction for only 3 min the reflections of cell 1 disappeared and a complete diffractogram of sodium cellulose (Na-cell) 1 was observed. But the intensity ratio of the two main reflections of Na-cell 1, $I_{(020)}/I_{(220)+(2\bar{2}0)}$, was 1.6. The value was higher than the normal value, about one, and increased to about two with the reaction time. Fig. 2 shows X-ray diffractograms of the cellulose regenerated from the mercerized membranes shown in Fig. 1 with water and dried.

Despite the high conversion to Na-cell by the previous 3 min treatment, regeneration gave cell 1 containing only a very small amount of cell 11. The ratio of cell 11 in the regenerated cellulose increased with the time of mercerization, complete conversion occurring after 480 hours. The belief that the mercerization of bacterial cellulose is difficult due to its high crystallinity was due to this regeneration of cell 1. But even after a 3 min reaction time, the regenerated cell 1 did not show the ionic-plava orientation of ($1\bar{1}0$). However, the regenerated cell 11 did show this orientation for all reaction times. Fig 3 shows the change in the amount of Na-cell 1 and regenerated cell 1 and cell 11 with respect to time. They were measured by the integral intensity of their X-ray reflection and expressed in relative fractions of their maximum values. The amount of preformed Na-cell 1 increased with reaction time to a maximum value at 120 hours. But the amount of cell 1 continued to decrease and of cell 11 to increase for 480 hours. After 3 min

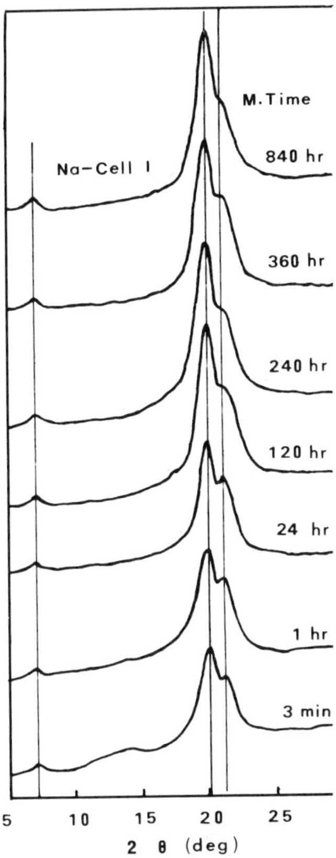

Fig. 1. X-ray diffractograms of bacterial cellulose membrane mercerized with 18% NaOH at 20 C for 3 min to 840 hr, taken in wet state. M. Time: Mercerization time.

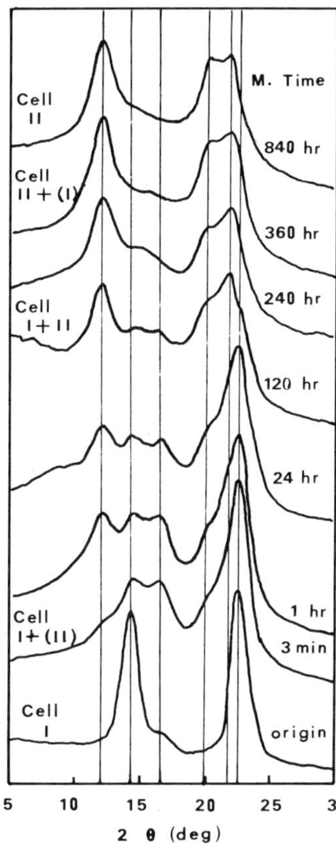

Fig. 2. X-ray diffractograms of regenerated cellulose from the mercerized bacterial membranes shown in Fig. 1.

mercerization the amount of Na-cell was about 60% of the maximum value, and a large amount of cell 1 was regenerated. Cell 1 was still regenerated, even after the maximum formation of Na-cell 1, clearly showing that a Na-cell 1 conversion is possible.

In our previous work [7,8] we found that Na-cell 1 prepared under conditions of low swelling could be converted into cell 1 with hot water. This Na-cell 1 differed from that prepared either from cell 1 by the usual method or from cell 11 by any method by the intensities of its meridional reflections. The two types were classified as Na-cell 1_1, and Na-cell 1_{11}. We assumed that Na-cell 1_1 and 1_{11} have "bent" and "bent and twisted" chain conformations respectively. When Na-cell 1 is decomposed with cold water it regenerates cell 1. Na-cell 1_{11} always regenerates cell 11, irrespective of the method.

Our results now show that the mercerization of bacterial cellulose gives initially a high proportion of Na-cell 1_1, due to

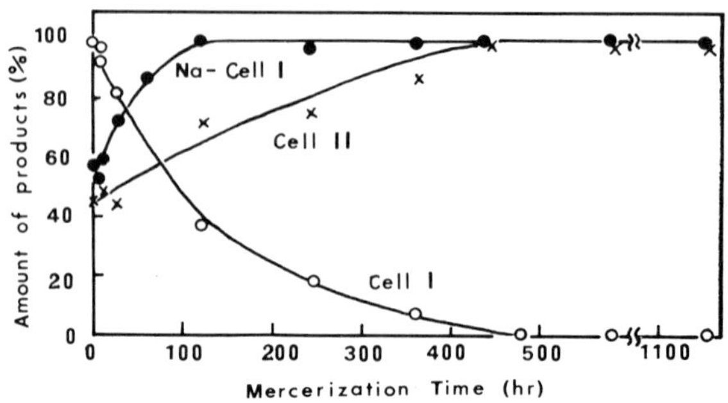

Fig. 3. Changes of amounts of the performed Na-cell and regenerated cellulose I and II from that with respect to the mercerization time.

the high crystallinity giving low swelling conditions. But this Na-cell 1_1, formed in the early stages of the reaction, gradually changes to Na-cell 1_{11}. The "bent" conformation changes to the "bent and twisted" form by relaxation of the chains with time. This should occur without major morphological change; any such major changes should have occurred at the time of the initial formation of Na-cell 1_1.

According to the mechanisms proposed by Kalpak et al and Sarko et al, Na-cell 1 with antiparallel chain structure is formed by the combination of chains from different microfibrils, each having a parallel chain structure but of different polarity. In this case all Na-cell 1 should be regenerated into cell 11. The present results show doubt on this mechanism. It is possible to consider Na-cell 1_1 and Na-cell 1_{11} having parallel and antiparallel chains respectively, but it is very unlikely that the Na-cell 1_1 to Na-cell 1_{11} conversion can be due to combination of chains from different microfibrils.

In our previous paper [9] are considered that bacterial membrane cellulose with uniplanar (110) orientation transformed into cell 11 without reorientation, and we are therefore critical of Kalpak's mechanism. However, this argument is no longer valid, since we have now shown that cell 1, regenerated from Na-cell 1_1, loses its (1$\bar{1}$0) orientation. This orientation appears to be required during the Na-cell 1_1 to Na-cell 1_{11} conversion. The cause of high intensity ratio of (020) to (220) + (220) of the Na-cell 1 is questionable, and may not be due to orientation. Fig 4 shows X-ray diffractograms of valonia cellulose mercerized, taken in the wet state. When mercerization was with 18% NaOH for 2 min a diffractogram of Na-cell 1 appeared but that of cell 1 was still clearly present. Moreover, the cell-1 retained its uniplanar orientation, though the pattern is less obvious as one proceeds from the origin. The mixed X-ray pattern of cell 1 and Na-cell 1 did not change significantly for at least 240 hours.

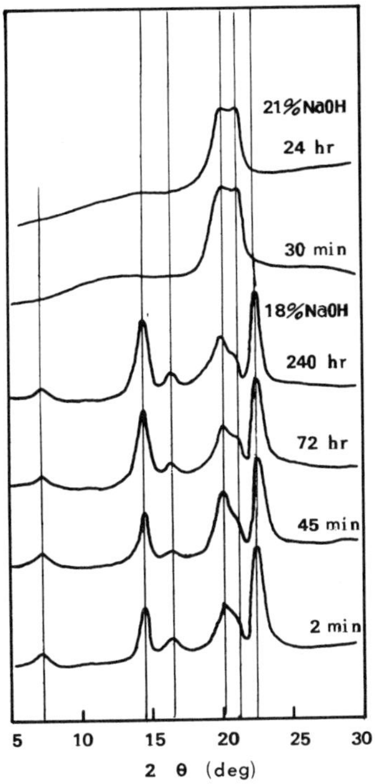

Fig. 4. X-ray diffractograms of valonia cellulose membrane mercerized with 18% or 21% NaOH at 20°C for various time.

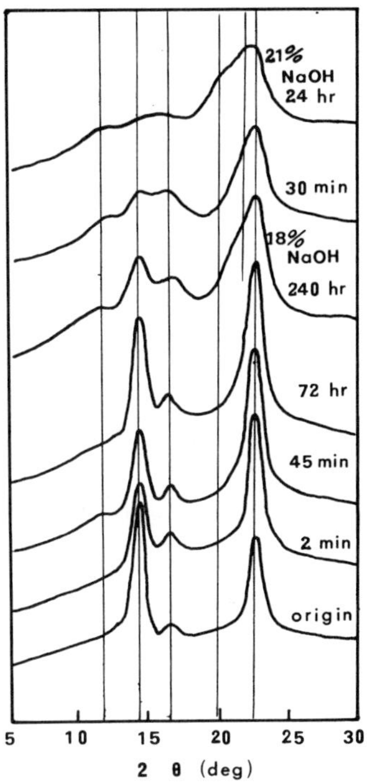

Fig. 5. X-ray diffractograms of regenera cellulose from the mercerized valonia membrane shown in Fig. 4.

However, on mercerization with 21% NaOH a diffractogram showing only Na-cell 1 was obtained after 30 min. The intensity ratio of the two main reflections of the Na-cell 1 was the standard value. There is a critical concentration of NaOH aq. solution for mercerization. It seems that the concentration for valonia cellulose is very high. Fig. 5 shows X-ray diffractograms of the regenerated cellulose from the mercerized valonia membrane shown in Fig. 4. For the sample with 18% NaOH for 2 min a complete pattern of the cell 1 with the uniplanar orientation was obtained. Most of the cell 1 should be unreacted cell 1 but part of it should be regenerated from Na-cell 1_1. For 18 % NaOH and 240 hr the pattern was mostly of cell 1 but the crystallinity of it decreased and (1$\bar{1}$0) of cell 11 at 2θ=12 increased a little. The sample regenerated after the mercerization with 21 % NaOH for 30 min showed a mixed pattern of cell 1 and 11. But the greater part of the mixture was cell 1 which lost orientation. The 24 hr sample showed a little increase in cell 11, and a lot of cell 1 remaining though the crystallinity of it decreased. The results suggested that most of the Na-cell 1 from valonia by the conditions was Na-cell 1_1. This resulted from higher crystallinity of valonia and the relatively lower concentration of NaOH solution used as

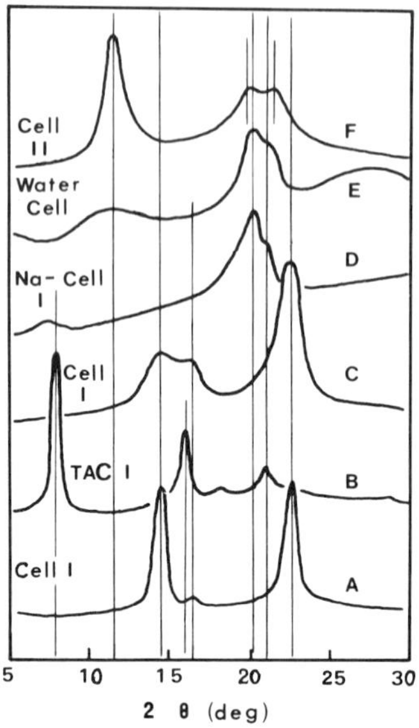

Fig. 6. Mercerization of the membrane without the uniplanar orientation of ($1\bar{1}0$). A: Bacterial cellulose, B: Acetylated A, C: Saponified B, D: Mercerized C, E: Decomposed D with water, F: Air dried E.

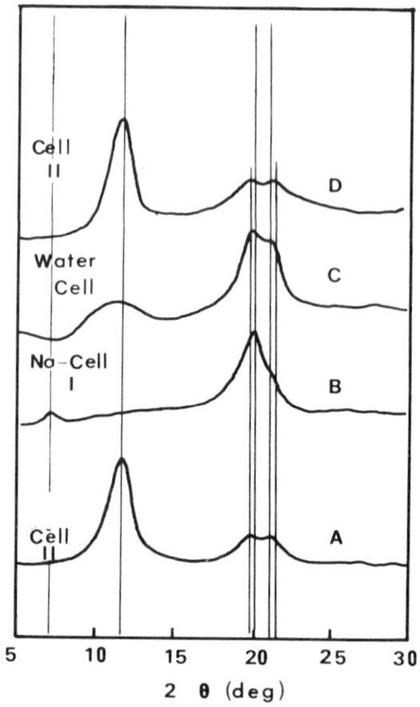

Fig. 7. Mercerization of a cellophane film with the orientation. A: Original cellophane, B: Mercerized A, C: Decomposed B with water, D: Air dried C.

compared with the critical concentration for valonia. In mercerizations of plant cellulose a lower concentration of NaOH solution used had a tendency to yield Na-cell 1_1 [8].

To investigate the secondary uniplanar orientation of regenerated cell 11, mercerization of membranes having no orientation was examined. Bacterial cellulose membrane was acetylated in the solid state, and followed by recrystallization and saponification. Fig. 6 shows their X-ray diffractograms. The regenerated cellulose was cell 1 which had lost orientation. The regeneration of cell 1 from the recrystallized triacetate was the characteristic result of bacterial cellulose. The regenerated cell 1 was mercerized with 18 % NaOH solution for 24 hr and followed by washing with water at 20°C and air-drying. X-Ray diffractograms taken at each stage are shown in Fig. 6. They showed that the secondary uniplanar orientation was formed at the stage of air-drying. Using cellophane film with the uniplanar orientation the same experiments were carried out, and the same results were obtained (Fig. 7).

It is known that ($1\bar{1}0$) planes, sheet plane of molecular chains, form interchain hydrogen bonds. Cell 11 should show greater tendency than cell 1 because OH groups of cell 11 stick out of the plane more than that of cell 1. It was considered that the crystallites were attracted to each other by their ($1\bar{1}0$)

planes to form hydrogen bonds during the drying process. It should bring about the secondary orientation of $(1\bar{1}0)$. For films which had various orientations, the secondary orientation of $(1\bar{1}0)$ by the processs was examined. Secondary orientation did not always occur and required films to have some initial orientation.

REFERENCES

[1] J. Hayashi, A. Sueoka and S. Watanabe, Nipon Kagaku Kaishi, 153, 160, 594, 1345 and 1587 (1973).
[2] S. Watanabe, J. Hayashi and S. Masuda, Kogyo Kagaku Zasshi, 73, 2058 (1970), Nippon Kagaku Kaishi, 948 (1974).
[3] A. Sarko, Tappi, 61, 59 (1978).
[4] F.J. Kolpak and J. Blackwell, Macromolecules, 9, 273 (1976).
[5] A. Sarko, T. Okano and H. Nishimura, Int. Dissolving Specialty Pulps Conf., Boston, 1983, TAPPI Proceedings, p.225.
[6] J.R. Colvin, Can. J. Microbiol. 11, 641 (1965).
[7] J. Hayashi, T. Yamadi and S. Watanabe, Seni Gakkaishi, 30, 190 (1974), ibid 31, (1975), Nippon Kagaku Kaishi, 1967 (1974).
[8] J. Hayashi, T. Yamada and K. Kimura, J. App. Polym. Sci., Sympo. 28, 713 (1976).
[9] J. Hayashi, Int. Sympo. on Wood and Pulping Chem., Tsukuba, 1983, Proceedings (1) 30.

Chapter 6

Cellulose - How much do we know about its structure ?

Anatole Sarko – Department of Chemistry and Cellulose Research Institute, State University of New York, Syracuse, New York 13210, U.S.A.

INTRODUCTION

Although cellulose was discovered nearly 150 years ago, its structure has remained a subject of widespread interest to this day. The cause of this interest has been, in large measure, the structural complexity of native celluloses -- from their long-chain, polymeric chemical structure to their morphological features of crystallinity and aggregation into a fiber. As a result, structural study of cellulose has had to be carried out on at least three separate levels: 1) chemical structure, 2) physical structure -- conformation and crystalline packing, and 3) fiber structure and supermolecular morphology. The chronology of the study of cellulose structure has roughly followed this order, with its chemical structure having been established, beyond dispute, long ago (see, for example, ref. 1), while studies of its crystallinity and morphology are still in progress. Our understanding of the latter two structural levels has increased significantly only during the past 10-15 years, coinciding with the development of new methods of structure analysis. Of major importance among the latter has been the combining of x-ray diffraction analysis with computer-based, stereochemical modeling [2,3]. The recent applications of the solid-state, Cross Polarization/Magic Angle Spinning (CP/MAS) NMR method have, likewise, been a source of new information [4]. In this context, it is interesting to note that cellulose has often been among the first substances to be studied by a new method. This was the case more than 60 years ago after the discovery of x-ray diffraction, and is equally true now after the development of the CP/MAS NMR technique. The combined impact of diffraction and NMR analysis, aided by modern molecular modeling, has been dramatic and we are now beginning to understand the cellulose structure on all of its levels in considerable detail. This is not to say that controversy does not remain: it is still there and it fuels continued interest in the structure of cellulose.

Finally, the study of the biosynthesis of cellulose is proceeding with considerable vigor. This research has also produced findings that have an important bearing on our understanding the structure of native celluloses [5].

In the following, a brief review and interpretation of the current knowledge of the physical and morphological structure of cellulose will be presented. Aspects of the crystalline structure, transformations between different crystal forms, the process of mercerization, and fiber morphology are reviewed. The discussion is based principally on the work done in the author's laboratory at Syracuse, using methods of x-ray diffraction and stereochemical modeling.

POLYMORPHY OF CELLULOSE CRYSTAL STRUCTURES

It has been known for a long time that cellulose crystallizes in four different forms -- the polymorphs known as celluloses I, II, III, and IV [1]. Of these, only cellulose I occurs naturally, accounting for practically all native cellulose structures seen in terrestrial and aquatic plants, as well as those produced by microorganisms. Cellulose II is the normal conversion product obtained through mercerization or solubilization-regeneration of native and other celluloses, while celluloses III and IV are obtained, respectively, by liquid ammonia and heat treatments of other polymorphs. Further distinction for celluloses III and IV can be made on the basis of the source polymorph used, viz., III_I and IV_I for starting with cellulose

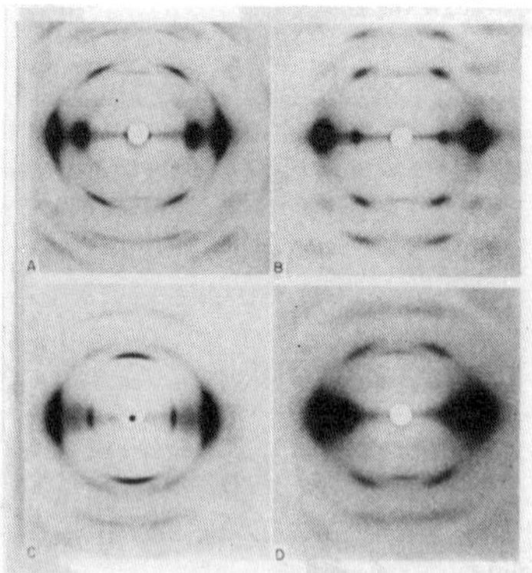

Fig. 1. X-ray diffraction diagrams of celluloses I, II, III, IV (A–D).

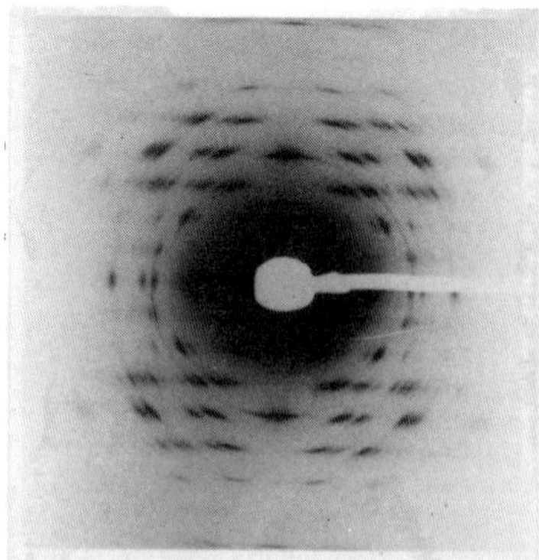

Fig. 2. Electron diffraction diagram of cellulose I of *Valonia ventricosa* [9].

I, and III_{II} and IV_{II} for starting with cellulose II. All cellulose polymorphs are quite crystalline and exhibit a marked degree of fibrous orientation of the crystalline domains, as shown by their x-ray diffraction diagrams (cf. Fig. 1). The crystal structures of celluloses I, II, III_I, IV_I, and IV_{II} have been analyzed in detail by x-ray diffraction; the findings are briefly described below.

Cellulose I

Cotton, ramie, wood, and other fibrous celluloses of medium crystallinity exhibit essentially the same diffraction pattern (cf. Fig. 1) which is consistent with a two-chain, monoclinic unit cell with dimensions a = 7.78, b = 8.20, c (fiber repeat) = 10.34 Å, and angle γ = 96.5°. Some very highly crystalline celluloses -- typified by that of the alga *Valonia ventricosa* -- yield a considerably better resolved but a very slightly different diffraction diagram (cf. Fig. 2) that shows evidence of a larger, eight-chain unit cell [6]. The dimensions of this, also monoclinic cell are twice as large as those of the two-chain unit cell: a = 15.76, b = 16.42, c = 10.34 Å, and angle γ = 96.8° [7]. The two-chain unit cells of other native celluloses can very nearly be considered as sub-cells of the *Valonia* unit cell [7,10].

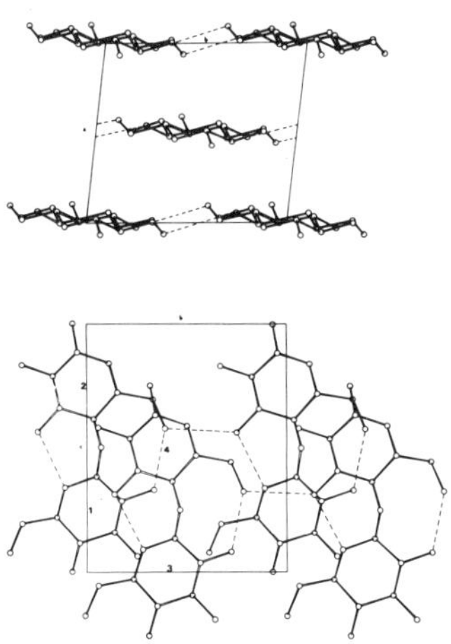

Fig. 3. Projections of ramie cellulose I unit cell onto: *a–b* (top) and *b–c* (bottom) planes [8]. (Courtesy of American Chemical Society).

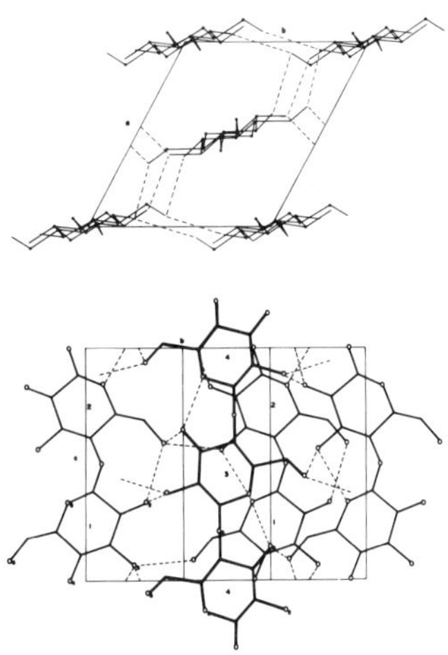

Fig. 4. Projections of cellulose II (Fortisan) unit cell onto: *a–b* (top) and *b–c* (bottom) planes [12]. (Courtesy of American Chemical Society).

The diffraction analysis of the two-chain cellulose I structure of ramie points rather firmly to a *parallel-chain* structure [8]. In the latter, the corner and center chains of the unit cell possess the same orientation along the *c* axis and are organized into sheet-like structures that parallel the *b-c* plane of the crystal lattice. As shown in Fig. 3, the corner and center sheets exhibit an identical intra- and intermolecular hydrogen-bond structure, but are offset from one another by a distance of approximately 1/4*c*. There is no hydrogen bonding linking the sheets.

The *Valonia* crystal structure, although not analyzed to the full extent possible, appears to be well represented by the same type of parallel-chain structure [7,10]. The detailed features of the *Valonia* structure and its differences relative to other celluloses must, however, await its full analysis. Since its unit cell contains eight chains, the complete analysis will be a difficult undertaking.

The reader may be reminded that the first reasonable crystal structure proposed for cellulose I -- by Meyer and Misch -- was based on *antiparallel* chains [11]. This proposal was not based on firm *evidence, but* the question of chain polarity in cellulose I has been a matter of controversy ever since. However, with the completion of several detailed crystal structure analyses of cellulose I, by independent research groups [7,8,10], it now appears highly probable that the structure of the native polymorph is, indeed, based on a parallel-chain structure. As shown below, this conclusion is further substantiated by x-ray diffraction analyses of other polymorphs, as well as by other evidence.

Cellulose II

When any of the celluloses is mercerized by alkali or solubilized and regenerated from solution, the end result is always the crystal structure of cellulose II. The crystal structures of both mercerized and regenerated forms of cellulose II have been determined in good detail by x-ray diffraction analysis, and both show the same features [12,13]. As can be seen in Fig. 4, the two most prominent features of the crystal structure of cellulose II are its *antiparallel-chain* structure and extensive, three-dimensional *hydrogen bonding.* The latter distinguishes it from all other cellulose polymorphs -- whether parallel or antiparallel -- and is undoubtedly the primary reason for the stability of cellulose II and its possession of the global minimum energy structure among all crystalline celluloses. Despite its three-dimensional hydrogen bonding, the crystal structure of cellulose II is still marked by sheet formation. In this case, however, the sheet structure can be thought to be present in two planes: the *b-c* plane common to all polymorphs, as well as that paralleling the *110* crystallographic plane.

Celluloses III_I and III_{II}

Treating any cellulose with liquid ammonia (or some amines) results in the structure of cellulose III. Cellulose III_I is obtained, in this fashion, from cellulose I and cellulose III_{II} is obtained from cellulose II. The crystal structure of cellulose III_I has been determined by x-ray diffraction [14], and is shown in a-b projection in Fig. 5. The structure is very similar to that of cellulose I -- in its features of parallel chains, sheet structure, and lack of inter-sheet hydrogen bonding. Because both the x-ray diagrams and unit cells of celluloses III_I and III_{II} are very similar, it is highly likely that their crystal structures are also very similar -- with the exception of the polarity of chain packing. Both cellulose III_I and III_{II} are unstable polymorphs and revert easily to their parent structures (III_I to I, and III_{II} to II) simply by heating in water. Since the parent structures differ in chain polarity, it is inescapable that celluloses III_I and III_{II} must also differ from one another in the same fashion.

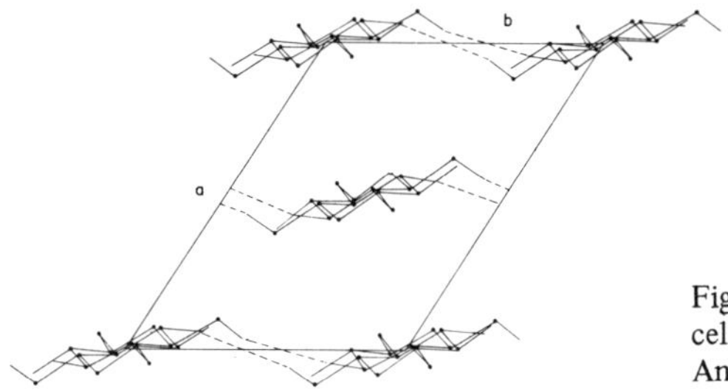

Fig. 5. Projection of cellulose III_I unit cell onto *a–b* plane [14]. (Courtesy of American Chemical Society).

Celluloses IV_I and IV_{II}

These polymorphs are most easily prepared by heating celluloses III_I and III_{II}, respectively, in glycerol to 260^0 C. The crystal structures of both have been determined and are shown in a-b projection in Fig. 6 [15]. The similarities between the two structures, as well as to polymorphs I and III, are obvious -- sheet structure, lack of three-dimensional hydrogen bonding, etc. The two polymorphs, IV_I and IV_{II}, possess nearly identical unit cells, but the important difference is, again, the chain polarity -- parallel in IV_I, antiparallel in IV_{II}.

It has been proposed that the cellulose IV polymorphs, being of low degree of crystallinity, are not true polymorphs but are simply mixtures of other crystalline forms. That this is not so has been proven by recording an electron diffraction diagram of cellulose IV_{II} [16]. This diagram clearly shows the presence of a unique cellulose IV crystal lattice.

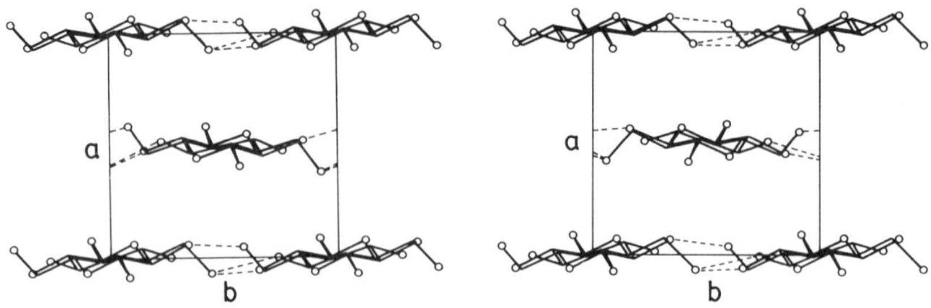

Fig. 6. Projections of cellulose IV_I (left) and IV_{II} (right) unit cells onto a–b plane [15]. (Courtesy of National Research Council of Canada).

CELLULOSE ACETATES

The distinction of cellulose polymorphs on the basis of chain polarity extends to their derivatives, notably cellulose triacetate (CTA). When cellulose I is carefully acetylated under non-swelling conditions, a parallel-chain CTA I structure results (cf. Fig. 7) [17]. On the other hand, when acetylation of cellulose I is carried out under swelling or solubilizing conditions, or cellulose II is acetylated with any method, or CTA I is treated with superheated steam, an antiparallel CTA II results (cf. Fig. 8) [18]. Upon deacetylation in the solid state, CTA I reverts to cellulose I and CTA II reverts to cellulose II [19].

Because there is no possibility of hydrogen bonding in CTA, its crystal structures need not preserve the same sheet formation that is characteristic of pure cellulose polymorphs. Although CTA I does show a type of sheet structure (in the $1\bar{2}0$ crystallographic plane), there is no evidence of it in CTA II, and the two polymorphs differ considerably in this regard. It is conceivable that the absence of hydrogen bonding may also allow the formation of other types of crystalline CTA polymorphs than have heretofore been characterized, as polymorphy of cellulose and other polysaccharide derivatives is well documented. For example, it is known that under certain conditions a CTA I-like x-ray diagram is obtained from heterogeneously

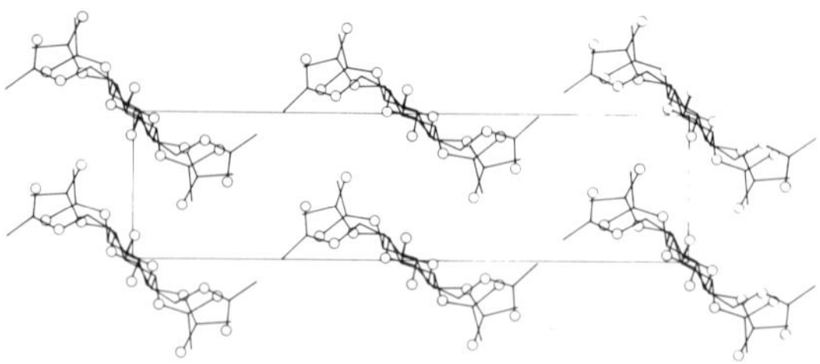

Fig. 7. Projection of CTA I unit cell onto a–b plane [17]. (Courtesy of Butterworths Scientific).

acetylated cellulose II [20]. An antiparallel CTA I-like structure is certainly possible based on stereochemical model analysis [17].

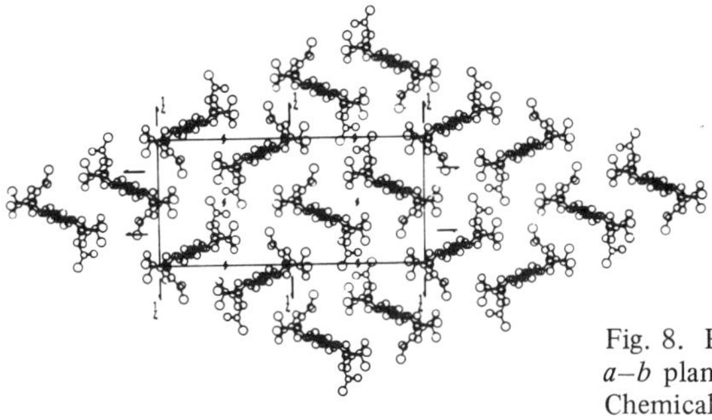

Fig. 8. Projection of CTA II unit cell onto *a–b* plane [18]. (Courtesy of American Chemical Society).

TRANSFORMATIONS AND CHAIN POLARITY

The combined results of crystal structure analyses of all cellulose polymorphs and derivatives completed to date strongly suggest that the cellulose structures fall into only two families: one based on parallel chains and the other on antiparallel chains. The native cellulose I is typical of the parallel-chain family, while cellulose II is the most stable structure within the antiparallel family. In both families, the chain conformations are only minimally different. The same is essentially true of the packing of the chains into sheet-like formations. The main difference between the two families resides in the packing of the sheets: the same polarity in the parallel-chain structures, opposite polarity in the antiparallel-chain structures (cf. Figs. 3 and 4). Transformations between the different structures reflect the degree of similarity (or its absence) in the various structures of the two families. As shown in Fig. 9, depicting the general scheme governing the interconversions of cellulose structures, transformations within each family are reversible. Some of the reversible transformations proceed easily, as for example, I <-> III_I, I <-> CTA I, II <-> III_{II}, and II <-> CTA II. Other such transformations, notably I <-> IV_I and II <-> IV_{II}, proceed only with great difficulty. The reasons for the latter are not clear. As

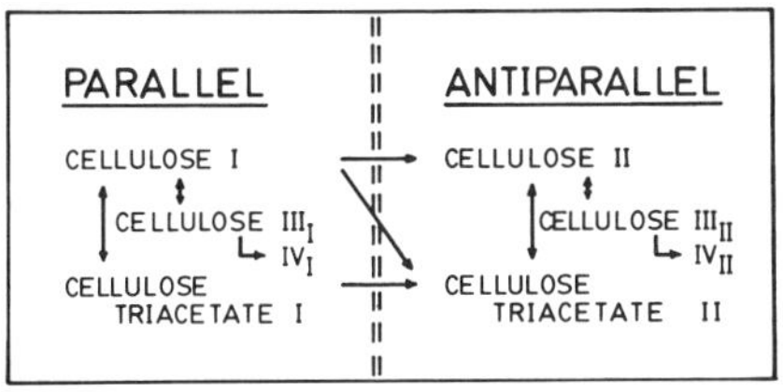

Fig. 9. Interconversions of cellulose polymorphs (including cellulose triacetate). (Courtesy of Gordon & Breach).

shown in Fig. 10, packing energy contour maps of both parallel and antiparallel families predict the existence of energy minima that correspond well with the known cellulose polymorphs. The energy maxima separating the different minima do not provide any clues why some transformations should proceed more easily than others. Obviously, a clearer understanding of the transformation dynamics is needed.

It is also well known that the parallel-to-antiparallel transformations are irreversible. As reference to Fig. 10 shows, the reasons for the irreversibility of the I -> II transformation may be entirely thermodynamic in origin -- cellulose II is the more stable structure of the two. Whether the same circumstance governs the CTA I -> CTA II transformation, or whether other factors such as entropy, kinetics, or statistical considerations come into play, are not yet known. Kinetic considerations are clearly important in the biosynthesis of cellulose I, as on thermodynamic grounds this polymorph should never come into existence. As shown by the brilliant work of R.M. Brown, Jr. and his collaborators, in the biosynthesis of cellulose I by the bacterium <u>Acetobacter xylinum</u> the aggregation of cellulose chains into a microfibril proceeds nearly simultaneously with the synthesis [5]. This directional process

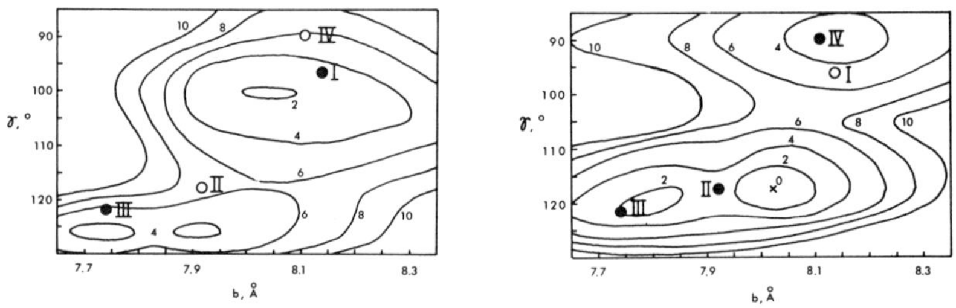

Fig. 10. Contour maps of predicted packing energies of parallel (left) and antiparallel (right) celluloses, as a function of unit cell parameters b and δ. (● represent known polymorphs, ○ unknown polymorphs). (Courtesy of John Wiley & Sons).

of synthesis-aggregation leads naturally to crystallization into a structure that resides in the nearest accessible energy minimum -- cellulose I. Conversely, if the bacterial cellulose synthesis was a process in which cellulose chains remained individually in solution for a length of time prior to crystallization, it is highly probable that cellulose II would be the end product.

MERCERIZATION AND STRUCTURAL TRANSFORMATIONS

When cellulose I is mercerized with alkali (or other chemical agents), the end product is always cellulose II. This process can be very fast (a matter of seconds) or as shown in a series of recent studies, it can be slowed down to weeks or months [21,22]. X-ray diffraction analysis of the intermediate structures observed during the slowed-down, alkali mercerization of ramie has produced interesting new information on the various structures of cellulose, the probable mechanism of mercerization, and the morphology of the cellulose fiber.

Initially, it was clearly demonstrated that the crystallinity of cellulose is not destroyed during the entire mercerization process -- beginning with cellulose I and ending with cellulose II. Instead, a series of apparently crystal-to-crystal phase transformations take place, without the occurrence of intervening amorphous stages. These phase transformations may involve up to five hydrated crystalline forms of cellulose, four of which also contain alkali. The transformations, shown in Fig. 11, are entirely reproducible, and as demonstrated by the x-ray diffraction diagrams shown in Fig. 12, produce very well oriented, fairly highly crystalline structures. These structures, termed "alkali-celluloses" (or "Na-celluloses"), had already been observed by x-ray diffraction as early as the 1930's, but the reproducibility of their transformations had previously not been attained [23].

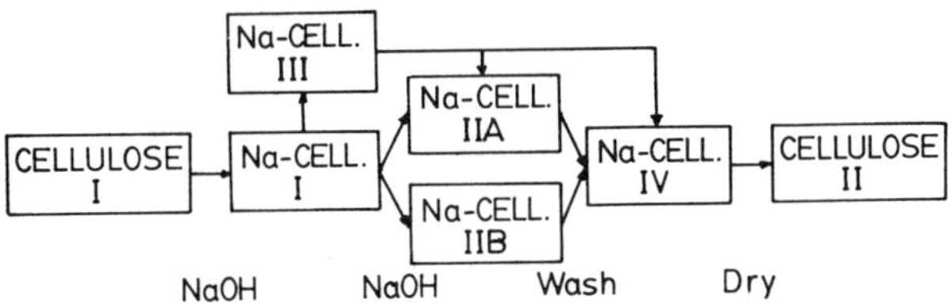

Fig. 11. Transformation of cellulose I to cellulose II through crystalline Na-celluloses during controlled mercerization [21]. (Courtesy of John Wiley & Sons).

Of the five alkali-cellulose structures in Fig. 11, three have been analyzed in detail by x-ray diffraction -- Na-celluloses I, IIB, and IV [24,25]. Features of their structures are illustrated in schematic form in Fig. 13. A number of interesting observations have emerged during the course of these analyses. First, the initial step in the transformation sequence, from cellulose I to Na-cellulose I, is irreversible, suggesting that a chain polarity change already takes place in this step. This is substantiated by the demonstration of <u>mixed</u> crystal structures, composed of unconverted cellulose I and Na-cellulose I, found at all times before the completion of the transformation. A mixed structure, e.g., composed of approximately 50% each of cellulose I and Na-cellulose I, upon quick wash-dry treatment yields a <u>mixture</u> of cellulose I and cellulose II crystal forms in roughly the same proportions [21]. This suggests very strongly that the Na-cellulose I component of the initial mixture, representing converted cellulose I, is the precursor of cellulose II and is thus already an antiparallel structure. Further support for this comes from the observation that cellulose II also gives rise to Na-cellulose I, but that transformation is reversible [21]. The final measure of agreement is found in the results of the crystal structure analysis which is definitely pointing towards an antiparallel Na-cellulose I structure [24].

The second interesting observation made in the course of this study concerns the structures of Na-celluloses IIA and IIB. Both form only after all cellulose I (or II) have been completely converted to Na-cellulose I and there is more than a certain minimum amount of NaOH present. The x-ray diffraction

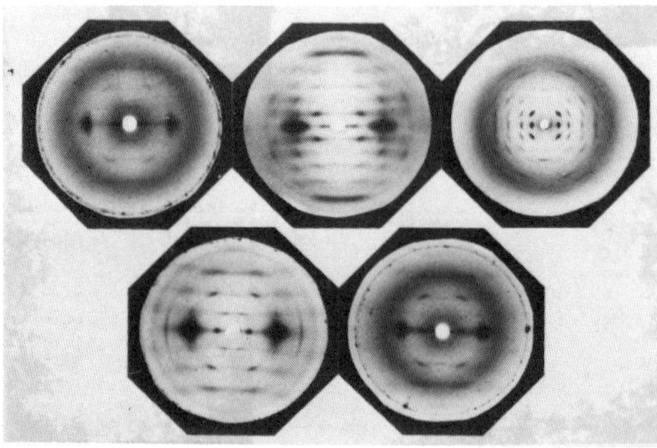

Fig. 12. X-ray diffraction patterns of Na-celluloses shown in Fig. 11. Top, left to right: Na-celluloses I, IIA, IIB; bottom, left to right: Na-celluloses III, IV [21]. (Courtesy of John Wiley & Sons).

diagrams of both Na-celluloses IIA and IIB show evidence of a fiber repeat of approximately 15 Å, longer than the usual 10.34 Å repeat of the pure cellulose forms. As shown in Fig. 13 B, the chain conformation consistent with this repeat is a three-fold helix, considerably different from the planar and ribbon-like conformation of the usual two-fold helix of pure celluloses. The three-fold helical conformation is completely devoid of intra- and intermolecular hydrogen bonds normally seen in cellulose crystal structures, suggesting that the primary attractive interactions present in these structures are those between Na^+ ions and the cellulose hydroxyl groups [25]. Support for this comes from CP/MAS NMR results that indicate a strong attraction between Na^+ ions and certain hydroxyl groups [26].

Finally, another interesting observation concerns the locations of the "guest" ions or molecules in the Na-cellulose complexes. As can be seen in Fig. 13, both in Na-celluloses I and IV the Na^+ ions in the former and water molecules in the latter are located <u>within</u> the sheet structure paralleling the <u>b-c</u> unit cell plane. This type of insertion of the guest ions or molecules destroys the intermolecular hydrogen bonds in the plane of the sheet. At first glance, it may seem more plausible that Na^+ or water would be situated between the planes thus preserving the two-dimensional hydrogen bonding of the sheets. That this does not take place argues for the importance of hydrophobic attractions between cellulose chains, directed roughly at right angles to the plane of the ribbon-like conformation of the chain. In fact, all cellulose structures possessing this conformation show a "stacking" of chains in this direction and an absence of hydrogen bonds along it (cf. Figs. 3-6, 13).

PROBABLE MERCERIZATION MECHANISM AND FIBER STRUCTURE

The transformation sequence of Na-cellulose structures shown in Fig. 11 suggests a model for the structural changes that take

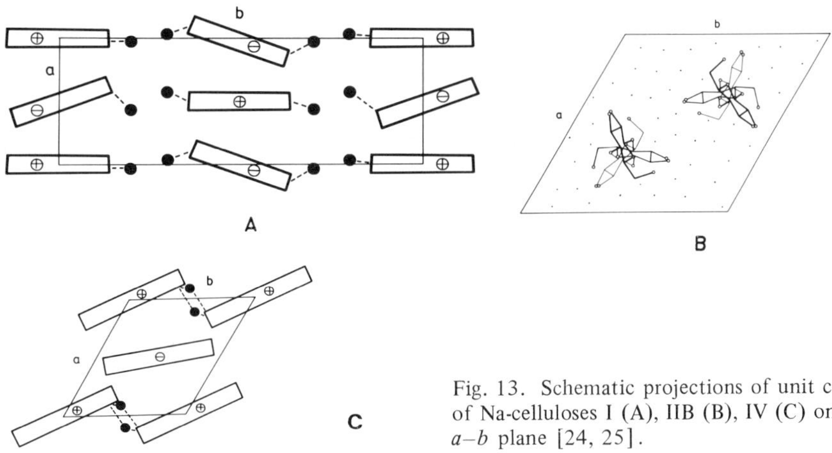

Fig. 13. Schematic projections of unit cells of Na-celluloses I (A), IIB (B), IV (C) onto *a–b* plane [24, 25].

place during mercerization, as illustrated in Fig. 14 [22]. According to this mechanism, NaOH enters the fiber structure in its amorphous regions -- the interfaces between the single-crystalline microfibrils. The initial conversion to Na-cellulose I occurs in these regions, followed by the conversion of cellulose I crystallites. After the conversion to Na-cellulose I has been completed, the structure of the latter ex-

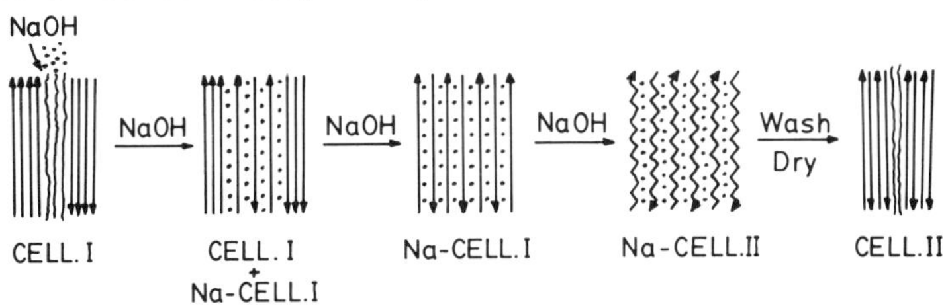

Fig. 14. Probable model of structural changes occurring during controlled mercerization of cellulose I [21]. (Courtesy of John Wiley & Sons).

pands, takes up more NaOH (and water), and converts from a two-fold helical conformation to a three-fold one of Na-celluloses IIA or IIB. Washing out the NaOH results in a hydrate structure (Na-cellulose IV) which upon drying converts to cellulose II.

To test the validity of this mechanism model, a series of crystallite size measurements were made during the cellulose I -> Na-cellulose I transformation [22]. As shown in Fig. 15, the results are in very good agreement with the proposed mechanism. The Na-cellulose I crystal phase is seen to begin forming very early during the transformation, long before the average crystallite size of cellulose I begins to diminish. At the same time, the average size (diameter) of Na-cellulose I crystallites remains constant at approximately 35 Å, suggesting that the amorphous, interfacial regions between the crystallites in ramie are of this order of magnitude. Only after all of the amorphous regions have been converted to Na-cellulose I, will the conversion of the main bulk of cellulose I crystallites take place, resulting in the final Na-cellulose I structure in which the average crystallite size is approximately 50 Å. The

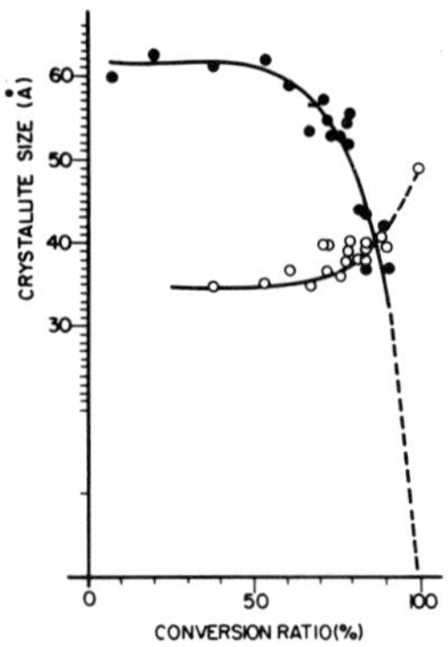

Fig. 15. Crystallite sizes of ramie cellulose I (●) and Na-cellulose I (○) during the first conversion step shown in Fig. 11 [22]. (Courtesy of John Wiley & Sons).

time course of this conversion was followed carefully, determining not only the average crystallite sizes, but also their size distribution. As shown in Fig. 16, the initial conversion rate is quite rapid, slowing down only after the amorphous phase has been converted (cf. also Fig. 15). During the rapid transformation, crystallite size distribution changes, indicating that along with the amorphous regions, the smallest crystallites of cellulose I are also being converted to Na-cellulose I [22]. The beginning of the slow rate of conversion coincides with the end of the conversion of these small crystallites. All of these observations support the mechanism shown in Fig. 14. They further suggest that these processes occur on a molecular and crystallite scale as shown in Fig. 17.

A model of the fiber structure that is consistent with this mechanism is shown in Fig. 18. A single fiber of ramie is composed of many microfibrils whose average diameter is approximately 65 Å (cf. Fig. 15). Even though the chain directions in

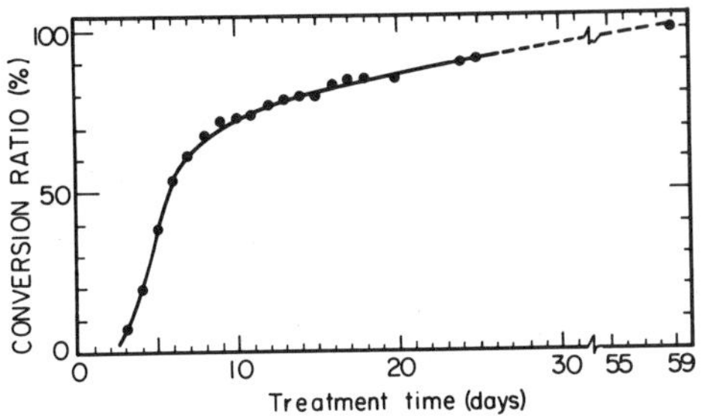

Fig. 16. Time course of the conversion shown in Fig. 15 [22]. (Courtesy of John Wiley & Sons).

each microfibril are parallel (because the latter are essentially single-crystalline in form) the assembly of the microfibrils into a macroscopic fiber cannot be expected to be anything but statistically random. Each fiber will thus contain roughly equal numbers of "up" and "down" oriented microfibrils. As a consequence, the regions between microfibrils will have chains oriented in both directions -- providing ideal sites for creating an antiparallel crystal structure with a minimum amount of lateral chain segment movement. In such a mechanism, there is no necessity to postulate major changes in the morphology of the fiber or having to "turn chains upside down". The same applies to the destruction of very small crystallites. In the latter, the surface-to-volume ratio is large, which al-

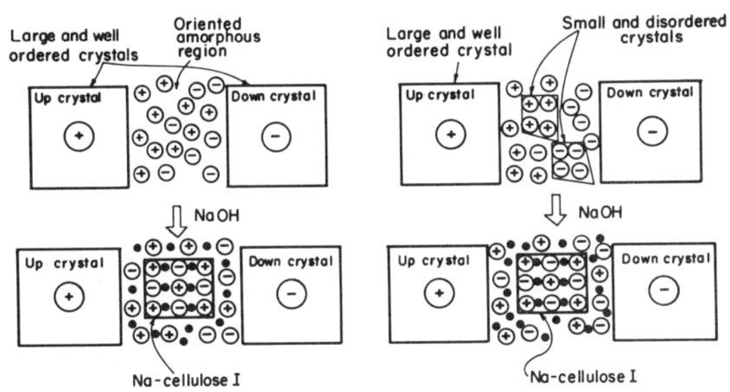

Fig. 17. Probable initial conversion mechanism of cellulose I to Na-cellulose I. Left: Amorphous regions; right: small crystallites [22]. (Courtesy of John Wiley & Sons).

most ensures that their surrounding amorphous structure contains chains of both orientations and therefore, suitable sites for the change to an antiparallel Na-cellulose I structure [24].

It is also possible that a parallel-chain Na-cellulose I (or one that is disordered) may be created from small cellulose I crystallites. Because such structures do not differ greatly from the antiparallel structure -- but are of higher energy -- it is not expected that they will remain in this form for very long, given the gel-like structure that prevails locally during the conversion. The presence of such a partially or substantially parallel-chain Na-cellulose I structure should be detectable by its reversion to cellulose I upon quick removal of NaOH. This reversion, in fact, has been claimed by Hayashi et al., in mercerizations that proceed at a fast rate [27]. Although this observation has not been duplicated in the author's laboratory, it is known that such Na-cellulose I structures, produced at a rapid rate, convert spontaneously to the usual, stable, antiparallel Na-cellulose I structure [28]. This observation agrees with theoretical predictions that a parallel-chain Na-cellulose I should be less stable than its antiparallel analog.

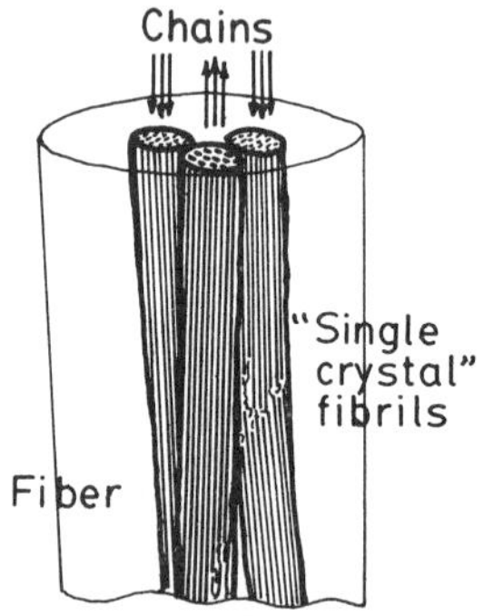

Fig. 18. Probable model of ramie cellulose fiber [21]. (Courtesy of John Wiley & Sons).

ACKNOWLEDGEMENT

This research has been sponsored by National Science Foundation under grants GP27997, CHE7501560, CHE7727749, CHE8107534, and PCM8320548; and by Hercules, Inc.

REFERENCES

1. R.H. Marchessault and A. Sarko, Adv. Carbohydr. Chem., 22, 421 (1967).
2. P. Zugenmaier and A. Sarko, in Fiber Diffraction Methods, A.D. French and K.H. Gardner, Eds., ACS Symp. Ser., 141, American Chemical Society, Washington, D.C., 1980, p. 225.
3. A. Sarko, in Cellulose: Structure, Modification, and Hydrolysis, R.A. Young and R.M. Rowell, Eds., John Wiley, New York, 1986, p. 29.
4. F. Horii, A. Hirai, and R. Kitamaru, in Polymers for Fibers and Elastomers, J.C. Arthur, Jr., Ed., ACS Symp. Ser., 260, American Chemical Society, Washington, D.C., 1984, p. 27.
5. C.H. Haigler, R.M. Brown, Jr., and M. Benziman, Science, 210, 903 (1980); R.M. Brown, Jr., Ed., Cellulose and Other Natural Polymer Systems, Plenum Press, New York, 1982.

6. J. Honjo and M. Watanabe, Nature (London), 181, 326 (1958).
7. A. Sarko and R. Muggli, Macromolecules, 7, 486 (1974).
8. C. Woodcock and A. Sarko, Macromolecules, 13, 1183 (1980).
9. R. Muggli and A. Sarko, unpublished diagram.
10. K.H. Gardner and J. Blackwell, Biopolymers, 13, 1975 (1974).
11. K.H. Meyer and L. Misch, Helv. Chim. Acta, 20, 232 (1937).
12. A.J. Stipanovic and A. Sarko, Macromolecules, 9, 851 (1976).
13. F.J. Kolpak and J. Blackwell, Macromolecules, 9, 273 (1976).
14. A. Sarko, J. Southwick, and J. Hayashi, Macromolecules, 9, 857 (1976).
15. E.S. Gardiner and A. Sarko, Can. J. Chem., 63, 173 (1985).
16. A. Buleon and H. Chanzy, J. Polym. Sci.: Part A-2, 18, 1209 (1980).
17. A.J. Stipanovic and A. Sarko, Polymer, 19, 3 (1978).
18. E. Roche, H. Chanzy, M. Boudeulle, R.H. Marchessault, and P. Sundararajan, Macromolecules, 11, 86 (1978).
19. E.S. Gardiner and A. Sarko, in Proceedings of the Ninth Cellulose Conference, A. Sarko, Ed., J. Appl. Polym. Sci.: Appl. Polym. Symp., 37, John Wiley, New York, 1983, Vol. 1, p. 303.
20. S. Watanabe, M. Takai, and J. Hayashi, J. Polym. Sci.: Part C, 23, 825 (1968).
21. T. Okano and A. Sarko, J. Appl. Polym. Sci., 29, 4175 (1984); ibid., 30, 325 (1985).
22. H. Nishimura and A. Sarko, J. Appl. Polym. Sci., in press (1987).
23. J.O. Warwicker, R. Jeffries, R.L. Colbran, and R.N. Robinson, Shirley Institute Pamphlet No. 93, Shirley Institute, Didsbury, Manchester, England, 1966, p. 90.
24. H. Nishimura and A. Sarko, to be published.
25. T. Okano and A. Sarko, to be published.
26. H. Nishimura, private communication.
27. J. Hayashi, T. Yamada, and K. Kimura, in Proceedings of the Eighth Cellulose Conference, T.E. Timell, Ed., J. Appl. Polym. Sci.: Appl. Polym. Symp., 28, John Wiley, New York, 1976, Part II, p. 713.
28. J. Hayashi, private communication.

Chapter 7

Characteristics of heterogeneously acetylated cellulose

Stephen E. Doyle, Anne Marie Gibbons and Richard A. Pethrick – Department of Pure and Applied Chemistry, University of Strathclyde, Cathedral Street, Glasgow G1 1XL, Scotland, UK.

INTRODUCTION

Cellulose acetate formation is a complex process involving first of all activation of wood or cotton by treatment with acetic acid, followed by prolonged acetylation. The initial treatment with acetic acid requires careful control of the moisture level and the contact time in order to ensure that the optimum conditions are achieved during the reaction.The viscous solution of cellulose triacetate formed is usually precipitated by addition of water or a dilute solution of acetic acid and the product has the form of flakes or a fluffy powder. During the acetylation process the original structure of the cellulose is destroyed and a new less ordered structure is generated.

In Cellucon 84, several papers reviewed the structure of cellulose [1 - 3] and we presented our initial observations on the synthesis and properties of fibrous cellulose acetate. In this paper we will consider some of the factors which influence the formation of fibrous cellulose acetate and the type of structure found in this material. Cellulose exhibits a variety of levels of structure. At a molecular level, cellulose - poly-cellobiose is composed of 1,4-linked anhydro β-D-gluco-pyranose rings, each having a chair conformation.Interaction between the pendant hydroxyl groups associated with carbon atom 3 and the ring oxygen atom leads to the development of a two-fold helical structure, figure(1). The cellulose molecules can either present a hydrophobic or a hydrophilic face to its neighbours and as a result they will tend to organize themselves into stacks held together by Van der Waals forces. The result is an aggregate structure in which the external surfaces are saturated with readily accessible hydroxyl groups. Most of the reactions involving cellulose use an initial process whereby the primary cellulose chains are solvated. This process is usually performed either by attaching suitable solvating groups to the hydroxyl hence forcing the chains apart or by degradation of the cellulose chains themselves. It is however possible that with careful control of the reaction the surface hydroxyl groups can be reacted allowing solvation of the stack without the need to degrade the chain structure.

The stacks of cellulose molecules are themselves organized into the elementary fibril structures which in turn form micro-fibrils, fibrils and ultimately the fibre, figure(2). Chemical modification of cellulose is therefore subject to accessibility of the reaction sites and hence many of the processes will contain an element of morphological control.

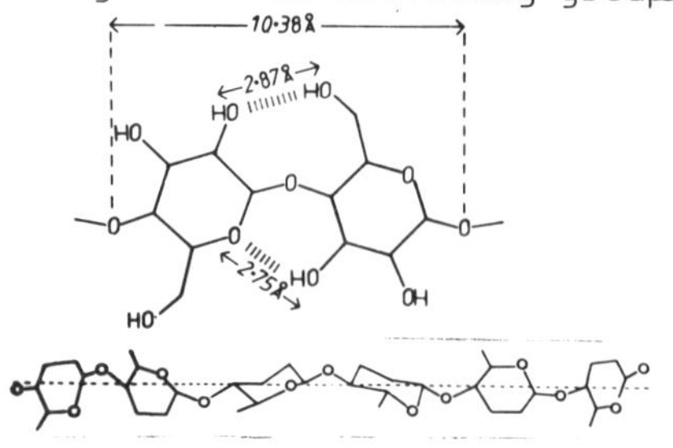

Fig. 1. 1,4β-D-Glucopyranose

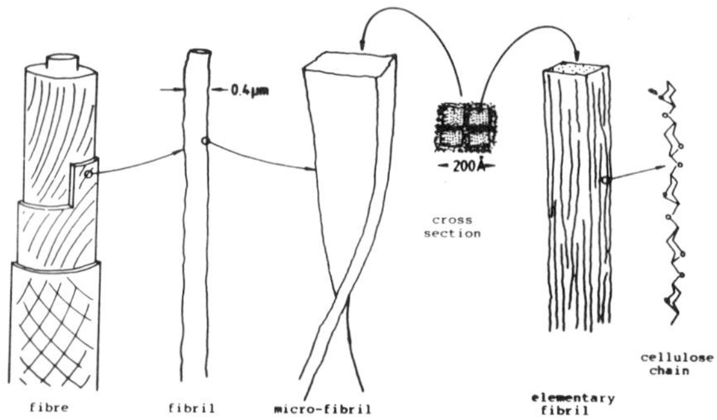

Fig. 2. Morphology of a cellulose fibre.

Natural cotton fibre will have an outer cuticle composed of lignins, waxes and other non-cellulosic material. Washing the cotton linters with dilute sodium hydroxide allows removal of the outer skin without influencing the structure of the fibres forming the main stem of the linters. Mono-fibrillar cellulose has an outer layer, known as the primary wall, consisting of cellulose fibrils interwoven in a cross-hatched fashion, figure (3). Inside this outer layer are a number of spiral wound layers culminating in an innermost pipe like structure in which the cellulose chains are orientated parallel to the stem axis.

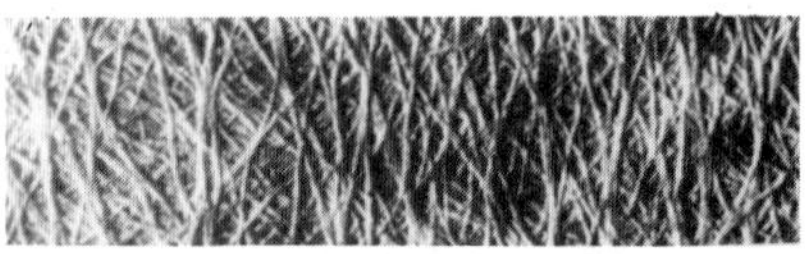

Fig. 3. Surface of linters.

This sturcture is typical of cotton linters and other monofibrillar cellulose materials.
A summary of the various levels of organization found in the cellulose fibre are presented in figure(4). In this paper we are concerned with the changes which occur in the cellulose fibres when they are subjected to acetylation.

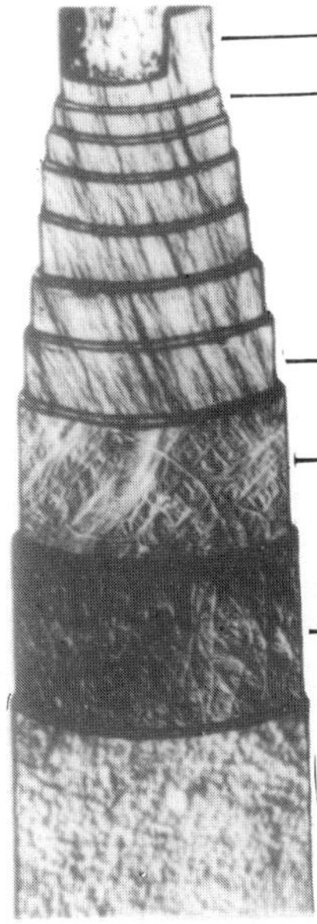

Fig. 4. Cotton linters

EXPERIMENTAL

Synthesis of fibrous cellulose acetate

A series of samples with varying degrees of acetylation were prepared from cotton linters by heterogeneous acetylation using the method of Tanghe[4]. A 10g portion of linters was placed in a 250 ml conical flask and a mixture of glacial acetic acid (80 ml), toluene (120 ml) and 72% perchloric acid (2 ml) added. The reaction mixture was vigorously shaken for a few minutes and then immediately poured back into the flask containing the linters. This procedure minimises the possibility of a high initial concentration of acetic anhydride contacting fibres closest to the neck of the flask and producing material with a higher than average degree of acetylation. The flask was stoppered and maintained at 303 K for a period between 5 minutes and several hours; the mixture being agitated periodically. Reaction was stopped by removal of the linters and washing with ethanol and then with water. The washing was continued until one drop of 0.05 M sodium hydroxide was sufficient to impart a faint pink colour to a 100 ml aliquot of the washing water containing phenolphthalein. The acetate was then washed a final time in ethanol and subsequently dried in vacuo over night at 333 K. The degree of acetylation was determined using a wet analysis method [5].

X-ray diffraction measurements

The method used has been described previously [6]. The residual scatter in the camera was determined using a silver foil reference. Corrections were not applied for Lorentz or polarization factors since these effects will be common for all the samples. The percentage crystallinity was determined using the method of Hermens and Weidinger [6] whereby the ratio of crystalline to amorphous material was determined by drawing a smooth curve through the minima of the principal diffraction features and areas above and below the curve were compared over an angular range of 2θ from 4 to 41° (d = 23 to 2.2 Å).

Electron microscopy

Samples of the polymer mounted on aluminium stubs and coated with gold were examined using a Philips PSM 100 scanning electron microscope. The micrographs were recorded photographically.

RESULTS AND DISCUSSION

Electron microscopy and X-ray diffraction measurements

Electron micrographs of unsubstituted cotton linters and a sample with DOS of 2.5 are shown in figure(5). Acetylation of the linters leads to an opening of the fibre structure with associated rupture of the internal fibrillar morphology. Normal commercial cellulose triacetate would have been degraded to a powder at this level of DOS. It is clear that the heterogeneous reaction used in this study leads to a very fibrous product, strongly resembling the form of the original cellulose linters.

(a) (b)

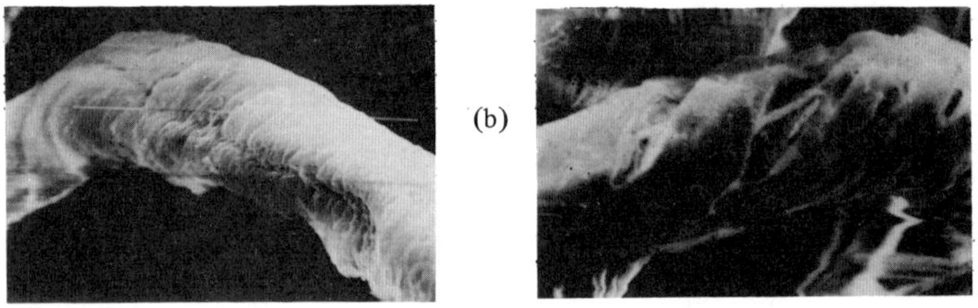

Fig. 5. Electron micrographs of cellulose – (a) and acetylated product with DOS of 2.5 – (b).

X-ray diffraction measurements on these samples have been reported elsewhere[7]. The unmodified cellulose linters were relatively long - 5 mm and therefore easily aligned, so that the radial spread of reflections observed was due to misalignment of the microfibrils in the fibre and not the sample. Due to the short length 1-2 mm and the brittle nature of the cellulose triacetate a poorer orientational pattern was obtained. The most obvious change occurring during acetylation was the development of a lattice spacing at 11.7 Å. Analysis of the X-ray data indicates that the acetylation process leads to an increase in the amorphous content of the material, however the generation of a crystalline peak associated with cellulose triacetate was very apparent. A plot of the intensity of the peak associated with the 11.7 Å spacing, indicates that the acetylation process used in this study allows retention of a significant degree of crystalline order in the fibre. A comparison of the electron micrographs and also the X-ray diffraction data indicate that at both the molecular and macroscopic level order is retained.

Influence of contact time and catalyst level on the morphology

It was observed in our initial studies on the heterogeneous acetylation of cellulose that the nature of the fibre formed was a function of the contact time and also the level of catalyst used in the synthesis. A more complete description of these studies will be published elsewhere[8], however it is relevant to present the main conclusions here.

i) Effects of contact time

The DOS of the samples were found to depend on two factors for a constant catalyst level; these were the amount of the anhydride used in the reaction and the contact time. In all cases it was found that an excess of the anhydride was necessary to achieve a particular level of acetylation. This implies that the acetylation process is an equilibration reaction. The rate at which equilibration is achieved depends on the DOS required and also the level of catalyst. For a constant level of catalyst the lower DOS samples were equilibrated faster than those with high DOS values. In most cases an equilibrium value had been achieved after 10 hours , however in some cases it was almost 20 hours before true equilibration was observed.

An optical examination of the fibres obtained using different reaction conditions indicated that once equilibration had occurred then the main effect was to shorten the fibre length. After two hours a broad distribution of fibre lengths would be observed ranging from values close to those of the original linters - 4 to 5 mm down to 0.1mm. As the contact time was increased so the average length of the fibre becomes shorter and values of the average were about 0.4 mm for a contact time of 47 hours.

It is apparent that during the acetylation a certain amount of degradation of the cellulose fibres occurs, the extent of the degradation depending upon the contact time and anhydride level.

ii) Influence of catalyst level

It has been observed that the DOS achieved with a given anhydride content depends on the level of perchloric acid used in the synthesis. The optimum degree of acetylation is achieved with a level of 2 ml of catalyst. It is also interesting to examine the way in which the fibre length varies with the level of catalyst. Increasing the level of catalyst from zero to 2ml leads to a shortening of the mean length of the fibre from a value of about 2mm to a value of 1mm. Further increase in the catalyst level leads to a suppression of the degradation and the observation that the length of the fibres once more increase. This observation implies that it is the anhydride which produces the degradation and not the perchloric acid catalyst.

Conclusions

In this paper we have summarized some of our observations on heterogeneously acetylated cellulose. Very high levels of substitution can be achieved whilst retaining a high level of ordered structure. Acetylation does lead to degradation but using this approach it is possible to generate a material with a

DOS of almost 3 whilst still retaining a fibre structure and at a molecular level a degree of crystallinity.

ACKNOWLEDGEMENTS

The authors wish to acknowledge the support of the MOD(Waltham) and the ROF(Waltham Abbey) for this study. The help and guidance of Drs F S Baker and T.J.Lewis are as always gratefully acknowledged.

REFERENCES

1) Hans Krässig, Cellulose and its Derivatives, Ed. J.F.Kennedy, G.O.Phillips,D.J.Wedlock and P.A.Williams,Ellis Horwood, Chichester, 3 (1985)
2) M.Lewin, Cellulose and its Derivatives, Ed.J.F.Kennedy, G.O. Phillips, D.J.Wedlock and P.A.Williams Ellis Horwood, Chichester 27 (1985)
3) J.O.Warwicker, Cellulose and its Derivatives, Ed. J.F.Kennedy G.O.Phillips, D.J.Wedlock and P.A.Williams Ellis Horwood Chichester, 3 (1985)
4) L.J.Tanghe, Methods in Carbohydrate Chemistry and Biochemistry Vol III, R.L.Whistler (Ed) Academic Press (1963) 193.
5) BS 2880: 1957 Methods of Testing Cellulose Acetate Flake.
6) P.H.Harmens and A.Weidinger J.Appl. Phys 19 491 (1948)
7) S.E.Doyle and R.A.Pethrick, J.Appl. Polym. Sci. (in press)
8) A.M.Gibbons and R.A.Pethrick (unpublished data).

Chapter 8

Molecular ordering in nitrocellulose

C. Viney and A. H. Windle – Department of Metallurgy and Materials Science, University of Cambridge, Pembroke Street, Cambridge CB2 3QZ, U.K.

INTRODUCTION

Studies of the molecular ordering in native cellulose I [1], regenerated cellulose II [2] and mercerised cellulose [3] are facilitated by the high degree of crystallinity in these materials. In particular, sharp x-ray diffraction peaks are obtained, which allow relatively easy determination of the unit cell.

Incompletely substituted cellulose derivatives, such as commercial grades of nitrocellulose (NC), are considerably less crystalline. Instead of thinking in terms of a unit cell, it is more appropriate to consider the structure in terms of local molecular conformation and correlation. Such cellulose derivatives share many characteristics common to liquid crystal-forming polymers [4,5]: the molecular chains are rigid or semi-flexible, have a high aspect ratio, and are unable to crystallise fully in three dimensions. In liquid crystalline random copolymers, for example, the molecules are specifically designed to discourage crystallinity. There is still, however, the possibility of crystal-like entities forming on a *local* scale, particularly in annealed specimens, as identical random sequences on adjacent chains match laterally [6]. The development of such local crystallinity in NC will be discussed below.

The detailed molecular structure of NC fibres can be expected to depend on supramolecular ordering dictated by the cellulose source organism. For this reason, we have developed a technique for drawing fibres of reconstituted NC, having a high degree of molecular alignment. Thus structural aspects which are not a direct consequence of the original positioning of molecules can be elucidated. Introducing molecular orientation into the reconstituted NC aids interpretation of the corresponding wide angle x-ray diffraction patterns, since the scattering can then be resolved easily into inter- and intra-molecular components.

EXPERIMENTAL

The mesogenic characteristics of NC molecules result in anisotropic (lyotropic liquid crystalline) solutions forming in particular solvents [4,7]. The Flory interaction parameter [8] between solvent and polymer can be modified by changing the temperature or by adding non-solvent [9], and the equilibrium phases formed can be plotted as a function of composition on a ternary diagram [10]. Our results for the system NC (DS=2.72) / tetrahydrofuran (THF; solvent) / ethanol (EtOH; non-solvent) are reproduced in Fig.1. The broken lines indicate the existence of a crystallosolvate phase; the kinetics of its formation are such

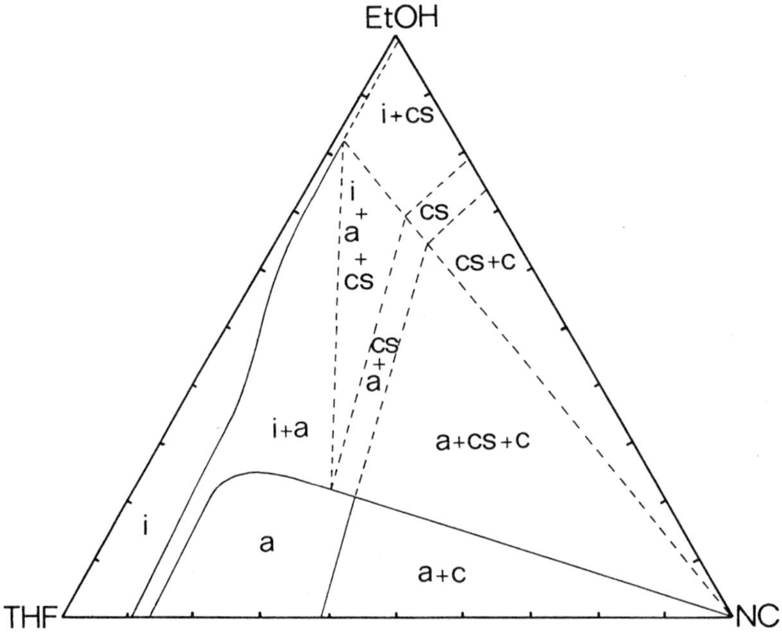

Fig. 1. Room temperature section through the phase diagram for NC/THF/EtOH. Adapted from ref. 7.
i = isotropic; a = anisotropic (mesophase); c = crystalline; cs = crystallosolvate.

that it has only been accessed by evaporating volatiles from the isotropic phase produced by compositions plotting in the three-phase field on the diagram [7]. A similar diagram has been obtained for NC with a DS of 2.32, but with no evidence of any crystallosolvate formation.

Anisotropic solutions lend themselves well to fibre drawing, since molecules are already aligned into domains, and it is only necessary to achieve mutual alignment of the domains during drawing [11]. Fig.2 shows schematically the apparatus used to

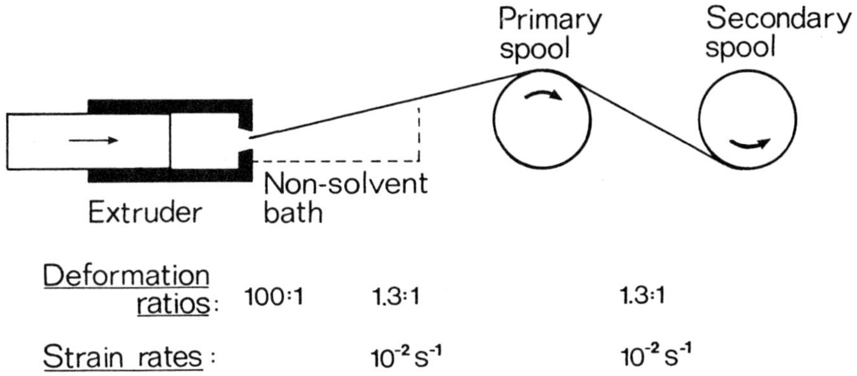

Fig. 2. Schematic side-view of apparatus used to extrude and draw fibres of reconstituted NC from NC/THF/EtOH compositions. Parameters characterising the various stages of the process are given.

draw fibres. Fibres drawn from initially single phase anisotropic solutions were extruded into EtOH (giving improved cohesivity on drawing), while those drawn from biphasic (isotropic+anisotropic) solutions were extruded into air. The former route generally gave fibres with greater overall perfection of molecular ordering. The latter route gave less perfect ordering, since the product contains regions of imperfectly aligned isotropic phase, but the *conditions* of ordering immediately prior to drawing are more exactly defined. We are confident that the product does not contain a significant amount of crystallosolvate. There is no weight loss on heating the fibres to 150°C (crystallosolvate decomposes rapidly at this temperature [7]), and there are no peaks assignable to crystallosolvate in the wide angle diffraction patterns of the reconstituted fibres.

A microbeam camera was used to obtain wide angle x-ray diffraction patterns from single fibres. Images were digitised with an image processing system [12] which reads data from the x-ray film via a scanning microdensitometer. Equatorial and meridional sections through the data are then easily obtained.

RESULTS AND DISCUSSION

Fig.3 shows diffraction data from 4 specimens: nitroramie (DS=2.35), drawn reconstituted nitroramie (DS=2.35), nitroviscose (DS=2.48), and drawn reconstituted nitroviscose (DS=2.48). The reconstituted specimens were drawn from biphasic

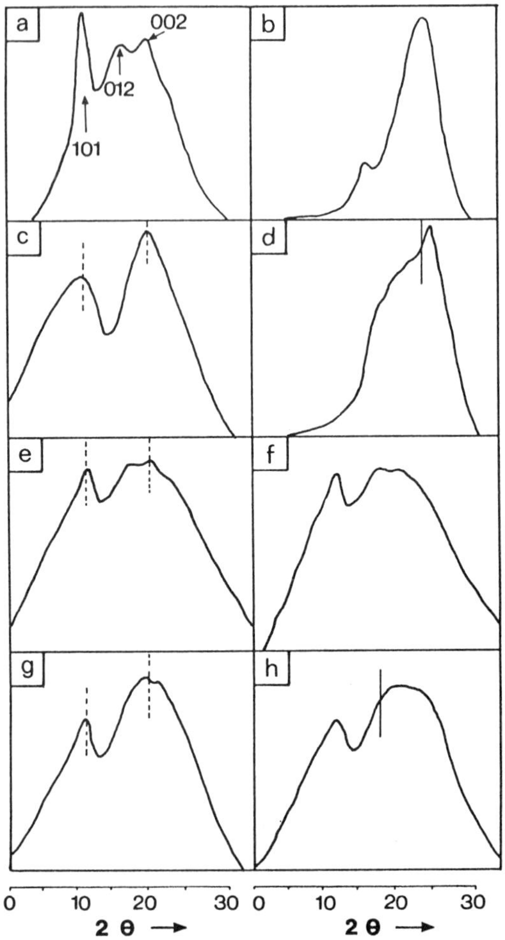

Fig. 3. Sections through microbeam x-ray diffraction data. Left: equatorial; right: meridional. (a), (b): nitroramie, DS = 2.35 (c), (d): drawn reconstituted nitroramie, DS = 2.35 (e), (f): nitroviscose, DS = 2.48 (g), (h): drawn reconsituted nitroviscose, DS = 2.48 Vertical broken lines in (c,e,g) denote the positions of (101) and (002) maxima as determined from (a). Vertical solid lines in (d) and in (h) denote the positions of the strongest reflections in (b) and in (f) respectively.

(isotropic + anisotropic) solutions. The absence of x-ray alignment in the nitroviscose is surprising in view of the high degree of alignment evident in unnitrated material. We were less successful in introducing alignment into reconstituted nitroviscose than we were in aligning reconstituted nitroramie; this is most likely a consequence of the lower molecular weight of the nitroviscose.

An immediate conclusion which derives from the equatorial data in Fig.3, is that the spacing between NC molecules, measured from as-nitrated material, represents an equilibrium value: the positions of the (101) and (002) reflections do not depend on the history of the specimens investigated. Native cellulose, in contrast, undergoes changes of several per cent in the unit cell dimensions normal to the chain direction when mercerised or reconstituted [1-3].

It was not possible to distinguish layer lines in the diffraction patterns from either as-nitrated or reconstituted material. One cannot therefore measure a helix repeat distance directly. Only nitration levels approaching the trinitrate give sufficient crystallinity for such a measurement to be made. However, the position of the most significant peak on the meridian is displaced to higher angles when nitroramie is dissolved and reconstituted (Figs 3b and 3d), and there is an indication that the same is also true for nitroviscose (Figs 3f and 3h). This suggests another departure from the behaviour of native cellulose, where the repeat distance along the chain is hardly affected by mercerisation / reconstitution. We suggest that the NC molecules in solution, afforded greater freedom to achieve their equilibrium conformation, adopt a tighter helix than in the as-nitrated state, promoted by the possibility of forming stronger nitrate-hydroxyl hydrogen bonds. Detailed computer modelling of possible conformations is in progress. Diffraction patterns calculated from the models will be compared with two-dimensional diffraction data collected in symmetrical transmission.

Some of the x-ray reflections are sharper than might be expected from a system of molecules in which the monomers occur in random sequence along the chains. Fig.4 shows equatorial sections of the x-ray patterns obtained from nitrocotton (DS=2.72), both for polymer drawn from biphasic isotropic + anisotropic solution and from single phase anisotropic solution. The highest degree of orientational order, and the greatest lateral extent of ordered regions, is evident in the later case, where the molecules are presumed to have had a high degree of *local* alignment prior to drawing. Whether or not such molecules can form lateral matches as a result of containing identical random sequences of monomers, raises statistical questions. An analysis for random copolymers consisting of two monomer types has been developed [6,13], where laterally matched sequences of the random chains are called "non-periodic layer" (NPL) crystallites. If we extend the model to include *three* monomer

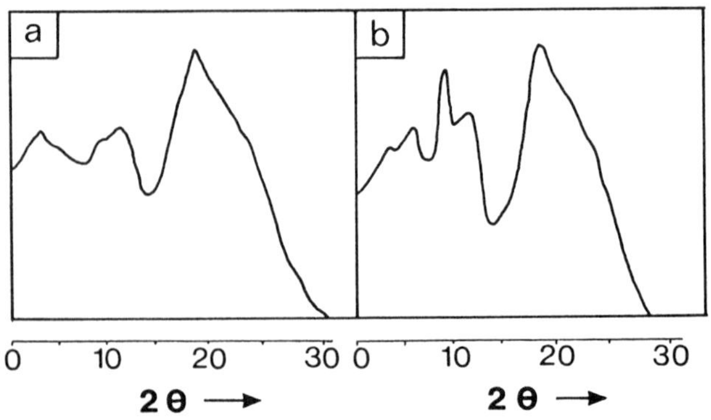

Fig. 4. Equatorial sections through microbeam x-ray diffraction data: a) Fibre drawn from two-phase NC (DS = 2.72)/THF/EtOH b) Fibre drawn from single-phase NC (DS = 2.72)/THF/EtOH, through EtOH.

types, we at least have a basis for describing NC with a DS of 2.62 or higher, the monomers in this case being [14]: 2,6-dinitrate, 3,6-dinitrate, and 2,3,6-trinitrate. The model can then be developed as follows:

A DS of 2+x (greater than 2.62) requires a probability x of a monomer being triply nitrated, while we assume that each of the doubly nitrated monomers has a probability of (1-x)/2 of occurring. Assuming these monomer types to be of equal length and randomly distributed along the chains, we "built" 100 molecules, each 100 units long, using a desk-top microcomputer. The first molecule is taken to be fixed in position. The second is placed next to it and shifted longitudinally, until the best (i.e. the longest) match between the chains is found. The second chain is then fixed in position, the third is placed next to it, and the process is repeated. In this way, a "primary match" may be propagated laterally for a distance of several chains, until a "better" match is nucleated at some other point along the current chain. A "nucleation weighting" can be introduced into the model, allowing for the fact that a crystallite, once nucleated, will continue to propagate laterally in preference to a new NPL crystallite being nucleated somewhere else along the chain. The conditions of molecular mobility necessary to allow searching for the best match, are most likely to be met in practice if the chains are locally aligned and in a shear flow environment. Such conditions approximate to our drawing of fibres from single-phase anisotropic NC/THF/EtOH compositions.

Fig.5(a-c) shows the extent of the primary matches predicted by our computer model, for nucleation weightings of 1:1, 2:1 and 5:1 respectively. The percentage of monomer units involved in the matches is 13.2%, 11.3% and 10.9% for the three cases respectively, giving a rough estimate of the degree of

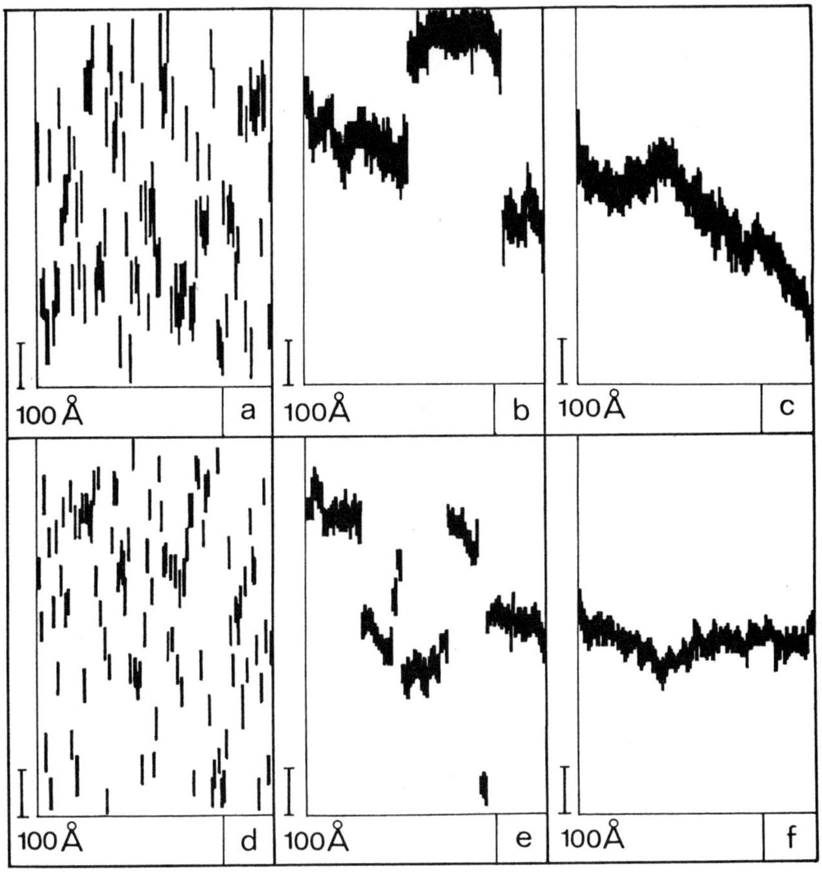

Fig. 5. The extent of "primary matches" derived from the computer model for NC. Each frame is 100 monomer units high and 100 chains wide. (a) – (c) DS = 2.72; (d) – (f) DS = 2.45. (a,d) nucleation weighting 1:1; (b,e) nucleation weighting 2:1; (c,f) nucleation weighting 5:1.

crystallinity. If we neglect the small number of unnitrated and 6-mononitrated residues in polymer having a DS of 2.45, the predictions of the model are as shown in Fig.5(d-f); the corresponding percentages of monomers involved in primary matches are now 8.3%, 6.9% and 6.2%.

The simple model therefore demonstrates that significant lateral order can form in a system comprised of entirely random chains. In practice, the percentage of monomer units involved in local reinforcing units could be higher than suggested by the above figures: while the primary matches may fix the relative register of neighbouring chains, smaller secondary matches between neighbouring chains can then also be identified. Refinement of the model will also need to take account of "blockiness": there is evidence that the sequence of monomer types is not entirely random [14].

ACKNOWLEDGEMENTS

This work has been carried out with the support of Procurement Executive, Ministry of Defence. We thank Derek Cosgrove for nitrating our samples, Drs T.J. Lewis and F.S. Baker for helpful discussions, and Simon Hanna for refinements to the computer model.

REFERENCES

1. Gardner, K.H. and Blackwell, J., 1974, *Biopolymers*, 13, 1975
2. Kolpak, F.J. and Blackwell, J., 1976, *Macromolecules*, 9, 273
3. Kolpak, F.J., Weih, M. and Blackwell, J., 1978, *Polymer*, 19, 123
4. Gray, D.G., 1983, *Journal of Applied Polymer Science: Applied Polymer Symposium*, 37, 179
5. Gray, D.G., 1985, *Faraday Discussions of the Chemical Society*, 79, 257
6. Windle, A.H., Viney, C., Golombok, R., Donald, A.M. and Mitchell, G.R., 1985, *Faraday Discussions of the Chemical Society*, 79, 55
7. Viney, C. and Windle, A.H., 1986, *Liquid Crystals*, in press
8. Flory, P.J., 1953, *Principles of Polymer Chemistry* (Cornell University Press, Ithaca), pg.509
9. Flory, P.J., 1956, *Proceedings of the Royal Society (Lond.)*, A234, 73
10. Papkov, S.P., 1984, *Advances in Polymer Science*, 59, 75
11. Prevorsek, D.C., 1982, *Polymer Liquid Crystals*, Ed: Ciferri, A., Krigbaum, W.R. and Meyer, R.B. (Academic Press, N.Y.), Chap.12.
12. Saxton, W.O. and Koch, T.L., 1982, *Journal of Microscopy*, 127, 69
13. Hanna, S. and Windle, A.H., 1986, to be submitted
14. Clark, D.T., Stephenson, P.J. and Heatley, F., 1981, *Polymer* 22, 1112

Chapter 9

Structural models of cellulose derivatives in the crystalline and liquid crystalline state

P. Zugenmaier – Institut fur Physikalisch Chemie der TU Clausthal D-3392 Clausthal-Zellerfeld, FRG.

ABSTRACT

The conformation of single chain molecules in the crystalline state are presented for some cellulose esters that were produced by a heterogeneous and homogeneous reaction, respectively. The determination of the conformation was based on X-ray analysis of crystalline fibres from which the packing of the chain molecules can also be concluded. Comparison of the X-ray patterns of variously derived cellulose leads to some new insights into the crystalline packing of cellulose chains.

If cellulose derivatives as cellulose tricarbanilate or ethyl cellulose are dissolved in high concentrations in various solvents, cholesteric liquid crystalline phases are observed. They are characterized by the pitch of the twisted nematic sheets that form helicoidal structures. The pitch with handedness of the cholesteric structures have been determined by different methods and found to be a function of temperature, concentration, solvent or solvent mixture and molecular mass. The flow behaviour of liquid crystalline cellulose is strongly dependent on structural features of the cholesteric phases.

INTRODUCTION

Structural models are a prerequisite for the understanding of mechanisms and interactions occurring

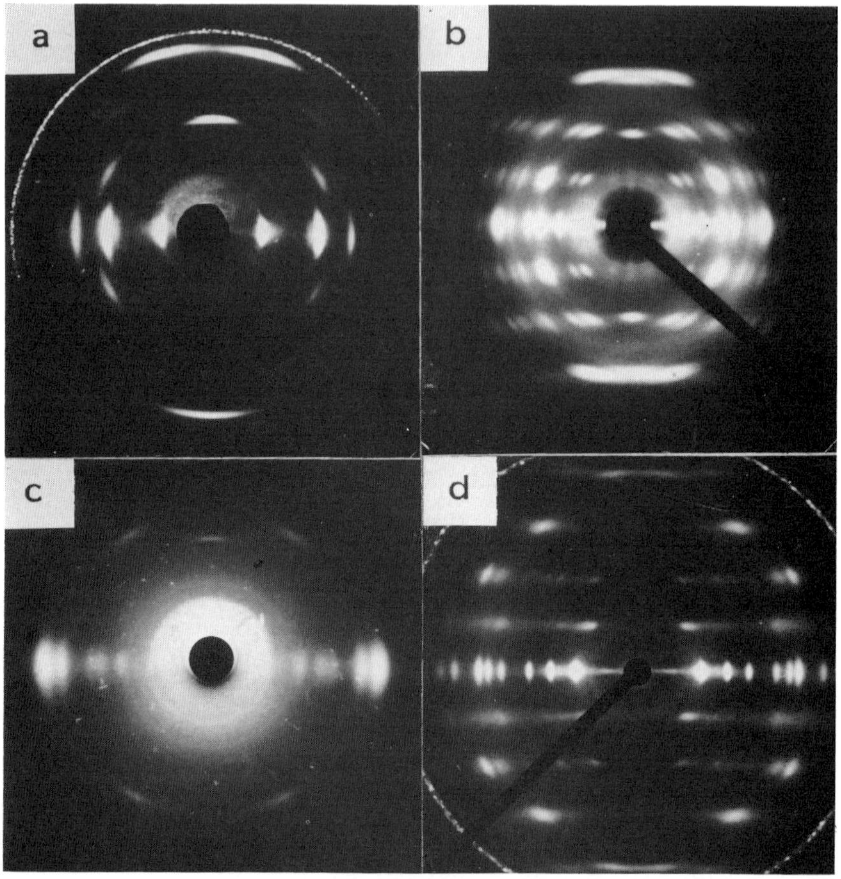

Fig. 1. Fibre X-ray diagrams of differently prepared cellulose derivatives (fibre axis vertical): (a) CTA I tilted, (b) CTA II, (c) CTP I tilted, (d) CTP II.

in solutions and the solid state. The evaluation of such models for the lyotropic liquid crystalline state of cellulose derivatives has to be developed in a two steps procedure: (i) A model for a single chain conformation is needed. (ii) The packing of single chains to form a lyotropic cholesteric liquid crystalline phase has to be evaluated.

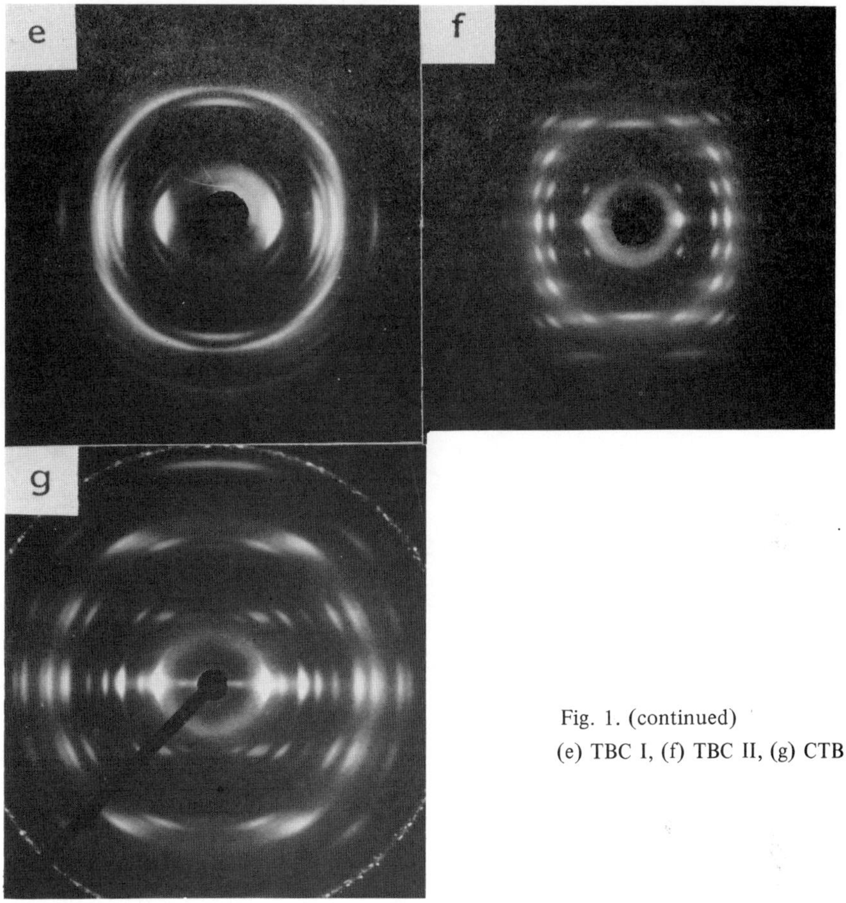

Fig. 1. (continued)
(e) TBC I, (f) TBC II, (g) CTB II.

A single chain conformation of a polymer can best be determined by X-ray analysis in the crystalline state combined with a computer based program of minimization of potential energy. The interaction with solvent, as experienced in the lyotropic liquid crystalline state, can then be approximated by studying the solvent complexed crystalline state. Conformational investigations of single chains with X-ray techniques are not possible in the cholesteric phase.

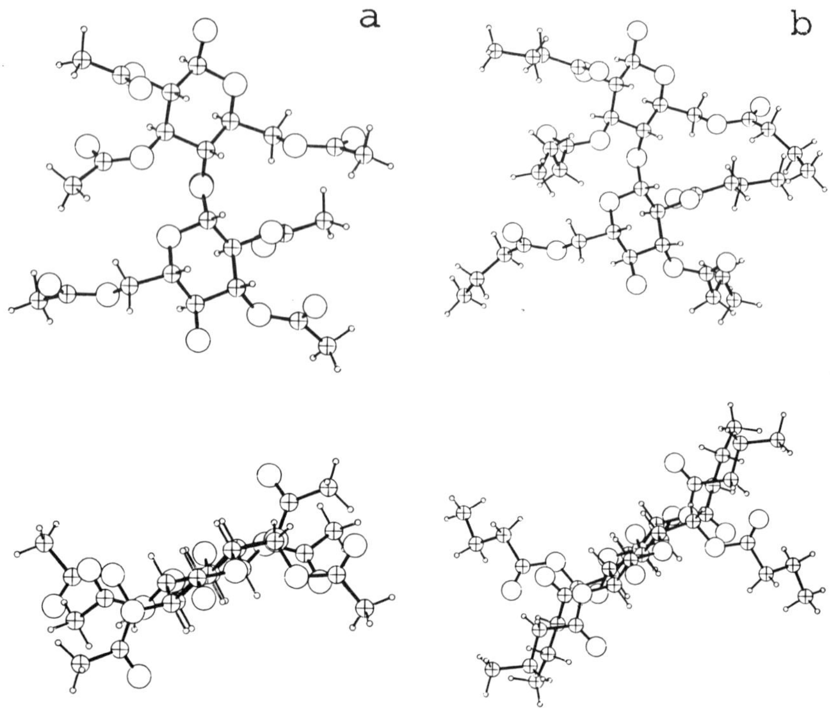

Fig. 2. Repeat unit of the conformation of cellulose derivatives in two projections: (a) CTA, (b) CTB II.

The packing of single chains in highly concentrated solutions that form a cholesteric phase is described by a model of twisted nematic sheets with a certain repeat distance, the so called pitch. Other parameters for a complete description are necessary, as e.g. the order parameter, but will not be considered in this short paper. The pitch and order parameter can be determined by optical means.

CRYSTALLINE STATE

The experimental data for the determination of a single chain conformation in the crystalline state can be obtained from fibre X-ray patterns as presented for a few cellulose derivatives in Fig. 1. Systematically absent meridional reflections lead to the selection of the symmetry elements along a

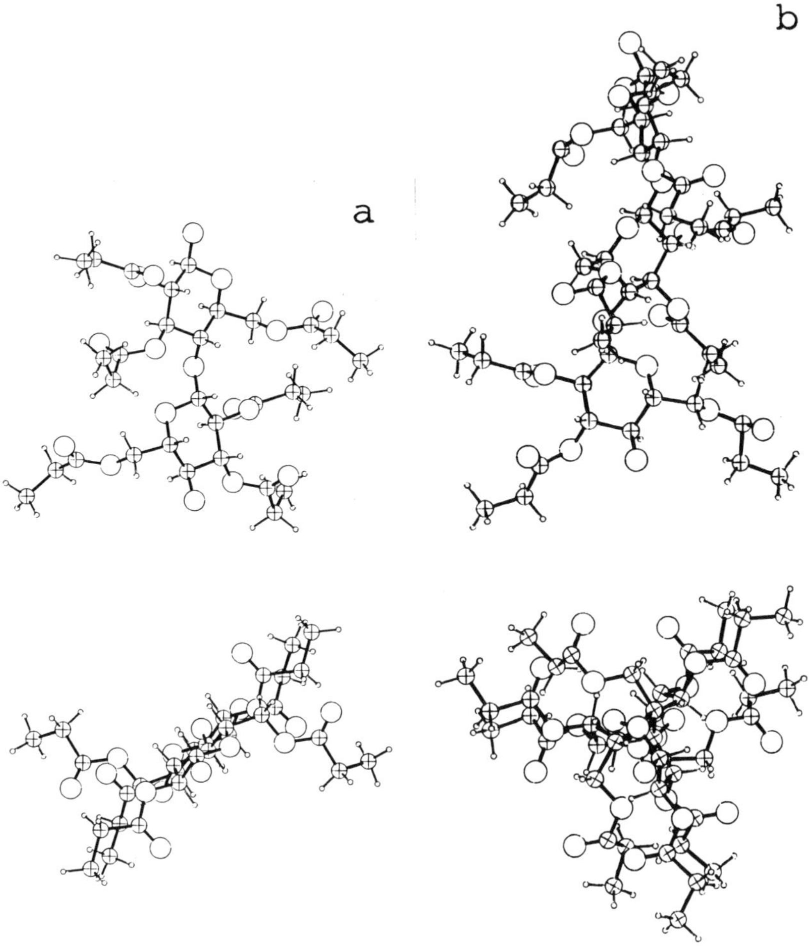

Fig. 3. Repeat unit of the conformation of cellulose derivatives in two projections: (a) CTP I, (b) CTP II (3/2 helix).

single chain, and the spacing between layer lines is used to determine the repeat unit or pitch of the chain conformation. The chain symmetry and the pitch imply strong constraints in a computer aided conformational analysis, and a low energy conformation of a single chain is easily available and depicted for the cellulose derivatives of Fig. 1 in Figs. 2 - 4. Two types of conformations can generally be destinguished for cellulose and cellulose derivatives and

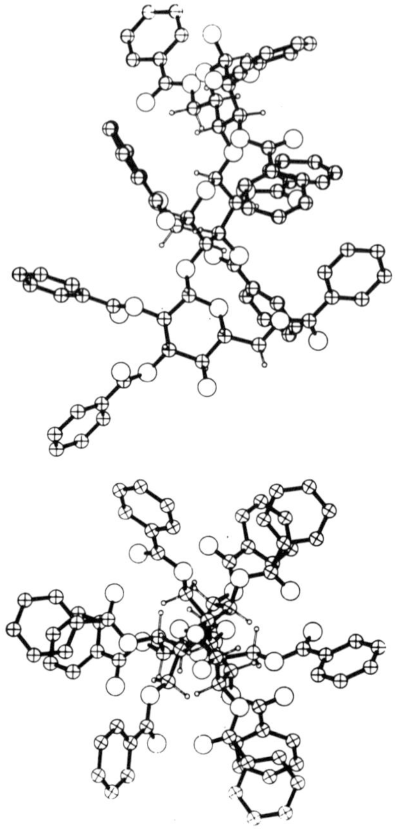

Fig. 4. Repeat unit of the conformation of TBC in two projection.

have been presented in previous publications [1-3]:

(i) A sheet like conformation of 2/1 helices is observed in many cellulose polymorphs. This conformation is also found in heterogeneously and homogeneously acetylated cellulose (CTA I, CTA II; The Roman numerals I and II are used to characterize the heterogeneous and homogeneous products, respectively.) and is depicted in Fig. 2a. The conformation of heterogeneously prepared cellulose tripropionate (CTP I). is shown in Fig. 3a, cellulose tributyrate (CTB II) in Fig. 2b. Trimethyl cellulose, etc. are other examples of this conformation [1-4].

(ii) A rod-like conformation of different left-handed (3/2, 5/3, 8/5) helices is present in tribenzoyl cellulose (TBC I, TBC II; Fig. 4), cellulose tripropionate (CTP II; Fig. 3b), triethyl cellulose and in cellulose derivative/solvent complexes as cellulose triacetate/nitromethane,etc. [1-3].

In this paper, the conformations of some newly derived cellulose structures are presented and discussed in detail, and conclusions will be drawn regarding the packing of cellulose molecules. The experimental procedures for the derivations will be discussed elsewhere [5]. Fibres of cellulose derivatives from Ramie were produced in the usual manner [3] and are represented in Fig. 1 with the corresponding conformational models in Figs. 2 - 4. Most 2/1 helical conformations show a small deviation from an exact 2_1 screw axis, because a meridional reflection on the third layer line is detected,disturbing a 2_1 symmetry element. This deviation seems to be necessary to relieve the somewhat short contact of the two H-atoms (H1 ... H4) centered by the bridge oxygen (O4) [6]. However, models with different hydrogen bonding schemes for distinguished different polymorphs cannot be discussed in cellulose derivatives structures as presented for cellulose in ref. 6.

The results of a structural study on heterogeneously and homogeneously derived crystalline cellulose samples of Ramie can be summarized as follows [5]: CTA I and CTA II (fibre X-ray pattern in Fig. 1a, b) form an approximate 2/1 helix (Fig. 2a) with different deviations from the exact symmetry. They are differently packed as indicated by different intensity distributions in the X-ray diagrams of Fig. 1a, b. CTA I is metastable and can be converted into CTA II. Cellulose tributyrate (CTB II; Fig. 1g) can also be approximated by a 2/1 helix (Fig. 2b) but one of the three substituent ester side groups at O3 deviates from an all trans conformation for the C-atoms [7] to accommodate for reasonable intra-molecular distances between atoms of adjacent residues and is easily recognized in Fig. 2b. Cellulose tripropionate (CTP I; Fig. 1c) forms a close to 2/1 helix (Fig. 3a) with a similar deviation at the ester group at O3 as CTB II. However, the stable polymorph CTP II (Fig. 1d) accommodates a left-handed 3/2 helix (Fig. 3b) to minimize the potential energy of short contacts between consecutive rings. If the ester side groups are bulkier as in tribenzoyl cellulose, both polymorphs TBC I (Fig. 1e) and TBC II (Fig. 1f) exhibit a 3/2 helix (Fig. 4). The unit cell and packing of chains is the same for both polymorphs as indicated by almost the same intensity distribution of the reflections in the X-ray diagrams. The apparent differences in arcing of the reflections in Fig. 1e,f result from a better orientation of the TBC II fibre.

It can be concluded by evaluating the results of the newly determined cellulose structures: They all fall into the established types of chain conformations with very extended chains of about 5 Å rise per residue. However, chain packing considerations lead to new insights concerning the chain polarity of heterogeneously and homogeneously produced fibres from Ramie cellulose. The chain polarity has to be the same for TBC I and TBC II and consequently for all other cellulose derivatives and cellulose I, cellulose II as proposed in a paper more than a decade ago [8]. It has to be concluded, that CTA I and CTP I cannot be readily transformed into the stable polymorph CTA II and CTP II, because of constraints implied by the morphology of the native microfibril and cannot be overcome with the introduction of small substituents. Such constraints can be visualized e.g. as a twist of the microfibril as a whole or as a bundle or pairs of chains, which might lead to a "non-crystallographic" arrangements of chains and cannot completely be described with a unit cell concept. It has at the present not been included in the discussion of chain polarities in native cellulose I [9,10]. As cellulose derived by a homogeneous reaction is considered to pack in an antiparallel fashion in the crystalline state, cellulose I from Ramie and the heterogeneous derivatives thereof have to form the same chain arrangements.

LIQUID CRYSTALLINE STATE

The conformational analysis provided a single chain structure in the crystalline state. The interaction with solvent, as present in the lyotropic liquid crystalline state, can be mimicked by consideration of the solvent complexed crystalline state. The sheet-like conformation of CTA is converted to a rod like one when nitromethane is built into the lattice, and the rod like conformation remains unchanged for cellulose tricarbanilate (CTC) with various solvents added [3]. The basic model for a single cellulose backbone chain in liquid crystalline phases has to be considered as a stiff, rod-like chain, including the results of measurements for chain rigidity of

CTC in solution [11].

Lyotropic liquid crystalline phases of cellulose derivatives exhibit cholesteric helicoidal structures, which consist, according to a model of de Vries [12], of twisted nematic sheets of single molecules with a pitch p, the size of which normally lies in the order magnitude of the wavelength for visible light. Closely related to the pitch are unusual optical properties, which can be used for its determination [3,13] : (i) Circularly polarized light of a selective wavelength λ_0 and handedness ($\lambda_0 = \bar{n} \cdot p$; $\bar{n}$ mean refractive index of a nematic plane) is reflected at the nematic sheets of the helicoidal structure of opposite handedness. (ii) Surface aligned cholesteric phases show a very high rotatory power with an anomalous optical rotatory dispersion at the selective wavelength λ_0.

The pitch for the right-handed helicoidal structures of highly concentrated CTC in ethyl methyl ketone (MEK) or 2-pentanone (2-P) - 0.6 g/ml CTC/MEK or CTC/2-P form easy recognizable liquid crystalline phases - increases with increasing temperature or concentration and decreases for the left-handed helicoidal structures of ethyl cellulose (EC) / glacial acetic acid (GAA) with increasing temperature or concentration [3,13]. An unusual positive birefringence was observed for CTC/MEK or CTC/2-P, a negative one for EC/GAA.

The lyotropic system EC/dichloroacetic acid (DCA) forms a right-handed cholesteric helicoidal structure in contrast to the left-handed structure of the system EC/GAA. A mixed solvent system EC/GAA-DCA at a certain ratio should lead to a cancellation of the twisting angle for the nematic sheets and result in a "quasi-nematic" phase. Such phases are of commercial interest for spinning fibres. The flow behaviour of such a "quasi-nematic" solution shows maximum shear thinning at low shear rates when the whole mixing range of the two solvents is considered. The results are discussed in details in ref. [14].

CONCLUSIONS

The model of de Vries and the equations derived for cholesteric structures of small molecules also hold for cellulose derivatives. Good agreement is obtained for the size of the pitch determined from optical measurement compared with the direct Grandjean line method [3,13].

It appears at the present quite common that crystalline structures of a cellulose derivative with the same unit cell are obtained through different derivation procedures. Tribenzoyl cellulose produced in a heterogeneous or homogeneous reaction shows the same unit cell in the crystalline state, and cellulose triacetate fibres obtained through a spinning process from a liquid crystalline solution [15] or heterogeneous acetylation of cellulose II samples [16] lead to comparable X-ray patterns of heterogeneously derived cellulose I of Ramie and can be indexed with a common unit cell [5].

The size of a unit cell is determined by the packing interactions of atoms of different chains. It seems unlikely that interactions of various helical conformations as a 3/2 helix for TBC or a 2/1 helix for CTA should be the same regardless of a parallel or antiparallel arrangement of chains. A single polarity of chain packing for cellulose and cellulose derivatives has to be concluded from this experimental evidence.

REFERENCES

1. P. Zugenmaier, J. Appl. Polym. Sci.: Appl. Polym. Symposium 37, 223 (1983).
2. P. Zugenmaier in "Polysaccharides", ed. W. Burchard, Springer-Verlag Berlin 1985, 260.
3. P. Zugenmaier in "Cellulose: Structure, Modification and Hydrolysis", eds. R. A. Young and R. M. Rowell, John Wiley & Sons, New York, 1986 221.
4. P. Zugenmaier and A. Kuppel, Colloid & Polymer Sci. 261, 231 (1986).
5. H. Steinmaier and P. Zugenmaier, to be published.
6. R. H. Atalla, Proceedings of the International Symposium on Fiber Science and Technology, Hakone, Japan, Aug. 1985, 86.
7. U. Eggert, Dissertation, Institut für Physikalische Chemie der TU Clausthal, D-3392 Clausthal-Zellerfeld, 1985.
8. A. Kuppel, E. Husemann, E. Seifert, P. Zugenmaier, Kolloid-Z. u. Z. Polymer 251, 432 (1973).
9. A. Sarko and R. Muggli, Macromolecules 7, 486 (1974).
   C. Woodcock and A. Sarko, Macromolecules 13, 1183 (1980).
10. K. H. Gardner and J. Blackwell, Biopolymers 13, 1975 (1974).
11. A. K. Gupta, J. P. Cotton, E. Marchal, W. Burchard, H. Benoît 17, 363 (1976).
12. Hl. de Vries, Acta Crystallogr. 4, 219 (1951).
13. U. Vogt, Dissertation, Institut für Physikalische Chemie der TU Clausthal, D-3392 Clausthal- Zellerfeld, 1985.
14. P. Zugenmaier, P. Haurand, Carbohydr. Research, in press.
15. E. J. Roche, J. P. O'Brien and S. R. Allen, Polym. Commun. 27, 138 (1986).
16. S. Watanabe, M. Takai, J. Hayashi, J. Polym. Sci. 23C, 825 (1968).

Chapter 10

DSC Studies on monovalent and divalent cation salts of carboxymethylcellulose in highly concentrated aqueous solutions

Kunio Nakamura – Industrial Research Institute of Kanagawa Prefecture.
Taksuko Hatakeyama – Research Institute for Polymers and Textiles, Tsukuba.
Hyoe Hatakeyama – Industrial Products Research Institute, Yatabe, Tsukuba.

INTRODUCTION

The interaction between water molecules and hydrophilic polymers including polyelectrolytes has been studied by many researchers [1]. Recently, we have studied the thermal and physical properties of systems consisting of water and hydrophilic polymers and water and polyelectrolytes using differential scanning calorimetry (DSC) and nuclear magnetic resonance (NMR) [2-9]. It has been ascertained that water associated with hydrophilic groups exhibits properties different from free water and that nuclear magnetic relaxation times of bound water are markedly different from those of free water.

Carboxymethylcellulose (CMC) is one of most common polyelectrolytes which have been commercialized and numerous applications in various industries have been found. However, the interaction between water and metal salts of CMC in a highly concentrated aqueous solution has not yet been sufficiently clarified. Therefore, in this paper special attention has been placed on clarifying the thermal properties of the water-CMC systems with various counter ions and water contents (weight of water/weight of dry sample, g/g) ranging from 0.3 to 4.0.

EXPERIMENTAL

Samples

Carboxymethylcellulose (CMC) samples, having monovalent

counterions such as Li^+, Na^+, Cs^+, divalent counterions such as Mg^{2+}, Ca^{2+}, Ba^{2+}, and a trivalent counterion Al^{3+}, were used. Degree of substitution (DS), molecular weight (Mw) of each monomer unit, and ionic radius (Ri) and atomic weight (Aw) of each counterion are listed in Table 1.

Table 1. Specific valves of CMC samples.

Conterion	DS	Mw	Ri (A)[10]	Aw
Li^+	0.61	201	0.68	7
Na^+	0.95	210	0.95	23
K^+	0.59	219	0.59	39
Cs^+	0.63	282	1.69	133
Mg^{2+}	0.59	203	0.65	24
Ca^{2+}	0.57	206	0.99	40
Ba^{2+}	0.56	232	1.35	137
Al^{3+}	0.57	200	0.50	27

Differential Scanning Calorimetry (DSC)

A DuPont 910 DSC equipped with cooling apparatus was used. A DSC sample was prepared in the following way. Dry CMC about 5 mg was weighed in an aluminium sample pan and an excess amount of water was added with a microsyringe. Then the sample pan was allowed to stand for a few minutes in order to obtain a certain amount of water and was weighed again after the sample pan was sealed. Water content (Wc) of the sample was calculated from the following equation: Wc, (g/g) = [weight of added water, (g)]/[weight of dry sample, (g)].

DSC curves were obtained in the temperature range from -130 °C to 50 °C. The scanning rate was 10 °C/min. Phase transition temperature and enthalpy of transition of sorbed water were calibrated using pure water as a standard. After a DSC measurement, the sample pan was again weighed to confirm that no weight loss had taken place during the scan. The bound water content was calculated by the method reported previously [8].

RESULTS AND DISCUSSION

Figure 1 shows DSC heating curves of the water-NaCMC system having various amounts of sorbed water. The DSC heating curve shows the glass transition (Tg) at about -90 °C, the cold crystallization (Tcc) peak at about -40 °C and the melting (Tm) peak at about 0 °C. The peak temperature of melting increases as Wc increases and levels off when Wc exceeds 2.0. Tg decreases rapidly with the increase of Wc and

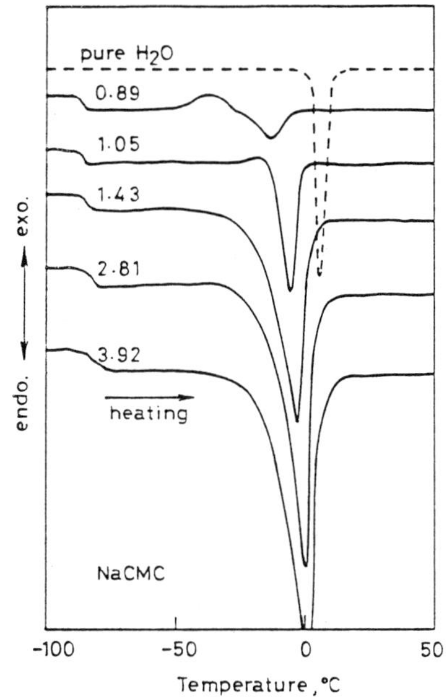

Fig. 1. DSC heating curves of the water-NaCMC system.

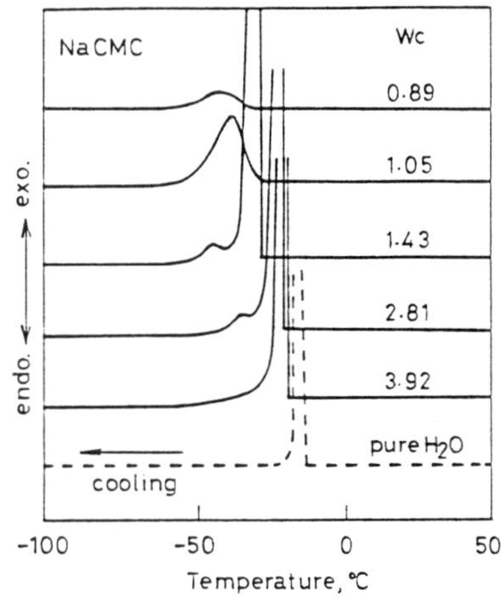

Fig. 2. DSC cooling curves of the water-NaCMC system.

reaches almost the constant value of -98 ^{o}C when Wc exceeds 1.0. However, any first order transition was not observed when Wc was less than 0.6. In the case of the water-CMC systems with other monovalent counterions of Li^+, K^+ and Cs^+, similar DSC curves were observed.

Figure 2 shows DSC cooling curves of the water-NaCMC system having various water contents. In general, when the water-NaCMC system is cooled from 50 ^{o}C to -100 ^{o}C, two exothermic peaks are observed in a DSC curve of a sample having Wc more than 1.1. One is a sharp peak (Tc_I) observed at about -30 ^{o}C in the DSC curve of the sample having Wc=1.43. This peak shows the crystallization of water. The other is a broad peak (Tc_{II}*) observed at about -50 ^{o}C in the same DSC curve. This peak shifts to a higher temperature range with the increase of water content and approaches the crystallization temperature of pure water. However, this peak shifts to a lower temperature range with the decrease of water content and disappears at a Wc less than ca. 0.7.

Figure 3 shows phase transition temperatures of water-CMC systems at various water contents. The phase transition temperatures are dependent on different kinds of counterions.

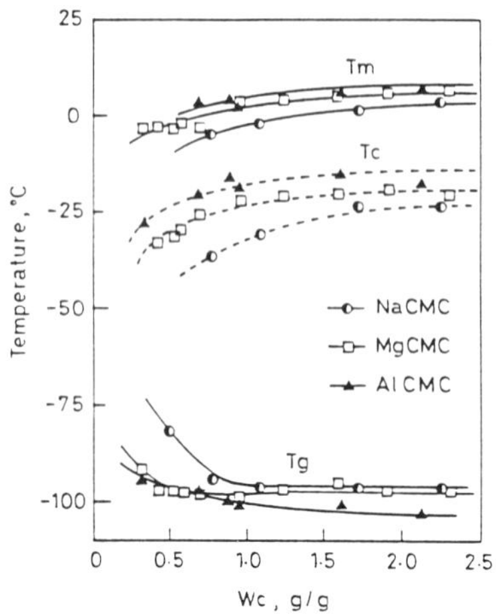

Fig. 3. Relationship between phase transition temperatures of the water-Metal-CMC systems and water content.

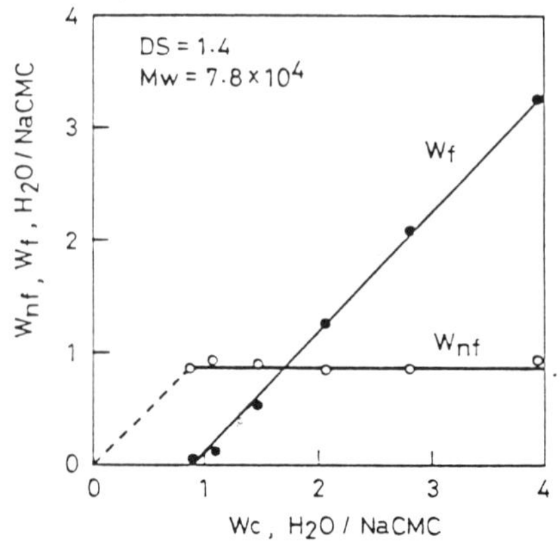

Fig. 4. Change of Wf and Wnf of the water-NaCMC system with water content.

As shown in the figure, Tm and Tc increase in the order $Al^{3+}>Mg^{2+}>Na^{+}$. This fact suggests that the interaction between water molecules and counterions decreases in the order $Na^{+}>Mg^{2+}>Al^{3+}$.

Generally the glass transition of the water-CMC system appears at a temperature lower than that of the first order transition. This is quite reasonable, as the glass transition of the system appears even with the presence of very small amounts of water which cannot crystallize. Although the melting and crystallization temperatures of the water-CMC system increase with the increase of Wc, the glass transition temperature decreases with increasing Wc when the value of Wc is lower than ca. 1.0, and the Tg reaches almost constant value which is dependent on the counterions of CMC in the Wc range higher than ca. 1.0.

The glass transition of CMC and pure water was not observed in the temperature range in which this experiment was carried out. Therefore, it is quite appropriate to consider that the combination of CMC with water causes a glassy state in the system.

In the case of the crystallization of water in the water-CMC system, the amount of water calculated from the enthalpy of crystallization is less than the amount of added water. The amount of water corresponding to the difference between the added water and that calculated from the enthalpy must be present. It is clear that water cannot move to the outside of a sample pan in our experimental conditions, as the weight of the sample pan before and after the DSC measurement was the same. Accordingly, the only possibility is that the water combined very tightly with the CMC and its counterion. This water is categorized as non-freezing water.

Such non-freezing water does not seem to have any kind of crystalline structure, as it shows no first order transition. The amount of freezing bound water which shows Tc_{II}^{*} in Figure 1 is too small to calculate, and can therefore be included in the amount of freezing water (Wf).

Accordingly, we have categorized water in the water-CMC system into two different kinds: freezing water (Wf) and non-freezing water (Wnf). By the subtraction of Wf from Wc, the amount of non-freezing water (Wnf) was calculated as follows:

$$Wnf = Wc - Wf \qquad (1)$$

The change of Wf and Wnf with Wc in the water-NaCMC system is shown in Figure 4. The amount of Wnf maintains almost constant value with increasing Wc. It is reasonable to assure that the extended Wnf line (broken line in Figure 4) intersects with the longitudinal axis at 0.92, since freezing water begins to appear at 0.92.

As shown in Figure 5, the Wnf value increases with the increase of degree of substitution (DS). This means that Wnf is dependent on DS. Therefore, we used the Wnf value of the sample with DS at about 0.6 as the reference standard for the comparison among CMC's having various counterions.

Figure 6 shows the relationship between Wnf and the radius of counterions. In the case of counterions having the same valency, the Wnf value decreases linearly with the increase of radius of counterions. In connection with the

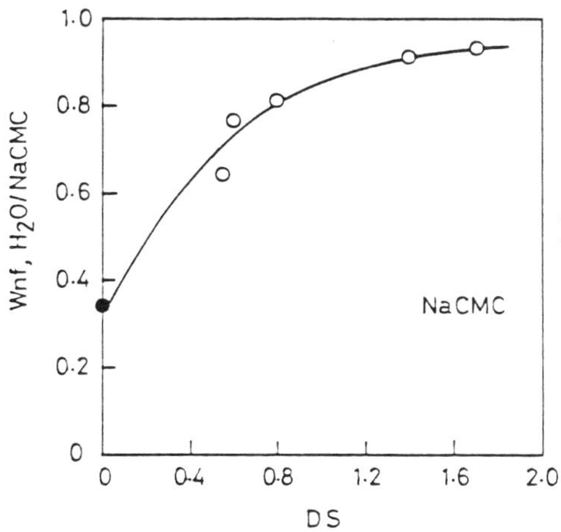

Fig. 5. Relationship between Wnf and degree of substitution of CMC in the water-NaCMC system.

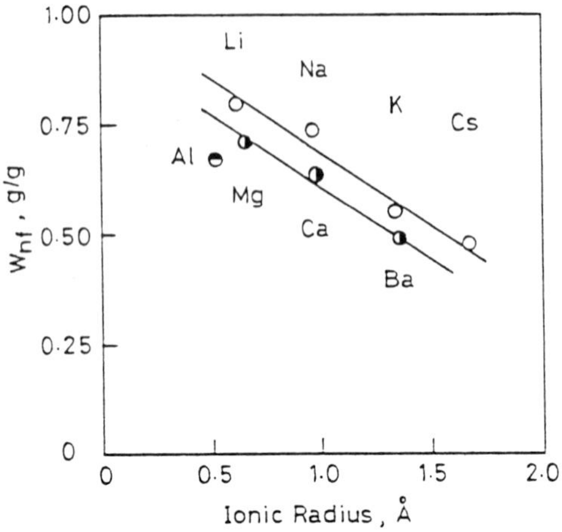

Fig. 6. Relationship between Wnf and ionic radius of the water-Metal-CMC systems.

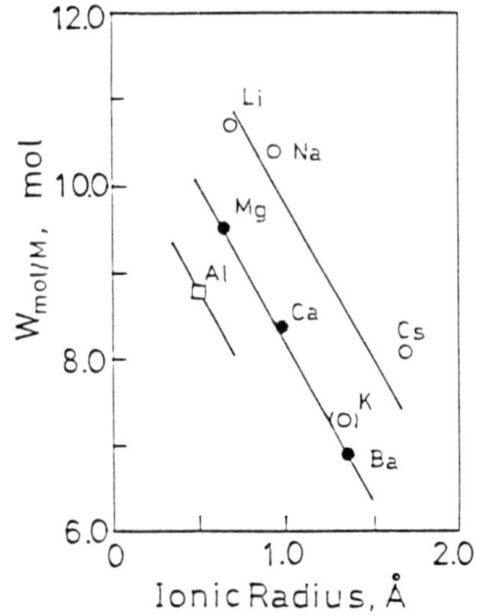

Fig. 7. Relationship between the moles of non-freezing water and ionic radius of the water-Metal-CMC systems.

above results, the sorption isotherms of the metal salts of CMC were obtained at 20 ^{0}C for the purpose of comparison. The results showed that the water increased in the order of Li^{+} $>Na^{+}>K^{+}>Cs^{+}$ for monovalent ions, $Mg^{2+}>Ca^{2+}>Ba^{2+}>Al^{2+}$ for divalent and trivalent ions. The results agree with the relatioinship shown in Figure 6. Therefore, it is quite reasonable to consider that the hydration of counterions in CMC has a close relationship with the ionic radius.

The number of moles of non-freezing water bound to a glucose residue in CMC can be calculated by the following equation:

$$Wmol/Gu=Wnf \times Mw/18 \quad (2)$$

where Wmol/Gu is the mole of non-freezing water sorbed onto a glucose unit, Mw is the molecular weight of a glucose unit of CMC, and 18 is the molecular weight of water.

It is approppriate to consider that the OH group in the metal salt of CMC binds the same amount of non-freezing water as natural cellulose does [5,9]. This suggests that the amount of non-freezing water around each metal ion of CMC is calculated by the following equation:

$$Wmol/M=[(Wmol/Gu)-(3-a)]/a \quad (3)$$

where Wmol/M is the mole of non-freezing water around a metal ion of CMC, and a=DS.

The calculated value of Wmol/M is shown in Figure 7. The number of water molecules around K^+ is extraordinarily small. Except for this value of water molecules around K^+, however, it may in general be said that the results show the presence of a linear relationship between Wmol/M and the radius of counterions. Therefore, it may not be wrong to consider that the size of the counterion of CMC strongly affects the hydration of water molecules around metal salts of CMC, and this effect decreases almost linearly with the increase of the ionic radius of a counterion.

REFERENCES

[1] S. P. Rowland ed., "Water in Polymers", ACS Symposium Series, **127**, American Chemical Society, Washington D.C., 1980.
[2] T. Hatakeyama, K. Nakamura and H. Hatakeyama, Netsusokutei, **6**(29), 50 (1979).
[3] K. Nakamura, T. Hatakeyama and H. Hatakeyama, Polymer, **22**, 473 (1981).
[4] H. Hatakeyama and T. Hatakeyama, Makromol. Chem. **182**, 1655 (1981).
[5] K. Nakamura, T. Hatakeyama and H. Hatakeyama, Textile Res. J., **51**, 607 (1981).
[6] K. Nakamura, T. Hatakeyama and H. Hatakeyama, Polymer J., **15**, 361 (1983).
[7] K. Nakamura, T. Hatakeyama and H. Hatakeyama, Polymer, **24**, 871 (1983).
[8] T. Hatakeyama, K. Nakamura, H. Yoshida and H. Hatakeyama, Thermochimica Acta, **88**, 223 (1985).
[9] H. Hatakeyama, H. Iwata, T. Hatakeyama, in "Cellulose and Its Derivatives", (Eds. J. F. Kennedy, G. O. Phillips, D. J. Wedlock and P. A. Williams), Ellis Horwood, Chichester, 1985, pp. 255.
[10] John A. Dean ed., "Lange's Handbook of Chemistry, 13th Ed.", McGraw-Hill, New York, 1985.

Chapter 11

Viscoelastic relaxations of cellulose acetate in the solid state

M. Pizzoli, M. Scandola and G. Ceccorulli – Centro di Studio per la Fisica delle Macromolecole del CNR and Istituto Chimico "G. Ciamician" University of Bologna, Via Selmi 2, 40126 Bologna, Italy.

INTRODUCTION

The relevance of dynamic mechanical spectroscopy in the study of molecular relaxations is well known [1,2]. The technique allows one to reveal not only the glass to rubber transition - that involves the cooperative motion of several chain segments (40-50 chain atoms) - but also much more restricted motions that occur in the macromolecule at a local level. Such motions give rise to absorption peaks in the damping factor (loss tangent, tan δ) and to a more or less evident decrease of the storage modulus (E'). The temperature and frequency where the relaxations are located depend on the energetic and volumetric requirements of the motion of the particular molecular entity involved.

The present paper discusses both recent results by the Authors [3-5] and new unpublished data on the viscoelastic relaxations of Cellulose Acetate and is mainly concerned with the problems connected with the molecular attribution of the dynamic mechanical absorption peaks.

EXPERIMENTAL

A series of Cellulose Acetate (CA) samples of varying degree of substitution (DS from 1.8 to 2.9) kindly supplied by Dr. R.T. Bogan of Eastman Kodak (Eastman Chemicals Division, Kingsport, Tennessee) were investigated.

Dynamic mechanical measurements were made by means of a Dynamic Mechanical Thermal Analyser (DMTA, Polymer Laboratories) from -150 to 230°C at a heating rate of 3 deg/min, over a range of selected frequencies from 0.33 to 30 Hz. The samples, in the form of small injection moulded bars [3], were tested by using either the single or the dual cantilever geometry. Some samples were hydrated by careful equilibration in controlled R.H. atmospheres, while the reference "dry" samples were stored in a desiccator under vacuum over P_2O_5 up to constant weight. The calorimetric measurements were carried out by means of a Du Pont Differential Scanning Calorimeter (9900 Thermal Analysis System) at a heating rate of 20 deg/min.

RESULTS AND DISCUSSION

Figure 1 shows the typical dynamic mechanical spectrum (at a frequency of 3 Hz) of a sample of Cellulose Acetate (DS 2.4) stored under normal room conditions. In the loss tangent (tanδ) curve three absorption peaks α, β and γ in order of decreasing temperature and a shoulder (β*) are clearly observed. The main α transition is characterized by a very steep drop of the dynamic

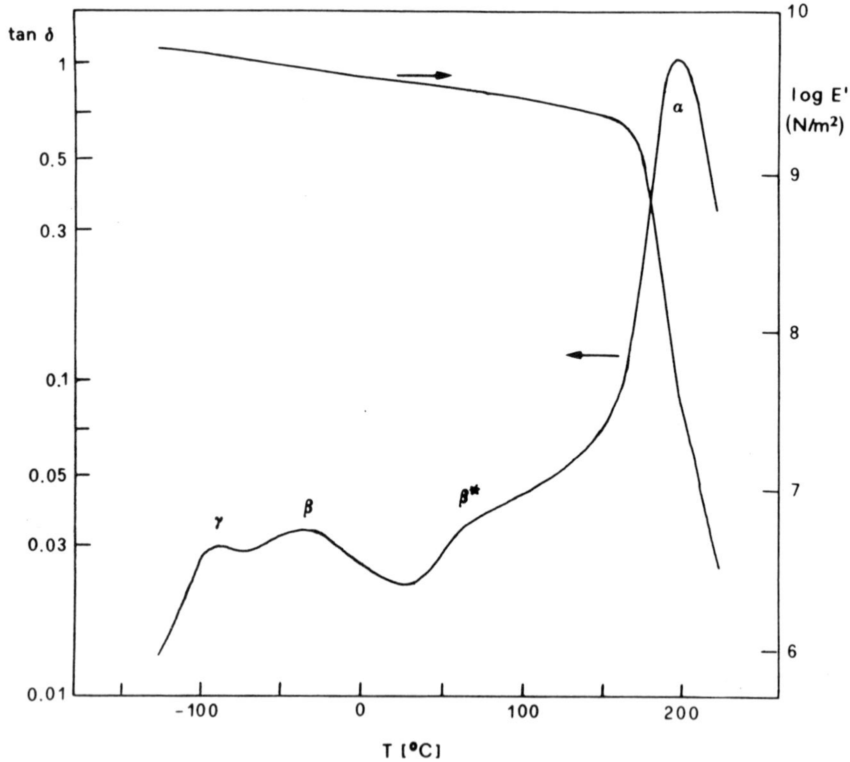

Fig. 1. Dynamic mechanical properties of Cellulose Acetate (DS 2.4) at 3 Hz.

storage modulus E' and is confidently assigned to the glass-to-rubber transition, its temperature location being in excellent agreement with the calorimetric Tg. The β^* shoulder, that can be reversibly eliminated from the spectrum by means of heat treatments, has been shown to be due to absorbed moisture [3]; this assignment is corroborated by both calorimetric and thermogravimetric results. Also the very low temperature γ relaxation is related to the presence of absorbed water in Cellulose Acetate: heat treatments that alter the amount of absorbed moisture induce variations of peak intensity and temperature [3]. The γ absorption is absent in the "dry" CA samples and has been attributed to the motion of water molecules associated with the macromolecules through polar interactions [5].

As regards the β relaxation, its assignment to a specific molecular motion is still controversial. Previous attributions [6-9] to motions involving the acetate side chains have not been confirmed by recent work by the Authors [3-5] and do not seem very likely in consideration of the temperature location and activation energy of the relaxation. Moreover, the parameters that affect the remaining relaxations of the spectrum (i.e. heat treatments and absorbed water) do not alter appreciably the β dispersion.

In order to investigate the possible involvement of the acetate group in the motion responsible of the β relaxation, the present paper takes under consideration the influence of a different parameter: the degree of substitution, i.e. the average number of ester side chains per glucopyranose ring. A series of five CA samples of varying degree of substitution (1.8 - 2.9) have been examined and the dynamic mechanical spectra of three of them are shown in Figure 2. The curves reported are relative to "dry" samples and the absence of any γ relaxation from the low temperature region of the spectrum - independently of the DS value - confirms the water-related character of the γ relaxation. Two further points are evident in Figure 2: (a) the insensitivity - within the limits of accuracy of the instrumentation - of the β absorption to the degree of substitution, contrasted by (b) a remarkable dependence on DS of both temperature (T_α) and intensity of the α relaxation. The decrease of T_α with increasing ester substitution is expected [8,10] and is found to be paralleled by an analogous decrease of the calorimetric glass transition temperature.

It appears that - apart from possible effects of different thermal history during the moulding procedure - the degree of substitution also influences the crystalline fraction of the partially crystalline polymer Cellulose Acetate, especially at high DS values. This feature is evidenced by the decrease of intensity of the α peak with increasing DS as well as by comparison of the modulus drop at the transition shown in Fig.1 for CA 2.4 and in Fig.2 for CA 2.9. The storage modulus of CA

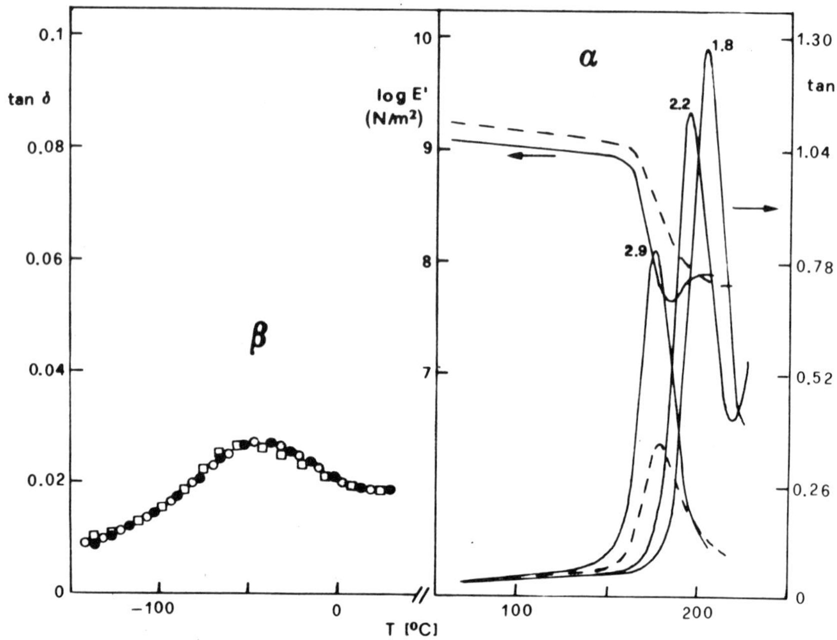

Fig. 2. Influence of the degree of substitution on the dynamic mechanical spectrum of dry Cellulose Acetate. β region, DS: □ 1.8, ● 2.2, ○ 2.9, α region, DS: number on curve; dotted curves refer to 2nd run on CA 2.9.

2.9 shows a much smaller drop associated with the α transition than the 2.4 sample and, unlike all other samples examined, goes through a minimum at $T>T_\alpha$. This latter observation suggests that the polymer chains in the amorphous regions of CA 2.9, after mobilization at the glass transition, undergo partial crystallization, the crystalline packing of the macromolecules being promoted by the regularity of an almost complete ester substitution. This result clearly indicates that only partial crystallization of CA 2.9 occurred on cooling in the mould after sample extrusion. The occurrence of crystallization after Tg in the first run is confirmed by a subsequent measurement on the same specimen (dotted curve in Fig.2) that shows a remarkable decrease of intensity of the α relaxation, i.e. a decrease of the amorphous fraction.

We recall that the dynamic mechanical curves shown in Figure 2 refer to "dry" samples. All samples investigated, irrespective of DS, absorb some humidity from the ambient when stored at normal room conditions and show a low temperature γ relaxation in their dynamic mechanical spectrum corresponding to the one seen in Figure 1 for CA 2.4. In this connection we emphasize that the influence of DS on the amount of crystallinity of Cellulose Acetate carries as a consequence an inherent difficulty in comparing the viscoelastic spectrum of hydrated samples having different DS. In fact, being commonly assumed that absorbed water primarily solvates the amorphous regions of partially crystalline polymers, it is practically impossible to convert the amount of absorbed water into significant concentration units in the absence of reliable information on the percentage of crystallinity of the sample.

As already pointed out, the region of the viscoelastic spectrum where the secondary β relaxation occurs appears to be insensitive to the average number of acetate side groups per sugar residue, at least in the range of DS presently investigated. This result is in clear disagreement with the view that the β relaxation is primarily due to the motion of the acetate side chains; on the other hand, preliminary results on a polysaccharide whose structure is closely related to Cellulose - Chitosan, i.e. deacetylated Chitin - supports the different opinion that the β relaxation is more likely due to some local motion of the polysaccharide main chain, as previously suggested by the Authors [3,5]. As a matter of fact Chitosan, whose molecular structure differs from that of Cellulose only by the substitution of -OH with $-NH_2$ on C2, shows a secondary relaxation below room temperature the origin of which cannot obviously be attributed to non-existing side chains.

In support of the existence of a relaxation associated with some local motion of the polysaccharide main chain we recall a NMR study by Froix and Nelson [11] where a relaxation process occurring in dry Cellulose at 370K (90 MHz) with an activation energy of 17.8 Kcal/mole was attributed to "partial hindered rotation of the Cellulose rings in the amorphous regions of the polymer". Assuming that the peak temperature of the relaxation process of Cellulose follows an Arrhenius-type equation over a wide range of frequencies, extrapolation of the above data to 3 Hz yields a peak centered at about $-60^{\circ}C$, a temperature comparable with the location of both the β relaxation of CA and the low temperature relaxation of dry Chitosan.

Although a definitive molecular attribution of the β relaxation of Cellulose Acetate is not yet possible, the above results altogether demonstrate that also natural polymers such as polysaccharides may show, as previously found for the protein elastin [12], relaxations in the glassy state related to local motions of the main chain, a feature that is frequently observed with linear synthetic polymers [13].

ACKNOWLEDGEMENTS

This work was partially supported by a grant of the Italian Ministry of Education

REFERENCES

[1] Murayama T., "Dynamic Mechanical Analysis of Polymeric Material" (1978) Elsevier, N.Y.
[2] Meier D.J., "Molecular Basis of Transitions and Relaxations" (1978) Gordon and Breach, N.Y.
[3] Scandola M. and Ceccorulli G., Polymer (1985) 26, 1953
[4] Scandola M. and Ceccorulli G., Polymer (1985) 26, 1958
[5] Ceccorulli G., Pizzoli M. and Scandola M., Polymer Comm. (1986), in press
[6] Mikhailov G.P., Artyukhov A.I. and Borisova T.I., Vysokomol. Soed. (1967) A9, 2401
[7] Mikhailov G.P., Artyukhov A.I. and Shevelev V.A., Vysokomol. Soed. (1969) A11, 553
[8] Mooroka T., Norimoto M., Yamada T. and Shiraishi N., Wood Research (1983) 69, 61
[9] Mooroka T., Norimoto M., Yamada T. and Shiraishi N., J. Appl. Polym. Sci. (1984) 29, 3981
[10] Seymour R.W., Weinhold S. and Haynes S.K., J. Macromol. Sci. Phys (1979) B16, 337
[11] Froix M.F. and Nelson R., Macromolecules (1975) 8, 726
[12] Pezzin G., Scandola M. and Gotte L., Biopolymers (1976) 15 283
[13] McCrum N.G., Read B.E. and Williams G., "Anelastic and Dielectric Effects in Polymeric Solids" (1967) Wiley, N.Y.

Chapter 12

Crystalline polymorphism of cellulose triacetate

M. Takai, K. Fukuda, M. Murata and J. Hayashi – Department of Applied Chemistry, Faculty of Engineering, Hokkaido University, Sapporo, Hokkaido, 060 Japan.

ABSTRACT

It was proved that cellulose triacetate (CTA) I can be formed completely by the heterogeneous acetylation of prehydrolyzed Fortisan of cellulose II as well as cellulose I when the swelling in the pretreatment media is low. In connection with the crystalline structures for cellulose I and cellulose II reported recently, the general concept of the dependency of CTA on cellulose structure, namely that CTA I is formed from cellulose I and consists of parallel structure, and CTA II from cellulose II and of antiparallel structure, is contradicted. Transformation of CTA I to CTA II can be obtained by superheated-steam treatment with retention of fibrous form. This transformation was examined by treating both poorly ordered (amorphous) and well-ordered (crystalline) CTA with the steam. The former is almost completely transformed from CTA I to CTA II, but the latter showed about 40% conversion from CTA I to CTA II. In addition, it is more difficult to get the transformation from bacterial CTA I to CTA II by the same treatment. Nevertheless CTA I from cellulose II of mercerized bacterial cellulose is easy to transform to CTA II with the treatment, regardless of the crystallinity of CTA I. At present the difference in conversion to CTA II between poorly-ordered, well-ordered CTA I and bacterial CTA I can not be explained. It may be associated not only with the effect of superheated steam on swelling, but also with the different patterns of hydrogen bonding between ramie and bacterial celluloses. Another difference in transformation behaviour of CTA I from cellulose II is based on the conformational change in cellulose structure from a "bent" form to a "bent-twisted" form.

INTRODUCTION

It is widely accepted that two stable polymorphic crystalline structures exist in cellulose triacetate (CTA): CTA I and CTA II (1, 2). The x-ray diffraction patterns of them can be distinguished from each other. In general, CTA I is obtained by the fibrous acetylation of cellulose I, while CTA II arises if CTA I is dissolved or swollen in solvent. In addition, heterogeneous acetylation of cellulose II yields CTA II. Both amorphous and crystalline CTA I are always saponified to cellulose I, while both forms of CTA II are saponified to cellulose II. Therefore it is said that CTA I is the derivative of cellulose I, and CTA II is that of cellulose II. In x-ray crystallographic studies of CTA I and CTA II, the structural model of CTA I has a parallel-chain packing and CTA II an antiparallel structure. In the previous paper, it was proved that CTA I could be formed by the heterogeneous acetylation of cellulose II when the swelling in the media was slight. Also the transformation of CTA I to CTA II was obtained by superheated steam treatment without any dissolving process. Recently, there has been some confusion in the transformation scheme between celluloses and CTAs. The purpose of the present paper is to establish the scheme of the dependency of CTA structure on cellulose structure and process history.

EXPERIMENTAL

Cellulose Samples Valonia cellulose (macrophysa), bacterial cellulose (harvested from Acetobacter xylinum), ramie, and Whatman cellulose powder (CF 11) were utilised as sources of cellulose I. Mercerization for the above cellulose sources was carried out with 18% NaOH, for 72hr at room temperature. After thoroughly washing with water until neutral and then drying, they were used as cellulose II. Fortisan prehydrolyzed by 2.4N-HCl at 100°C for 3hr was also utilized as a source of cellulose II.

Acetylation Fibrous acetylation involved pretreating the cellulose materials in glacial acetic acid at 20°C for 24hr, or in an aqueous solution of 40% potassium acetate at 50°C for 1 hr. Then the pretreated materials were acetylated by using 1% of nascent sulfonic acetic acid as a catalyst in a mixture of 25 parts of acetic anhydride and 75 parts of benzene at 20°C for 48hr. The products were washed with benzene to be free from acid and then immersed in 75% neutralized ethyl alcohol at 50°C for 1hr to remove sulfonic acetic acid thoroughly. Finally they were washed exhaustively with distilled water and then dried at room temperature in a vacuum desiccator over P_2O_5.

Heat or Superheated-Steam Treatment. Heat treatment of the fibrous or powdery product was carried out with silicone oil at 210°C for 15 min. Superheated steam treatment was done in an autoclave at 170°C for 2hr.

X-Ray Analysis X-ray diffractograms of the products were obtained by the reflection method with an automatic diffractometer of nickel-filtered Cu K_a radiation in the w-2θ scanning mode.

RESULTS AND DISCUSSION

CTA I Derived from Cellulose II by the Fibrous Acetylation

In our previous paper (3), it was proved that CTA I could be formed by the heterogeneous acetylation of cellulose II when swelling in the pretreatment media was slight. When mercerized celluloses were acetylated by the pretreatment with glacial acetic acid, they gave mixed patterns of CTA I and CTA II(3). However normal rayon or Fortisan fiber was hardly acetylated by the pretreatment with glacial acetic acid. The acetylation requires a severe pretreatment(4). Thus the acetylated rayon always had a CTA II pattern. Sakurada et al.(5) noticed that the reactivity of rayon toward acetylation was markedly enhanced by acid hydrolysis. Whatman cellulose powder or acid-prehydrolyzed Fortisan fiber was subjected to a comparatively mild pretreatment with glacial acetic acid, followed by acetylation. Highly acetylated products were obtained from Whatman cellulose powder (DS=2.80) and prehydrolyzed Fortisan (DS=2.90). Figure 1 shows x-ray diffractograms of these triacetates before and after heat treatment. They are generally sharper than those of the triacetate from unhydrolyzed celluloses. Whatman cellulose powder yields the typical CTA I pattern in which the diffractions characteristic of CTA I are at 11.5A (2θ=7.7°) and 5.5A

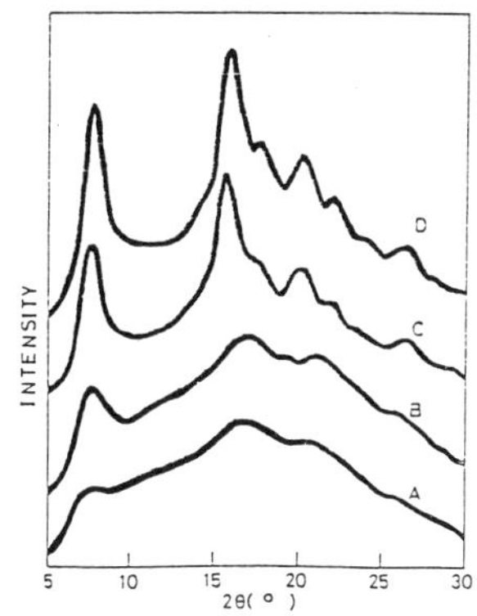

Fig. 1. X-ray diffractograms of CTA.
Amorphous CTA from
(A) prehydrolyzed Fortisan
(B) Whatman cellulose powder,
(C) and (D) Heat-treated CTA from (A) and (B), respectively.

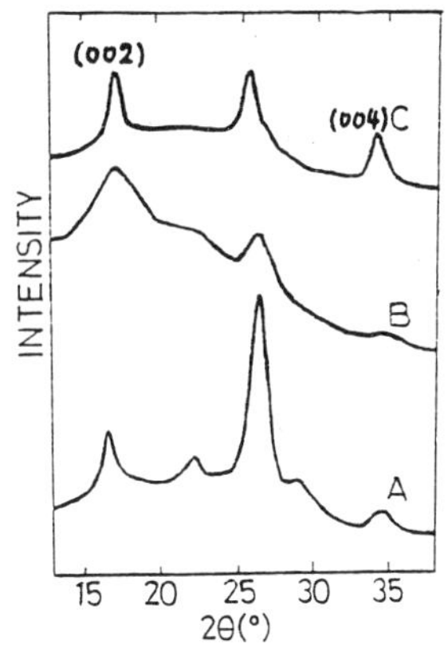

Fig. 2. Meridional X-ray diffractograms of CTAI and CTAII.
CTAI: (A) ramie
(B) mercerized ramie
CTAII: (C) Arnel.

(16.1°). It becomes clear that cellulose II Fortisan yields CTA I and that there is no substantial difference between mercerized cellulose and rayon. This fact contradicts the previously held view that CTA I is yielded from cellulose I and CTA II from cellulose II. In the molecular and crystal structure analysis of CTA II, Dulmage(6) proposed an antiparallel structure for it, and Roche et al.(7) recently confirmed it. Stipanovic et al. (8) have similarly analyzed CTA I obtained from ramie cellulose, and its structure is proposed to be based on parallel chains. If this is true, CTA I can never be derived from cellulose II which consists of antiparallel chains. However, as mentioned above, a typical CTA I is obtained from cellulose II, prehydrolyzed Fortisan with high crystallinity. At least, it seems that a parallel arrangement is not necessarily the correct structure for CTA I. In our previous papers(9,10), the main distinction between CTA I and CTA II was not the matter of chain polarity, but of the orientation of acetyl groups and of the chain packing. The crystal structure of CTA I is based on the sheet-like structure of the $(1\bar{1}0)$ plane of the original cellulose, which is densely packed by methyl or carbonyl groups. The shape of the CTA I lattice markedly resembles that of cellulose II, whereas the chain packing within the sheet is similar to that of cellulose I. On the other hand, the crystal structure of CTA II is also basically a sheet-like structure, but van der Waals contacts are within the sheet because of the different orientation of the acetyl groups. The sheets interlock each other and are difficult to slide over each other. Figure 2 shows meridional x-ray diffraction traces of CTA I and CTA II. The peak intensity ratio (002) to (004) of CTA I from ramie is ca. 2.0, whereas that of CTA I from mercerized ramie is as high as 6.0. On the other hand, for CTA II from Arnel (commercial cellulose triacetate fiber), the peak intensity of (002) is about the same as that of (004). Such results would seem to indicate that CTA I from cellulose I and CTA I from cellulose II should be distinguished from the point of difference in intensity of meridional reflection. Thus they are denoted as CTA I_I and CTA I_{II}, respectively.

Transformation of CTA I to CTA II without Dissolution It is possible to obtain transformation of CTA I to CTA II by the treatment with superheated steam without any dissolving process. Such a treatment can be regarded as a combination of swelling and heat treatment. Both crystalline and amorphous CTA I were treated with superheated steam at 170°C for 2hr in an autoclave. The transformation of CTA I to CTA II results in the change of the x-ray diffraction patterns as shown in Fig. 3. Both acetylated Whatman cellulose powder and acetylated Fortisan, which are amorphous, were almost completely transformed from CTA I to CTA II. However acetylated bacterial and Valonia celluloses were hardly transformed to CTA II by steam treatment, regardless of their crystallinity. On the other hand, the acetylated cellulose powder and Fortisan precrystallized to CTA I by the heat treatment showed about 40% conversion from CTA I to CTA II in the former and complete conversion in the latter as observed in a sharper pattern. This fact explains the considerable difference in the transformation between CTA I_I and CTA I_{II}. CTA I_{II} is always easy to transform to CTA II by the steam treatment, regardless of its crystallinity. As to CTA I_I, the ease of the transformation must be concerned, not only with the crystallinity before the steam treatment, but also with the species of native cellulose sources such as bacterial or Valonia cellulose, and ramie or Whatman cellulose powder. In the former two, the CTA I is resistant to the transformation to CTA II, while in the latter two the extent of transformation depends upon their CTA I crystallinity. The effect of the selective uniplanar orientation in the former case was studied. The cellulose sample with no preferential orientation saponified from bacterial cellulose triacetate was reacetylated using the above conditions. As the mixed pattern of CTA I and CTA II shows in Fig. 4, the transformation to CTA II was slightly improved by the process, but most was recrystallized to CTA I. Therefore the preferential orientation is not the main reason for the resistance to transformation. The reason why they show the difference is still not completely elucidated at present. Reasons for the differences have been proposed by Atalla et al.(11) based on comparisons of the solid state ^{13}C-NMR spectra for the native celluloses from bacteria or Valonia and from higher green plans. They also identified cellulose I_α and I_β. The different hydrogen bonding should have an influence on the acetylation and the subsequent steam treatment for the transformation. Thus we denote their acetyl derivatives as CTA $I_{I\alpha}$ and CTA $I_{I\beta}$ respectively.

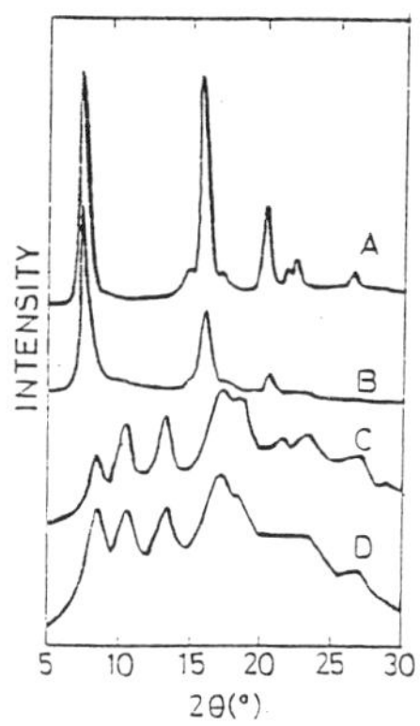

Fig. 3. X-ray diffractograms of CTAs with superheated steam treatment:
(A) Valonia cellulose
(B) bacterial cellulose
(C) prehydrolyzed Fortisan
(D) Whatman cellulose powder

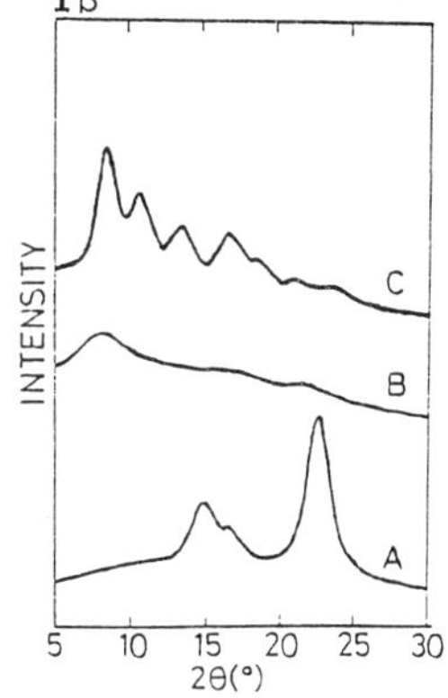

Fig. 4. X-ray diffractograms of reacetylated bacterial cellulose with no preferential orientation saponified from the CTA:
(A) saponification product
(B) acetylated product of (A)
(C) superheated-steam treatment of (B).

CTA $I_{I\alpha}$ is difficult to transform to CTA II, but CTA $I_{I\beta}$ is easy to transform to CTA II. In contrast with this, the mercerization process results in the native celluloses from bacteria or Valonia and from higher green plants becoming similar. The acetylated products of mercerized bacterial cellulose or mercerized ramie cellulose give rather hazy CTA I patterns (Fig.5), although the two equatorial diffractions can be recognized at 11.5 and 5.5A. Both of them are almost completely transformed from CTA I to CTA II by the steam treatment. This means that there is no difference in the transformation between CTAs from mercerized bacterial cellulose and from mercerized ramie. This is because the mercerization process causes the change in chain conformation from the "bent-twisted" type by rearranging hydrogen bonds.

Saponification Some idea of the stability of the structures of CTA I and CTA II and further proof of their dependency on the initial cellulose structure can be obtained from saponification experiments. Creely and Conrad(12) concluded that if CTA I, formed by fibrous acetylation of cellulose I, is treated with formic acid in aqueous concentrations greater than 72% and then heated, the product always has a CTA II structure. Therefore, CTA I can be transformed to CTA II without loss of fibrous form. They also concluded that upon saponification CTA II formed by treatment with less than 75% formic acid is always converted to cellulose I, while that with more than 75% formic acid is converted to a mixture of cellulose I and II. In our previous paper(2), we concluded that there is a general trend for the saponification products to approach cellulose II as the crystallinity of CTA increases. Regardless of the

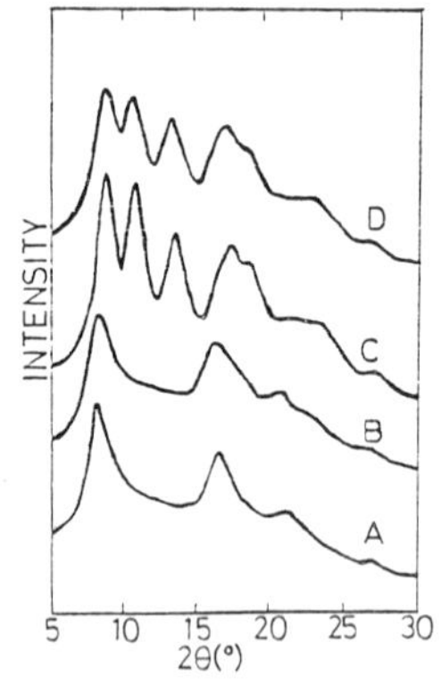

Fig. 5. X-ray diffractograms of CTAs from mercerized celluloses after heat treatment from
(A) mercerized bacterial cellulose
(B) mercerized ramie
(C) and (D) after superheated-steam treatment for (A) and (B), respectively.

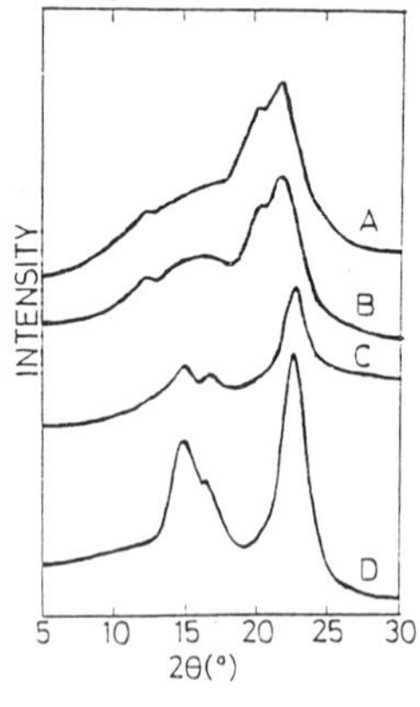

Fig. 6. X-ray diffractograms of saponification products of heat-treated CTA
(A) Whatman cellulose powder
(B) ramie (C) Valonia cellulose
(D) bacterial cellulose.

initial cellulose, highly ordered CTA I and CTA II having a complete crystalline structure are almost entirely converted to cellulose II. This result disagrees with the previous observation that CTA I saponifies to cellulose I and that CTA II saponifies to cellulose II. There is also some confusion in chain polarity; according to the molecular and crystal structure analysis, CTA I is a parallel chain structure, while CTA II an antiparallel chain structure. If this is true, CTA I can never convert to cellulose II of antiparallel structure, and CTA II produced by swelling CTA I in 72-75% formic acid will never saponify to cellulose I of parallel structure. However, in this study, crystalline CTA I from Valonia or bacterial cellulose is always saponified to cellulose I regardless of its crystallinity, but the preferential orientation disappears on saponification (Fig.6). This is a different finding from the previous results in which crystalline CTA I from ramie or cotton cellulose is essentially saponified to cellulose II with increasing crystallinity of CTA I. In addition, when less crystalline CTA I and CTA II were saponified, the products gave mixed patterns of cellulose II and cellulose IV. The same result was obtained for saponified product from acetylated mercerized bacterial cellulose. As yet we can not adequately explain the variation of the content of cellulose IV contained in the saponification products. It will, however, be associated with the crystallinity of CTA and saponification conditions, for the content of cellulose IV decreases as the CTA become highly ordered and the saponification temperature decreases.

REFERENCES

1. K. Hess, C. Trogus, Z. Physik. Chem., B5, 161(1929).
2. B. S. Sprague, J.L. Riley, H.D.Noether, Textile Res.J., 28, 275(1958).
3. S. Watanabe, M. Takai, J. Hayashi, J. Polym. Sci., Part C, 23, 825(1968).
4. M. Saito, T. Kasai, Sen-i Gakkaishi, 19, 465(1963),
5. I. Sakurada, Y. Sakaguchi, Y. Tanaka, Sen-i-Gakkaishi, 14, 900(1958).
6. W. J. Dulmage, J. Polym. Sci., 26, 277(1957).
7. E. Roche, H.D. Chanzy, M. Boudeulle, R. H. Marchessault, P. Sundarajan, Macromolecules, 11, 86(1978).
8. A. J. Stipanovic, A. Sarko, Polymer, 19, 3(1978).
9. M. Takai, H. Watanabe, J. Hayashi, S. Watanabe, Bulletin of the Fac. Engn., Hokkaido Univ., No.82, 93(1976).
10. M. Takai, H. Watanabe, J. Hayashi, S. Watanabe, Bulletin of the Fac. Engn., Hokkaido Univ., No.82, 109(1976).
11. R. H. Atalla, D. L. VaderHart, Macromolecules, 17, 1465 (1984).
12. J. J. Creely, C.M. Conrad, Textile Res. J., 35, 184(1965).

Chapter 13

Cadoxen and solutions of cellulose in cadoxen

Malcolm B. Huglin and Stephen J. O'Donohue – Department of Chemistry and Applied Chemistry, University of Salford, Salford M5 4WT, UK.

ABBREVIATIONS, SYMBOLS AND CONDITIONS

The following abbreviations are adopted in the text:

AQ/CAD cadoxen-water mixture, c(g/cm^3) concentration, C (mol/cm^3) concentration, CAD cadoxen, CEL cellulose, en ethylene diamine, $\bar{G}$(s^{-1}) rate of shear, GPC gel permeation chromatography, M relative molar mass, n refractive index, v(cm^3/g) specific volume of solution, v_o (cm^3/g) specific volume of solvent, $\bar{v}_2$ partial specific volume of cellulose in solution, W weight fraction of Cd in cadoxen, W_2 weight fraction of cellulose in solution, η (cP) absolute viscosity, η_{sp} and η_r specific and relative viscosities, [η] (cm^3/g) intrinsic viscosity, ϕ_1 volume fraction of $[Cd(en)_3]^{2+}$ in cadoxen, γ_1 (cm^3/g) coefficient of selective adsorption of $[Cd(en)_3]^{2+}$ to cellulose, κ (ohm^{-1} cm^{-1}) specific conductance, λ (nm) wavelength in vacuo, μ (J/mol) chemical potential, ρ (g/cm^3) density.

Unless specified otherwise, the overall concentrations of en and NaOH in CAD are 28% wt and 0.35 mol/dm^3 respectively and in AQ/CAD the ratio CAD/water = 1:1 (vol). A temperature of 25°C applies to the reported measurements and data.

INTRODUCTION

GPC allows changes in M with time to be followed conveniently during degradation of CEL. This technique requires a sensitive detecting device for the concentration of CEL as well as a calibration via a [η]/absolute M relationship. None of the solvents for CEL [1-4] is simple in nature and many suffer from disadvantages such as (i) colour, (ii) oxidative instability, (iii) dissolving power limited to low concentrations of CEL (dependent also on M). Despite item (iii) for it, CAD remains the most widely used of these solvents. There are, in fact, several cadoxens differing in composition, but there is a dearth of reported physicochemical data on them and on CEL/CAD solutions. Here attention is focused mainly on fundamental aspects of the solvent, the solutions and modes of effecting dissolution. Absolute values of M via light scattering for acid degraded CEL will be given in a separate report.

Cadoxens comprise $[Cd(en)_3](OH)_2$ in aqueous en. It is usual, but not essential, to incorporate NaOH. CEL dissolves within a rather restricted range of composition, viz. W ∽ 0.040–0.075, en ∽ 27–32% wt, water ∽68–72% wt. The designation AQ/CAD applies to CAD diluted with water and also to the solvent medium after dilution of a solution of CEL in CAD. It is emphasised that, because CAD itself is highly aqueous, it is the concentrations of Cd, en and sodium hydroxide which are reduced significantly in AQ/CAD, whereas the concentration of water is increased comparatively slightly.

PREPARATION AND ANALYSIS OF CADOXEN

CAD was prepared according to Brown [5]; freshly distilled en (b.p. 117–118°C) was essential. Except for a small number of experiments, sodium hydroxide was then added [1]. Final storage was in the dark at 4°C.

The content of total en (complexed + free) was determined by titration [6]. Five separate procedures were used for the concentration of cadmium, viz: (a) complexometric titration vs. EDTA using eriochrome black or, for a sharper end-point, a mixture of this with methyl red as indicator [7], (b) gravimetry using ammonium thiocyanate and pyridine [6], (c) atomic absorption spectroscopy after initial linear calibration with cadmium iodide solutions within the same range of molar concentrations as the cadoxens, (d) via precise measurements of ρ (Anton Paar digital densimeter model DMA-55), which afforded linear depend-ences on W in CAD and AQ/CAD, (e) similarly, with a Pulfrich refractometer linearity was found between n and W. All procedures except for (b) were fast, only (d) and (e) were non-destructive and only (a) and (b) did not require any previous calibration.

PROPERTIES OF CADOXEN AND AQUEOUS CADOXEN

Table 1 summarises some physico-chemical data obtained on cadoxens of known cadmium content. The linear dependences quoted were obtained by using 6 different cadoxens of W = 0.02–0.06. Because of its high conductivity, NaOH was excluded for measurements of κ. For comparison, measurements were also made on a simple cadmium salt, cadmium chloride, in aq. en (28% wt). At a particular W, the value of κ decreased in the order AQ/CAD > CAD > $CdCl_2$. The values of η were obtained from flow times in an Ubbelohde viscometer in conjunction with the flow time, density and absolute viscosity of water and the density of CAD. For the most common CAD (W = 0.05), η is about twice as large as for the corresponding AQ/CAD (W = 0.025), which in turn is about twice as large as η for water. Hence for viscosity of CEL solutions it is clearly quicker to adopt AQ/CAD rather than CAD.

Table 1. Dependence of properties (P) of CAD and AQ/CAD on concentration of Cd. For ρ, n and η: $P = a_0 + a_1 W$. For K: $P = a_0 + a_1 C + a_2 C^2$ where C = concentration of Cd in mol/cm^3.

P	Coefficients	CAD	AQ/CAD
ρ	a_0/g cm^{-3}	1.0076	1.0001
	a_1/g cm^{-3}	1.4	1.4×10^{-4}
η	a_0/cP	2.7	1.49
	a_1/cP	22.7	8.7
n(λ=436)	a_0	1.3913	1.3651
	a_1	0.26	0.21
n(λ=546)	a_0	1.3823	1.3586
	a_1	0.28	0.21
n(λ=578)	a_0	1.3815	1.3563
	a_1	0.27	0.25
κ	a_0/ohm^{-1} cm^{-1}	3.14×10^{-5}	1.00×10^{-3}
	a_1/ohm^{-1} cm^2 mol^{-1}	155	257
	a_2/ohm^{-1} cm^5 mol^{-2}	-6.48×10^4	0

STABILITY OF CADOXEN AND AQUEOUS CADOXEN

Instability in CAD is manifested by development of a yellow coloration (oxime formation [1]) and/or precipitation of a white solid. Access to sunlight and storage at room temperature are to be avoided. We have kept CAD successfully for up to 6 months in the dark and below 4°C. Stability during actual use at room temperature is the relevant factor and for this purpose it was found essential to incorporate NaOH. Dilutions with water in excess of the normal 1:1 (vol) ratio in AQ/CAD have no deleterious effect, but at very high dilutions W becomes too small for the product to act as a solvent for CEL. Water-miscible organic liquids should not, in general, be added to CAD, since they induce turbidity and precipitation. This effect is pronounced with even small additions of acetone and ethanol, whilst rather higher contents of iso-propanol can be tolerated. However, higher contents of added glycerol for fractionation of

CEL have been reported [8].

Occasionally an unusual effect was noted for solutions or, as indicated next, in the solvent alone. When the liquid was blown through the orifice of a pipette or sucked rapidly by vacuum up the capillary tube of a viscometer, a white precipitate appeared due to decomposition of the solvent. Because CAD is pumped through narrow tubes in a GPC apparatus, the effect seems unlikely to be attributable directly to pressure. One may conjecture that, under a sudden pressure change, there is a transient localised rise in temperature.

Regarding temperature alone the findings are more conclusive. When a CAD of a particular W is kept at a temperature T, precipitation occurs rather abruptly after a time t. The higher the value of T the smaller is t. Similarly, for cadoxens of different W at a fixed T, the larger the value of W the smaller is t. These experiments were also conducted by maintaining CAD at T in a UV cell. The UV maximum absorption peak for CAD occurs at 235 nm, but with time at slightly elevated temperature this is shifted to 239 nm, which is probably due to replacement of en ligands by hydroxyl groups or water. However, as seen in Fig.1, there is no change in absorption (which would otherwise have permitted a kinetic analysis of the decomposition process). After the previously noted interval t, the absorption rises abruptly and precipitation occurs. The washed, dried white precipitate arising as decomposition product of CAD was identified as $Cd(OH)_2$ via X-ray powder diffraction. It is indicated thus that, in addition to the recommendations on storage, it is necessary to use CAD at temperatures not exceeding ca. 30°C and to utilise CAD of W not greater than ca. 0.06.

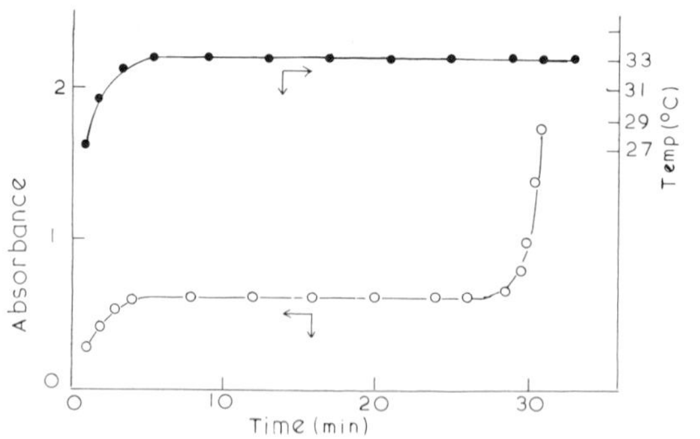

Fig. 1. Absorption at 239 nm for a sample of cadoxen (W = 0.065) held at 33°C. Actual temperature of liquid in the cell was obtained with a thermocouple and is indicated in the upper curve.

DISSOLUTION OF CELLULOSE

Regenerated film ($M \sim 5 \times 10^4$) and undegraded K-linters ($M \sim$

8×10^5) were used. The former dissolved without difficulty at room temperature. For K-linters any turbidity persisting even after prolonged stirring could be dispelled immediately by the addition of a trace of water to give clear solutions. Several types of measurement on solutions of linters did not afford as high a level of precision as those on solutions of regenerated film. This could be rectified if the K-linters were subjected to a form of regeneration, i.e. if the CEL was precipitated from CAD solution into water, washed, dried and redissolved in CAD. Quantitative recovery of CEL from solution by precipitation was found to require at least a 10-fold excess of water.

Preparation of a solution of CEL in AQ/CAD can be achieved by (a) dissolution in CAD followed by addition of water, (b) immersion of CEL in water followed by addition of CAD, (c) dissolution of CEL in a pre-formed AQ/CAD mixture. In (b) and (c) the contact with water is sufficient to overcome hydrogen-bonding in CEL, but there is restriction to specific binding with $[Cd(en)_3]^{2+}$ ions because of the low content of Cd. The highest volume fraction of water in the resultant AQ/CAD to yield solubility is ca. 0.25. For a typical CAD of W = 0.05 this means that W is reduced to 0.75 x 0.05 = 0.0375, which is less than the threshold value of 0.040 for solubility (see Introduction). In contrast, procedure (a) yields solutions of CEL for volume fractions of water not only up to the commonly adopted value of 0.50 but also up to 0.75. At this latter dilution the value of W is reduced to 0.012, i.e. well below the threshold value. However, it is important to distinguish between the Cd content in the final diluted solution and the more crucial value, which is the Cd content in the the actual dissolution medium. We have verified that, for a low volume fraction of water (0.20), procedures (a), (b), and (c) all yield the same value of [η].

Fig.2 illustrates the effect on [η] of dilution via (a). AQ/CAD is seen to be a thermodynamically better solvent than CAD alone. Indeed only a 2% content of water is sufficient to increase [η] markedly over its value in CAD alone. There is no evidence of significant change in [η] within the region 20%-60% water and it would appear, consequently, that adoption of exactly 50% in AQ/CAD is not too critical for viscometric studies. Higher values of [η] in AQ/CAD than in CAD were observed not only for solutions of regenerated CEL but also for all samples obtained from controlled degradation of K-linters.

Failure to take account of any shear effect can, in principle, lead to erroneous values of [η]. Corrections have been found necessary, for example, for solutions of tricarbanilated CEL of high M [9]. One of the largest values of [η] in the present study is yielded by slightly degraded K-linters in AQ/CAD, the value in CAD being somewhat lower. In neither case did finite

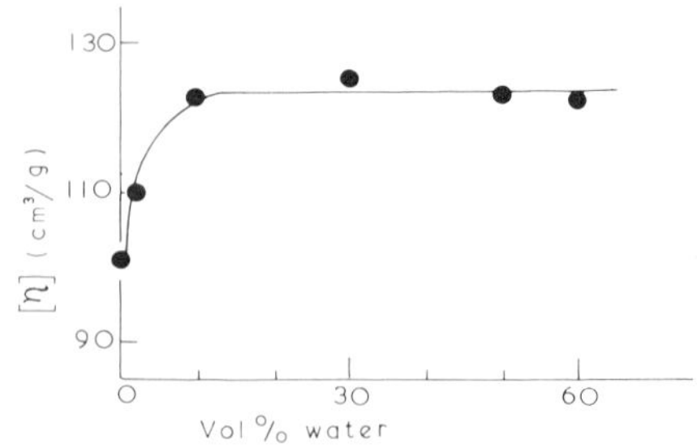

Fig. 2. Intrinsic viscosity of regenerated cellulose in cadoxen and in cadoxens diluted with water (T = 25°C).

changes in shear rate affect [η] and it is concluded that values of [η] yielded directly in a single bulb viscometer are correct if [η] < ca. 600 cm^3/g. Figure 3 illustrates estimation of [η] for slightly degraded linters in CAD via a double extrapolation procedure to zero concentration and zero $\bar{G}$. Coincidence between the values of [η] according to the equations of Huggins and Kraemar is also demonstrated.

STABILITY OF SOLUTIONS OF CELLULOSE

CEL solutions were prepared in three different cadoxens having W = 0.041, 0.051 and 0.056. Solutions in the corresponding AQ/CAD were also made. As indicated before, the latter yielded somewhat larger values of [η] than the former group, but within each group [η] was sensibly independent of W. After 8 weeks' storage at 4°C the measured values of [η] indicated some slight degradation of the CEL corresponding to a decrease of ca.10% in M. Generally solutions can be used within a few days of preparation and hence degradation may be considered negligible.

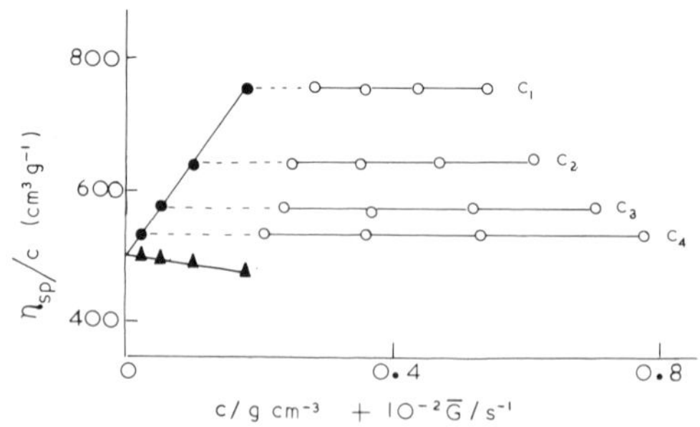

Fig. 3. Determination of intrinsic viscosity for slightly degraded K-linters in cadoxen (W = 0.051) at 25°C. Data relate to 4 concentrations ($c_1 - c_4$) and 4 rates of shear $\bar{G}$. Via Huggins equation – open circles and filled circles as values extrapolated to $\bar{G}$ = 0. Via Kaemar equation [(1/C ln n_r vs. c] – points omitted to avoid overcrowding and only values extrapolated to $\bar{G}$ = 0 shown as filled triangles.

OTHER PROPERTIES OF SOLUTIONS OF CELLULOSE

The partial specific volume $\bar{v}_2$ of regenerated CEL was determined in 6 different media, viz for W = 0.041, 0.051 and 0.056 in CAD and half these values in AQ/CAD. In each system ca. 7 different values of W_2 were used and the specific volume v of the solution measured. $\bar{v}_2$ was obtained from the linear plots according to eq.(1),

$$v = v_1 + (\bar{v}_2 - v_1)W_2 \qquad (1)$$

There was no influence of the solvent medium on $\bar{v}_2$, the average value being 0.680 ± 0.005 cm^3/g. Hermans [10] has analysed reported densities and refractive indices of several sources of CEL and has obtained a mean value of 58.3 cm^3/mol for the molar refraction at λ = 589 nm. This lies in close accord with the value of 58.2 cm^3/mol at λ = 578 nm which we have derived via the Gladstone-Dale relationship [11] from values of $\bar{v}_2$, the specific refractive increment of CEL solutions and n.

Specific refractive index increments dn/dc of CEL solutions were measured in a Brice-Phoenix differential refractometer calibrated with aq. KCl [11]. No dependence of dn/dc on W_{23} was observed. In CAD of W = 0.051 values of 0.090 and 0.089 cm^3/g were obtained at λ = 436 and 546 nm respectively, whilst in the corresponding AQ/CAD values of 0.095 cm^3/g were obtained at both wavelengths. These values relate to constant composition, but it is well established that the appropriate quantity for light scattering in multicomponent systems is dn/dc at constant chemical potential, i.e. $(dn/dc)_\mu$. This is obtained at dialysis equilibrium of diffusible low molar mass species. By means of our earlier procedure [12] a solution was prepared of CEL in CAD of W = 0.051. The subsequent solution in AQ/CAD was separated from AQ/CAD by a Visking membrane (note that the cellulosic membrane becomes soft and swollen, but does not dissolve, since the medium was AQ/CAD and not CAD). Refractometric measurements were made after attainment of dialysis equilibrium (48 hr). The resultant value of $(dn/dc)_\mu$ = 0.189 cm^3/g was found at both values of λ.

In eq (2) $dn/d\Phi_1$ represents the variation of refractive index of AQ/CAD with volume fraction of the complex ion $[Cd/en)_3]^{2+}$. Its value was estimated via the data in Table 2 to be 0.177.

$$\gamma_1 = [1/(dn/d\Phi_1)][(dn/dc)_\mu - (dn/dc)] \qquad (2)$$

The value of γ_1 is thereby found to be + 0.53 cm^3 of $[Cd(en)_3]^{2+}$ per gram of CEL, and the positive sign signifies preferential adsorption of this species (as opposed to desorption, which would be indicated by a negative sign). Invoking M = 162 for the repeating unit in CEL this corresponds to 0.62 mol of complex ion selectively adsorbed per monomer unit. Molecular models were used to demonstrate that this finding is sterically feasible if 0.62 is approximated to 2/3, i.e. if 2 complex ions are involved in binding to 3 repeating units via the hydroxyl groups. Conductivity measurements on CEL solutions may also be used to quantify selective adsorption [13]. Here it will suffice to indicate such assumption has a physically meaningful connotation rather than being merely a thermodynamic parameter related to activity coefficient. Adsorption of some of the ions to the polymer should reduce ionic mobility and hence κ although there is no change in the total number of ions. The value of κ in CAD or AQ/CAD was found to decrease linearly with concentration of dissolved CEL. Thus for a CEL concentration of 1×10^{-2} g/cm^3, the value of κ is ca. 6% smaller than that in the solvent alone. It is possible that in addition to dn/dc, dκ/dc may prove an alternative means of detecting the concentration of CEL.

ACKNOWLEDGEMENT

The authors thank the Procurement Executive, Ministry of Defence for financial support and for a maintenance grant (to SJO).

REFERENCES

1. D Henley, Arkiv Kemi, 18 (1961) 327.
2. K F Gadd, Polymer, 23 (1982) 1867.
3. M Terbojevich, A Cosani, G Conio, A Ciferri and E Bianchi, Macromolecules 18 (1985) 640.
4. R B Seymour, E L Johnson and G A Stahl in "Macromolecular Solutions" (R B Seymour and G A Stahl, Editors) Pergamon Press, London (1982) p.90.
5. W Brown, Svensk Papperstidn., 70 (1967) 458.
6. A I Vogel "Textbook of Quantitative Inorganic Analysis", Longman, London and New York (1978).
7. H Vink, Makromol. Chem. 76 (1964) 66.
8. L S Bolotnikova, S N Danilov and T I Samsonova, J Appl. Chem. (USSR), Eng. Trans. 34 (1961) 2443.
9. J T Guthrie, M B Huglin, R W Richards, V I Shah and A H Simpson, Europ. Polym. J. 11 (1975) 527.
10. P H Hermans "Contribution to the Physics of Cellulose Fibers", Elsevier, New York (1946) p.207.
11. M B Huglin in "Light Scattering from Polymer Solutions" (M B Huglin - Editor), Academic Press, London (1972) Chapt. 6.
12. J T Guthrie, M B Huglin and G O Phillips, Makromol. Chem. 149 (1971) 309.
13. M B Huglin, S J O'Donohue and P M Sasia - in preparation.

Chapter 14

Advances in the understanding of the nitration of cellulose in mixed acids

Hugh S. Munro and Robert D. Short – Department of Chemistry, University of Durham, South Road, Durham DH1 3LE, U.K.

ABSTRACT

The role of the nitronium ion in the nitration of cellulose in mixed acids is discussed on the basis of studies using ESCA and Raman spectroscopy. That the nitronium ion is the species responsible for nitration is supported by nitrations in potassium nitrate/sulphuric acid solutions and nitronium ion salts. The extent of nitronium ion formation and denitration capabilites in the acid mixes are considered with respect to the final degree of substitution.

INTRODUCTION

Despite the numerous applications in which cellulose nitrates have been used (e.g.lacquers, packaging, munitions *etc.*) a clear unambiguous understanding for the degree of nitration of cellulose obtained in mixed acids has still to emerge. Studies aimed at probing the relationship between acid mix composition and the final degree of substitution have resulted in two very different models [1] :-

a) The micro and macroscopic structure of cellulose gives rise to inaccessible regions where some of the hydroxyl groups are not available for nitration. The extent of accessibility is dependent on the nitrating mix composition.

b) Nitration of cellulose is a reversible process. For example, if a 2.6 D.O.S. cellulose nitrate is immersed in an acid mix capable of only nitrating to a D.O.S of 2.0 then denitration of the 2.6 D.O.S. material is observed. The maximum D.O.S obtained in a given acid mix is representative of the equilibrium between the nitration and denitration reactions. The position of this equilibrium is dependent on the acid mix composition.

It may be the case that a combination of both models applies.

Before the relative merits of the models can be discussed in terms of determining the dominating process for a limiting D.O.S. it is necessary to understand the chemistry of the nitrating reaction. In particular the identification of the species responsible for the nitration of the cellulose in a mixed acid is required. Earlier studies have not identified this species. The aim of this paper is to discuss the identification of this agent as a result of studies using ESCA and Raman spectroscopy and to examine some consequences of its identification.

Electron Spectroscopy for Chemical Applications (ESCA) has been shown to be a very powerful technique for the examination of the composition of surfaces. It has been successfully applied to the determination of the D.O.S. in cellulose nitrates [2,3,5]. Due to the heterogenuous nature of for example the nitration of cellulose, the surface is of the material is affected first. Reaction at a surface is not constrained to the same diffusion and degradation problems as the bulk material and as a consequence a greater variety of reaction conditions can be studied if characterisation of the surface is possible. Thus the monitoring of surface D.O.S. by ESCA enables further insights into the nitration of cellulose to be obtained. Raman spectroscopy is a powerful tool for the characterisation of acid mixes [6].

RESULTS AND DISCUSSION

In technical acid mixes the nitronium ion (NO_2^+) (a powerful nitrating agent) is known to be present[1]. However, Miles [1] concluded that NO_2^+ was not the nitrating agent for two main reasons. First cellulose can be nitrated in acid mixes in which NO_2^+ is not spectroscopically detected and secondly nitration can take place in the vapour of nitric acid at low pressures. The formation of NO_2^+ in the vapour phase is improbable. However this latter argument may be invalid if condensation of the vapour occurs before reaction then autoionisation of the acid may occur.

$$2HNO_3 \rightarrow H_2NO_3^+ + NO_3^-$$

$$H_2NO_3^+ \rightarrow NO_2^+ + H_2O$$

For the former, that NO_2^+ is not spectroscopically observed does not necessarily mean that NO_2^+ is absent. Further, there may be more than one species capable of nitrating cellulose, although it is worth noting that most of the nitrating media used in organic chemistry produce NO_2^+ as the nitrating agent[4].

On examining the D.O.S. obtained for acid mixes of fixed nitric variable sulphuric/water composition (Table 1) it is observed that as the sulphuric acid content decreases the D.O.S. decreases. The Raman spectra for the corresponding range of acid

Table 1. Acid mix composition and D.O.S. obtained

% composition by weight			D.O.S	
H_2SO_4	HNO_3	H_2O		
75	22.5	2.5	2.5	Mix 1
70	22.5	7.5	2.5	
65	22.5	12.5	2.4	Mix 2
60	22.5	17.5	2.1	Mix 3

mix compositions are shown in Figure 1. The bands centred at 1400, 1320, 1050 and 966 cm^{-1} can be assigned to NO_2^+, HNO_3, HSO_4^- and H_2SO_4 respectively [6].

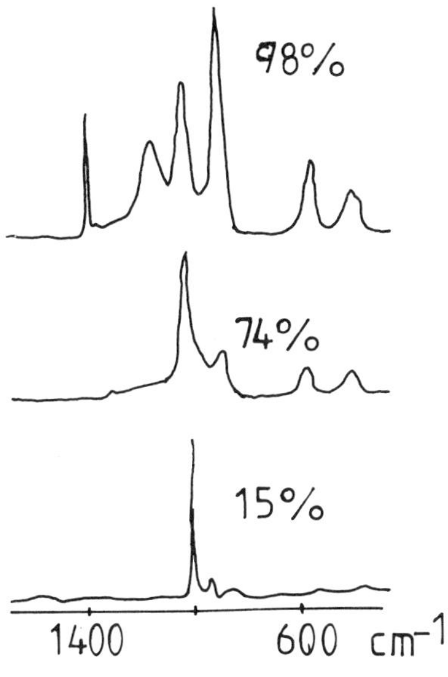

Fig. 1. Raman Spectra of mixed acids.

It can be clearly seen that as the D.O.S. of the cellulose nitrate increases so does the intensity of the nitronium ion band relative to the nitric acid band. Thus as the concentration of sulphuric acid in the mix increases the conversion of nitric acid to the nitronium ion is enhanced. As the nitronium ion [4] is an acknowledged nitrating agent in so many aspects of organic chemistry it would seem strange that if in these present mixes this species did not play an important role.

Further supporting evidence that the nitronium ion does nitrate cellulose arises from other reactions in which it is known to be present. For example, the nitrogen species present in potassium nitrate solutions of different sulphuric acid content have been identified [7]. Only in solutions containing greater than 82% sulphuric is the nitronium ion observed (Figure 2). When cellulose is immersed in these solutions nitration occurs only when the nitronium ion is present. Similarily immersion of cellulose in a saturated nitronium ion salt (NO_2^+ BF_4^-) for two minutes results in nitration on the surface to a D.O.S. of 2.82. These results clearly indicate that when NO_2^+ is present it will nitrate cellulose. Consequently in technical acid mixes where this species is nearly always observable in the Raman spectra, it will have a significant role to play.

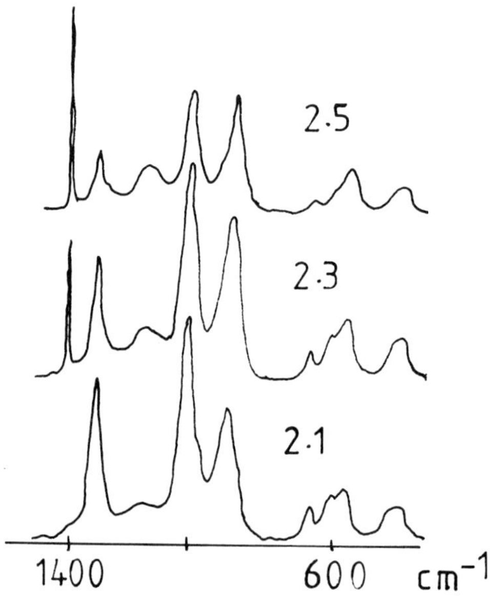

Fig. 2. Raman Spectra for KNO_3/H_2SO_4 mixes.

There are two main problems in interpreting the nitronium ion as the effective nitrating agent in nitric/sulphuric acid/water mixes. The first arises in cases where the nitronium ion concentration, as observed by raman spectroscopy, and sulphuric acid contents of the acid mix are very high yet the D.O.S. obtained is lower than would be anticipated. The second arises in cases where the nitronium ion is not observed spectroscopically, yet nitration to reasonable level occurs.

Table 1 indicates the first complication. The D.O.S for a mix with fixed nitric acid content and increasing sulphuric plateaus. Raman spectra indicate that for the acid beyond the maximum D.O.S. the nitronium ion content is slightly higher than that for the maximum. As we have alluded to above it would have been expected that the greater the nitronium ion concentration the higher the D.O.S.. However, we have not accounted for, if any, the possibility that sulphuric acid may in fact act as a denitrating agent. Miles [1] concluded that denitration of a cellulose nitrate in sulphuric acid was not possible. The limit on this observation being that in very concentrated sulphuric acid solutions the cellulose nitrate is destroyed. Thus the observation for denitration using bulk techniques is not possible. By utilising the surface sensitivity of ESCA and the heterogenuous nature of the reaction, it is possible to observe(by appreciably reducing the reaction time)that denitration of cellulose nitrate does indeed occur when the sulphuric acid content (relative to water) is greater than 80%. This concentration is comparable to that for the acid mix which gives rise to a lower than anticipated D.O.S..

When, for an acid mix of fixed nitric composition, the sulphuric content is reduced and the water content increased, the intensity of the nitronium ion band decreases relative to the nitric increases and the D.O.S. drops. However, on continuing this trend there comes a point whereby the acredited nitrating species is no longer spectroscopically observable yet a reasonable final D.O.S. is obtained. This point generally is outside of the technical acid composition region. Is the nitronium ion still responsible for nitration or is some other species now operable? This question has yet to be fully resolved. However, it is interesting to compare the time dependence of nitration for acid mixes which give rise to similar final D.O.S.'s but for which the nitronium ion is either just observable in the raman spectrum or is not.

The D.O.S. obtained in the first second and after 40 minutes are compared in Table 2. It can be clearly seen that when NO_2^+ is definitely present the nitration reaction at the surface is extremely rapid (1s) and when it is not the initial D.O.S. is very low. This time dependence distinguishes the two conditions. The slow nitration time for the latter case is also observed in very dilute nitronium ion salt solutions where the nitronium ion is the only nitrogen species present and is not observed in the Raman spectrum. This may suggest that the nitronium ion is involved in the nitration of cellulose in acid mixes where it is not spectroscopically observable. Further work is required in this area.

To summarise, the change in the D.O.S. in acid mixes of fixed nitric content but variable sulphuric/water content may be described as follows. At low sulphuric concentrations the extent of nitronium ion formation is low (i.e. not spectroscopically observable) and results in a relatively low D.O.S. with a slow reaction time for reaching the final degree of nitration. As the sulphuric content is increased the conversion of the nitric to NO_2^+ increases (spectroscopically observable). The reaction

Table 2. D.O.S. obtained after 1s and 40 minutes for different acid mixes.

ACID MIX % Nitric/Sulphuric/Water		D.O.S 1second	D.O.S 40minutes
1. 25.3 / 59.4	/ 15.3	0.1	1.8
2. 22.5 / 61	/ 16.5	1.0	2.0
3. 22.5 / 75	/ 2.5	2.1	2.4

In acid mix 1 NO_2^+ is NOT detectable.
In acid mix 2 NO_2^+ is just detectable.
In acid mix 3 NO_2^+ is very detectable.

time is significantly reduced and the D.O.S. increases. (It may also be the case that the denitration reaction, that is generally associated with the nitric acid, will also be less favoured as the effective concentration of molecular nitric present in the mix is reduced). Although the effective nitrating power of the mix (as determined by the amount of NO_2^+ present) is increased at very high sulphuric acid concentrations the observed D.O.S. does not continue to rise as denitration due to the sulphuric becomes significant.

Although the above discussion relates the D.O.S. with nitronium ion concentration and hence suggestive of a balance between nitration and denitration reactions as the controlling factor on the maximum degree of nitration (i.e. equilibrium) it does not exclude the possibility that some hydroxyl groups are in fact inaccessible (micro accessibility).

EXPERIMENTAL

The preparation of cellulose nitrates in mixed acids has been previously described [5].

Cellulose Whatman No.1 papers were immersed in saturated KNO_3/H_2SO_4 mixes for two minutes, longer times resulted in the dissolution of the cellulose.

The nitronium ion salt, $NO_2^+BF_4^-$ was dissolved in sulpholan (99%, Aldrich Chem Co.). Whatman No.1 papers were immersed in a saturated solution of the nitronium ion salt at 24°C for 2 minutes. The nitration was performed under a dry nitrogen atmosphere. Before analysis the papers were soxhlet extracted with hexane for >2 hours to remove the sulpholan adhering to the surface.

Core level spectra were obtained on a Kratos ES300 electron spectrometer employing $Mg_{K\alpha}$ x-rays. Spectra were peak fitted using the software on the Kratos DS300 data system on which spectra were stored. Area ratios are within ± 5%. Raman spectra were obtained on a Varian Cary 82 Raman spectrometer using an argon laser (514.5 nm).

ACKNOWLEDGEMENTS

We wish to thank R.A.R.D.E for the provision of a research studentship to R.D.S.

REFERENCES

1. Miles F.D. "Cellulose Nitrate ", Interscience, London 1953.
2. Clark D.T., Stephenson P.J., Polymer Communications, 23,1034 (1982).
3. Clark D.T., Stephenson P.J., Polymer Communications, 23,1295 (1982).
4. Urbanski T., " Chemistry and Technology of Explosives ", 4,Pergamon Press, London 1984.
5. Clark D.T. and Fowler A.H.K., Polymer Communications, 24,140 (1983).
6. a) Ingold C.K., Millen D.J., Poole H.G., J.Chem. Soc. 2576 (1950)
   b) Hogget J.G., Moodie R.B., Penton J.R. and Schofield K., "Nitration and Aromatic Reactivity ", Cambridge University Press,Cambridge 1971.
7. Bayliss N.S., Watts D.W., Aust. J. Chem., 16,943 (1963).

Chapter 15

Preparation of butyl carbamates of wood

H. West and W. B. Banks – Department of Forestry & Wood Science, University College of North Wales.

GENERAL EXPERIMENTAL TECHNIQUE

Throughout this study, specimens of Scots pine (Pinus sylvestris) sapwood were used. They were cut to the following dimensions 15 mm tangential, 15 mm radial, 5 mm longitudinal. The short longitudinal dimension was chosen to ensure easy penetration of reagent throughout the cellular structure. Prior to use the specimens were lightly sanded to remove loose fibre, and then solvent extracted by refluxing with a toluene/ethanol/acetone (4 : 1 : 1 v/v) mixture for 4 - 6 hours to remove non structural substances (extractives). The specimens were then oven dried at 105 $\pm$ 2°C for about 16 hours and stored over anhydrous phosphorus pentoxide until required for treatment.

Reactions with n-butyl isocyanate reagent were carried out by adding diluent solvent, catalyst (where used) and weighed specimens to a stainless steel pressure reaction vessel. The vessel and contents were brought quickly to reaction temperature by plunging for a short period under controlled conditions into an oil bath maintained at a fixed temperature, about 150°C above reaction temperature. This was to minimise the temperature rise period and hence ensure reaction conditions as nearly isothermal as possible. After reaction the vessels were cooled rapidly by immersing in cold flowing water. It was established that warm up and cool down periods did not exceed 5 mins.

The reacted blocks were cleaned by reflux in acetone for 4 hours to remove unreacted reagent and catalyst. Reflux solvent was then allowed to evaporate and the blocks were finally heated strongly (105°C for 16 hrs) to ensure minimal solvent retention. Specimens treated in this way were observed to lack the infra red absorption band at 2230 cm^{-1} characteristic of the N = C = 0 group. It can be assumed therefore, that the prescribed clean up procedure, effectively removes unreacted isocyanate.

THE NON CATALYSED REACTION

It is known that aliphatic isocyanates react relatively easily with alcoholic hydroxyl under homogeneous conditions, (see for example Kirk-Othmer, 1981)

$$R - OH + O = C = N - R \longrightarrow R - O - \overset{\overset{\displaystyle O}{\|}}{C} - \overset{\overset{\displaystyle H}{|}}{N} - R$$

Rowell and Ellis (1979) have demonstrated also that wood hydroxyl will react with a low molecular weight isocyanate (CH_3-N=C=O) without catalyst. In preliminary studies we have found that n-butyl isocyanate reacts only very slowly with oven dried wood in the absence of catalyst even under quite severe conditions. For example reaction with a 6% solution of butyl isocyanate in toluene for 16 hours at 140°C led to a weight increase of about 4.5%. This is equivalent to reaction of only some 3% of the hydroxyl groups present in wood.

It is well established, that the reactivity of cellulose varies depending on the extent to which the fibrillar aggregations are disrupted (see for example, Krassig, 1985). The suggestion is made that in non swelling systems reaction starts at fibril surfaces and proceeds via less ordered non crystalline zones into tightly hydrogen bonded crystalline regions, by a bond unzipping process. Where the cellulose is swollen, reaction can occur more rapidly, because the swelling agent (aqueous base or organic swelling liquid) opens up the structure, allowing easy access to non crystalline regions within the fibrillar aggregate. With powerful swelling agents, the crystalline zones become accessible, allowing a faster reaction.

Results from our preliminary studies with butyl isocyanate, support these ideas. Where wood is swollen by neutral reagents (acetone and methyl acetate) reaction under similar conditions to those described above (6% butyl isocyanate, 16 hrs. 140°C) gave rise to the following results (Table 1).

Table 1. Wood – butyl isocyanate reaction data in swelling and non swelling systems

Reagent	Vol. Swell relative to water	Weight Gain %	-OH sites reacted %
6% butyl isocyanate in toluene	0%	4.5 ± 0.5	3%
6% butyl isocyanate in acetone	50	20.5 ± 0.6	15%
6% butyl isocyanate in methyl acetate	50	18.0 ± 0.4	13%

It is important to note, that the increased rate of reaction observed is in the absence of any chemical catalysis and that the extent of cell wall swelling induced by the swelling liquids used is relatively small (only about 50% of that achieved in water).

THE BASE CATALYSED REACTION

The homogeneous isocyanate - hydroxyl reaction has been extensively investigated, forming as it does, the basis of much urethane chemistry. In heterogeneous systems, the reaction has been used by Rowell & Ellis(1979), Rowell (1980), Ellis & Rowell (1984) and Rowell & Ellis (1984) in a series of studies concerned with the modification of wood to impart dimensional stabilisation and decay resistance. However little work is reported concerning the mechanism and kinetics of the heterogeneous reaction. We have studied the reaction between butyl isocyanate and wood, under pyridine catalysis,in some depth. Data relating to this study are given in Table 2.

Table 2. Reaction data for pyridine catalysed, wood n-butyl isocyanate reactions (Reagent, 10% n-butyl isocyanate, 90% pyridine (Vol:vol) Volumetric wood swell∿85% of that in water

Temperature (^{o}C)	Rate Constant (hr^{-1})
60	1.06×10^{-3}
70	1.88 "
80	3.40 "
90	6.33 "
100	10.5 "
110	16.0 "
116	23.5 "

From the data of Table 2, it is seen that wood is substantially swollen in the pyridine - butyl isocyanate reagent. An Arrhenius plot of these data yields an activation energy of ~ 60 kJ/mole which suggests that reaction to weight gain 20% (~15% of hydroxyl groups reacted) would occur in 30 min with a 6% BuNCO soln. in pyridine @ 140°C. This is in contrast to the swollen/non catalysed systems where reaction to about 20% weight gain took 16 hours and to the non swollen and non catalysed system where reaction for 16 hours caused only 4.5% weight gain. Hence it seems that the reaction is strongly influenced both by accessibility of functional groups in wood (the swelling effect) and by chemical, in this case base, catalysis. This view is reinforced by the data of Table 3, which summarise the results of reactions between wood and n-butyl isocyanate in the presence of pyridine and triethylamine catalysis.

Table 3. n-Butyl isocyanate – wood reaction with pyridine and triethylamine catalyst.

Reagent	Catalyst pKa	Volumetric wood swell %*	Reaction condition	Weight gain %
1 pt butyl isocyanate 2 pts triethylamine (vol : vol)	10.8	0	115°C : 8hrs	3.4
1 pt butyl isocyanate 2 pts pyridine	5.2	70%	115°C : 8 hrs	53.4

* water = 100%

CATALYSIS IN SWOLLEN AND NON SWOLLEN WOOD

The data for the reactions outlined above, suggest strongly that effective reaction requires the wood to be swollen and a chemical calalyst to be present. Systematic preliminary studies of the effect of these two factors have been carried out in a series of reactions between oven dry and pre swollen (solid emulsion) wood with butyl isocyanate in the presence of several catalysts. The wood solid emulsions were prepared by saturating specimens with water and then replacing the water with several changes of dry acetone. Solvent exchange was repeated with fresh batches of dry solvent until the acetone in contact with wood specimens no longer reacted with sodium metal to liberate hydrogen. By this means the wood cell wall was retained in a swollen state, substantially in the absence of water or other hydroxy reagents liable to react with the isocyanate in competition with the wood cell wall. Throughout this series of reactions, isocyanate reagent was supplied in about 8 fold molar excess (based on total wood hydroxyl) and the reactions were conducted at 100°C for 2 hours. The reagent mixtures (50 ml) contained 25 ml butyl isocyanate, and were molar in catalyst, diluent solvent being toluene. Four replicate wood specimens were used in each reaction. Reaction data are given in Table 4 below.

Specimens heated under identical conditions with solvent/catalyst only, showed small weight decrease (~0.5%). This is probably due to removal of small amounts of cell wall and extractive material. Hence the true weight gains due to reaction are slightly higher than those reported in Table 4.

Table 4. Reaction data for swollen and non swollen wood reacted with n-butyl isocyanate in the presence of several catalysts.

Catalyst	Catalyst pKa	Non swollen mean weight increase (1 s.d)	Swollen mean weight increase (1 s.d)
None	-	2.9 (0.4)	8.8 (0.7)
Pyridine	5.2	0.7 (0.1)	10.3 (0.7)
Triethylamine	10.8	2.7 (0.5)	19.0 (0.7)
D.A.B.C.O.*	8.6 (3.0)	3.0 (0.2)	55.8 (0.3)
D.B.T.D.A.*	-	4.5 (0.5)	44.1 (1.0)

* D.A.B.C.O. = 1, 4 di azo bicyclo [2,2,2, octane]

D.B.T.D.A. = di butyl tin di acetate

The data show convincingly that none of the catalysts are effective with wood in a non swollen state. Simply activating the wood by swelling leads to a significant increase in the extent of substitution achieved under the specified reaction conditions. The effectiveness of pyridine used as a combined reagent solvent/catalyst is certainly due to a combination of its powerful wood swelling and base catalytic properties. Where wood is pre swollen, more powerful bases (triethylamine and D.A.B.C.O.) show a more powerful catalytic effect. This is in agreement with observations made on homogeneous urethane reactions(Hostettler

& Cox 1960). Similarly the organo tin compound (D.B.T.D.A.), known to be a very efficient homogeneous urethane reaction catalyst, shows little catalytic effect on non swollen wood, but a very marked catalysis in the systems where wood is activated by swelling.

DISTRIBUTION OF URETHANE PRODUCT WITHIN THE WOOD CELL WALL

Each of the three major classes of wood cell wall polymer (cellulose, hemicelluloses, lignin) contain reactive hydroxyl groups. From an analysis of reaction profiles for wood reacted with butyl isocyanate under pyridine catalysis, it seems that lignin reacts more rapidly, but to a smaller extent than polysaccharide (Fig 1).

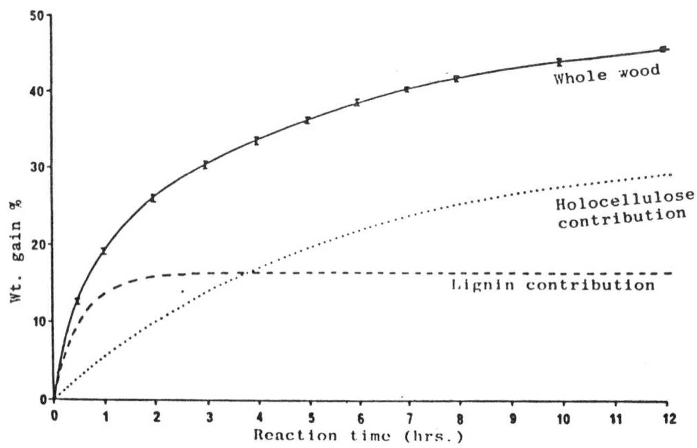

Fig. 1. Wood – butyl isocyanate reaction data.

This idea is supported by results from studies of the reaction of isolated holocellulose (produced by brown rot organisms) and lignin (West and Banks, 1986). Precise reasons for the apparently greater ease of reaction of lignin are not known. It may be due simply to a lower extent of intra and inter molecular hydrogen bonding in the relatively rigid lignin network, compared to that in fibrous and gelatinous polysaccharide. The observation is in agreement with that of Rowell (1980)

Within the polysaccharide fraction, reaction may occur in hemicelluloses, in non crystalline regions in cellulose and in crystalline cellulose. It is suggested by Krassig (1985), that reaction (acetylation) of isolated cellulose fibre, begins at surfaces and less crystalline regions, proceeding gradually into crystalline zones by a hydrogen bond "unzipping" process. We have observed the X ray diffraction pattern for wood cell wall polysaccharide reacted to various levels of substitution with n-butyl isocyanate. Our data are summarised in Fig. 2.

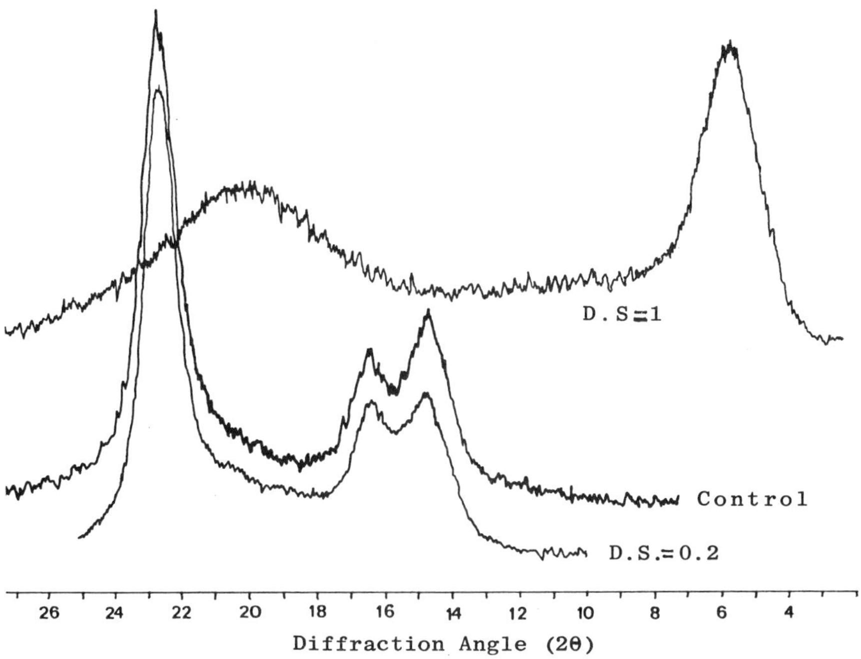

Fig. 2. X-ray diffractograms of treated wood.

The diffractograms for the unreacted material, and that reacted to a degree of substitution ~ 0.2 are virtually superimposable. This suggests strongly, that up to this degree of substitution, reaction is confined to non crystalline regions (hemi celluloses or less ordered cellulose regions). At the higher degree of substitution however reflections in the 0.4 - 0.6 nm region are severely attenuated, whilst a new peak corresponding to a lattice spacing of ~ 1.5 nm appears. This spacing is in the region observed in base swollen cellulose 101 plane (Howsman and Sisson, 1954). It may be therefore that as reaction proceeds to a high degree of substitution, the crystal structure is forced into a form resembling that of base swollen cellulose, and held in this condition after removal of base, by the ordered bulking effect of bonded urethane groups in crystalline regions. Krassig (1985) describes such processes as being generally applicable to substitution reactions carried out on fibrous cellulose.

REFERENCES

Ellis,W.D.,& Rowell,R.M.,(1984) Wood & Fibre Sci. 16,349-356

Hostettler,F.,& Cox,E.F.(1960) Industrial & Engineering Chemistry 52, 609-610.

Howsman,J.A.,& Sisson,W.A.,(1954) in Cellulose and Cellulose Derivatives eds. E.Ott, H.M.Spurling and M.W.Grafflin. Interscience.

Kirk-Othmer (1981). Encyclopedia of Chemical Technology,3rd Edition. John Wiley & Sons.

Krassig,D.H.(1985). Structure of Cellulose and its relation to properties of cellulose fibres. In Cellulose and Its Derivatives, eds. Kennedy,J.F.,Phillips,G.O.,Wedlock,D.J. and Williams,P.A. Ellis Horwood Ltd.

Rowell,R.M.(1980) Wood Science 13, 102-110.

Rowell,R.M. & Ellis,W.D. (1979) Wood Science 12,52-58

Rowell,R.M. & Ellis,W.D. (1984) Wood & Fibre Sci.16, 257-267.

Chapter 16

The topochemistry of aspen sulphonation

R. P. Beatson – Pulp and Paper Research Institute of Canada.

SUMMARY

Small sticks of aspen, <u>Populus tremuloides</u> were treated with 1 mol L^{-1} sodium sulphite solutions of pH 7 at 120°C for up to 240 minutes. The extent of sulphonation of the middle lamella cell corner region, fibre wall and vessel wall was measured by using a transmission electron microscope fitted with an energy dispersive X-ray analyser.

The sulphur content of the lignin in the cell corner region of the middle lamella and the vessel wall was twice the level found in the fibre wall. The distribution of sulphur matched the distribution of guaiacyl units.

INTRODUCTION

The sulphonation of lignin plays an important role in all sulphite pulping processes (Glennie, 1971). Recently, it has been found that sulphonation of lignin and carboxylate formation are factors that determine the strength properties of the chemi-mechanical pulps obtained on refining sulphite-treated aspen chips (Heitner and Atack, 1983). However, despite the obvious importance of the sulphonation of hardwood lignin it has been a rather neglected subject.

During sulphite treatments in the pH range of 4 to 7, lignin in birch, maple and aspen has been found to sulphonate to a lesser degree than in softwoods (Lindgren and Saeden, 1951). This led to the conclusion that these species contained approximately half of the number of reactive 'A-groups' that are found in spruce lignin. Recent studies (Beatson, Heitner, Rivest and Atack, 1985) show that for aspen, this difference occurs in the very early stages of sulphonation corresponding to the sulphonation of guaiacyl units bearing free para-hydroxyls, the so called 'X-groups' of spruce lignin. In the same study it was also noted that although the kinetics of sulphonation of aspen lignin were the same as reported for spruce lignin (Beatson, Heitner and Atack, 1984), the maximum extent of sulphonation of aspen lignin was about 50 percent of that of black spruce lignin.

Marth 1959 investigated the sulphonation of lignin during treatment of aspen veneer, 2.5 mm thick, with sulphite solutions in the pH range of 4.2 to 9.0. This treatment entailed a warm-up period of 90 minutes to 170°C followed by periods of 30 to 600 minutes at temperature. As part of the study, sections of sulphite treated chips were stained with p,p'-azodimethylaniline; a dye that turns red on contact with sulphonated lignin. The red-colouration was found to progress from the middle lamella inward to the fibre lumen. The walls of the vessels and ray cells became pink early in the cook whereas the libriform cell walls took longer to show a pink colour. The pink colour in the libriform cell was most pronounced at pH 4.5.

These observations were taken to indicate the flow of sulphite liquor through the chip. The liquor was thought to contact and sulphonate the vessels and rays cells first and then penetrate the libriform fibres through cell wall diffusion. Little information was given on the early stages of aspen sulphonation and the experimental design did not permit a clear decision as to whether sulphonation was controlled by liquor penetration or by differences in the chemical reactivity of the lignin in the various regions of the wood.

The present work was designed to determine quantitatively the distribution of sulphur in aspen chips that had been treated for short times with sulphite liquors after a thorough impregnation. These are the conditions encountered during the sulphite treatment stage of modern chemimechanical pulping processes. Quantitative data was obtained by using a technique developed during a study of the topochemistry of black spruce sulphonation (Beatson, Gancet and Heitner, 1984). This technique involves counting the X-rays emitted from sulphur atoms in various regions of sulphite treated wood sections when they are bombarded with electrons in a transmission electron microscope.

EXPERIMENTAL

Sulphite Treatment of Aspen

Aspen (Populus tremuloides) wood chips were broken into small sticks in a Waring blender. About 7.3 g O.D. of the disintegrated chips were fully soaked with water and then immersed for 24 h at room temperature in 220 mL of a solution of 104 g L^{-1} of sodium bisulphite and 33.3 g L^{-1} of sodium hydroxide. The impregnated chips and sulphite liquor were placed in a 250 mL stainless steel bomb and immersed in an oil bath heated to 155°C. After 13 minutes the contents of the bomb reached 120°C and cooking was continued at this temperature for known times. The bombs were cooled rapidly and the wood sticks removed. The sticks were washed by repeated soaking in deionized water until no more sulphite was eluted as determined by a permanganate test.

A few sticks were removed for use in preparation of thin sections. The remainder were used for chemical analysis. The acid insoluble and acid soluble lignin were determined by CPPA Technical Section standard method G9 and TAPPI useful method UM250 respectively. The sulphur content was determined by CPPA Technical Section standard method G28. Yields of the treatments were determined by parallel experiments. The results are shown in Table 1.

Determination of Sulphur Concentrations in different Morphological Regions

The relative sulphur concentrations in the middle lamella, fibre and vessel walls were measured with a transmission electron microscope fitted with an energy dispersive X-ray analyser. Details of the procedure are given in a previous report [Beatson, Gancet and Heitner, 1984].

Table 1. Sulphur content of aspen treated at 120°C with pH 7 sulphite solution having a total SO_2 concentration of 1 mol L^{-1}.

Treatment Time (minutes at 120°C)	Yield (percent)	Lignin Content (percent)	Lignin Losses (percent O.D. wood)	Sulphur Content of Treated Wood
13	94.7	21.8	0.8	0.10
70	88.4	21.3	1.3	0.19
240	85.6	19.1	3.5	0.30

Lignin content of original wood: 22.6 percent.

RESULTS AND DISCUSSION

Distribution of Sulphur

Figure 1 depicts the relative sulphur concentrations in different regions of a 0.15 μm transverse section of aspen which had been treated with sodium sulphite to a bulk sulphur contentof 0.3 percent. The points represent the position of the 0.2 μm diameter electron beam used for the 100 second irradiation of the sample. It is apparent that the sulphur concentration in the vessel wall is about one third of that in the cell corner region of the middle lamella. The sulphur concentration in the fibre wall is even lower than in the vessel wall. The outer area of the fibre wall has a sulphur concentration about one seventh that in the cell corner. The sulphur concentration in the fibre wall drops to zero as the lumen is approached. The low X-ray counts in the lumen and the inner regions of the fibre wall arise mainly from background radiation.

Table 2 compares the sulphur concentrations in the middle of the fibre wall, vessel and cell corner. Results are given for aspen which had been treated for three different times at 120°C with pH 7 sulphite solutions 1 mol L^{-1} in total SO_2. It is apparent that the sulphur concentration in S_2 layer of the fibre wall is on average about 11 percent of that in the cell corner region of the middle lamella. The value of 11 percent is a maximum since up to 50 percent of the low X-ray counts in the fibre wall may arise from background radiation. Thus, the sulphur content of the fibre wall may be as low as 5 percent of that in the cell corner. This is in marked contrast to our findings with black spruce where the concentration of sulphur in the fibre wall was about thirty percent of that in the cell corner (Beatson, Gancet and Heitner, 1984). The sulphur concentration in the vessel wall of aspen is much higher than in the fibre wall being about 33 percent of that in the cell corner.

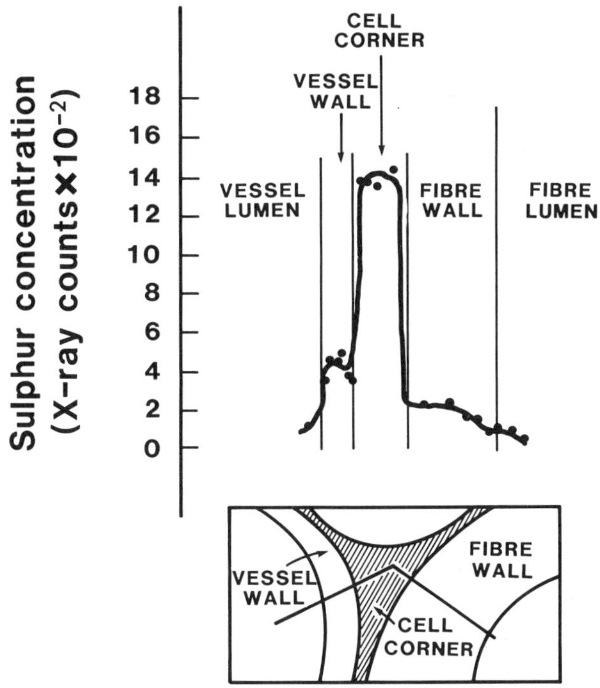

Fig. 1. The relative sulphur concentration in different regions of a section of sulphonated aspen as given by the X-ray counts in the sulphur K_α window. Sulphite treatment was with 1 mol L^{-1} sodium sulphite at pH 7 and 120°C for 240 minutes.

The low degree of sulphonation of the fibre wall indicates that sulphonation can play only a minor role in softening the fibre wall of aspen during chemimechanical pulping. Apparently most of the fibre softening and resulting increase in bonded area, caused by sulphite treatment during chemi-mechanical pulping of aspen, must arise from the formation of carboxylic acid groups in the fibre wall (Heitner and Atack, 1983).

It is reasonable to assume that the distribution of lignin in aspen is similar to that in birch (Fergus and Goring, 1975). Using the data for birch, the relative sulphur contents of the lignin in each region can be calculated as shown in Table 2. The low concentration of sulphur in the fibre wall lignin relative to that in the cell corner and vessel wall lignin, when combined with the observation that maximum extent of sulphonation of aspen lignin is only 50 percent that of black spruce lignin (Beatson, Heitner and Atack, 1984), indicates that the aspen has less reactive sites in its fibre wall lignin.

Table 2. Sulphur content of lignin in morphological regions of aspen

Morphological Region		Ratio of Sulphur Concentrations	Ratio of Lignin Concentrations*	Ratio of Sulphur Contents of Lignin
Fibre	S_2	1	1	1
	ML_{cc}	9	4.5	2.0
Vessel	S_2	3	1.4	2.1

* Fergus and Goring, 1970.

Nature of the Reactive Site in Aspen Lignin

Recent studies on the kinetics of aspen sulphonation (Beatson, Heitner, Rivest and Atack, 1985) have shown them to mirror the kinetics of black spruce sulphonation. Thus, as black spruce contains only guaiacyl lignin, this strongly implies that only guaiacyl units in aspen are sulphonated. It has been demonstrated previously, by u.v. microscopy (Musha and Goring, 1975), and fibre fractionation (Hardell, Leary, Stoll and Westermark, 1980) that the syringyl moieties in hardwoods are located mainly in the fibre secondary wall whereas the lignin in the cell corner and vessel walls (Walter, Harkin and Kirk, 1974) is composed mostly of guaiacyl units. Thus, the distribution of sulphur in aspen lignin matches the distribution of guaiacyl residues providing further evidence that only the guaiacyl units react.

No 'a priori' reason for the lack of sulphonation of syringyl units is apparent in the results of studies on the sulphonation of vanillyl and syringyl alcohol (Janson and Sjöström, 1965). However, it has been shown that at pH 7 lignin is sulphonated via quinone-methides which are only formed if the free p-phenolic groups are present (Beatson, Heitner and Atack, 1984). Thus, a possible reason why the syringyl moieties in aspen fail to sulphonate is that nearly all their phenolic groups are involved in ether linkages (Nimz, 1973; Robert and Gagnaire, 1981). Indeed, it has been shown (Chang, Cowling, Brown, Adler and Miksche, 1975) that cellulolytic enzyme lignin and milled wood lignin from sweetgum have phenolic hydroxy contents about 67 percent lower than those from spruce. A carbon-13 NMR study (Nimz, Robert, Faix and Nemr, 1981) has also shown that milled wood lignins from the hardwoods beech, maple, oak and cherry also have lower phenolic hydroxy contents than softwood lignins.

Thus, there is evidence that only the guaiacyl units in aspen lignin can be readily sulphonated. This hypothesis explains not only the topochemistry of the sulphonation but is consistent with the kinetics and mechanism of lignin sulphonation and the low free phenolic content of hardwood lignins. Further evidence for slow sulphonation of syringyl lignin in aspen can be found in the low yields of syringaldehyde on nitrobenzene oxidation of spent liquors from the early stages of sulphite cooking of aspen (Marth, 1959; Stone, 1955).

The observation that lignin sulphonation in the cell wall occurs mainly in the outer layers implies that the guaiacyl units are localised in this region. This is not totally surprising when one considers the sequence of lignification in wood cells.

Lignification begins in the cell corner and outer regions of the S_1 layer and proceeds into the S_2 layer of the secondary wall (Wardrop, 1965 and 1971; Saka and Thomas, 1982). Also, in ring porous angiosperms such as aspen, formation and lignification of the vessel walls preceeds that of the xylem elements (Kozlowski, 1971; Evert and Kozlowski, 1967). Thus, the distribution of sulphur and guaiacyl units follows the sequence of lignin deposition. This implies that the precursors for formation of syringyl lignin are only present during the later stages of lignification. One can surmise that the absence of these precursors during the early stages of lignification results from the same factors that govern the lack of syringyl lignin in softwoods (Higuchi, 1980). That is, an absence of ferulate-5-hydroxylase or a lack of activation and/or reduction of sinapic acid.

Distribution of Guaiacyl and Syringyl Units

It is possible to estimate the guaiacyl distribution in aspen from the relative sulphur contents of the lignin by making a few reasonable assumptions. The first assumption is that syringyl moieties are absent from the cell corner region of the middle lamella. This is generally the case for hardwoods (Fergus and Goring, 1970; Saka, Hosoya and Goring, 1983).

The other assumptions are that the guaiacyl moieties in the fibre and vessel walls sulphonate to about a 30 percent higher degree than in the cell corner, by analogy with results on black spruce (Beatson, Gancet and Heitner, 1984; Whiting and Goring, 1982), and also that 10 percent of the lignin in all regions exists as p-hydroxyphenyl moieties (Smith, 1955).

The distribution of syringyl and guaiacyl units in aspen, estimated from the sulphur distribution, are shown in Table 3 along with the data for birch obtained by bromination/EDXA (Saka, Hosoya and Goring, 1983) and by U.V. microscopy (Fergus and Goring, 1970). There is general agreement in that vessel wall lignin contains more syringyl units than lignin in cell corner region of the middle lamella, but much less than lignin in the fibre wall.

It should be noted that although the conclusions on the distribution of syringyl and guaiacyl units are consistent with much of the published literature they contradict the conclusions by D.E. Marth in 1959, that for aspen the middle lamella was rich in syringyl units and the fibre wall rich in guaiacyl units.

Table 3. Distribution of guaiacyl and syringyl units in aspen and birch.

		Guaiacyl/Syringyl/p-Hydroxyphenyl (percent)		
		Aspen	Birch*	
		(from sulphur distribution)	(from bromine distribution)	(from U.V. spectral analysis)
Fibre	S_2	32/58/10	12/88/0	20/80/0
Vessel	S_2	66/24/10	88/12/0	100/0/0
ML_{cc}		90/ 0/10	91/9/0	100/0/0

* Saka, Hosoya and Goring, 1983.

Also, our results disagree with those recently obtained by infra-red and methoxyl analysis on tissue fractions from hardwoods (Obst, 1982). These techniques failed to produce evidence for a discrete guaiacyl lignin associated with a cell type or morphological region and led the author to state that 'discrete syringyl and guaiacyl homopolymers, if present, would have to be thoroughly mixed and uniformly distributed throughout the wood'.

CONCLUSIONS

The distribution of sulphur in different morphological regions of sulphite treated aspen (Populus tremuloides) has been determined, with reasonable accuracy, by use of a transmission electron microscope fitted with an energy dispersive X-ray analyser.

The sulphur content of the lignin in the middle lamella, cell corner region, and the vessel wall was found to be about double that in the fibre wall. The low degree of sulphonation of the fibre wall indicates that the sulphite treatment at pH 7 is not an effective method of softening fibres. Thus, sulphonation of aspen chips at pH 7 will not increase significantly the bonded area and tensile strength of aspen chemimechanical pulps.

For aspen treated with sulphite at pH 7 and 120°C, the distribution of sulphur matches the distribution of guaiacyl moieties. This observation, combined with the previous report that the kinetics of sulphonation of aspen lignin mirror those of spruce, indicates that only guaiacyl units in aspen lignin can be sulphonated. The lack of sulphonation of syringyl units is apparently a result of the formation of quinone-methides being blocked by involvement of the p-hydroxy group in ether linkages.

Since sulphonation occurs only at guaiacyl lignin units it is possible to estimate the guaiacyl content of lignin in different regions of aspen from the sulphur distribution. The middle lamella, cell corner region, and the vessel wall are guaiacyl rich whereas most of the syringyl units are found in the fibre wall.

ACKNOWLEDGEMENTS

The author wishes to thank Drs. C. Heitner, J.F. Revol and S. Saka for helpful discussions and Mr. N. Muradali for expert technical assistance.

REFERENCES

Beatson, R.P., C. Gancet and C. Heitner, 1984. The Topochemistry of Black Spruce Sulfonation. Tappi Journal, 67 (3), 82-85.

Beatson, R.P., C. Heitner and D. Atack, 1984. Factors Affecting the Sulphonation of Spruce. Journal of Pulp and Paper Science, 10 (11), J12-J17.

Beatson, R.P., C. Heitner, M. Rivest and D. Atack, 1985. Sulphite Treatment of Aspen. Factors Affecting the Generation of Carboxylate and Sulphonate Groups. Paperi ja Puu 67 (11), 702-708.

Chang, H-L., E.B. Cowling, W. Brown, E. Adler and G. Miksche, 1975. Comparative Studies on Cellulolytic Enzyme Lignin and Milled Wood Lignin of Sweetgum and Spruce. Holzforschung 29, 153-159.

Evert, R.F. and T.T. Kozlowski, 1967. Effect of Isolation of Bark on Cambial Activity and Development of Xylem and Phloem in Trembling Aspen. American Journal of Botany, 54, 1045-1055.

Fergus, B.J. and D.A.I. Goring, 1970. The Distribution of Lignin in Birch Wood as Determined by Ultraviolet Microscopy. Holzforschung 24, 118-124.

Glennie, D.W., 1971. Lignins. p. 597-637, K.V. Sarkanen and C.H. Ludwig, (eds.), Wiley Interscience, New York.

Hardell, H-L., G.J. Leary, M. Stoll and U. Westermark, 1980. Variations in Lignin Structure in Defined Morphological Parts of Birch. Svensk Papperstidning, 83 (3), 71-74.

Heitner, C. and D. Atack, 1983. Ultra-High-Yield Pulping of Aspen, Effects of Ion Content. Pulp and Paper Canada, 84 (11), T252-T257.

Higuchi, T., 1980. Biochemistry of Lignification. Wood Research 66, 1-16.

Janson, J. and E. Sjöström, 1965. Note Concerning the Sulphonation Rates of Vanillyl and Syringyl Alcohol. Acta Chemica Scandinavia 19 (2), 525-527.

Kozlowski, T.T., 1971. Growth and Development of Trees, Vol. II, p. 26, Publ. Academic Press, New York and London.

Lindgren, B.O. and U. Saeden, 1951. Sulphonation and Dissolution of the Lignins of Monocotyledons and Dicotyledons with Sulphite Solutions at pH 4-7. Svensk Papperstidning 54, 795-799.

Marth, D.E., 1959. Studies on the Lignin Fraction of Aspenwood Pulps Produced by Sulphite-Bisulphite Cooking Liquor Systems. Tappi, 42 (4), 301-308.

Musha, Y, 1970. Distribution of Lignin in Aspen Wood. M.Sc. Thesis, McGill University, Department of Chemistry.

Musha, Y. and D.A.I. Goring, 1975. Distribution of Syringyl and Guaiacyl Moieties in Hardwoods as Indicated by Ultraviolet Microscopy. Wood Science and Technology, 9, 45-58.

Nimz, H., 1973. Chemistry of Potential Chromophoric Groups in Beech Lignin. Tappi 56 (5), 124-126.

Nimz, H.H., D. Robert, O. Faix and M. Nemr, 1981. Carbon-13 N.M.R. Spectra of Lignins, 8, Holzforschung 35, 16-26.

Obst, J.R., 1982. Guaiacyl and Syringyl Composition in Hardwood Cell Components. Holzforschung 36, 143-152.

Robert, D. and D. Gagnaire, 1981. Quantitative Analysis of Lignins by C^{13} NMR. Proceedings of the Ekman-Days 1981, Stockholm, June 9-12, Volume 1, p. 86-88.

Russ, J.C., 1974. X-ray Microanalysis in the Biological Sciences. Journal of Submicroscopic Cytology 6, 55-79.

Saka, S., S. Hosoya and D.A.I. Goring, 1983. The Distribution of Lignin in Hardwood as Determined by Bromination with TEM-EXDA. In Preprints of The 1983 International Symposium on Wood and Pulping Chemistry, Vol. 1, 24-29, Tsukuba Science City, Japan, May 23-27.

Saka, S. and R.J. Thomas, 1982. A Study of Lignification in Loblolly Pine Tracheids by the SEM-EDXA Technique. Wood Science and Technology 16, 167-179.

Smith, D.C.C., 1955. 'p-Hydroxybenzoate Groups in the Lignin of Aspen (Populus tremula). Journal of the Chemistry Society, 2347-2351.

Stone, J.E., 1955. A Study of the Lignin Removal During a Neutral Sulphite Cook of Aspen. Tappi 38, 610-612.

Walter, K.E., J.M. Harkin and T. Kent Kirk, 1974. Guaiacyl Lignin Associated with Vessels in Aspen Callus Cultures. Physiol. Plant. 31:140-143.

Wardrop, A.B., 1965. Cellular Differentiation in Xylem. p. 61-97 in 'Cellular Ultrastructure of Woody Plants', Ed. W.A. Côté Jr., Proceedings of the Advanced Science Seminar, Pinebrook Conference Center, Upper Saranac Lake, New York, Sept. 1964. Publ. Syracuse University Press.

Wardrop, A.B., 1971. Lignins. p. 19-41, K.V. Sarkanen and C.H. Ludwig, (eds.), Wiley Interscience, New York.

Whiting, P. and D.A.I. Goring, 1982. Relative Reactivities of Middle Lamella and Secondary Wall Lignin of Black Spruce Wood. Holzforschung 36, 303-306.

Chapter 17

Liquid ammonia as a solvent for the chemical modification of wood and wood products

S. G. Graham, D. J. Harman and J. R. Holker – Materials Science Business Centre, Shirley Institute, Didsbury, Manchester M20 8RX, England.

INTRODUCTION

The overall objective of this project was to assess the largely unexplored potential of liquid ammonia as a solvent to replace or complement water and hydrocarbons in the chemical treatment of wood to impart various properties, e.g. dimensional stability, flame-retardancy, water-repellency, and resistance to micro-organisms. If successful, it might offer some long-term technical, economic, and environmental benefits over established processes. Thus, the solvent power of liquid ammonia for many organic and inorganic substances would enable a wider range, and novel types, of reagents to be applied to wood, whilst its strong swelling action on cellulose should assist their deeper, more uniform penetration. Ammonia is in plentiful supply from non-petrochemical sources, a possible strategic advantage, and should be easily recoverable in a suitably designed closed circuit plant, a factor that may assume importance as antipollution legislation becomes increasingly stringent.

Although liquid ammonia treatment of wood has been explored in several laboratories, little attention appears to have been given to its use as a solvent for additives. In the textile area, a few companies treat cotton fabric with liquid ammonia to facilitate subsequent finishing. Some of the physical effects of ammonia on wood have been reviewed by Bariska and

Schuerch /1/ in an article on softening and forming processes and there are references to the pretreatment of wood with ammonia in order to render it more accessible to chemical modification. Barkhuysen /2/ has investigated the dyeing of cotton from liquid ammonia, whilst Brederick and Pfundtner /3/ have attempted to apply creaseproofing finishes. Standard texts list many inorganic compounds with a high solubility in ammonia, including several ammonium salts and other nitrates and bromides, whilst ammonia has also been extensively used as a solvent for organic reactions, e.g. those reactions of acetylene promoted by sodamide.

RESULTS AND DISCUSSION

Many compounds of known or potential value for the treatment of wood were screened qualitatively or semi-quantitatively for solubility in liquid ammonia, to supplement already published information. Those soluble compounds of more immediate interest included many phenols, mercaptobenzthiazole, some organotin derivatives, alkyl phosphates and a phosphite, ammonium salts (chloride, bromide, thiocyanate, tetrafluoroborate, phosphite, and sulphamate), and certain self-curing organosilanes. Simple aryl phosphates, ammonium phosphates, and many copper salts (acetate, chromate, citrate) were at most only sparingly soluble.

When cubes of red or white pine that had been stained with Erythrosin B for periods up to 6 hr at -33°C in liquid ammonia were split along the grain, lateral penetration was seen to be negligible and penetration through the end-grain never exceeded about 3-5 mm; pre-drying of the cubes had no significant effect. Modest improvements, largely in end-grain penetration, were observed if the temperature of treatment was lowered to -63°C or if the samples were purged with carbon dioxide. The solvent-saturated wood had swollen by about 2% along the grain and about 10% in both radial and tangential directions. Subsequent removal of ammonia, however, brought an overall shrinkage of about 20% by volume and the densified cubes were considerably harder than the original blocks.

Cubes dyed under pressure (10 bar, 25°C; 22 bar, 60°C) showed complete penetration along the grain, and much greater, although still incomplete, lateral penetration. Not surprisingly, however, the incidence of cracking, shrinkage, and distortion was more pronounced.

The resistance of wood to penetration by even small solute molecules was confirmed quantitatively with solutions (20 g/1, or 2.5 g/100g) of ammonium chloride in liquid ammonia. It was first established that a Scots pine sapwood block (20mm x 20mm x 80mm) absorbed some 50-55% w/w, or 40-45% v/v of ammonia during 4 hr at -33°C; addition of the solute did not significantly affect this value. A simple calculation, therefore, indicates that the wood should pick up about 1.25% w/w ammonium chloride. However, uptake at atmospheric pressure was less than half of this value at -33°C, but closely approximated to it at -63°C (Table 1). There was a massive increase in uptake of ammonium chloride, to ca 4%, when impregnation was conducted under pressure. Practical difficulties prevented measurement of the sorption of solvent under pressure, but the amount of solute taken up indicates that it was at least 150% w/w, i.e. about three times the amount absorbed at -33°C. To account for the low uptake of solute at -33°C, it is suggested that a substantial proportion of the ammonia may be absorbed from the saturated vapour, by analogy with the moisture regain of cellulose, which rises rapidly as the relative humidity approaches 100%.

Table 1. Sorption of ammonium chloride from liquid ammonia.

Solvent	NH_4Cl (g/1)	Time (h)	Temp (°C)	NH_4Cl (g/100g wood)
Liquid Ammonia	20	2	40 (15 bar)	4.44
	20	2	20 (10 bar)	4.15
	20	4	-33	0.54
	20	4	-63	1.10
Water	20	4	100	1.10
	20	4	20	0.34
	20	4	0	0.37
	200	4	100	4.84

The dominance of end-grain penetration into Scots pine sapwood blocks was illustrated by sorption values for blocks with different end-grain:lateral surface ratios and with their end-grain optionally sealed with an epoxy resin. A rough calculation indicated that diffusion of ammonia (g/cm^2 wood surface) was of the order of five times greater via the end-grain than laterally; similar comparisons could not be made for pressure impregnation, because the severe conditions caused cracking and lifting of the seals.

Preferential yet limited end-grain penetration was also demonstrated on cubes (25 mm side) of white pine treated with pentachlorophenol (PCP) at 1, 5, 10 and 20% concentration in liquid ammonia at -33°C and then sliced across the grain into eight 3mm thick sections. Analysis of each section (Table 2) revealed a dramatic fall in PCP content only a few mm into the cube, especially at the highest solute concentration. Once again, the total uptake fell far short of the roughly calculated values for wood saturated with liquid ammonia.

Table 2. Sectional analysis of PCP-treated white pine.

Section	PCP content ($mg.cm^{-3}$)			
	1% PCP	5% PCP	10% PCP	20% PCP
1	1.11	4.96	6.89	20.88
2	0.42	1.68	3.07	14.46
3	0.17	0.62	0.40	1.70
4	0.13	0.44	0.22	-

Scots pine sapwood blocks (20 x 20 x 80 mm) were next impregnated with sodium pentachlorophenate or mercaptobenzthiazole (10g/1) in liquid ammonia, over periods of one, two and four hours. One set of samples was subjected to direct pressure impregnation by allowing the liquid ammonia to heat up to 30°C and a pressure of 11.5 bar. The solute contents of the treated wood (Table 3) showed large variations between replicates, even though samples with similar grain and growth rate were selected. This alone did not ensure uniformity, and the variations must in part be attributed to the heterogeneous nature of wood. The most obvious conclusion is that treatment under pressure again favoured far superior uptake of the reagent. Under atmospheric pressure, assuming a 50% w/w uptake of ammonia, solutions containing 10g/1 should afford a solute content in the treated wood of about 0.6%. Under pressure at 30°C, the uptake of ammonia must have been 250% w/w or more, to account for a content of over 3%.

Table 3. Uptake of PCP and MBT by Scots pine sapwood.

t(h)	PCP (g/100g wood)	MBT (g/100g wood)
1	0.18	0.31
2	0.38	0.04
4	0.11	0.10
1 (30°C)	3.32	3.44

The solubility of ammonium chloride, sulphamate, and fluoborate in liquid ammonia prompted a comparison of their effectiveness as non-permanent, i.e. water-extractable, flame-retardants when applied at ambient pressure from this solvent and from water to several wood species; dibutyl phosphite was included in the trial as a simple organophosphorus flame-retardant. To obviate problems associated with limited penetration, and for convenience in subsequent assessment by the Critical Oxygen Index method (ASTM-D2563), the wood was treated in the form of veneers. No attempt was made to determine the salt content of the products.

Table 4. COI values for FR-treated veneers

Reagent	Beech	Oak	Scots Pine	Teak	Walnut
Untreated	21.0	21.0	20.3	21.1	20.9
NH_3 only	19.7	19.8	20.4	20.3	20.3
2% NH_4BF_4 (1 h)	20.4	20.1	20.5	21.6	22.3
2% NH_4BF_4 (4 h)	23.6	21.0	21.9	21.5	22.5
4% NH_4BF_4 (4 h)	22.8	22.5	22.6	22.2	24.2
10% NH_4BF_4 (1 h)	28.3	31.9	24.5	28.5	28.4
10% NH_4BF_4 (1 h)*	23.5	24.1	22.4	21.9	22.8
10% $NH_4SO_3NH_2$ (1 h)	30.1	33.2	27.8	31.0	
10% NH_4Cl (1 h)	26.7	33.3	26.8		28.0
10% NH_4Cl (1 h)	23.2	25.2	24.0	22.5	23.7
10% HP (O) $(OBu)_2$(1 h)	23.1	27.3			

* From water

These treatments produced an immediate improvement in COI (Table 4) which generally increased as concentrations were raised. At a given concentration and exposure time, aqueous treatments caused significantly less improvement in COI values than those carried out using ammonia. No conclusions could be drawn about the relative effectiveness of the four reagents, but there were distinct increases for different wood species, oak proving particularly responsive.

Ammonium chloride and ammonium bromide were compared in closer detail in experiments also designed to assess any influence non-cellulosic wood substances, e.g. oils, resin, lignin, and any ammonia-soluble polysaccharides, might have on flammability. First, the ability of liquid ammonia to extract non-cellulosic substances was amply demonstrated in the cumulative results of four successive 1 h treatments of Scots pine veneer at $-33^{o}C$, with intermediate drying at $100^{o}C$ _in vacuo_; the following mean values again showed very wide variations between the ten individual samples :

time (h)	1	2	3	4
wt.loss (%)	4.1	7.1	9.9	10.9

Dimensional changes accompanying the 3 hr weight loss showed far better replication, with average values of 3.6% shrinkage in length and 13.3% in width.

Table 5. Flammability of Scots pine veneer treated with NH4Cl and NH4Br.

Treatment	Net wt. gain(%)	Halide (%)	COI (%)
Control	-	-	18.9
NH_3 only (1h)	-8.6	-	20.4
In NH_3, $-33^{o}C$, 1h			
10% NH_4Cl[1]	5.3	6.7	26.1
10% NH_4Cl, (leached)[2]	6.5	5.8	26.4
20% NH_4Cl	11.7	8.8	27.8
10% NH_4Br	4.0	8.3	27.4
10% NH_4Br (leached)	5.8	8.4	27.3
20% NH_4Br	10.9	13.1	28.3
20% NH_4Br (leached)	15.8	16.0	27.9
In H_2O 1h			
10% NH_4Cl, $20^{o}C$	3.1	5.0	21.2
10% NH_4Cl, $100^{o}C$	4.3	10.7	23.7
10% NH_4Cl, $100^{o}C$ (leached)	6.4	5.9	27.6
10% NH_4Br, $20^{o}C$	3.2	5.6	24.1
10% NH_4Br, $100^{o}C$	6.8	9.8	25.2
10% NH_4Br, $100^{o}C$ (leached)	-	13.9	28.1

Notes : (1) Saturated solution + solid
(2) Pre-leached for 3 x 1 h in liquid NH_3; wt gains based on extracted weights.

The COI values (Table 5) indicate that removal of soluble materials prior to treatment in ammonia had only a slight influence. Preleaching did, however, result in distinctly (3-4%) higher COI values for veneer treated with either ammonium chloride or ammonium bromide in water at 100°C.

CONCLUSIONS

Liquid ammonia dissolves a variety of organic and inorganic reagents with the ability to improve the microbiological resistance, flame-retardancy, water-repellency, and other properties of wood. It was, however, necessary to remove or highly compress occluded air in order to obtain a reasonable degree of penetration of solutes into all but the thinnest samples of wood. This was achieved most conveniently by working under autogenous pressure at ambient temperature or above.

Limited treatments of small pine blocks with pentachlorophenol and mercaptobenzthiazole (two common microbiocides) and of veneers with ammonium chloride and ammonium bromide, confirmed the value of pressure impregnation, although there was often very poor agreement between replicate samples, perhaps reflecting the highly heterogeneous nature of wood. Veneers of several wood species, impregnated at atmospheric pressure with simple flame-retardants, showed marked increases in COI - to values over 30% in some cases.

Experience strongly suggests that, given the right conditions for impregnation, substantial amounts of both established and novel additives could be taken up by timber; wood pulps, chip and veneers, with their high specific surface, could probably be treated satisfactorily, even at atmospheric pressure.

REFERENCES

1. Bariska, M and Schuerch, C., 'Wood Technology: Chemical Aspects', A.C.S. Symposium Series No.43, 1977.

2. Barkhuysen, F.A., S.A.W.T.R.I. Techn. Rep., Vols. 406 and 427, 1978.

3. Bredereck, K and Pfundtner, R., Textilveredlung, 1975, 10, No. 3, 92-100.

Chapter 18

Influence of additives on the alkaline degradation of cellulose

R. Evans, A. F. A. Wallis and R. H. Wearne – Division of Chemical and Wood Technology C CSIRO, Private Bag 10, Clayton, Victoria 3168, Australia.

ABSTRACT

When digestions of cotton cellulose with soda and kraft liquors at 170°C were carried out in the presence of 9,10-anthraquinone (AQ), increased yields of residual cellulose with higher carboxyl contents, but with lower DPs, resulted. However, AQ in its reduced form had no effect on soda-dithionite cooks of cellulose. These observations are consistent with oxidation of cellulose end groups to aldonic acid groups by AQ, and AQ-assisted cleavage of glucosidic bonds in the cellulose. It is proposed that the latter reaction proceeds by oxidation of the hydroxyls along the cellulose chain, and bond cleavage by a β-elimination mechanism. In support of this proposition, AQ and sodium anthraquinone-2-sulphonate were found to oxidise cyclohexanol and cyclohexanemethanol under soda pulping conditions.

INTRODUCTION

The discovery that catalytic amounts of 9,10-anthraquinone (AQ) as well as many of its derivatives aid the alkaline pulping of wood [1,2,3] has led to the commercial acceptance of the soda-AQ and kraft-AQ pulping processes [4]. Adding AQ to pulping liquors increases pulp yields and delignification rates. The soda-AQ process also offers a sulphur-free alternative to kraft pulping, while allowing the development of new sulphur-free chemical recovery technologies.

To explain the effectiveness of the small amounts of AQ used, typically 0.1% AQ based on oven-dry wood, a cyclic mechanism has been proposed [5,6,7] in which AQ is reduced by the polysaccharides and the reduced AQ species is oxidised back to AQ by the lignin. The higher pulp yields are due in part to the oxidation of the terminal aldehyde groups of the polysaccharides to aldonic acid groups by AQ, thereby stabilising the polysaccharides towards endwise depolymerisation (peeling) reactions [8]. Although this oxidation is well documented, the effect of AQ on the random alkaline cleavage of glycosidic bonds during alkaline pulping is less well understood.

Arbin et al. [9] found that AQ assisted the random alkaline scission of glucosidic bonds in amylose, both in the presence and absence of oxygen. For alkaline degradation of cellulose at 170°C, although AQ promotes the random cleavage in the presence of oxygen [10], the effect of AQ and its derivatives without oxygen is less clear. Whereas Carlson and Samuelson [10] found that adding 5% AQ (cellulose basis) to soda cooks of cotton cellulose under nitrogen gave increased yields of residual cellulose with slightly lower viscosities (indicative of chain cleavage), other studies showed that addition of 0.1% AQ and its derivatives [11] or tetrahydro-AQ [12] under nitrogen gave increased yields and slightly higher viscosities. As part of a study of factors influencing the strengths of soda-AQ pulps, we re-examined the effect of AQ and other additives on the alkaline degradation of cotton cellulose. A study of reactions of polysaccharide model compounds under soda-AQ pulping conditions was also carried out.

EXPERIMENTAL

The cellulose sample used was bleached cotton linters with an α-cellulose content of 99.3%, and DP_w 5340. Viscosities were measured as 0.5% cuene solutions by Tappi method T230 om-82. The carboxyl contents of the celluloses were obtained by the methylene blue method [13].

DP values were measured by high performance size exclusion chromatography (HPSEC) of the tricarbanilate derivatives on a series of three Showdex columns, KF-806, 805 and 804, with tetrahydrofuran as the eluent. The cellulose tricarbanilates were prepared by a modification of the method of Schroeder and Haig [14], in which the cellulose samples were reacted with phenyl isocyanate at 80°C with volumes of pyridine and reaction times depending on the molecular weight of the sample.

The alkaline treatments of the cellulose were carried out by placing 4.0 g of bleached cotton linters and 200 mL 1M sodium hydroxide in 500 mL steel vessels. One series of cooks was performed in an oxygen atmosphere, and all other cooks were in a nitrogen atmosphere with care taken to remove oxygen from the vessels. Various additives were introduced, and the bombs were rotated, six at a time, in a hot air bath at 170°C. After cooling and filtration, the treated pulps were kept at 60°C with a solution of 0.1M sodium dithionite and 0.25M sodium hydroxide for 10 min, and then filtered and washed with water. The procedure was repeated continually until the red colour of reduced AQ from the AQ-treated pulps was no longer evident. As additives, AQ (208 mg, 1 mmol), 9,10-diacetoxyanthracene (DAA) (294 mg, 1 mmol) and sodium dithionite (582 mg, 3.3 mmol) were used. The kraft liquor had 25% sulphidity and the effective alkali was equivalent to 1M sodium hydroxide. In another series of cooks, 5 g samples of cellulose and 50 mL 1M sodium hydroxide with 5.2 mg AQ as the additive were placed in 180 mL vessels.

Cellulose models and quinones (AQ and sodium 9,10-anthraquinone-2-sulphonate (AMS)) were added to 4 mL 1M sodium hydroxide in 10 mL glass tubes which were sealed under nitrogen. The tubes were placed in 20 mL steel vessels and heated in a rocking air bath at 170°C for various periods of time. The contents were acidified and extracted with dichloromethane.

Cyclohexanol (10 mg, 0.1 mmol) was reacted with AQ (25 mg, 0.12 mmol) or AMS (37 mg, 0.12 mmol). The reaction mixture was analysed by gas chromatography (GC) on a BP20 capillary column (Scientific Glass Engineering, Melbourne) with anisole as an internal standard.

Cyclohexanemethanol (11.5 mg, 0.1 mmol) was reacted with AQ (50 mg, 0.24 mmol) or AMS (74 mg, 0.24 mmol). The reaction mixture was methylated with boron trifluoride-methanol and analysed by GC on a BP20 capillary column with 1,2-dimethoxy benzene as the internal standard.

RESULTS AND DISCUSSION

Alkaline degradation of cellulose

Alkaline cooks with cotton cellulose at 170°C were carried out and compared with alkaline cooks containing AQ and DAA as additives. Two charges of AQ were used, 5% and 0.1% based on cellulose; the larger charge was used in an attempt to simulate the soda-AQ pulping of wood, during which AQ would be continually regenerated. The data on the treated celluloses are presented in Table 1.

Digestion of cotton cellulose with 1M sodium hydroxide in an oxygen atmosphere gave a cellulose residue which was low in both yield and viscosity, indicating that severe degradation of the cellulose had taken place. Adding AQ to these cooks

Table 1. Alkaline digestion of cotton cellulose at 170°C.

Cook	Additive	Time at 170°C (h)	% yield		Cuene viscosity (mPa.s)		DP_w		DP_n		carboxyl groups per molecule	
			no additive	additive	no additive	additive	no additive	additive	no additive	additive	no additive	additive
soda (oxygen atmosphere)	5% AQ	0	87.5	92.8	4.54	3.27	1018	666	442	321	0.82	1.04
		0.5	79.3	86.4	3.19	2.64	644	396	299	215	0.83	0.97
		1.0	69.6	84.1	2.57	2.34	434	344	223	198	0.72	1.07
soda	5% AQ	0	97.7	98.1	27.9	18.9	3778	3363	1407	1232	0.61	0.58
		0.5	96.5	97.4	12.2	9.43	2753	2353	1056	907	0.46	0.59
		1.5	92.5	96.1	6.08	5.37	1595	1324	667	559	0.41	0.54
soda	0.1% AQ	0	97.2	97.9	26.6	24.7	3956	3670	1492	1293	0.94	0.69
		0.5	92.9	94.5	12.8	11.1	2472	2474	1002	989	0.60	0.62
		1.0	92.8	93.2	7.38	5.78	1668	1652	731	723	0.54	0.53
kraft	5% AQ	0	97.8	97.6	27.6	24.4	3688	3430	1353	1281	0.63	0.68
		0.5	96.8	97.4	10.9	9.7	2274	2166	948	903	0.49	0.58
		1.5	92.8	96.6	5.26	5.07	1502	1393	668	613	0.45	0.58
soda-dithionite	5% AQ	0	97.7	97.6	27.6	22.8	3835	3510	1472	1304	0.79	0.80
		0.5	96.9	96.3	15.3	13.7	3174	2702	1297	1144	0.61	0.54
		1.5	95.2	95.6	7.90	7.95	1904	1923	796	791	0.50	0.41
soda-dithionite	7.35% DAA	0	98.3	98.4	23.2	19.5	3529	3160	1341	1172	0.45	0.56
		0.5	97.3	97.1	11.6	11.9	2638	2483	1100	966	0.57	0.58
		1.5	94.7	92.3	5.18	5.36	1378	1403	625	611	0.43	0.43

resulted in higher yields and lower viscosities than for the control soda cooks. All subsequent digestions were performed under nitrogen.

When soda cooks of cellulose were carried out under nitrogen in the presence of either 5% or 0.1% AQ, higher yields and lower viscosities than for the soda cook resulted. The magnitude of the effects was greater for the larger AQ charge. Addition of 5% AQ to cellulose digestions with kraft liquor gave less viscosity decrease than when 5% AQ was added to the soda cooks. These observations on cotton cellulose corroborate those of Carlson and Samuelson [10] and contrast with those in which viscosity increases were noted [11,12].

The DP values of the treated celluloses, determined by HPSEC of tricarbanilate derivatives, reflect the cuene viscosity values. In addition, the HPSEC results indicate that the loss in DP of the cellulose in the AQ cooks took place across the whole range of molecular weight. Thus the observed differences in the yields and viscosities of the treated pulps are not a consequence of local effects of a narrow molecular weight fraction.

The number of carboxyl groups per cellulose molecule is higher for the AQ-treated celluloses than for the celluloses treated without AQ. This and the observations above may be explained by oxidation of cellulose by AQ. The increase in yield and carboxyl content of AQ-treated celluloses may be due to end-group oxidation to aldonic acid groups [8], thus stabilising the cellulose towards the peeling reaction. The decrease in viscosity of the celluloses after treatment with AQ may result from oxidation of the hydroxyl groups at C-2, C-3 and C-6 along the cellulose chain to carbonyl functions, rendering the glucosidic bonds more labile to cleavage by alkali through a β-elimination mechanism. This is in accordance with the proposal of Gierer _et al._ [15] that AQ pretreatment of pine wood oxidises the polysaccharides, thus facilitating their degradation during kraft pulping.

To test whether the action AQ exerts on cellulose under soda pulping conditions is oxidative, a series of cooks was performed in the presence of the reducing agent, sodium dithionite, which converts AQ to the hydroquinone in the alkaline solution [16]. In this case, no yield improvement or difference in the number of carboxyl groups ensued from addition of 5% AQ to the cooks, although the reduction in viscosity of the AQ-containing cooks was still evident. The latter observation could be due to incomplete conversion of the solid AQ to the soluble hydroquinone by the dithionite, allowing some cellulose oxidation to occur. To overcome this, the 5% AQ was replaced by an equivalent amount of DAA, which would hydrolyse to the hydroquinone under the alkaline conditions. In this case, there was no yield increase on addition of the DAA, and only a small viscosity reduction early in the cook. Thus the effect of AQ on the alkaline degradation of cellulose appears to be due to oxidation. To gain further insight into the nature of this oxidation, the reactions of AQ with model alcohols were studied.

Reaction of quinones with alkaline solutions of alcohols

The principal reaction of AQ with polysaccharides under soda pulping conditions is oxidation of aldehyde end groups to aldonic acid groups [8]. That this reaction proceeds through an aldosulose intermediate was confirmed by Vuorinen [17], who showed that the initial oxidation product of **D**-glucose with the soluble quinone AMS under alkaline conditions at 50°C was **D**-glucosone. AMS and AQ have also been found to oxidise lignin model benzyl and cinnamyl alcohols to carbonyl functions under alkaline conditions [18,19].

In the present study, reactions of AMS and AQ with cellulose models under alkaline conditions was undertaken to determine whether the quinones oxidise hydroxyl groups at C-2, C-3 and C-6 along the cellulose chain, and thus make the cellulose more labile towards random alkaline cleavage of glucosidic bonds. Monohydric alcohols were chosen as models because their reaction products would be considerably simpler than those from carbohydrates. Cyclohexanol (Ia) and cyclohexanemethanol (Ib) as dilute solutions in alkali were used as models of secondary and primary hydroxyl groups respectively in cellulose, and were reacted with both the soluble quinone AMS and the insoluble AQ.

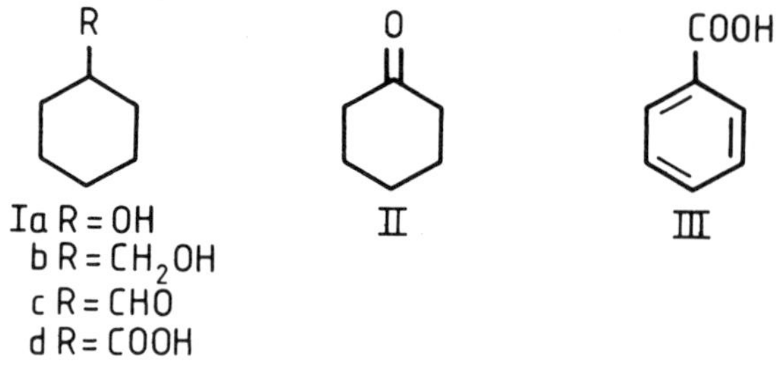

A solution of cyclohexanol (Ia) in 1M sodium hydroxide was unchanged after treatment under nitrogen at 170°C for 6 h (Figure 1). When an equimolar amount of AMS was added to the reaction mixture, the cyclohexanol was oxidised to cyclohexanone (II), attaining a yield of 62% after 6 h (GC analysis). Adding the insoluble AQ to the reaction also gave cyclohexanone as the sole reaction product, but in the lower yield of 18% after 6 h.

Fig. 1. Alkaline reactions of quinones with cyclohexanol.

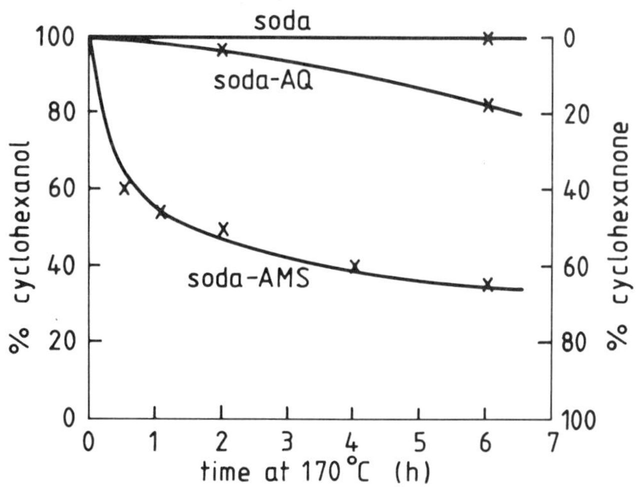

The primary alcohol, cyclohexanemethanol (Ib), was stable after reaction with 1M sodium hydroxide at 170°C under nitrogen for 4 h, giving only a small amount of the oxidised product, cyclohexanecarboxaldehyde (Ic) (Table 2). However, in the presence of 2.4 molar equivalents of AMS, 46% of the starting material was transformed to products, among which were the aldehyde (Ic) and acid (Id) as well as benzoic acid (III). Under soda-AQ conditions, the alcohol was oxidised only to a limited extent.

The results show that the quinones are more reactive to the secondary alcohol Ia, a model for the C-2 and C-3 hydroxyls in cellulose, than the primary alcohol Ib, a model for the C-6 hydroxyl in cellulose. Similar oxidation of the hydroxyls along the cellulose chain could lead to a significant loss in DP.

Table 2. Alkaline reactions of quinones with cyclohexanemethanol.

Cook	% yield			
	Ib	Ic	Id	III
soda 4 h	100	0.6	–	–
soda-AMS 4 h	54	0.9	10.6	11.1
soda-AQ 6 h	99	1.7	tr.	–

REFERENCES

1. Holton, H., Pulp and Paper Mag. Can., **78** (10), T218 (1977).
2. Nomura, Y., Wakai, M. and Sato, H., Japan Patent 43403 (1976); Chem. Abstr., **85**, 65105 (1976).
3. Farrington, A., Nelson, P.F. and Vanderhoek, N., Appita, **31** (2), 119 (1977).
4. MacLeod, J.M. and Clayton, D.W., Paper, **197** (4), 32 (1982).
5. Fleming, B.I., Kubes, G.J., MacLeod, J.M. and Bolker, H.I., Tappi, **61** (6), 43 (1978).
6. Algar, W.H., Farrington, A., Jessup, B., Nelson, P.F. and Vanderhoek, N., Appita, **33** (1), 33 (1979).
7. Iiyama, K., Kulkarni, A.G., Nomura, Y. and Nakano, J., Mokuzai Gakkaishi, **24** (10), 766 (1978).
8. Lowendahl, L. and Samuelson, O., Tappi, **61** (2), 19 (1978).
9. Arbin, F.L.A., Schroeder, L.R., Thompson, N.S., and Malcolm, E.W., Tappi, **63** (4), 152 (1980); Cellul. Chem. Technol., **15** (5), 523 (1981).
10. Carlson, U. and Samuelson, O., Svensk Papperstidn., **82** (2), 48 (1979).
11. Paik, K., Nonglim nonjip, **23**, 15 (1983); Chem. Abstr., **101**, 232111 (1984).
12. Satoh, A., Nakano, J., Ishizu, A., Nomura, Y. and Nakamura, M., Kami Pa Gikyoshi, **33** (6), 410 (1979).
13. Wilson, W.K. and Mandel, J., Tappi, **44** (2), 131 (1961).
14. Schroeder, L.R. and Haig, F.C., Tappi, **62** (10), 103 (1979).
15. Gierer, J., Kjellman, M. and Noren, I., Holzforschung, **37** (1), 17 (1983).
16. Sawicki, E., Stanley, T.W. and Hauser, T.R., Anal. Chem. **30** (12), 2005 (1958).
17. Vuorinen, T., Carbohydr. Res., **116**, 61 (1983).
18. Gratzl, J.S., EUCEPA Symposium Manuscripts, Vol. II, No. 12, June 2-5, 1980, Helsinki, Finland.
19. Apfeld, P.B. and Dimmel, D.R., J. Wood Chem. Technol., **2** (3), 269 (1982).

Chapter 19

Degradation of cellulose triacetate in acetylation media

P. Howard and S. U. Malook – Chemistry Department, University of Surrey, Guildford, Surrey.

Acetylation of cotton linters using sulphuric or perchloric acid inevitably gives a product of lower molecular weight due to chain scission. Since the catalyst has a dual role in respect of the acetylation and degradation processes, the overall catalytic mechanism is complex and presently not fully understood. Both processes are obscured by the heterogeneous nature of the conditions, particularly in the early stages. A useful way of studying the degradation process under controlled homogeneous conditions, in the absence of acetylation, is to compare degradation rate constants for cellulose triacetate in various acetic anhydride-solvent mixtures.

EXPERIMENTAL

Materials

Cellulose triacetate (CTA) samples were prepared from cotton linters and had an acetic acid yield of at least 62.3%. Details of CTA preparation, method of analysis, molecular weight characterisation, and purification of solvents are given elsewhere [1-3]. In the present work number-average molecular weights of CTA samples used in the three chloroalkane solvents studied were 81100 ($CHCl_3$), 91140 (CH_2Cl_2) and 92607-112035 ($C_2H_2Cl_4$).

Procedure

A weighed amount of vacuum dried CTA was dissolved in the chloroalkane solvent to give a stock solution which was kept at the degradation working temperature for at least three days before use. The catalyst was weighed into a volumetric flask, in a dry-box, to give a 25 ml stock solution in a weighed amount of acetic anhydride. A measured volume was then added to a second volumetric flask, a little of the chloroalkane solvent added and the stoppered flask brought to the working temperature. A measured amount of the CTA stock solution was transferred to this flask and the solution immediately made up to the mark with thermostatted solvent. After shaking, to give a completely homogeneous solution, about 10 ml was filtered (No. 3 sinter) into an Ubbelohde suspended-level viscometer in the thermostat bath.

The zero-time for commencement of degradation was taken as that at which the CTA stock solution was added to the catalyst solution. Viscometer flow-times for the solution of CTA in the chloroalkane-acetic anhydride-catalyst solution were then recorded at times extending over 150 min. A specially designed viscometer head, connected to solvent traps and drying tubes, prevented solution evaporation and atmospheric moisture contamination during the degradation runs.

Limiting viscosity numbers (LVN), obtained from the solution flow-times, were used to calculate the number-average molecular weight of CTA at successive degradation time intervals as described below. Solvent flow-times were measured for each chloroalkane-acetic anhydride composition (without catalyst) at each working temperature.

All degradation runs were carried out with a CTA concentration of 0.800 g dl^{-1} and an acetic anhydride concentration of 2.12 mol dm^{-3}.

LVN Determinations

These were calculated from viscosity numbers (η_{sp}/c) measured at each degradation reaction time using the single-point viscosity equations proposed by Maron [4] (chloroform) or Solomon and Gotesman [5] (methylene chloride and tetrachloroethane). These relationships gave LVN values correct to about 2% over the range of temperatures studied.

Molecular weight at each reaction time

This was obtained from the calculated LVN using previously determined Mark-Houwink constants, using different number-average molecular weight CTA samples, in the corresponding solvent mixtures at each temperature.

RESULTS

Treatment of kinetic data

This was based on the theoretical treatment of random chain degradation [6,7] and later discussed by Grassie [8]. It is assumed here that each oxygen link between glucopyranose CTA units has the same probability of scission, in which case the first-order kinetic equation for degradation is given by

$$\frac{1}{M_t} = \frac{1}{M_o} + \frac{kt}{288 \cdot 2} \tag{1}$$

where M_o and M_t are the number-average molecular weights of CTA initially and after time t, respectively. The rate constant, k, represents the fraction of oxygen links broken in unit time and 288.2 is the mass of a single glucopyranose triacetate unit. Rate constants were obtained from slopes of the linear plots of $1/M_t$ or $(M_o-M_t)/M_tM_o$ versus t. Where curved plots were obtained (discussed later) the rate constant was evaluated from the initial slope (t=o) for the 'best-fit' polynomial equation obtained by least squares regression analysis. Choice of polynomial order was based on lowest standard deviation from experimental points consistent with closest agreement with the theoretical intercept (Equation 1).

Perchloric and sulphuric acid catalysed degradation

Typical kinetic plots are shown in Figures 1 and 2 for methylene chloride solutions with perchloric and sulphuric acid catalysts, respectively.

The effect of temperature on rate constant was satisfactorily expressed by the Arrhenius equation. The activation energy, E_a, was evaluated from good Arrhenius plots obtained for each of the chloroalkane solvents. Degradation studies completed in tetrachloroethane [3] now enables a direct comparison to be made between chloroform, methylene chloride, and tetrachloroethane as solvents. Rate constants and activation energies are shown in Tables 1 and 2 for the three chloroalkanes using perchloric and sulphuric acids, respectively.

DISCUSSION

The linear kinetic plots for perchloric acid catalysed degradation generally found (Figure 1) support the proposed first-order rate equation, although slight curvature was observed at higher temperatures. We believe the curvature to be due to a slow secondary side-reaction between catalyst and acetic anhydride, which is more marked for the sulphuric acid catalyst. Evaluation of initial rate constants (t=o) thus seems justified in these cases. Non-linear plots were observed for sulphuric acid catalyst at all temperatures in chloroform, methylene chloride and tetrachloroethane.

Perchloric acid is a much faster catalyst than sulphuric acid, comparable rates of degradation only being found with a perchloric acid concentration about one tenth of that for sulphuric acid. For perchloric acid catalyst, rates of degradation decreased in the order chloroform > methylene chloride > tetrachloroethane at all temperatures reported (Table 1). This was true for sulphuric acid catalyst up to 25°C, above which the order would appear to be chloroform > tetrachloroethane > dichloromethane (Table 2).

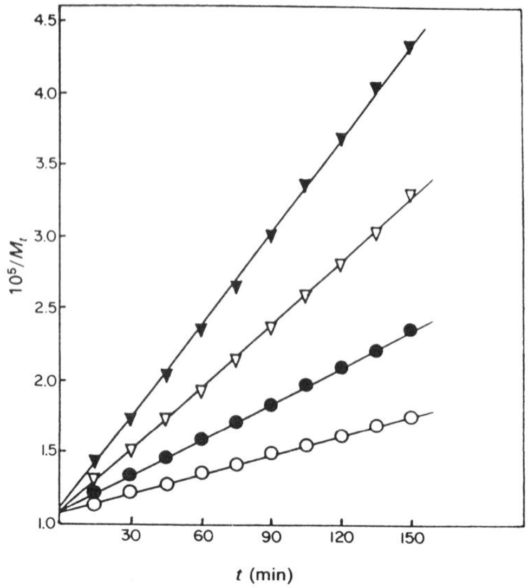

Fig. 1. Kinetic plots for perchloric acid catalysed degradation of CTA in dichloromethane (2.12 mol dm^{-3} Ac_2O) at 18°C. Values given $\times 10^{-3}$ mol dm^{-3} $HClO_4$ (○) 1.65; (●) 3.30; (△) 5.40; (▲) 7.20.

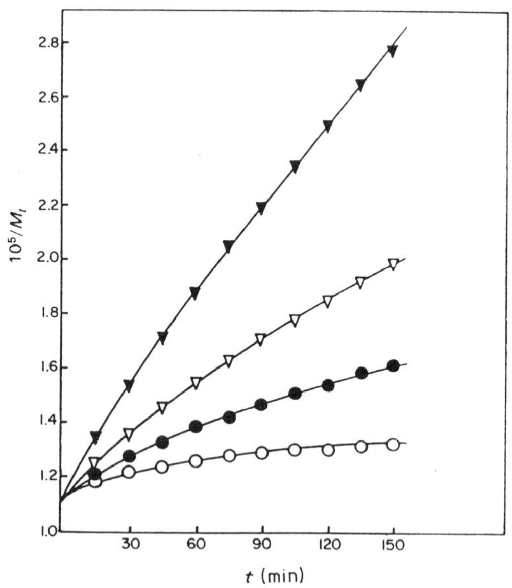

Fig. 2. Kinetic plots for sulphuric acid catalysed degradation of CTA in dichloromethane (2.12 mol dm^{-3} Ac_2O) at 18°C. Values $\times 10^{-2}$ mol dm^{-3} H_2SO_4: (○) 2.00; (●) 3.00; (△) 4.00; (▲) 6.00.

Table 1. Degradation of CTA in Chloroalkane Solvents $[HClO_4] = 1.65 \times 10^{-3}$ mol dm^{-3} $[Ac_2O] = 2.12$ mol dm^{-3}.

Solvent	$10^5 k/\text{min}^{-1}$				Ea/kJ mol^{-1}
	15°	20°	25°	30°	
Chloroform	1.89	3.39	5.14	9.02	74.1
Dichloromethane	0.827	1.54	2.86	4.22	78.6
Tetrachloroethane	0.223	0.405	0.732	1.26	83.7

Table 2. Degradation of CTA in Chloroalkane Solvents $[H_2SO_4] = 1.69 \times 10^{-2}$ mol dm^{-3} $[Ac_2O] = 2.12$ mol dm^{-3}.

Solvent	$10^5 k/\text{min}^{-1}$				Ea/kJ mol^{-1}
	15°	20°	25°	30°	
Chloroform	4.98	7.33	11.1	15.6	55.7
Dichloromethane	0.820	1.23	1.93	2.70	58.4
Tetrachloroethane	0.431	0.875	1.50	3.17	94.7

Reaction mechanism

Scission of the CTA chain under anhydrous conditions in the presence of acetic anhydride is thought to occur by acetolysis involving the acetic anhydride. First-order kinetics is observed since the acetic anhydride concentration is in large excess of the oxygen linkage concentration. A proposed mechanism is as follows:

$$HClO_4 \rightleftharpoons H^+ + ClO_4^- \qquad (2)$$

$$HClO_4 + Ac_2O \rightleftharpoons Ac^+ + ClO_4^- + AcOH \qquad (3)$$

$$H_2SO_4 \rightleftharpoons H^+ + HSO_4^- \qquad (4)$$

$$H_2SO_4 + Ac_2O \rightleftharpoons Ac^+ + HSO_4^- + AcOH \qquad (5)$$

Effective catalytic species for degradation are suggested to be H^+ and Ac^+ (acetylium ion CH_3CO^+). Thus the degradation rate constant would depend on the equilibrium concentrations of these species and their catalytic coefficients k_{H^+} and k_{Ac^+} as follows

$$k = k_{H^+}[H^+] + k_{Ac^+}[Ac^+] \qquad (6)$$

Equilibria (2)-(5) are assumed to be rapidly established and the relative $[H^+]$ and $[Ac^+]$ concentrations would depend on the relevant equilibrium constants. A reaction mechanism based on these equilibria would explain the difference in catalytic ability for the two acids. The magnitudes of the catalytic coefficients are at present unknown but it seems reasonable to assume that $k_{H^+} > k_{Ac^+}$. Clearly more work is required to elucidate these values.

Finally, the chain breaking process is believed to take place according to

$$-\mathrm{O}- \xrightarrow[\text{fast}]{H^+} -\overset{\overset{\displaystyle H}{|}}{\underset{+}{\mathrm{O}}}- \xrightarrow[\text{slow}]{Ac_2O} -\mathrm{OAc} + \mathrm{AcO}- + H^+ \qquad (7)$$

$$-\mathrm{O}- \xrightarrow[\text{fast}]{Ac^+} -\overset{\overset{\displaystyle Ac}{|}}{\underset{+}{\mathrm{O}}} \xrightarrow[\text{slow}]{Ac_2O} -\mathrm{OAc} + \mathrm{AcO}- + Ac^+ \qquad (8)$$

with the reaction between acetic anhydride and the charged intermediate complex being the rate determining step.

REFERENCES

1. M.A. Bhatti and P. Howard, Die Makromol. Chemie, 177, 101, (1976).
2. P. Howard and S.U. Malook, Polymer, 26, 297 (1985).
3. P. Howard and S.U. Malook, to be published.
4. S.H. Maron, J. Appl. Polym. Sci., 15, 282, (1961).
5. O.F. Soloman and B.S. Gotesman, Die Makromol Chemie, 104, 107, (1967).
6. W. Kuhn, Z. Phys. Chem., 159, 368, (1932).
7. A. Af. Ekenstam, Ber. Deut. Chem. Ges., 69, 549, 553 (1936).
8. N. Grassie, "Chemistry of High Polymer Degradation Processes", Butterworths, 1956.

Chapter 20

The effect of degree of substitution on diffusion and mixing in cellulose nitrate/solvent systems

U. P. Gandhi, P. J. F. Griffiths and G. S. Park – Department of Applied Chemistry, UWIST, PO Box 13, Cardiff CF1 3XE, Wales, U.K.

INTRODUCTION

The study of the diffusion of a range of solvents into cellulose nitrate films using an interferometric technique has been reported by Park and Said [1] and correlations between rates of diffusion and molecular size of diffusant were put forward. Values of the critical solvent concentrations required before interdiffusion can occur and the maximum mutual diffusion coefficients for the solvent/polymer systems were measured.

Polymers of only one degree of nitration (12.6%N) were studied. In order to further the study of diffusion in cellulose nitrate, values for both mutual diffusion coefficients and self-diffusion coefficients of solvents in polymers having a range of nitrogen contents have been measured using the interferometric technique of Robinson [2] and the radiotracer technique of Park and Tran van Hoang [3]

EXPERIMENTAL

Four cellulose nitrate samples having nitrogen contents of 11, 12.2, 12.6 and 13.4% were used. These polymers were purified by precipitation into water from a 5% w/v solution in acetone. The precipitation was repeated and the purified polymer was dried for 24 hours in vacuo at 35°C.

A 3% solution in acetone was used to cast films of thickness about 30μm on glass using the technique described elsewhere to give uniform sheets [4]. Residual traces of acetone were removed by leaching the films in water and then drying them for 24 hours in vacuo at 35°C. The films were then stored in vacuo over silica gel.

Thoroughly dried acetone of A.R. quality was used throughout.

The interferometric technique used by Park and Said [1] which followed closely that described by Robinson [2] was used to study the interdiffusion of acetone with the four cellulose nitrates and mutual diffusion cofficients were calculated from the concentration against distance data obtained using Matano's [5] method.

The purified 11, 12.6 and 13.4% N content cellulose nitrates with the plasticizer di(2-ethylhexyl)phthalate as solvent were used in the measurement of the solvent self-diffusion coefficients using the radiotracer technique described by Park and Tran van Hoang [3].

Films containing the required amounts of plasticizer and ones of similar concentrations containing carbon-14 labelled plasticizer were cast from solutions in dry acetone containing plasticizer and cellulose nitrate in the required ratios.

The appropriate amount of solution was poured into 30mm diameter glass rings standing on sheets of plate glass. After the solvent had been allowed to escape through a diffusion barrier [4] for a few days, the resulting plasticized films were removed, together with the casting

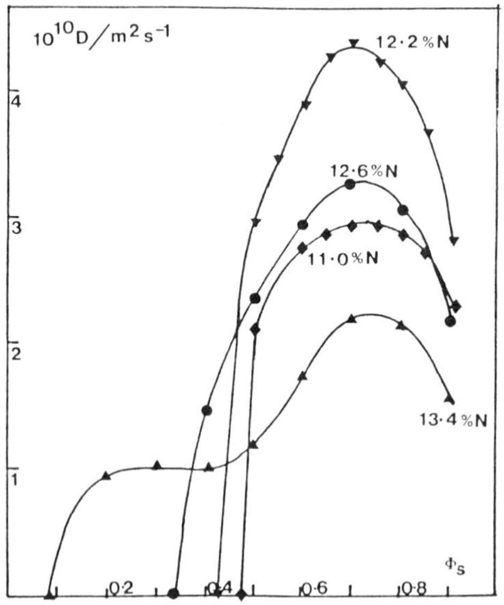

Fig. 1. Mutual diffusion coefficient, D, as a function of volume fraction Φ_s, for various cellulose nitrate-acetone systems at 25°C.

ring and kept in a vacuum desiccator, over silica gel, for 24 hours to remove the last traces of acetone. The films were cut with a specially sharpened cork borer to give discs of diameter about 18mm. The thicknesses of the discs were measured using a micrometer screw gauge on discs which were sandwiched between pairs of microscope slides.

The uniformity of distribution of the plasticizer was checked by infra-red and radioactive measurements.

INTERDIFFUSION IN THE CELLULOSE NITRATE/ACETONE SYSTEMS

Diffusion coefficient/concentration curves for the interdiffusion of acetone with 11, 12.2, 12.6 and 13.4% N content cellulose nitrate are shown in Fig 1. For each polymer, diffusion is zero until a certain solvent concentration, volume fraction Φ_S^0, is attained. Above this critical concentration there is a rapid increase in the value of the mutual diffusion coefficient to a maximum beyond which there is a slow decrease as the concentration approaches 100% solvent.

The data for the cellulose nitrates of four different extents of nitration show an interesting variation of the critical concentration, Φ_S^0, and maximum value D_{max} of the diffusion coefficient with the nitrogen content. Table 1 gives the values for these data.

Table 1. Diffusion coefficient maximum, D_{max}, and critical volume fraction of solvent, Φ_S^0, for diffusion to occur at 25°C. Acetone: Cellulose nitrate systems.

% N	$10^{10} \times D_{max}/m^2s^{-1}$	Φ_S^0
11.0	2.95	0.46
12.2	4.28	0.43
12.6	3.24	0.32
13.4	2.44	0.12

The values of Φ_S^0 decrease with increasing nitrogen content whereas those of D_{max} pass through a maximum at a nitrogen content of 12.2%.

The regular decrease of Φ_S^0 with increasing nitrogen content could be attributed to the concomitant decrease in extent of hydrogen bonding.

The observation of a maximum value for D_{max} may be associated with the compensating effects of the thermodynamic driving force for diffusion and the frictional resistance to molecular motion.

The frictional resistance, f, to interdiffusion would be expected to decrease as the extent of hydrogen bonding decreases with increasing nitrogen content so that D would be expected to increase. The diffusion coefficient, D, however, also depends on the relationship between the thermodynamic driving force and the concentration gradient (eq. 1)

$$D = (RT/f)\ d \ln a_s/d \ln \Phi_s$$

In this relationship a_s is the thermodynamic activity of the solvent at a volume fraction Φ_s. If, therefore, $d \ln a_s/d \ln \Phi_s$ decreases with increasing nitrogen content the two opposing effects would lead to a maximum value for D_{max}.

For the cellulose nitrate (13.4% N) (Fig. 1) it is seen that the diffusion coefficient/concentration relationship is different from those observed for the polymers of lower nitrogen content. The diffusion coefficient increases and then becomes steady before increasing again to a maximum and falling again at high acetone concentrations. The curves

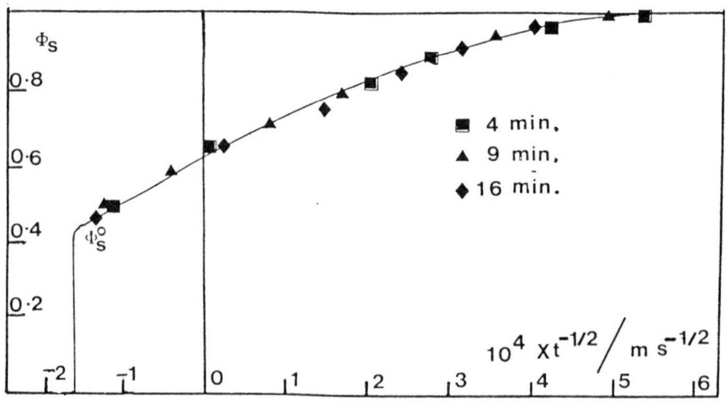

Fig. 2. Boltzman correlation between distance, X, from the original boundary and root time, $t^{1/2}$, for the 12.2% nitrogen content cellulose nitrate-acetone system at 25°C.

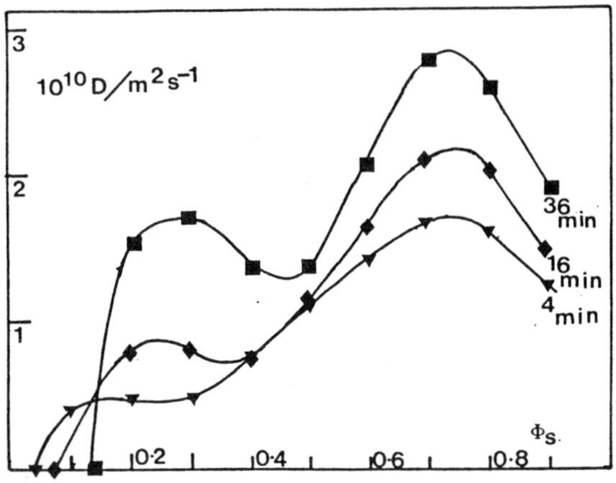

Fig. 3. The influence of time effects on the calculated mutual diffusion coefficient, D, at solvent volume fraction, Φ_s, for the 13.4% nitrogen cellulose nitrate-acetone system at 25°C.

for all the polymers are calculated from mean values of plots of concentration against the ratio of distance to the square root of time. For the lower nitrogen content polymers the curves are independent of time as is shown in Fig. 2 for the 12.2% N content polymer. The plots for the 13.4% N content polymer, however, show appreciable change of shape with increasing time and by taking the individual plots for 4, 16 and 36 min. the mutual diffusion coefficients in Fig. 3 were calculated using Matano's method [5].

An interesting feature of Fig. 3 and of the data obtained at other temperatures is the decrease of Φ_S^0 with time (and also with temperature). It is well known that the sorption of acetone by cellulose nitrate at low acetone concentrations is "non Fickian" so that time effects should be present even in the low nitrogen content polymers. These are not observed because the diffusion coefficients are effectively zero (below Φ_S^0) at these low concentrations and so their variation with time has no effect on the intermixing pattern. The low Φ_S^0 value for the 13.6% N polymer means that the diffusion coefficients in the time dependent region of concentration are now appreciably greater than zero. This affects the intermixing patterns so that the Boltzmann correlation between distance and root time no longer holds and Φ_S^0 decreases with time.

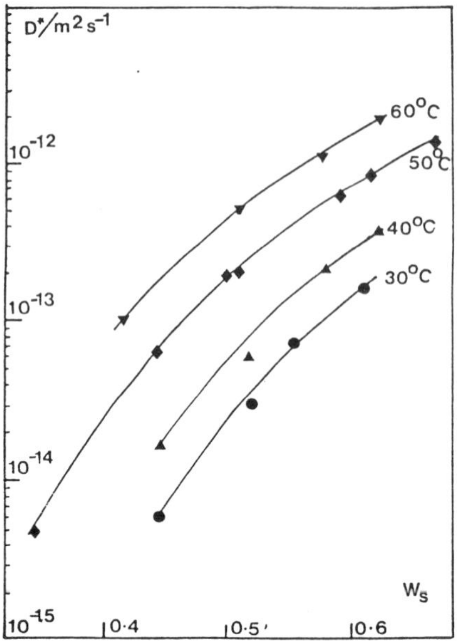

Fig. 4. Self diffusion coefficient, D*, of di (2-ethylhexyl) phthalate in 12.6% nitrogen content cellulose nitrate as a function of the weight fraction, W_s, of phthalate.

SELF DIFFUSION IN THE CELLULOSE NITRATE/PLASTICIZER SYSTEMS

In order to measure molecular mobility without the complications due to time effects on mixing and to obtain data at low mobilities, solvent self-diffusion coefficients D*, have been obtained for the system cellulose nitrate/ di(2-ethylhexyl) phthalate having nitrogen concents of 11%, 12.6% and 13.4% at temperatures of 30, 40, 50 and 60°C and at plasticizer concentrations of between 50phr (0.33 wt fraction plasticizer) and 160phr (0.62 wt fraction plasticizer).

Fig. 4 gives the diffusion coefficients for a range of plasticizer concentrations in the 12.6% N cellulose nitrate plotted against the weight fraction of plasticizer. The same general pattern is observed for all three polymers.

For each polymer at each temperature there is a pronounced change in diffusion coefficient with change in plasticizer concentrations. Fig. 5 shows the plasticizer diffusion coefficient/concentration relationship for each of the three cellulose nitrates at 50°C. The variations with plasticizer concentration differ significantly between the different nitrogen contents. It is of considerable interest that in both the high and the low nitrogen content polymers the diffusion coefficients are lower than in the 12.6% nitrogen polymer for a given concentration of plasticizer and depend more on the plasticizer concentration.

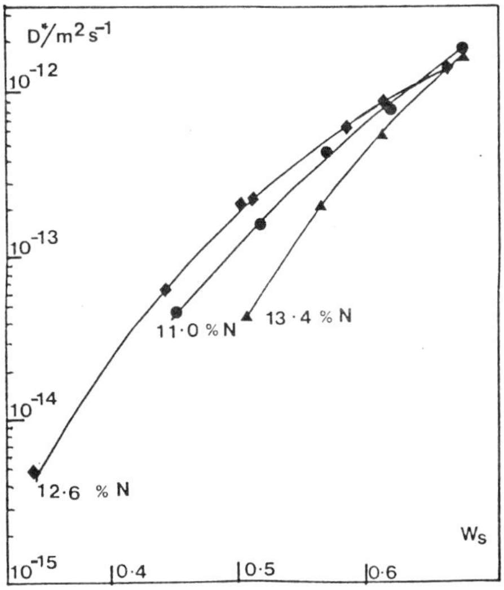

Fig. 5. Self diffusion coefficient, D*, of di (2-ethylhexyl) phthalate in cellulose nitrates of various nitrogen contents as a function of the weight fraction, W_s, of phthalate at 50°C.

Activation energies were determined from the variation of diffusion coefficient with temperature and Table 2 gives the activation energies at three plasticizer concentrations calculated for the temperature range 30–60° C. It is interesting to observe that the extent of change of activation energy with concentration increases in the order 13.4%N < 11% N < 12.6% N whilst the concentration dependence of the diffusion coefficients is in the reverse order. The activation energies are thus most dependent on concentration of plasticizer in those systems where the diffusion coefficients are least dependent on concentration, which are also the systems in which the diffusion coefficients are larger.

Table 2. Activation energies for the self diffusion of di (2-ethyl-hexyl) phthalate in cellulose nitrates containing 11.0, 12.6 and 13.4% N.

Plasticizer concentration wt fraction(phr)	Activation Energy/kJ mol^{-1}		
	11.0% N	12.6% N	13.4% N
0.5 (100)	82.0	83.1	86.7
0.556 (125)	71.6	72.6	79.5
0.6 (150)	66.3	67.	74.6

For all conditions of temperature and concentration, the self-diffusion coefficient D* of the plasticizer like the maximum of the mutual diffusion coefficient D_{max} of acetone in cellulose nitrate has a maximum value for the 12.2 to 12.6% N content polymer.

This content of nitrogen corresponds to a hydroxyl content of about 0.5 per glucose ring.

The way in which the critical concentration of acetone, Φ_S^0, changes with degree of substitution indicates that it is largely determined by the extent of interchain hydrogen bonding whilst the observation of a maximum of D* as well as of D_{max} at a hydroxyl content of about 0.5 per glucose ring suggests a maximum degree of randomness of structure at this composition with greater regularity for the dinitrate and the trinitrate.

ACKNOWLEDGEMENTS

With the exception of the 11% polymer the cellulose nitrates used were given to us by RARDE, Waltham Abbey. The ^{14}C labelled phthalate was made by Dr. J. Gilbert of the MAFF Norwich. The work was carried out with the support of the Procurement Executive, Ministry of Defence

REFERENCES

1. G.S.Park and M.M.Said, "Some aspects of Diffusion in Cellulose Nitrate" Chap. 30, p 337-344, in "Cellulose and its Derivatives" (J.F.Kennedy, G.O.Phillips, D.J.Wedlock and P.A.Williams, Editors). Ellis Horwood Limited, Chichester, 1985.
2. C. Robinson, Proc. Roy. Soc. A204, 549-569 (1951).
3. G.S.Park and Tran van Hoang, Euro. Polym. J., 15, 817-822 (1979).
4. J.Crank and G.S.Park, in "Diffusion in Polymers" (J. Crank and G.S.Park, Editors) Academic Press, London and New York, 1968, Chap. I.
5. C. Matano, Jap. J. Phys. 8, 109 (1932-3).
   See also J. Crank. The mathematics of diffusion, 2nd Edn, Clarendon Press, Oxford, 1975 p230-234.

Chapter 21

Properties of lyotropic liquid crystals of cellulose derivatives: Rheological properties of ethyl cellulose

S. Suto and M. Karasawa – Department of Polymer Chemistry, Faculty of Engineering, Yamagata University, Jonan 4-3-16, Yonezawa, Yamagata, 992, Japan.

ABSTRACT

Rheological properties of liquid crystalline ethyl cellulose (EC) solutions were determined by means of two viscometers (cone-plate and capillary types). The viscometric and elastic behavior of EC systems was similar to that of hydroxypropyl cellulose systems: 1) the concentration-temperature superposition for viscosity was valid, 2) the elastic parameter at die exit was related to that at die entrance. The viscometric behavior in the biphasic region was governed by the volume of the anisotropic domain.

INTRODUCTION

In our previous papers, rheological properties of liquid crystalline hydroxypropyl cellulose (HPC) were determined at both low (1,2) and relatively high (3,4,5) shear rates. Our major findings were that 1) concentration-temperature superposition for viscosity could be applied, 2) the concentration dependence of the elastic parameters was similar to that of the viscosity, 3) the elastic parameters at die entrance were related to the other elastic parameters at die exit.

In this article, ethyl cellulose (EC) was used. EC is of commercially greater importance than HPC because EC forms liquid crystals in suitable solvents and is not water-soluble. The rheological properties of the EC solutions were determined with two types of rheometer. We discuss whether the behavior observed for HPC systems is valid for EC systems or is fortuitous and also discuss the concentration-temperature superposition for viscosity in terms of varying volume of the anisotropic domain.

EXPERIMENTAL

Samples

Ethyl cellulose(EC) was obtained from Tokyo Kasei Co.Ltd.. The number-average molecular weight of EC was 1.8 x 10^4. EC powder was dried at 60°C in vacuo for about 24 hr before use. Reagent-grade m-cresol and benzyl alcohol (Wako Pure Chemical Ind.Ltd.) used were purified by distillation prior to use.

Preparation of concentrated solutions

A sample of 40 - 50 g of EC was weighed accurately into a glass-stoppered flask. A given volume of solvent was added from a volumetric pipette and the mixture was stirred for about a week and was then stored in the dark for about three months at room temperature. The concentration ranges of solutions were 21-50 wt% for EC/m-cresol system, and were 35 - 55 wt% for EC/ benzyl alcohol system.

Rheometry

A cone-plate type viscometer was used for low shear rates and a capillary viscometer for relatively high shear rates. The details of measurements are given elsewhere (1 - 5). The elastic parameters observed with capillary viscometer were Bagley coefficient (ν), entrance pressure drop (ΔP_{ent}) , and die swell (B). ν and ΔP_{ent} were parameters at die entrance and B was at die exit.

Results and discussion

1) Viscosity at a low shear rate (1 sec^{-1})

Figure 1 shows the temperature dependence of the viscosity for EC/m-cresol systems. The viscosity exhibited a minimum (η_{min}) and a maximum (η_{max}) (at temperatures designated T_{min} and T_{max}). This behavior was the same as that reported by us for HPC systems (1,2). The concentration-temperature superposition of the viscosity shown in Fig.1 is shown in Fig.2. This superposition could be applied to not only the systems in this study (m-cresol and benzyl alcohol) but the other 8 systems for HPC and EC which are not reported here. These results suggest that the concentration-temperature superposition is valid for all lyotropic liquid crystals of cellulose derivatives.

We try to explain the superposition in terms of varying volume of an anisotropic domain. A tentative picture of the temperature dependence of the viscosity is schematically represented in Fig.3. Log viscosity is plotted against 1/T. The plots are linear for a single phase region (isotropic and anisotropic) and can be expressed by an equation of the Arrhenius form, but are not linear for a biphasic region. As noted by Conio et al. (6), the iso-

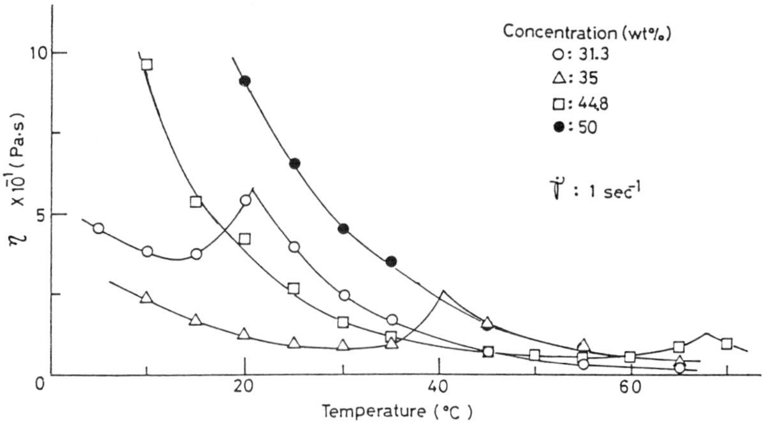

Fig. 1. Temperature dependence of viscosity for EC/m-cresol system.

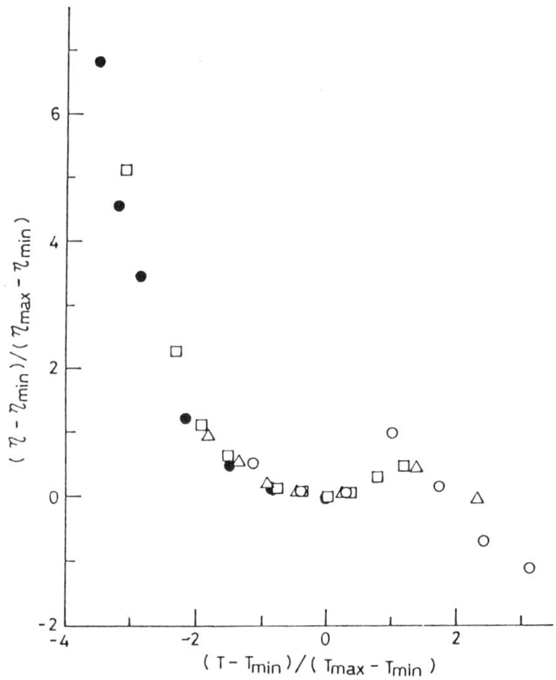

Fig. 2. Master curve for EC/m-cresol system; concentrations as for Fig. 1.

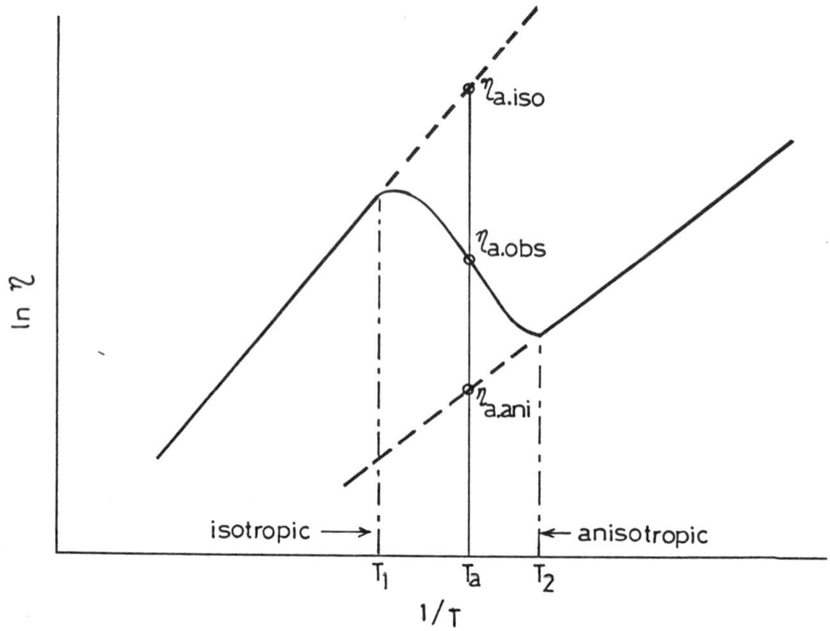

Fig. 3. Schematic picture of temperature dependence of viscosity for liquid crystalline solutions.

tropic phase begins to form at T_2 of Fig.3 and the isotropy letely ends at T_1. In other words, the picture in Fig.3 reflects the transformation from an anisotropic to an isotropic phase. Volume of the anisotropic phase begins to decrease at T_2, decreases with increasing temperature, and finally approaches zero at T_1. If the phase transformation does not occur, the plots of log viscosity against 1/T for the single anisot opic and isotropic phase should be linear over a wide range of temperature (indicated by the dashed curves of Fig.3).

The weight fraction of the anisotropic phase may be estimated from the viscosity curve presented in Fig.3, by assuming the following equation:

$$\eta_{obs} = \eta_{iso} \cdot W_1 + \eta_{ani} \cdot W_2 \qquad (1)$$

where η_{obs} is the viscosity observed in the biphasic region, η_{iso} and η_{ani} are the viscosity of the isotropic phase and that of the anisotropic one, respectively. They are estimated from the dashed curves of Fig.3. W_1 and W_2 are the weight fractions of the isotropic and anisotropic phases;

$$W_1 + W_2 = 1 \qquad (2)$$

Therefore, W_2 may be estimated from eqs. (1) and (2) by measuring η_{obs} and estimating η_{iso} and η_{ani}. To clarify the description, the estimation procedure of W_2 is given. At T_a the

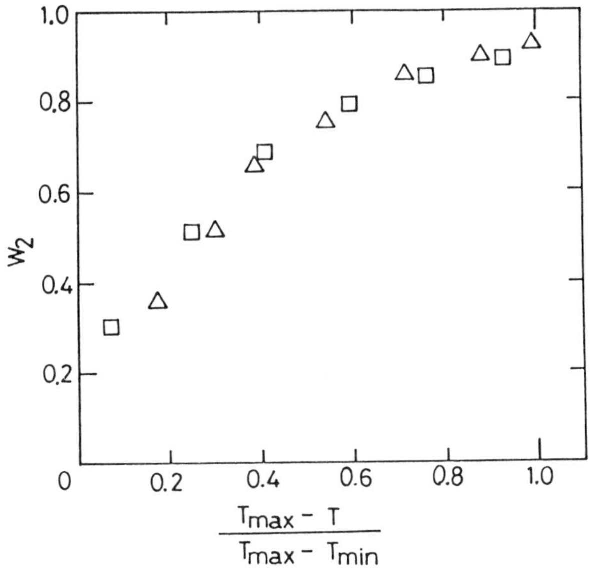

Fig. 4. Temperature dependence of weight fraction of anisotropic domain for EC/m-cresol system; concentrations as for Fig. 1.

viscosity observed is $\eta_{a.obs}$ and the viscosities estimated are $\eta_{a.iso}$ and $\eta_{a.ani}$. Then, the weight fraction of the anisotropic phase W_2 at T_a can be estimated. After the same manner as noted above, W_2 at given temperatures which are between T_1 and T_2 is estimated. The W_2 estimated for EC/m-cresol system is plotted against the reduced temperature in Fig.4. Fig.4 shows clearly that the viscometric behavior in the biphasic region is governed by the volume of the anisotropic domain. The volume of the anisotropic domain is a function of temperature and polymer concentration, and the volume estimated for each polymer concentration can be fitted to a master curve. A shape of the master curve tends to vary with the liquid crystalline systems. It may be due to the difference of the size of the anisotropic domain.

As noted above, we assumed the simplest equation (1) for estimating W_2, but the assumption had no theoretical basis. Our estimation of the W_2 is considered qualitatively reasonable, however, quantitatively questionable. Therefore, we need to confirm the validity of the eq. (1) or to derive an alternative equation.

2) Viscoelastic properties at relatively high shear rates

Fig.5 shows the polymer concentration dependence of the viscosity for EC/m-cresol system. This behavior was typical of the lyotropic liquid crystals. The elastic parameters at both die entrance and die exit exhibited the same behavior as the visco-

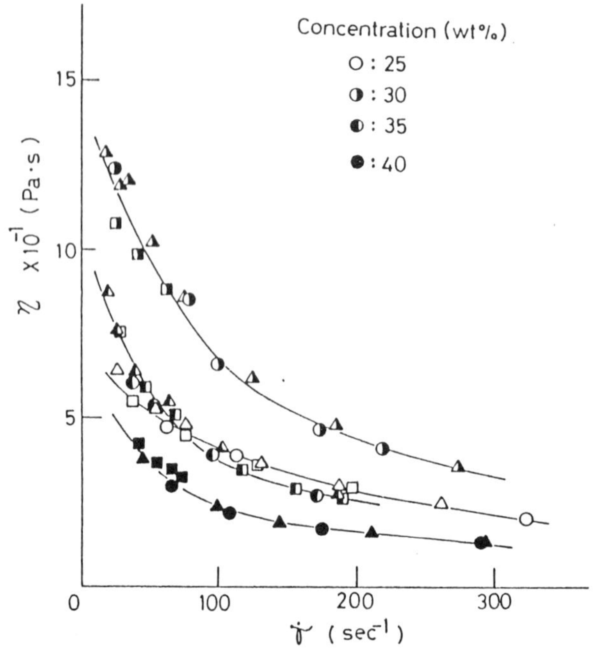

Fig. 5. Shear rate dependence of viscosity for EC/m-cresol system.

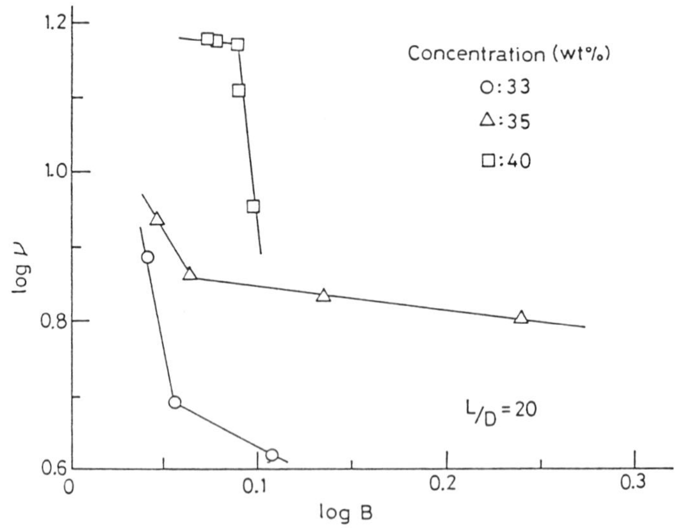

Fig. 6. Correlation between logarithm of Bagley coefficient ν and logarithm of die swell B for EC/m-cresol system.

sity with respect to polymer concentration. This finding was similar to that reported by us for HPC systems (3,4,5). However, it should be noted that the concentrations at the maximum and minimum viscosity are not identical to th se at the maximum and minimum elastic parameters strictly. Both viscosity and elastic parameters are dependent on shear rate, however, the shear rate dependence of the elastic parameters seems to be greater than that of the viscosity. Therefore, it is possible that different results could be obtained if comparisons were made at different shear rates.

Fig.6 shows that there is a following correlation between μ and B for EC/m-cresol system:

$$\mu \propto B^{n} \qquad (3)$$

There was found also to be a similar correlation between ΔP_{ent} and B:

$$\Delta P_{ent} \propto B^{n'} \qquad (4)$$

where n and n' are constants which depend on concentration. The findings for EC system were qualitatively similar to those obtained for HPC systems (3,4,5). Therefore, the behavior in this study seems to be general for all lyotropic liquid crystals of cellulose derivatives. However, our findings have no theoretical basis. We need to explain theoretically the anomalous behavior of the lyotropic liquid crystals.

CONCLUSIONS

1) The viscoelastic behavior of EC was almost the same as that of HPC. The viscometric behavior in the biphasic region was governed by the volume of the anisotropic domain.
2) The following correlations found for HPC systems were valid for EC system:

$$\mu \propto B^{n},$$

$$\Delta P_{ent} \propto B^{n'}.$$

These correlations seemed to be the rule for the cellulosic liquid crystalline solutions.

ACKNOWLEDGEMENT

The assistance of Mr.M.Ohshiro and of Mr.K.Obara in performing a number of experiments is acknowledged.

REFERENCES

1) Suto,S., et al., Kobunshi Ronbunshu, 43,231(1986)
2) Suto,S., et al., J.Polym.Sci.POlym.Phys.Ed., in print.
3) Suto,S., et al., Polym.Commun., 26,335(1985)
4) Suto,S., et al., Nihon Reoroji Gakkaishi, 14,87(1986)
5) Suto,S., et al., Polymer, in Press.
6) Conio,G., et al., Macromolecules, 16,1264(1983)

Chapter 22

The preparation and characterisation of polyisobutylene-grafted cellulose pulp

S. Coleman-Kammula and **H. Hulskers** – Koninklijke/Shell-Laboratorium, Amsterdam (Shell Research B.V.) Badhuisweg 3, 1031 CM Amsterdam, The Netherlands.

ABSTRACT

The preparation of polyisobutylene-grafted cellulose pulp involves activation of cellulose by conversion to cellulosates. The necessity of the activation step is demonstrated by reactions with model acid chlorides and anhydrides, which show that esterification via sodium cellulosate yields fibrous products with high levels of derivatization. Esterification of cellulose itself via the conventional pyridine-catalysed route leads to powdering of the pulp.

Reaction of anhydride-terminated polyisobutylene (PIB-MALA) with sodium cellulosate gives polyisobutylene-grafted cellulose products in high graft yields (110 %) and graft frequency (degree of substitution = 0.1). The yields depend on reaction time and on the amount and molecular weight of the PIB-MALA. The non-solubility of the grafted products in cuprammonium hydroxide, as well as analyses by infra-red and X-ray photoelectron spectroscopy, provide proof of grafting.

INTRODUCTION

Despite the extensive research in the area of cellulose graft copolymerization, reports of polyolefin grafting onto cellulose are scarce. This may be because conventional free-radical-initiated copolymerization is not effective for polymerizing simple olefins such as ethylene and propylene. Transition metal catalysts for polymerizing ethylene and propylene do not, when used with cellulose, result in polyolefin-grafted products (Kaminsky, 1984), as there are no direct mechanisms for formation of initiator sites on cellulose.

An alternative route to polyolefin-grafted cellulose is coupling of preformed polymer chains containing electrophilic end groups with the hydroxyl groups of cellulose. This allows the much desired control of graft length - a feature lacking in free-radical polymerization. However, reports of grafting by this route have also been limited, and here we envisage problems such as barriers to diffusion of the polymer chains into the cellulose pulp and the low number of end groups in relation to polymer chain length. Grafting onto cotton yarns by esterification with acid-chloride-terminated polystyrene in the presence of pyridine, which is the only known example in this category, resulted in low grafting percentages (Avny and Schwenker, 1967). We report here esterification of an activated form of cellulose pulp, i.e. sodium cellulosate. Our aim was not only to improve the grafting process but also to obtain a fibrous product, since for applications such as reinforcements in fibre-reinforced composites fibrous grafted products are desired. In order to find the best conditions for optimum derivatization without fibre destruction we first examined esterification of cellulose and cellulosate by model acid chlorides and anhydrides.

Esterification by model compounds

Via pyridine catalysis

Synthesis of cellulose mono-, di- and triesters, using excess acid chlorides (C_{10} and C_{16}) and pyridine at 150 °C, was first described in the literature early this century (e.g. Gault and Ehrman, 1923). Some of the products were found to be soluble in organic solvents and the higher esters were meltable and probably non-fibrous.

In our work, the reaction temperature was kept at 110 °C to minimize fibre degradation. The concentrations of acid chloride and pyridine were varied so that their effects on the fibre structure could be examined. With three equivalents of acid chloride per anhydroglucose unit (AGU), products with a high degree of substitution (DS = 2.5 to 3.0) are formed, which is similar to the reactions reported in the literature. The products are lumpy powders which do not disperse or swell in water.

At substantially lower concentrations of acid chloride (0.3 and 0.1 equivalents) esterification still occurs. The products did partially retain the original fibrous structure. DMAP, 4-(N,N-dimethylamino)pyridine, which is known to be a more efficient acylation promoter than pyridine, gave higher-DS products in which the fibre structure was retained. Reaction of cellulose with alkyl-substituted succinic anhydrides (R=CH_3 and R=$C_{12}H_{23}$) using pyridine as base, under conditions similar to the above, resulted in negligible derivatization.

Solid-state ^{13}C nmr, a relatively new tool in this area, was found to be useful for determining the degree of substitution of cellulose laurate: see Fig. 1, a-d.

In Fig. 1,a the ^{13}C nmr spectrum of cellulose (Whatman, fibre grade) itself is depicted and carbon resonances of the anhydroglucose units are indicated. In addition to these the spectrum shows aliphatic constituents containing up to 5 % carbon.

In Fig. 1,b and c, ^{13}C spectra of lauroyl-grafted cellulose material are shown. The aliphatic region (14-44 ppm) shows clearly the presence of paraffinic carbon chains in both spectra. Many of the individual carbons of the C_{11} chain can be distinguished, as is indicated in Fig. 1,c. The unresolved resonance of the ester carbonyl at 175 ppm could indicate the presence of different esters. On the basis of the carbon signal intensities of the alkyl chains and of the cellulose units, the DS values are estimated to be <0.01 and 0.3 for spectra b and c respectively.

The carbon resonances of the glucose units of the cellulose laurates are not significantly different from those of the underivatized cellulose. If preferential reaction at one of the three hydroxyls had occurred, one would expect those particular carbon resonances to have altered. This has been demonstrated in nitrosation of cellulose, where solid-state ^{13}C nmr has indicated preferential substitution at the C_6 hydroxyl group (Lippmaa et al., 1983). Therefore it is conceivable that, with the laurates, random distribution of lauroyl chains over all the three hydroxyl groups minimizes the observable changes in the spectra.

The ^{13}C spectrum in Fig. 1,d clearly indicates complete grafting of the cellulose hydroxyls to form a cellulose trilaurate. The DS value of 3.3 estimated by nmr integration indicates the possible presence of laurate residues due to adsorption.

From these results it may be concluded that solid-state ^{13}C nmr spectroscopy can determine the degree of substitution of cellulose derivatives. More work is necessary to establish the position of substitution conclusively.

Via sodium cellulosate

Sodium cellulosate

Formation of sodium cellulosate using cotton yarns and sodium methoxide in methanol (eq. 1) has been reported by Schwenker

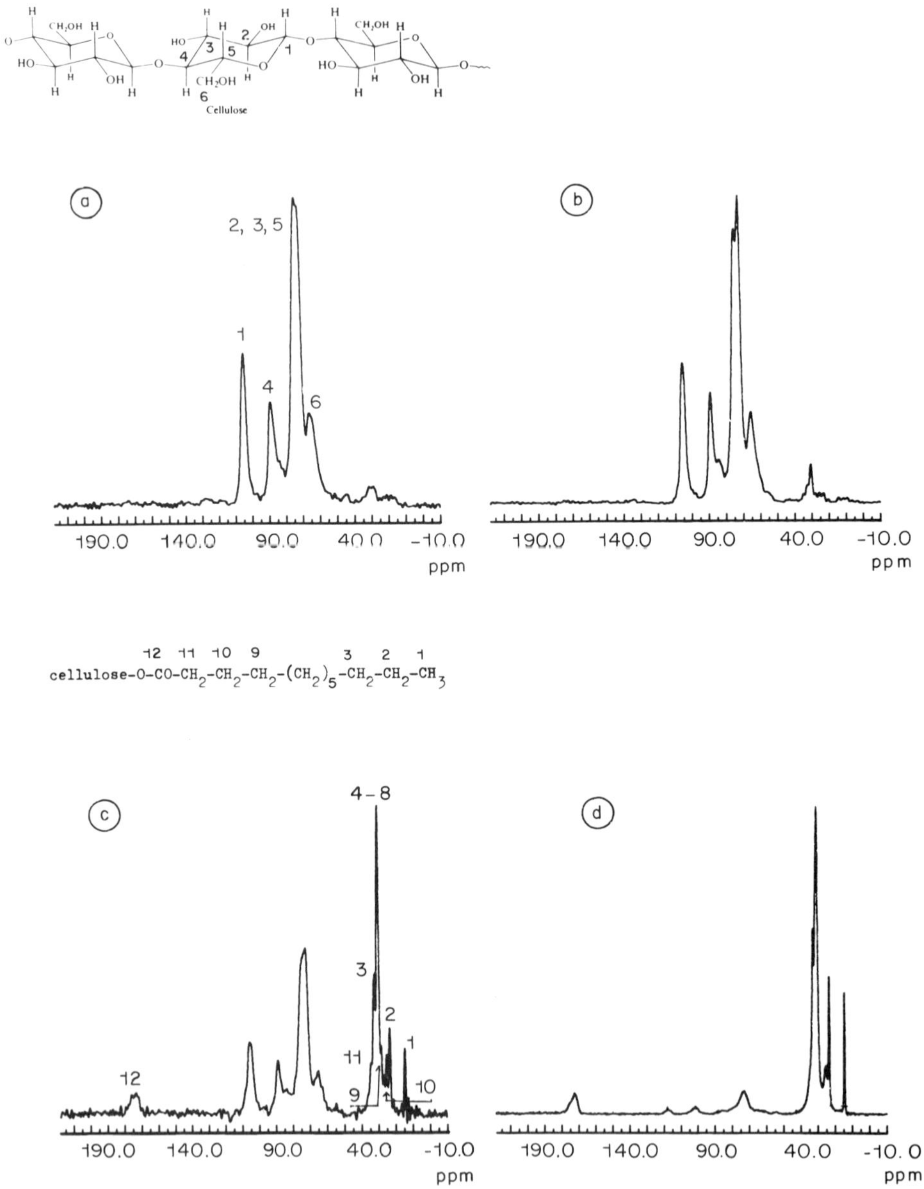

Fig. 1. Solid state ^{13}C nmr of cellulose and cellulose laurates. a = Whatman cellulose; b–d = cellulose laurate; b: DS $\leqslant$0.01, c: DS = 0.3, d: DS = 3.3.

et al. (1961). We examined briefly the effect of such a reac-

$$Cell - OH + NaOCH_3 \rightleftharpoons Cell - ONa + CH_3OH \quad (1)$$

tion on the fibre structure of lignin-free wood pulp. Sodium cellulosate was prepared using Whatman cellulose according to the procedure described by Schwencker et al. With 1N sodium methoxide, products containing 4.2 meq of sodium cellulosate per gram of cellulose (i.e. 0.7 Na^+ per AGU) were obtained. Microscopic examination of the cellulose regenerated from the cellulosate showed that little damage to the original fibre structure had occurred.

Since this intermediate had not yet been characterized by spectroscopic means we measured its infra-red (IR) spectrum. The IR spectrum of the purified reaction product (Fig. 2,b) showed absorptions at 1600 and 1450 cm^{-1} which we assign to NaO-cell bonds on the basis of IR data of model sodium alkoxides. By comparison, the spectrum of cellulose pre-swollen in sodium hydroxide and extracted with methanol (Fig. 2,c) shows only changes in cellulose morphology (Fig. 2,a) due to alkali treatment.

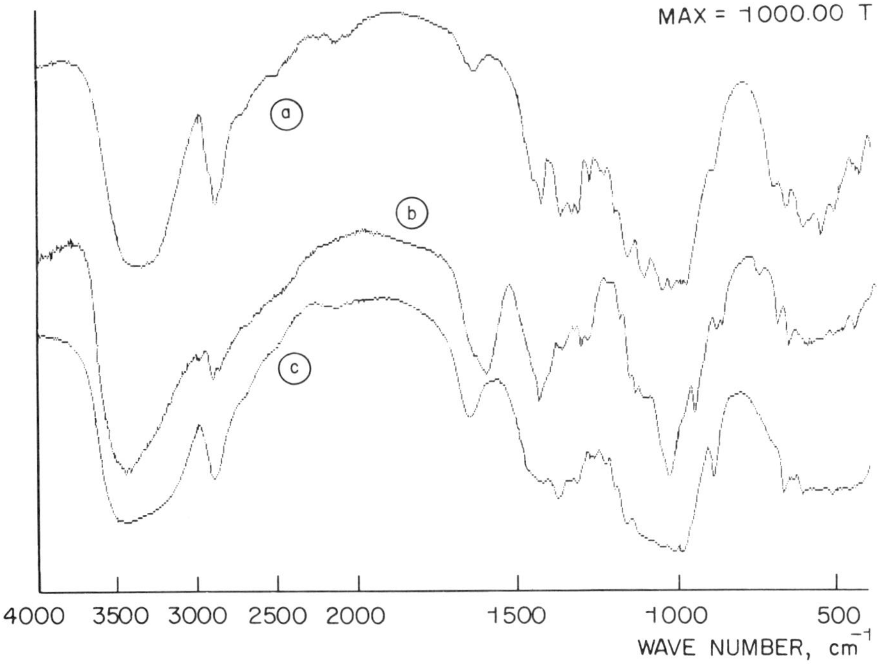

Fig. 2. IR spectra of (a) cellulose, (b) sodium cellulosate and (c) alkali-treated cellulose.

Esterification of sodium cellulosate

Sodium cellulosate (DS = 0.7) was reacted at 60 °C with varying amounts of lauroyl chloride in a minimum amount of toluene for 20 h. Addition of 0.6, 2.5 and 3.8 equivalents of acid chloride per AGU resulted in a DS of 0.25, 0.8 and 1.1 lauroyl groups per AGU respectively. Since the DS of sodium cellulosate was 0.7, this means that, under the conditions concerned, some esterification of the hydroxyl groups in additon to the alkoxy groups of the cellulose occurs.

In view of the availability of polymers functionalized by succinic anhydrides and the high reactivity of this functional group towards alcoholates, we have examined esterification by some model succinic anhydrides. Reaction of equimolar amounts of cellulosate and substituted succinic anhydrides, followed by acid washing, resulted in the half ester as evidenced by the IR absorbance at 1720-1695 cm^{-1}. The products were fibrous, bearing ester degrees of substitution in the region of 0.3.

These observations indicate that sodium cellulosate is a useful intermediate for grafting onto cellulose fibres via coupling with acid- and anhydride-terminated polymers. As an illustration, the preparation of polyisobutylene-grafted cellulose via coupling of succinic-anhydride-terminated polyisobutylene with sodium cellulosate is described.

POLYISOBUTYLENE-GRAFTED CELLULOSE (PIB-g-CELL)

The reaction (eq. 2) of sodium cellulosate (DS = 0.7) from Whatman cellulose with PIB-MALA (the chlorine-mediated adduct of maleic anhydride and polyisobutylene, supplied by Shell Chemicals UK; mol. wt. = 1000) dissolved in the minimum amount of toluene was examined over different reaction times and at different temperatures.

After filtration and washing with ethanol and 1N HCl, the products were extracted overnight in a soxhlet apparatus with

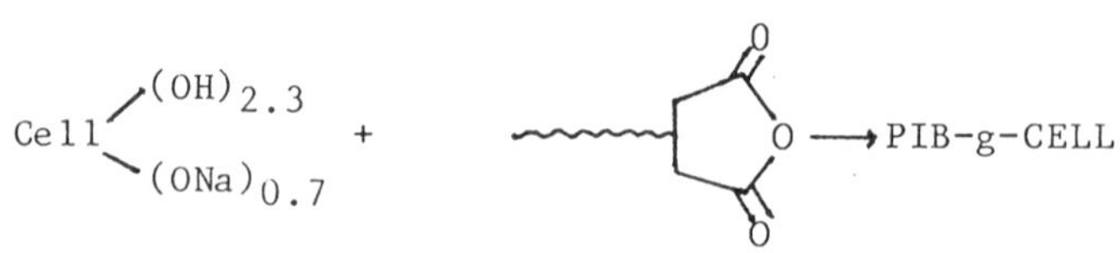

PIB-MALA (2)

cyclohexane in order to remove any adsorbed PIB-containing residues. Percent grafting and degree of substitution were calculated from weight gain.

Percent grafting varied from 50 to 113 (DS = 0.07 to 0.18) depending largely on the amount of PIB-MALA used and the reaction time, but hardly on temperature. This reflects the diffusion-controlled nature of this heterogeneous reaction.

The formation of grafted cellulose was confirmed by the behaviour in cuprammonium solutions. As expected, a freshly prepared solution of cuprammonium hydroxide dissolves Whatman cellulose completely. By comparison, only 40 % of the material from the grafting reaction was soluble. Regeneration of this soluble portion by acidification of the cuprammonium solution showed it to be mostly ungrafted cellulose. The insoluble 60 % of the material is the grafted copolymer.

Examination by scanning electron microscopy (SEM) of the products showed only partial breakdown of the fibres and the presence of islands of polyisobutylene on the cellulose surface.

Figure 3,a and b, which are IR spectra of PIB-g-CELL and PIB-g-CELL minus cellulose respectively, reveal both ester

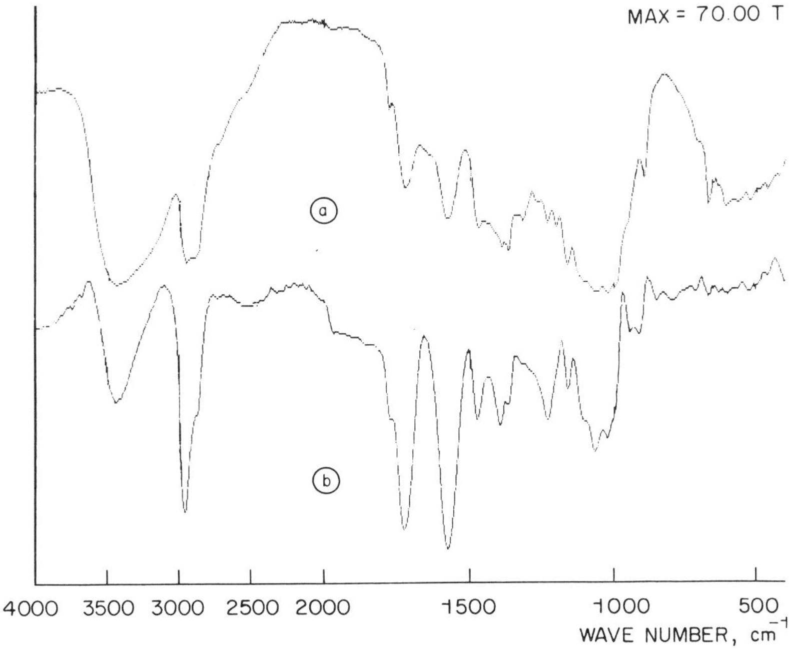

Fig. 3. IR spectra: (a) PIB-g-CELL and (b) the difference spectrum of PIB-g-CELL minus cellulose.

and carboxylic acid carbonyl groups, proving that the PIB chains are covalently bound to the cellulose as half esters.

CONCLUSIONS AND FINAL COMMENTS

The work we have described demonstrates that grafting via coupling of preformed polymers containing reactive end groups is a viable route, especially when it is mediated by sodium cellulosate. The reaction illustrated here, with succinic anhydride end group, can be successfully extended to other electrophilic groups such as epoxides, esters and silyl chlorides.

The polyisobutylene-grafted products synthesized by the present route are fibrous and lend themselves to applications such as reinforcements for thermoplastic matrices. The use of cellulose fibres (pulp) as reinforcements for thermoplastic matrices has been limited due to problems such as poor dispersion, poor wettability and insufficient fibre to matrix adhesion. Grafting of polymers onto the pulp is expected to minimize some of these problems.

This and other related applications are currently under investigation in our laboratory.

REFERENCES

Avny, Y. and Schwenker, R.F. (Jr.). Textile Res. J. 37 817 (1967).
Gault, H. and Ehrman, P., Compt. Rend. 177 124 (1923).
Kaminsky, W. Naturwissenschaften 71 93 (1984).
Lippmaa, E.T., Teeyaer, R.E., Zhbaukov, R.G., Pupko, O.A., Marchenko, G.N. and Zabelin, L.V. Dokl. Akad. Nauk SSSR 271 1159 (1983).
Schwenker, R.F. (Jr.), Kinoshita, T. and Beurling, D. J. Polymer Sci., 51 185 (1961).

GPC of cellulose derivatives on poly(styrenedivinylbenzene) packings in organic solvents

L. L. Lloyd and F. P. Warner – Polymer Laboratories Ltd, Essex Road, Church Stretton, Shropshire, SY6 6AX, U.K.

J. F. Kennedy – Research Laboratory for the Chemistry of Bioactive Carbohydrates and Proteins, Dept of Chemistry, The University of Birmingham, P.O. Box 363, Birmingham B15 2TT, U.K.

C. A. White – Chembiotech Ltd, Institute of Research and Development, Vincent Drive, Birmingham B15 2SQ, U.K.

INTRODUCTION

The gel permeation chromatographic (GPC) analysis of water soluble polysaccharides has been dominated by the hydrophilic bonded phase, silica-based packing materials [1,2]. For such materials to provide maximum resolution over a defined fractionation range it is essential for them to have as large a pore volume as possible and an accurately controlled distribution of pore sizes. As a consequence, the resulting packing materials are frequently susceptible to damage by shock or high pressure. In addition, being silica-based, dissolution of these materials in accelerated when used with eluents outside the pH range 2 to 7.

Hydrophilic polymer gels have been developed to provide water-compatible, polymer-based GPC packing materials [3,4] which are stable to eluents within the wider pH range of pH 2 to 12. Whilst being able to resolve materials with molecular weights up to about 30 million they have an inferior resolution power compared with the silica-based columns.

A range of highly crosslinked poly(styrene-divinylbenzene) packing materials have been developed [5] which provide rigid, macroporous resins that have greater stability over a wider range of pH and eluent type. Whilst initally developed for the GPC analysis of organic-soluble resins using eluents of medium polarity such as toluene and dichlorobenzene, the highly crosslinked nature of the resins gives wide solvent compatability and allows high polarity solvents such as dimethyl formamide, dimethyl sulphoxide and _N_-methyl pyrrolidinone to be used without loss of performance.

Gel permeation chromatographic analysis is a useful technique for the routine evaluation of cellulose derivatives since it provides a rapid method for the determination of molecular weight distribution. Such distributions are of more use in production control and end-use performance evaluation than are single point viscosity measurements for mean molecular weight evaluation [6]. However, there have been few reports of such analysis being applied to these materials using either aqueous [1] or organic solvent [7] systems. This paper describes the use of the macroporous poly(styrene-divinylbenzene) column packing materials for the molecular weight distribution analysis of cellulose derivatives using polar organic eluents.

EXPERIMENTAL

Equipment

The GPC system consisted of a pump (Knauer model 64), injector (Rheodyne 7125 equipped with a 200ul loop), column oven (Knauer model 60) and refractive index detector (Knauer model 98). In place of a chart recorder, a data station (Trivector Tri Lab II) was employed to convert the elution data into molecular weight distributions using the derived data from the calibration standards. All molecular weights are quoted in terms of polystyrene or polyethylene oxide/glycol standards as appropriate.

Columns

The following columns (300 x 7.5mm ID) were selected from the range of seven defined pore types and a mixed pore gel currently available as PLgel (Polymer Laboratories Ltd):

a) PLgel 10um 10^6 Å 30cm
b) PLgel 10um 10^6 Å 30cm
c) PLgel 5um Mix Å 30cm

Eluents

Tetrahydrofuran and dimethylformamide were hplc grade. Lithium bromide was analytical grade. All eluents were filtered (0.2 micron) and degassed prior to use.

Column calibration

Three column systems were established and calibrated using polyethylene glycol/oxide and polystyrene polymer standards (Polymer Laboratories Ltd).

System 1

A three-column system, comprising two columns of PLgel

10um 1 million Å and one column of PLgel 1 thousand Å, was eluted with tetrahydrofuran (1.0ml/min) at 20°C and calibrated using polystyrene standards (with a molecular weight range 1 thousand to 8.5 million). The calibration curve obtained is shown in Figure 1.

System 2

A single column of PLgel 5um Mix Å was eluted at 50°C with dimethylformamide containing lithium bromide (0.1% w/v, 1.0/min) to prevent aggregation of solute molecules. The system was calibrated using polyethylene glycol (molecular weight range 960 to 8.6 thousand) and polyethylene oxide (molecular weight range 18 thousand to 1 million) calibration standards (see Figure 2).

System 3

Two columns of PLgel 5um Mix Å were connected in series and eluted at 20°C with tetrahydrofuran (1.0ml/min). The system was calibrated using polystyrene (molecular weight range 1 thousand to 4.3 million) calibration standards (see Figure 3).

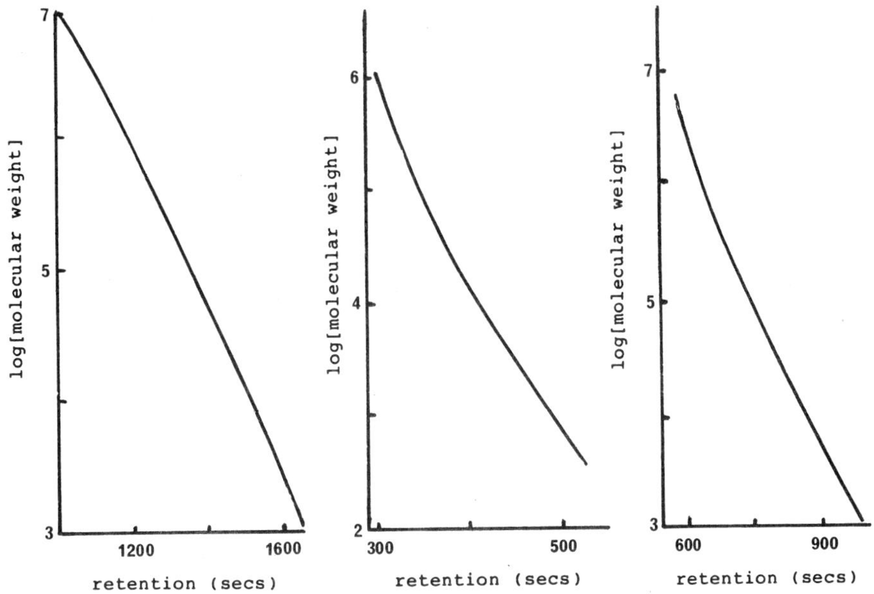

Figs. 1–3. Calibration curves for the Plgel column Systems 1 to 3 respectively (see text for details of column combinations, etc.).

ANALYSIS OF CELLULOSE DERIVATIVES

Nitrocellulose

The GPC analysis of nitrocellulose has been described [7] using large particle Styragel columns. The use of 8 to 10 columns (4 feet x 0.375 inch ID) operated in series at 1.0ml/min results in unacceptably long analysis times. With the availability of high performance, macroporous packings, such as PLgel, it is no longer necessary to use either the quantity of columns or large column dimensions. Using the three column combination described in System 1 the total run time is reduced to 30 minutes.

A single determination at 0.2%w/v concentration (200 ul injection) of cellulose nitrate was used to illustrate the separations which can be obtained (rather than the more tedious method of repeat analyses at successively lower concentration and extrapolation to zero concentration used to obtain accurate molecular weight values by overcoming concentration effects [7]). The effect of stabilizing a cotton nitrocellulose, by boiling in aqueous solutions of acid and alkali, on the molecular weight and its distribution can be observed in Figure 4. Not only has the peak molecular weight been reduced (from 393000 to 196000) but the dispersity (the ratio of weight average to number average molecular weights) has increased from 5.6 to 8.1. Nitrocelluloses with similar peak molecular weights (196000 for 12.6% N cotton-based and 146000 for 12.2%N wood-based materials) can be distinguished on account of the greater dispersity of the cotton-based materials (8.1 compared to 6.7) resulting from the greater degradation required for the cotton-based product (see Figure 5).

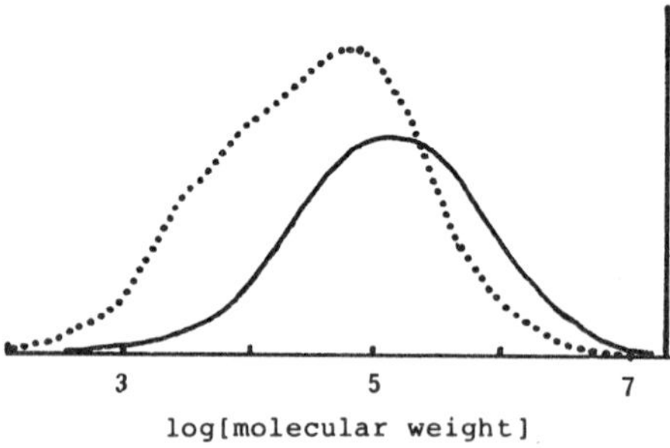

Fig. 4. GPC separation of unstabilized (——) and stabilized (.......) cotton nitrocelluloses using System 1.

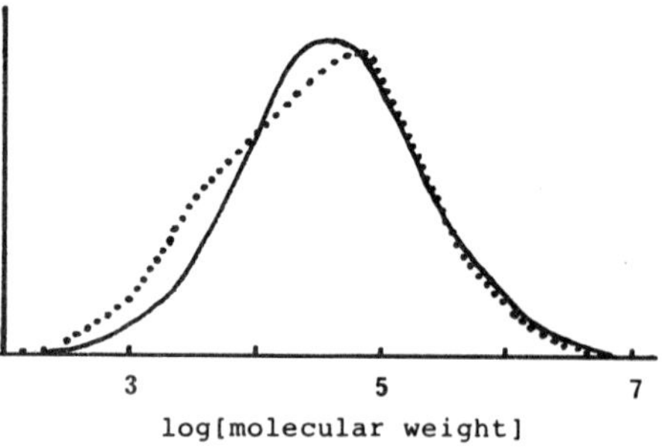

Fig. 5. GPC separation of cotton-based (.......) and wood-based (——) nitrocellulose of similar molecular weight using System 1.

Hydroxyethyl cellulose

The analysis of hydroxyethyl cellulose using a single mixed pore column eluted with dimethylformamide alone gave nonreproducible results and elution of material prior to the column exclusion limit. The addition of lithium bromide (0.1% w/v) to the eluent resulted in genuine GPC being obtained with reproducible results. It is considered that the lithium bromide prevents aggregation of the hydroxyethyl cellulose. Calibration of System 2 required the use of polar materials (polyethylene oxide/glycol standards) since polystyrene standards interact with the gel matrix under the elution conditions employed. The use of elevated temperature was made in order to lower the viscosity of the eluent, improve the mass transfer and also lower the pressure in the GPC system.

Analysis of samples (0.1% w/v, 100 ul injection volume) from three different grades of hydroxyethyl cellulose (graded according to viscosity) shows that the peak molecular weights follow the accepted grading order (see Figure 6) and therefore GPC can be used to evaluate the molecular weight distribution of such materials and correlate the data with the traditional viscosity-derived data. However, GPC does give more information than single point measurements (eg viscosity measurement) which can resolve the different end-use performances observed for samples of similar viscosity. Figure 7 illustrates the molecular weight distributions of two samples which have viscosities of 118 cp and 148cp under identical conditions. The higher viscosity material has the lower molecular weight values (number average, 24000; weight average, 113000; and peak value, 77000) compared with the values for the lower viscosity sample (27000, 140000 and 80000 respectively). The dispersity may be a more important parameter in this respect since the higher viscosity is obtained for the sample with lower dispersity (4.7 compared to 5.2).

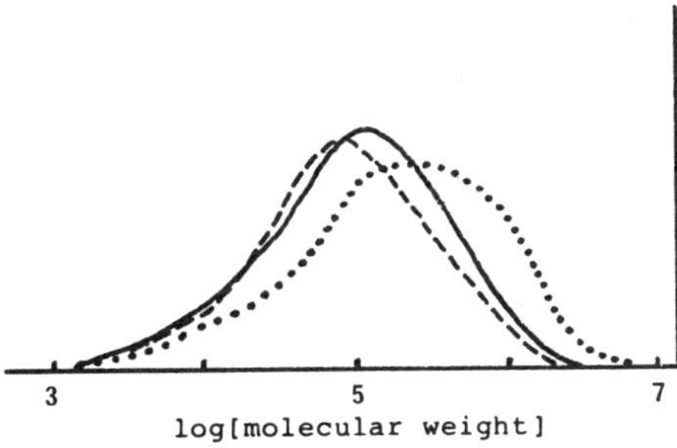

Fig. 6. GPC separation of hydroxyethyl cellulose of low (-------), intermediate (——) and high viscosity (.......) using System 2.

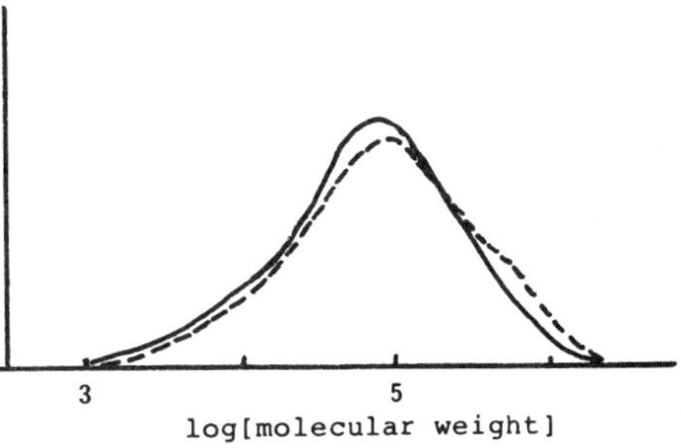

Fig. 7 GPC separation of two samples of hydroxethyl cellulose of similar viscosity using System 2. (——) 148 cp; (- - - -) 118 cp.

Cellulose acetate

A single sample of cellulose acetate (1.0% w/v) was analysed using System 3 (20 ul injection volume) to illustrate the application. This commercial grade of cellulose acetate had a uniform molecular weight distribution with the following values: number average, 43800; weight average 142661; peak molecular weight, 82813; and dispersity, 3.3 (see Figure 8).

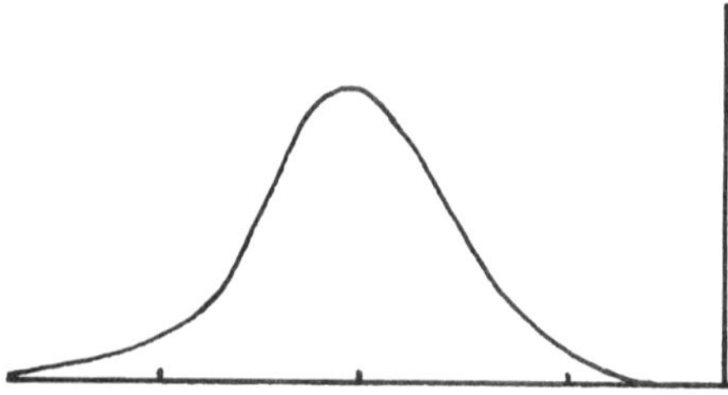

Fig. 8. GPC separation of cellulose acetate using System 3.

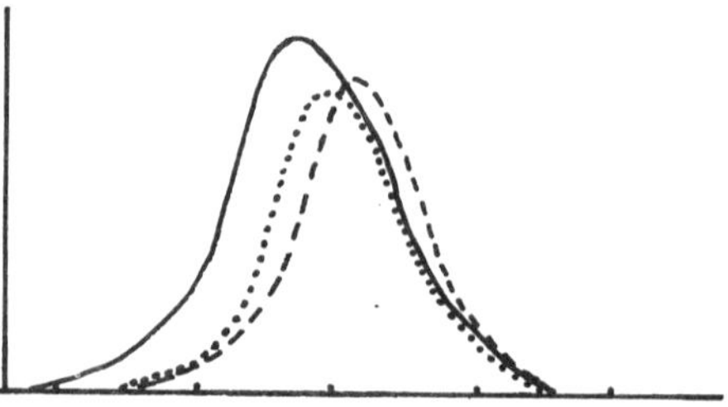

Fig. 9. GPC separation of identical ethyl celluloses of small (——), intermediate (------) and large (.......) flake size using System 3.

Ethyl cellulose

The analysis of ethyl cellulose for molecular weight distribution has been shown [8] to be important in the assessment of the suitability of the product for extrusion into tubular inhibitive coatings in rocket motors etc. It was reported that, in general, the samples with the broader molecular weight distributions gave fewer problems on extrusion with a smoother, more uniform surface. Three samples of ethyl cellulose, which were supposedly identical except for flake size, were analysed using System 3 (sample 0.2% w/v, injection volume 200 ul).The results obtained (see Figure 9) demonstrate that GPC can differentiate between materials reported to be very similiar. The material with the smallest flake size had the lowest molecular weight (number average, 24000; weight average, 13000; and peak, 54000) and greatest dispersity (5.4) which might imply that more degradation had occurred during production. The order of results did not follow the order of flake size however since the material with intermediate flake size had the largest molecular weight (number average, 72000; weight average, 214000; peak, 157000) and the sample having the largest flake size had intermediate molecular weight (number average, 56000; weight average, 170000; peak, 89000). The dispersity of the two samples with the higher molecular weights was found to be equal (3.0).

Cellulose tricarbanilate

Cellulose tricarbanilate [cellulose tri(N-phenyl carbamate] is a derivative formed by the reaction of phenyl isocyanate in pyridine and, due to the relatively mild reaction conditions, is reported to be formed without degradation of the

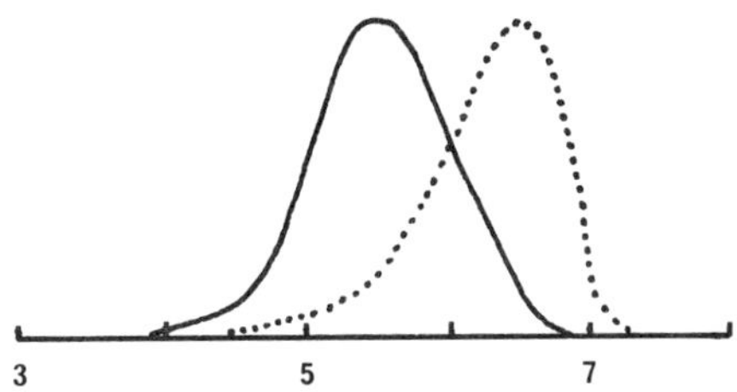

Fig. 10. GPC separation of tricarbanilates from different cotton celluloses using System 1. Sample A (——) has the following molecular weight values: number average, 211000, weight average, 623000; peak, 354000; dispersity, 3.0; and sample B (.......); number average, 1.34 million; weight average, 3.44 million; peak, 3.50 million; dispersity, 2.6.

parent cellulose [9]. As a result it has been suggested that this derivative offers the opportunity of studying the molecular weight distributions of cellulose samples without the need for eluents such as cadoxen which can degrade the sample or column matrices.

Analysis of two carbanilates prepared from different cotton celluloses (sample concentration 0.2% w/v, injection volume 200 ul) using System 1 shows that the system can analyse derivatives from celluloses of different molecular weight up to about 4 million (see Figure 10). It demonstrates that cellulose is not monodisperse as proposed by a number of workers but exists as a distribution of molecular weights with relatively low dispersities.

CONCLUSIONS

The examples quoted in this paper demonstrate the potential of the highly crosslinked poly(styrene-divinylbenzene) packings (PLgel) for the GPC analysis of a wide range of cellulose derivatives soluble in organic solvents. The rigidity of these column packings means that changes in eluent can be readily achieved without recourse to lengthy gradual composition changes required for other packings [8]. Thus, the number of columns required for analysing a wide range of derivatives can be minimized and the effect of different eluents readily studied.

ACKNOWLEDGEMENTS

The authors wish to thank those who supplied the derivatives used in this study.

REFERENCES

[1] H.G Barth and F.E. Regnier, J. Chromatogr.192 (1980) 275.
[2] Y. Kato, K. Komiya, H. Sasaki and T. Hashimoto, J. Chromatogr. 190 (1980) 297.
[3] T. Hashimoto, H. Sasaki, M. Aiura and Y. Kato, J. Polymer Sci. 16 (1978) 1789.
[4] Y. Kato, H. Sasaki, M. Aiura and T. Hashimoto, J. Chromatogr. 153 (1978) 546.
[5] Polymer Laboratories technical literature.
[6] M.G. Wirick and J.H. Elliot, J.Appl. Polymer Sci. 17 (1973) 2867.
[7] A.F. Cunningham, C. Heathcote, D.E. Hillman and J.I. Paul, In Liquid chromatography of polymers and related materials II, Chromatographic Science Series Vol 13' (eds J.Cazes and X. Delamore) Marcel Dekker, Basel, 1980, p173.
[8] A.P. Brookshaw, D.E. Hillman and J.I. Paul, Brit. Polymer J.5 (1973) 229.
[9] J.Danhelka and I. Kossler, J. Polymer Sci. 14 (1976) 287.

PART 2

BIOTECHNOLOGY

Physical, chemical and biochemical considerations in the biological degradation of wood

T. W. Jeffries – Institute for Microbial and Biochemical Technology, USDA Forest Service, Forest Products Laboratory.

INTRODUCTION

Microbial degradation of wood should be considered on two levels. From an organismic point of view degradation concerns the interaction of the cell with its microenvironment. At this level, degradation is strongly affected by diffusion of the degradative agents and the uptake of the oligomeric products. From a biochemical perspective, degradation concerns the molecular architecture of the biopolymers, the capacities of the enzymatic catalysts to bind to them and the catalytic turnover rates. Lignin reactivity is limited by the accessibility, heterogeneity, and stability of the polymeric linkages; cellulose reactivity is limited mostly by accessibility and crystallinity; and hemicellulose reactivity is limited by substitution.

The relationships among wood structure, chemistry and biological decomposition have been reviewed many times in articles and occasionally in complete texts (Nicholas, 1973; Loewus and Runeckles, 1977; Rowell, 1984; Higuchi, 1985). If one also considers the specialized texts and monographs on the structure and degradation of individual wood components such as cellulose (Hajny and Reese, 1969) and lignin (Kirk, *et al.*, 1980; Crawford, 1981; Kirk and Cowling, 1984) and volumes from symposia on biomass conversion (*e.g.*, Wilkie, 1975; Gaden *et al.*, 1976; Ghose, 1978; Hollaender, 1981), it is apparent that one cannot add much new detail to what has previously been said. Such, however, is not the purpose. My intention here is to resynthesize past knowledge with a few major recent findings and to relate tissue and polymer structure of wood to mechanisms for its biological and biochemical degradation.

GROSS CHEMICAL COMPOSITION

The chemical composition of wood varies from one species to another. Angiosperms generally contain appreciably more xylan and acetyl, and gymnosperms generally contain significantly more mannan and lignin. Other gross chemical differences exist. Specifically, North American angiosperms average 45% glucan, 18% xylan, 1.2% galactan, 0.6% arabinan, 2.4% mannan, 4.6% uronic anhydride, 4% acetyl, and 22% lignin. North American gymnosperms average 44% glucan, 7% xylan, 2.1% galactan, 1.6% arabinan, 10.4% mannan, 4.2% uronic anhydride, 1.4% acetyl, and 29% lignin (Pettersen, 1984). The structure of the hemicelluloses and lignins differ significantly between the two taxa. The chemical compositions of woods have been extensively reviewed by several authors (Pettersen, 1984; Sjöström, 1981; Fengel and Grosser, 1975; Rydholm, 1965; Browning, 1963). Aside from the presence or absence of extractives, which markedly retard biodegradation, the gross chemical composition of wood actually tells relatively little about its potential for degradation.

DISTRIBUTION OF WALL COMPONENTS

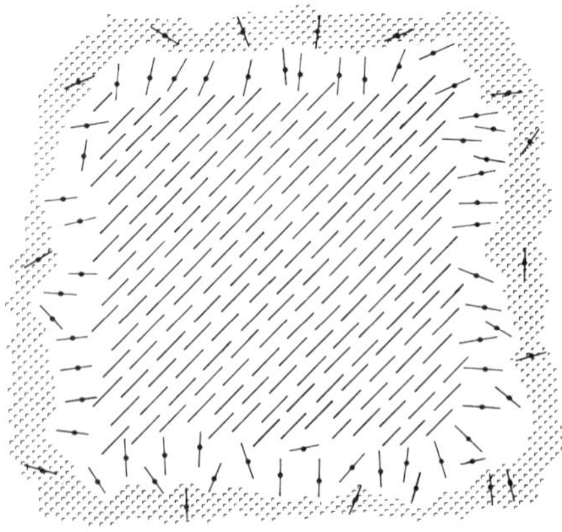

Fig. 1. Cellulose (–), hemicellulose (-•-) and lignin (⁛) complex in wall.

The three principal components of wood are found throughout the wall. Cellulose is organized into laminar crystallites which are bundled into the microfibrils. Each micro - fibril contains regions of amorphous cellulose interspersed and intertwined with hemicellulose (Fig. 1). The latter is in turn crosslinked to lignin. Most of the cellulose is in the secondary wall layers. Its concentration depends on the relative portions of hemicellulose and lignin. Distribution of hemicellulose within wood tissue is not homogeneous because vessels and fibers in angiosperms contain appreciably more cellulose while parenchyma cells contain large amounts of glucuro - noxylan (Meier, 1985). Presumably the higher hemicellulose content of parenchyma cells assists in distribution of nutrients. Pentosan content is higher in ray cells than in the total wood (Timell, 1967). In *Pinus sylvestris*, for example, tracheids contain more than twice as much mannose and less than half as much xylose as ray cells.

Qualitative localization of hemicellulose within the cell walls can be accomplished with selective stains and other techniques, but quantitative distribution is more difficult to establish (Meier, 1985). Hardell and Westermark (1981) removed individual cell wall layers by microdissection and isolated three fractions from tracheids of spruce: The (M + P) fraction had substantially higher arabinose and galactose content, but xylose and galactose varied little throughout the wall, indicating that the hemicellulose content in cell walls is proportional to the cellulose and not to the lignin.

Lignin is the encrusting material. It is found in the highest concentration in the middle lamella, but because the volume of the middle lamella is small, most of the total lignin is found in the secondary wall (Table 1).

Table 1. Distribution of lignin in two gymnosperm species (Saka and Goring, 1985).

Species	Morphological region[a]	Tissue volume[b]		Lignin			
				% total		Conc (g/g)	
		E	L	E	L	E	L
Black spruce	S	87	94	72	82	0.23	0.22
	ML	9	4	16	10	0.50	0.60
	ML_{cc}	4	2	12	8	0.85	1.00
Douglas fir	S	74	90	58	78	0.25	0.23
	ML	10	4	18	10	0.56	0.60
	ML_{cc}	4	2	11	6	0.83	0.90

[a]S= secondary wall, ML = middle lamella; ML_{cc}= middle lamella, corner region

[b]E= earlywood; L=latewood

The total amount of lignin in the secondary wall is greater in latewood than in earlywood, primarialy because of the increase in the thickness of the secondary wall layer. Even though the concentration of lignin increases in latewood middle lamella, the proportion of lignin found in this region decreases mostly because the middle lamella is much thinner.

COMPONENT CHEMISTRIES

Lignin is a stable, highly cross-linked aromatic macromolecule arising through dehydrogenative free radical polymerization of *p*-coumaryl, coniferyl and sinapyl alcohols. Gymnosperm lignin is made up principally of coniferyl alcohol (guaiacyl lignin) whereas angiosperm lignin is made up from approximately equal amounts of coniferyl and sinapyl alcohols (syringyl/guaiacyl lignin) (Freudenberg and Neish, 1968; Adler, 1977). *p*-Coumaryl alcohol is found principally as a precursor to the lignin of grasses and in reaction wood (Freudenberg and Neish, 1968). Although the lignin polymer can be described by a combination of three monomeric units, the large number of interunit linkages which arise through free radical polymerization leads to a great deal of structural complexity. A few prominent substructures comprise most of the interunit linkages (Fig. 2). By far the most common (40-50%) is the arylglycerol-β-aryl ether link (A). The next most common (≈ 10%) are the phenyl coumaran (B) and biphenyl (C) substructures. The biosynthesis and chemical characterization of lignin has been presented extensively elsewhere (Higuchi, 1985; Fengel and Wegener, 1984; Adler, 1977; Sarkanen and Ludwig, 1971; Freudenberg and Neish, 1968).

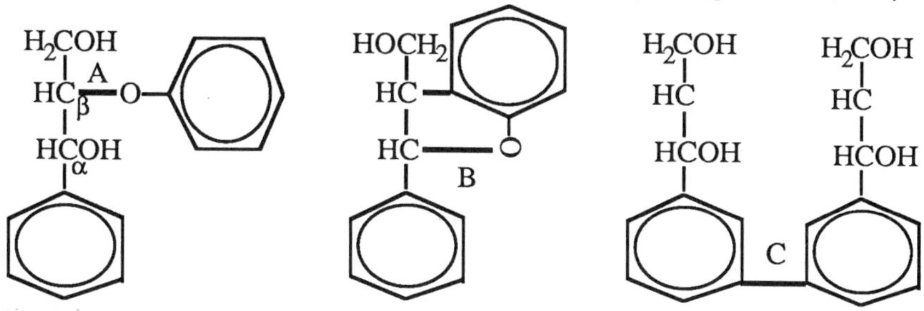

Fig. 2. Major interunit linkages found in lignin (Alder, 1977).

The relative stability of the chemical bonds determines the reactivity of the polymer as a whole. Although different bonds are reactive under different conditions, all of the linkages in lignin are stable in comparison to other biological polymers. Lignin is condensed by strong acid, but degradation occurs in strong alkali or in the presence of strong oxidizing agents. Any linkage to the α carbon will exhibit relative instability. In fact, the rate of alkaline degradation of open-chain *p*-hydroxybenzyl-aryl ethers is too fast to be measured (Miksche, 1980). *p*-Hydroxylphenylcoumaran structures (B) are also relatively easily degraded with cleavage of the Cα-O ether linkage and the formation of a double bond between Cα–Cβ, but the degradation rates of all the other linkages are much lower than these. The most stable structure is the aromatic ring. Probably because of the greater stability of the Cβ-O and biphenyl linkages, the lignin linkage most readily oxidized during biological degradation is the Cα-Cβ.

Lignin is not simply deposited between cell wall polysaccharides, but is chemically linked to the hemicelluloses with a frequency of about 1 per every 35 phenyl propane units (Obst, 1982). Direct linkages to glucan are relatively rare except at the non-reducing terminus. Carbohydrates are common contaminants in lignin preparations, even when highly purified; and degradation of lignocellulose by polysaccharidases leads to the isolation of polymeric lignin with carbohydrate attached (Adler, 1977). Methylation analysis has shown that the predominant link to carbohydrate is through arabinoxylan. Lignin bonding can occur at the C-2 and C-3 of the xylose monomer and at the C-5 of arabinose side chains (Minor, 1983). Arabinose, xylose and 4-O-methylglucuronic acid are the most common sugar moities to which lignin is cross-linked. Benzyl ester, benzyl ether and phenyl glycosidic linkages are probably involved (Fengel and Wegener, 1984).

Cellulosic microfibrils are the most conspicuous element of the plant cell wall. The β-(1,4)-D-glucose chains which make up the cellulose polymer can be arranged in both parallel and antiparallel crystallites; but based on what we now know about the biosynthetic (Hassid, 1969; Lloyd, 1984) and biodegradative (Chanzy and Henrissat, 1985) mechanisms and x-ray crystallographic studies (Gardner and Blackwell, 1974), the parallel arrangement is almost certainly the form found in nature.

The structure of cellulose is derived from its chemical character and its biochemical origin. Each chain of cellulose is held straight by O6-O2 and O3-O5 intrachain hydrogen bonds and adjacent chains are bonded into a flat sheet by O3-O6 hydrogen bonds. Adjacent laminar sheets are held together by van der Waals' forces (Gardner and Blackwell, 1974; Marchessault and Sundararajan, 1983). The cellulose chains are packed with the plane of the glucose ring oriented diagonally with respect to the flat planes of the microfibril (Caulfield, 1971).

Interspersed throughout the cellulose crystallite are amorphous regions in which crystallinity is very much reduced. These regions could result from the curvature of the cell membrane during cellulose synthesis. A crystallite forming as the cellulose synthetase complex moves across the surface of the cell membrane would eventually jut out away from the membrane too far for the complex to maintain order in it. An amorphous region would be formed between the crystallite and the complex, and a new crystallite would be organized out of the packing of the synthetase complex.

The resistance of the cellulose crystallite to hydrolysis is attributable to the rigidity of the rings held tightly by the crystal structure. The rotational energy barrier encountered in ring flexure is a fundamental factor in controlling the rate of hydrolysis (Harris, 1975), and the principle extends to conformation and substitution of the glycosides involved. If, for example, one compares the relative rates of hydrolysis for C1-β-methyl pyranosides and β-(1-4)-linked disaccharides, they fall in the same order as determined by the ring stability of the glycone (the non-reducing terminal glycoside which is released to form a reducing group) and not the aglycone (the reducing terminus) (Table 2).

The glycosidic linkages of a hemicellulosic polymer of glucose is more stable than that of mannose, and a polymer of mannose is more stable than one of xylose. The steric hinderance attributable to the crystallinity of the cellulose polymer combined with the greater stability of the bond, accounts for why hemicellulose is so much more readily hydrolyzed than cellulose. This picture is complicated somewhat, by the fact that glucuronosides – substituents in angiosperm xylans – are hydrolyzed at about one thirtieth the rate of glucosides (Harris, 1975).

Table 2. Relative hydrolysis rates of methyl β-pyranosides and β-(1,4)-linked disaccharides (Harris, 1975).

Methyl pyranoside of	Relative rate	Disaccharide	Relative rate
β-D-glucose	1.0	Cellobiose	1.5
β-D-Mannose	3.0	Mannobiose	2.7
β-D-Galactose	4.8	Lactose	3.6
β-D-Xylose	5.8	Xylobiose	11.0

The major hemicellulose of angiosperms is *O*-acetyl-4-*O*-methylglucuronoxylan. A second hemicellulose found in much lower proportions is glucomannan. In gymnosperms, the major hemicelluose is *O*-acetylgalactoglucomannans. Gymnosperms also contain arabino-4-*O*-methylglucuronoxylan which is similar to the polymer found in angiosperms (Timel, 1967; Whistler and Richards, 1970). The main chain of angiosperm xylan is composed of 6-membered pyranose rings in the chair conformation linked β-(1-4) as in the case of cellulose. Because xylose does not possess the C6 hydroxyl, O2-O6 intrachain and O6-O3 interchain hydrogen bonding is not present. In xylan, O3-O5' hydrogen bonding form the chain into a twisted ribbon with a repeat unit of xylotriose (Marchessault and Liang, 1962). Straight-chain, unsubstituted xylan is relatively insoluble. In wood cell walls, xylan possesses monomeric side branches consisting of D-glucuronic or 4-O-methylglucuronic acids linked α glycosidically to the O2 position with an average frequency of 1 per 10 xylose residues (Timell, 1967). Also, native xylan is frequently acetylated on the O3 or the O2 position with a frequency of about 7 in 10 xylose residues. This acetylation makes the polymer much more soluble in water (Biely, 1985; Whistler and Richards, 1970). The significance of such acetylation to enzymatic degradation is frequently overlooked because the acetyl groups are commonly removed during isolation. The main chain of the xylan polymer contains one β-(1,3)-linked branch point about every 75 residues (Whistler and Richards, 1970).

Glucomannans are the major hemicellulose of gymnosperms where they comprise about 15-17% of the total dry weight. In contrast, they are minor components of angiosperms, generally comprising only 3-5%. In angiosperms, they contain glucose and mannose most commonly in a 1:2 ratio linked in a β-(1,4) straight chain (Timell, 1967). In gymnosperms, they occur in close association with cellulose (Whistler and Richards, 1970). The majority of glucomannans contain a few galactose residues as non-reducing end groups. The ratio of galactose, glucose, and mannose in gymnosperms is commonly 2:10:30 (Timell, 1961). Arabinogalactans are found principally in larch wood, but they are also minor occasional components in angiosperms. The polymer consists of a β-(1,3)-galactose main chain with β-(1,6)-galactose and (1,6)-arabinose (furanose) side branches, and it is highly water soluble (Whistler and Richards, 1970). Although larch arabinogalactan is a minor and atypical hemicellulosic component of wood, it is frequently employed in xylanase assays, presumably because of its easy isolation and solubility. For the taxonomic distributions and structures of hemicelluosic polymers see specific treatments (McNeil, *et al.*, 1984; Stephen, 1983; Rees, 1977; Towle and Whistler, 1973; Timell, 1967; Whistler and Richards, 1970).

MICROSTRUCTURE AND POROSITY

Porosity and microstructure are critical in determining accessibility of enzymes to their polymeric substrates. The relationship between enzyme diffusion and substrate accessibility has been shown in a number of ways. For example, reducing particle size greatly increases surface area and speeds up the enzymatic hydrolysis of cellulose (Millet *et al.*, 1975). Grinding or cutting wood to open all the lumina to attack will expose considerably more surface than is available in whole wood. The total surface area reaches about 1 $m^2 g^{-1}$. Once the lumina are accessible, however, relatively little further increase in surface area occurs following further grinding, because the cell wall thickness is very small compared to its length (Stone *et al.*, 1969). Particle size is not the only factor important in enzymatic depolymerization. Enzymatic accessibility depends on the structure of the wall itself.

The pore volume of cell walls can be estimated from the water content. In the dry state, the wood cell wall is non-porous (Stone *et al.*, 1966). Green wood, however, is wet, so it follows that in its wet state, the wall is swollen above its dry volume by an amount equal to the volume of water it contains. At the fiber saturation point, *i.e.*, when the walls are completely saturated with water but before any free water is found in the cell lumens, the volume of water contained in the fiber is equal to the total pore volume. In the case of sprucewood, the total pore volume is about 0.48 ml g^{-1}, virtually all comprised of pores less than 50 Å and 60% less than 10Å in diameter (Stone and Scallan, 1968). The molecular pore distribution of individual wall components of native wood is poorly characterized. However, it is known that the cellulose crystallite excludes water by hydrogen bonding. The water content of lignin is also probably low by virtue of its aromatic structure. Hemicellulose, on the other hand, has a high affinity for water; so it seems likely that the bulk of the pore structure in wood cell walls will be found in this fraction.

A distinction should be made between the effects of substrate crystallinity and those of microporosity. Susceptibility of pure cellulose to enzymatic hydrolysis is inversely related to its degree of crystallinity (Sasaki *et al.*, 1979; Caulfield and Moore, 1974; Fan *et al.*, 1980). However, in any practical situation, enzymatic hydrolysis is not carried out with pure cellulose; so considerations other than crystallinity prevail. Recent studies by Grethlein (1985) using wood which had been partially hydrolyzed by dilute acid have shown a linear correlation between pore volume accessible to a protein with a nominal diameter of 51 Å and the initial rate of cellulose hydrolysis obtained with *Trichoderma reesei* cellulase. The crystallinity index of the substrate was not related to the rate of hydrolysis. Likewise, in an examination of enzymatic hydrolysis of a series of treated lignocellulosic substrates, the rate of cellulose hydrolysis did not correlate with crystallinity (Puri, 1984). Also, in a study of the initial hydrolysis rates obtained with *Clostridium* and *Trichoderma* cellulases acting on six different cellulosic substrates, Weimer and Weston (1985) concluded that the rate limiting factor for both cellulases was accessibility and diffusion rather than crystallinity.

The relationship between component microstructure and enzymatic degradation has been studied most extensively with cellulose. Stone *et al.* (1969) employed solute exclusion to measure the pore volume in cotton linters swollen to varying degrees in phosphoric acid. The swollen cellulose gels were hydrolyzed with *Trichoderma* cellulase having a reported molecular weight of 60 kd. They found a linear relationship between the initial reaction rate and the surface area within the gel which was accessible to a sphere of 30-40 Å dia., corresponding nominally to a protein of 40 to 70 kd (Stone and Scallan, 1968; Cowling and Brown, 1969; Weimer and Weston, 1985). However, neither the pores nor the enzymes are spherical.

Stone *et al.* (1969) visualized the pores in swollen cellulose not as circular in cross section, but as slit-like spaces between the cellulose lamellae. The elongated morphologies of cellulases have been recognized for some years, with the earliest work dating to that of Whitaker *et al.* (1954). Of 27 cellulases summarized by Cowling and Brown (1969), the ratio of length to width averaged 5.9 ± 0.33 and the apparent molecular weights ranged over 10 fold. One can therefore conclude that elongate shapes for cellulases are more characteristic than their sizes.

Even very small enzymes can possess relatively large binding sites. For example, lysozyme (molecular weight = 12.5 kd) is known from x-ray crystallographic studies to bind six pyranose rings (Imoto *et al.*, 1972), and indirect evidence has been given to show that a β-(1,3)-D-glucanase with a molecular weight of 27 kd has a binding site encompassing at least 14 β-(1,3)-D-glucose rings (Jeffries and Macmillan, 1981). Contrary to what one might expect, however, the groove for the binding site of lysozyme runs across the short axis of the enzyme. If endoglucanases follow a similar pattern, the portions of the enzyme sticking out on the sides adjacent to the binding site could be involved in separating laminar sheets of the cellulose crystallite.

DEGRADATION OF WHOLE WOOD

Microbial degradation of whole wood occurs principally by fungi. Although bacteria apparently possess much of the biochemical machinery necessary for biodegradation of the wood components, they do not form mycelia, and hence, cannot propagate through the tissue structure or translocate nutrients into the wood or from one region to another as decay progresses.

The various components found in different types of cells and within cell walls affect the degradation rates. Electron microscopic studies of wood degraded by white-rot fungi have shown that in some instances, parenchyma cells can be almost completely degraded, leaving vessels behind (Otjen and Blanchette, 1984). The different rates of degradation could be related to the different hemicellulose compositions of the two cell types. The relationship between wood structure and microbial degradation is somewhat ambiguous because of the many separate variables such as lignin content and composition, the presence of extractives, the density of the tissue structure, moisture content, and the relation of these variables to the invading organism (Wilcox, 1973). The most resistant wood species generally have high extractives contents and have been summarized (Scheffer, 1973), and Wilcox (1973) has summarized the organisms characteristic of wood decay.

Wood degradation is limited by diffusion. Diffusion of oxygen and CO_2 through the wood is restricted, and the interior of a decaying tree is essentially anoxic (Jensen, 1969). Heartrot fungi appear to possess special metabolic capabilities which enable them to grow under the restrictive conditions found in the center of a log (Highley and Kirk, 1979). Diffusion of gases occurs principally through the cell lumens. If the cells are filled with water, essentially no oxygen transport occurs. In fact, methanogenic bacteria – a taxonomic group intolerant of even traces of oxygen – have been found in the heart of decaying, water-saturated cottonwood, (Zeikus and Ward, 1974). Diffusion of gas out of the wood is limited and concentrations of CO_2 can be as high as 20% of the total gas volume in wood which is undergoing fungal decay. (Hintikka and Korhonen, 1970; Jensen, 1969). As decay progresses, porosity increases and oxygen availability increases as well.

The intracellular lumen and wall pits are particularly important in degradation because they provide access for the invading mycelium to pass from one cell to another. When wood is dried, the pit membrane becomes aspirated or forced against one side of bordered pits, thereby blocking the free passage of water, oxygen, and possibly mycelia. Pit aspiration has been proposed as a reason for the slower degradation of gymnosperms by white-rot fungi, but investigations have not supported this idea (Highley, 1978). Pits are not essential for the passage of fungi such as white- and brown-rotters which can produce bore holes perpendicular to the cell axis. Hubert (1924) suggested that this ability could distinguish between such decay fungi and various staining fungi which merely inhabit the wood. Both white-rot and brown-rot degradative fungi initially invade through the pits, whereas in later stages of decay they produce bore holes.

White-rot and brown-rot fungi are commonly indicated and studied for their abilities to decay whole wood. White-rot fungi are generally associated with decay of angiosperms and brown-rot fungi with the decay of gymnosperms (Cowling, 1961) with white-rot fungi showing the more pronounced host selectivity (Eslyn and Highley, 1976). White-rot fungi degrade cellulose and hemicellulose more or less at the same rates while lignin is degraded at the same rate or slightly faster. By comparison, brown-rot fungi utilize the cellulose and hemicellulose leaving the lignin demethylated, but otherwise largely undigested (Highley and Kirk, 1979). In both instances, the hyphae are initially concentrated in the ray cells and propagate into the center of the wood along the inside of the cell lumens. Degradation of the cell occurs from the inside out. The mycelia of both white-rot and brown-rot fungi are surrounded by a hyphal sheath (Palmer *et al.*, 1983a,b) which, in the case of brown-rot fungi, spreads away from the fungus and sometimes encompasses the cellulosic fibrils (Highley *et al.*, 1983a; Highley *et al.*, 1983b) It is not clear whether the sheath of white-rot fungi is involved in cellulose degradation, but both white-rot and brown-rot fungi produce a diffusible cellulose-degrading system because degradation occurs at some distance from the mycelia (Highley and Murmanis, 1984).

BIODEGRADATION OF LIGNIN

Biodegradation of lignin has been studied for more than 80 years with only incremental progress over much of the time (Crawford and Crawford, 1984). Early assays employed the disappearance of acid-insoluble Klason lignin. The development of synthetic ^{14}C-labeled lignins (Kirk *et al.*, 1975) enabled rapid radiorespirometric determination of lignin degradation under controlled conditions, and a series of experiments using this assay led to improved understanding of the cultural and nutritional conditions under which lignin biodegradation occurs.

In *Phanerochaete chrysosporium,* a co-metabolizable carbon source is essential for lignin biodegradation (Kirk *et al.*, 1976), and it occurs in response to nitrogen starvation (Keyser *et al.*, 1978). This indicates that the ligninolytic system is formed as a part of secondary metabolism in this organism. Carbohydrate starvation likewise leads to a rapid but transient onset of ligninolytic activity (Jeffries *et al.*, 1981). Elevated oxygen levels increase the rate of lignin biodegradation (Kirk *et al.*, 1978), through the production of hydrogen peroxide as an extracellular oxidant and the induction of ligninolytic activity (Faison and Kirk, 1983). The hydrogen peroxide is derived from the cometabolism of cellulose and/or hemicellulose by reactions which are as yet incompletely characterized.

The complexity of lignin prevents meaningful biochemical studies on the polymer itself; so concepts about biodegradative mechanisms are derived from studies with biodegradation or modification of model compounds and direct chemical characterization of products from wood decayed under ligninolytic conditions. It is not completely clear whether the reactions observed in oligomeric substrates are completely related to what occurs in the native lignin polymer. For example, cleavage of aromatic rings in monomeric aromatic compounds was demonstrated several years ago using cellular extracts of a lignin-degrading fungus (Buswell and Erickson, 1979), but degradation of aromatic rings of lignin is now known to occur in the polymer (Haider *et al.*, 1985). Some of the earliest studies on lignin models were done on dimeric compounds containing the DL-glycerol β-aryl ether linkage (Russell *et al.*, 1961), the choice being based on the fact that this linkage accounts for most of the interunit bonds in lignin (Adler, 1977).

One of the principal problems in employing model compounds in whole cultures is to carry out the cleavage of interest while preventing subsequent reactions which lead to degradation or polymerization. For example, guaicyl-glycerol-β-guaicyl ether (I) contains a phenolic hydroxy group and is readily degraded by white-rot fungi containing phenol oxidase while the non-phenolic veratryl glycerol-β-guaiacyl ether (II) is not degraded (Kirk *et al.*, 1968). The problem of sequential reactions is particularly acute because the initial reaction of interest is often rate-limiting. A major breakthrough in the design of model lignin substrates came with the use of the non-phenolic model compound 1,2-bis(3,4-dimethoxyphenyl)propane-1,3-diol (III). Biodegradation of III in ligninolytic cultures shows that primary cleavage occurs between Cα and Cβ with incorporation of molecular oxygen (Nakatsubo *et al.*, 1982). A recent review of how model compounds have elucidated the degradative pathway for lignin has been given by Higuchi (1985b).

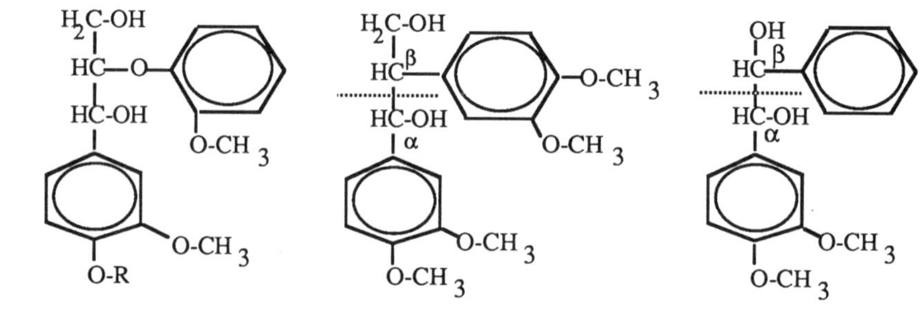

I R = H II R = CH_3 III IV

Enzymes have now been identified in the cell-free extracellular broth of ligninolytic cultures which will depolymerize spruce lignin and catalyze the degradation of lignin-related models (Tien and Kirk, 1983). Depolymerization is carried out by a 42 kd ligninase with a pI of 3.5 which requires hydrogen peroxide for activity. Ligninase was purified to homogeneity (Tien and Kirk, 1984) and shown to contain a single protoheme IX prosthetic group. Ligninase is similar to classical peroxidases in its resonance Raman spectra of the heme active site (Kuila *et al.*, 1985), and like peroxidase, H_2O_2 reacts stoichometrically with the heme to give an oxidized intermediate. Following reduction, the enzyme reacts with the lignin model compound (Tien *et al.*, 1986) by abstracting an electron from the aromatic ring thereby leaving a cation radical (Kersten *et al.*, 1985). Once the cation radical is formed, Cα cleavage occurs spontaneously (Hammel *et al.*, 1985). A number of subsequent reactions, such as demethyoxylation or Cα – Cβ cleavage, can follow – largely dependent on the nature of the substrate.

Subsequent to the enzymatic reaction, the radicals can undergo other spontaneous reactions with the incorporation of oxygen. These sequential reactions complicate interpretation of the stoichiometries, and it is necessary to employ model substrates which exhibit only the reaction of interest. One such compound is 1-(3,4-dimethoxyphenyl)-2-phenylethanediol (IV). Using this substrate, it was shown that substrate free radicals are formed as intermediate products (Hammel *et al.*, 1985). When acting on 1,4-dimethoxybenzene ligninase abstracts one electron at a time from the ring to produce cation radical intermediates which decompose through the addition of water and the removal of methanol groups to form *p*-benzoquinone (Kersten *et al.*, 1985). In recent studies of lignin model compounds representing the predominant arylglycerol-β-aryl ether substructure of lignin, Kirk *et al.* (1986a) showed that the major consequence of the ligninase activity is cleavage of the Cα- Cβ linkage. Cα oxidation is much less frequent and removal of aromatic alkyl ethers is a relatively minor reaction, although both types of reactions occur.

Other enzymes produced by *Phanerochaete* in liquid culture may play separate roles in biodegradation of lignin. For example, Huynh and Crawford (1985) have identified an extracellular H_2O_2-dependent aromatic methyl ether demethylase and an extracellular Mn^{++} dependent peroxidase which has also been observed by Glenn and Gold (1985). At least 13 different proteins are found in the extracellular broth of *Phanerochaete* cultures, 10 of which appear to be haemproteins (Kirk *et al.*, 1986b). When one also considers the possible roles of cellobiose:quinone oxidoreductase (Westermark and Eriksson, 1974; Buswell, Hamp and Eriksson, 1979) and phenol oxidase (Ander and Eriksson, 1976) – if not in the initial depolymerization reactions then perhaps in subsequent reactions to drive the depolymerization reactions forward – then it is clear that the biochemical explanation of lignin degradation is far from complete.

Direct chemical studies of lignin degradation products in decayed wood do not always point to the same mechanisms or pathways as are revealed by model compound studies. Notably, a more prominent role for ring cleavage is indicated than is observed with model compounds (Chen and Chang, 1985). It is important to remember that lignin exists in wood in quite a different environment from model compounds in defined culture media. The localized concentration of substrate is very much higher and the availability of water is very much lower in wood than it is in culture broths. Also, lignin is found closely associated with hemicellulosic materials, the degradation of which could almost certainly affect lignin depolymerization. The differences, however, might simply reflect substrate accessibility. In the polymer, the aromatic rings would be attacked with the formation of cation radicals. The radicals would migrate to cleave the relatively reactive Cα - Cβ linkages near the surface of the polymer. Those on the interior would be shielded from attack. The surface ring structures would subsequently be degraded by sequential electron abstraction.

Other aspects of lignin biodegradation such as the roles of oxidative species (Kirk *et al.*, 1984), details of biochemical mechanisms (Kirk, 1986), regulation (Kirk and Fenn, 1982) and bacterial lignin degradation (Crawford and Crawford, 1984) have been reviewed elsewhere.

BIODEGRADATION OF CELLULOSE

The biodegradation of cellulose has probably attracted more attention world-wide than any other single biochemical event outside the field of medicine, yet the mechanism of cellulose hydrolysis is still imperfectly understood. The cellulase complex from *Trichoderma reesei* has been studied most thoroughly, but it is not yet clear exactly how this complex works to depolymerize cellulose. The focus here is placed on the biochemical mechanisms involved in hydrolysis.

Reese *et al.* (1950) first postulated the C1 Cx hypothesis for cellulose hydrolysis to account for differences observed with crystalline and non-crystalline cellulose substrates. Since that time, the number of proteins known to catalyze cellulose depolymerization have proliferated, and they are now commonly classified into three groups (Table 3). Each of these is found in various isozymes, and their action on cellulose is synergistic. The cellobiohydrolases act on straight chain, unsubstituted cellulose and exhibit activity against crystalline cellulose. *Endo*glucanases will act on acid swollen straight-chain cellulose and on substituted cellulose such as carboxymethyl cellulose. β-glucosidases act on cellobiose and aryl β-glucosides and their activity decreases rapidly as the degree of polymerization (DP) increases.

Table 3. Enzymes of the cellulase complex.

Systematic name	Enzyme no.	Trivial names	Substrate and product
1,4-β-D-glucan cellobiohydrolase	(EC 3.2.1.91)	*exo*glucanase, cellobiohydrolase	crystalline cellulose cellobiose
endo-1,4-β-D-glucan 4-glucanohydrolase	(EC 3.2.1.4)	*endo*glucanse, β-glucanase	amorphous cellulose cellooligosaccharides
β-D-glucoside glucohydrolase	(EC 3.2.1.21)	cellobiase, β-glucosidase	cellobiose, triose glucose

As is shown in Table 4, the principal endoglucanase (EG I) of *T. reesei* is very active against carboxymethyl cellulose (CMC) and shows essentially the same specific activity against phosphoric acid swollen cellulose (PSC). In contrast, cellobiohydrolase I (CBH I) shows more than 60 times as much activity against PSC than with CMC. Both CBH I and EG I show activity against avicel, but whereas CBH I can degrade it completely, EG I will attack it only until the amorphous regions are consumed. Only the EG I will attack xylan and only β-glucosidase (βG I) will attack cellobiose or *p*-nitrophenyl glucoside (PNPG) (Shoemaker *et al.*, 1983).

Table 4. Specific activities of cellulases isolated from *T. reesei* (Shoemaker *et al.*, 1983).

	Specific activity (U mg^{-1})					
	Avicel	PSC	CMC	Cellobiose	PNPG	Xylan
CBH I	0.04	0.6	<0.01	<0.01	<0.01	<0.01
EG I	0.17	26.0	27	<0.01	<0.01	6.4
βG I	<0.01	<0.01	<0.01	31.4	19.5	<0.01

Reese (1974) identified five criteria by which cellulases could be distinguished. Each gives some insight into the hydrolytic mechanism (Table 5). The first of these, effect of DP, has already been discussed. The second, retention or inversion of anomeric configuration is directly related to the hydrolytic mechanism. The mechanism for glycolysis has been characterized best by x-ray crystallographic studies of lysozyme.

Table 5. Classification criteria for cellulases (Reese, 1974).

Criterion	Observation
Effect of degree of polymerization	*exo* have higher affinity on longer chain
Retention or inversion	*endo* retain; *exo* invert
Ability to transfer glycosyl	β-glucosidases are more able
Effect of lactone inhibitor	*endo* are more susceptible
Bond specificity	*exo* is more specific in its hydrolysis

Based on similar sequence homologies, it has been proposed that the catalytic sites of endoglucanases I and II (EG I, II), the cellobiohydrolase I (CBH I) and egg white lysozyme are similar (Paice *et al.*, 1984). Catalysis proceeds by an ion-pair mechanism. During binding, the C6 carbon of the glycone interacts with the enzyme to distort the glycone ring into a flattened half-chair. Catalyis is initiated by protonation from an acidic amino acid to the glycosidic oxygen. This brings about cleavage and leaves a positively-charged carbonium ion at the C1 of the glycone which is stabilized by ionic interaction with a negatively-charged amino acid residue on the opposite side of the ring. The glycone is finally released when the negatively-charged amino acid is displaced by a hydroxyl (Phillips, 1966; Eveleigh and Perlin, 1969). One feature of this mechanism is that it requires the substrate to sit in a groove in the enzyme, hence it might not be appropriate to the hydrolysis of a crystallite.

The anomeric configuration of the product depends on whether an enzyme-bound intermediate is formed. The presence of a bound intermediate is indicated by the ability of some glucanases to carry out transglycosylation reactions. Transglycosylation is most commonly seen with β-glucosidases but is also observed in endoglucanases to a lesser extent (Okada and Kisizawa, 1975). Cellobiohydrolases do not generally exhibit transglycosylation (Patil and Sadana, 1984; Reese, 1974). This observation combined with the fact that cellobiohydrolase does not retain the configuration of the C1 hydroxyl, indicates that no stable enzyme-bound intermediate is formed and anomeric inversion suggests that the hydroxyl displaces the glycosidic oxygen in a single step mechanism.

*Endo*glucanases do not exhibit a great deal of specificity with regard to the position of the linkage hydrolyzed, but they are specific with regard to the aglycone portion of the binding site, as indicated by the products formed through action on mixed-linkage glycans (Parrish and Perlin, 1960). Cellobiohydrolases are not active on mixed linkage β-glucans nor do they hydrolyze hydroxyl-substituted celluloses such as CMC.

The presence of the half-chair intermediate in hydrolysis by endoglucanases is indicated by the inhibitory activity of substrate analogs. One of the forms of D-glucono-1,5-lactone is very similar to the half-chair intermediate D-gluconopyranosyl cation formed during hydrolysis. D-glucono-(1,5)-lactone is a potent inhibitor of endoglucanases but is much less effective with cellobiohydrolase (Reese *et al.*, 1971). The glycolytic mechanism for *exo*-splitting hydrolases such as cellobiohydrolase is almost certainly different from that of *endo*-splitting hydrolases. Probably the cellobiose structural repeat unit of cellulose accounts for its appearance as a product.

In recent years, cellobiohydrolases of several fungi have been separated and characterized (Mishra *et al.*, 1983; Beldman *et al.*, 1985; Shoemaker *et al.*, 1983; van Tilbeurgh *et al.*, 1985) as two immunologically distinct proteins (Fägerstam and Pettersson, 1979) that show no sequence homology (Fägerstam and Pettersson, 1984). Amino acid sequence homology does, however, exist between CBH I and EG I (Shoemaker *et al.*, 1983; Paice *et al.*, 1984). CBH I and II are similar in that both show high affinity for crystalline cellulose and release cellobiose as their principal product. But CBH II shows greater specific affinity for cellobiose than CBH I (van Telbeurgh *et al.*, 1984). Surprisingly, cellobiohydrolases show little specificity for the 4-OH non-reducing terminal substituent, and they can be differientiated by their relative abilities to bind and cleave a series of methyl umbelliferol (4-O-MeUmb) substituted cellooligosaccharides. CBH I will hydrolyze 4-MeUmb-(Glu), 4-MeUmb-$(Glu)_2$ or 4-MeUmb-$(Glu)_3$. CBH II in contrast will not hydrolyze any model smaller than 4-MeUmb-$(Glu)_3$, and even then its specific rate is 10-fold lower than that of CBH I on this substrate. Based on these results, van Tilbeurgh *et al.* (1985) have proposed that CBH I and II are able to interact with different cellobiosyl groups within longer chains. This finding also supports a proposed mechanism for the synergism observed between these two enzymes.

Structural assymetry was first proposed by Fägerstam and Pettersson, (1980) as the basis for synergism between CBH I and CBH II. This notion has recently been given a more definite form by Wood *et al.*, (1986). When crystalline cellulose is used as a substrate, synergism between CBH I and CBH II is observed. However, when acid swollen cellulose is employed, synergism is negligible. Wood *et al.* (1986) have speculated that these two different types of enzymes might attack two different types of non-reducing end groups which should arise in the cellulose crystallite (Fig. 3). This sort of synergism is different from that which has been long recognized between *endo* - and *exo*glucanases and between cellobiohydrolase and β-glucosidase (Wood and McCrae, 1979; Kanda *et al.*, 1980; Henrissat *et al.*, 1985). The synergisms are related to the creation of new enzyme cleavage sites and the relief of feedback inhibition.

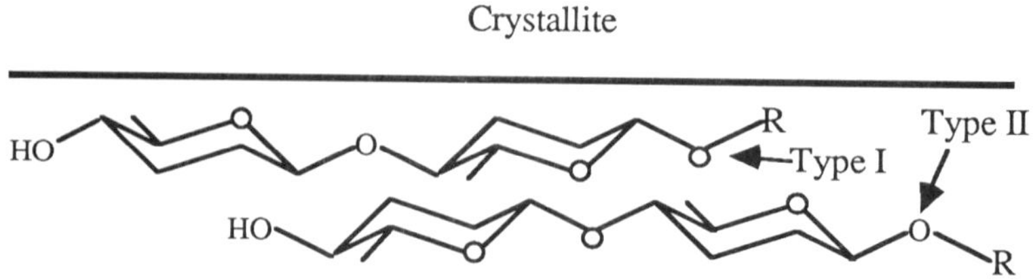

Fig. 3. Model of assymetric glycosidic linkages in cellulose crystallite.

Purified CBH I is capable of attacking and degrading microcrystalline cellulose from *Valonia* without any other enzyme present (Chanzy *et al.*, 1983). When the enzyme is labeled with colloidal gold, its attachment can be visualized. The enzyme binds to the edges rather than the face of the crystallite. (Chanzy *et al.*, 1984). The specificity of CBH I is shown strikingly in its unidirectional attack on the non-reducing end of the molecule, giving rise to a tapered point (Chanzy and Henrissat, 1985). The reducing end is left untouched, as are the kinks in the crystallite (Fig. 4). If, however, a purified endoglucanase is added, hydrolysis is initiated at the kinks and multiple unidirectional hydrolysis is observed.

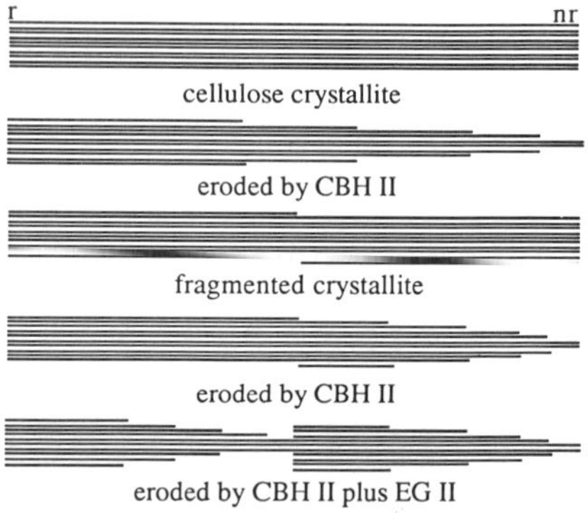

Fig. 4. Degradation of crystalline cellulose by purified cellulases.

When all of the cellulases are present, *Valonia* crystals undergo a delamination to form bundles of microcrystallite fibrils (Chanzy and Henrissat, 1983). A similar delamination has been observed with the purified bacterial cellulose from *Acetobacter xylinum* (White and Brown, 1981). Although no direct evidence is available, the delamination probably occurs between the sheets adjacent van der Waals-bonded sheets rather than between adjacent hydrogen bonded chains.

The distinction between CBH I and II is not confined to *T. reesei* . CBH I and II are also found in the cellulase complex of *Penicillium pinophilum* (Wood *et al.*, 1986) and in *Fusarium lini* (Mishra *et al.*, 1983). In the latter instance, CBH I was found to be active against D-xylan, whereas CBH II was not. Given the apparent importance of the C6 carbon in forcing the half-chair conformation during the hydrolysis of the glycosidic bond by lysozyme (Phillips, 1966), it is apparent that substrates lacking this substituent may not be appropriate for some cellulases.

Reese (1974) pointed out that most cellulases don't act on xylan. This was puzzling in view of the fact that xylan did not possess additional side groups which might cause steric hinderance. It is possible that they use a mechanism similar to lysozyme to bend the ring into a reactive conformation. It is now apparent that in some instances at least, highly purified cellulase preparations possess activity against xylan (Shoemaker *et al.*, 1983), albeit at an insignificant rate in comparison to a true xylanase.

BIODEGRADATION OF HEMICELLULOSE

Because of the diversity of hemicelluloses, it is not possible to survey here all of the different types known. Dekker and Richards (1976) have provided an extensive, useful compilation of this sort. Likewise, Reilly (1981), Woodward (1984) and Biely (1985) have recently reviewed the literature on microbial xylanases. Mannanases have been studied to a lesser extent, hence, the emphasis here is placed on xylanases.

The characterization of hemicellulase mechanisms presents numerous difficulties not encountered with cellulase because the substrate employed is heterogeneous and branched. Moreover, the literature is not definitive because substrates such as larch arabino - glucuronoxylan commonly used for hemicellulose assays – presumably because of their easy solubility – but they are not really representative of the bulk of hemicellulose found in wood. Selection of an appropriate substrate is critical because the apparent specific activity of a xylanase can vary greatly depending on the xylan preparation used for assay. Substrate variability is not confined to the type of xylan employed, because different batches of xylan from the same supplier can lead to apparent changes of as much as two-fold (Khan *et al.*, 1986). Native xylans tend to be highly acetylated. Acetyl groups increase solubility but decrease the rate of enzymatic depolymerization (Biely *et al.*, in press); so the method of substrate preparation can have a profound affect on the apparent activity.

Xylanases, like cellulases, can be classified into β-xylosidases, *exo*-β-xylanases, and *endo*-β-xylanases (Reilly, 1981; Frederick *et al.*, 1985). The endoxylanases can be further differentiated on the basis of their abilities to cleave at L-arabinose branch points or produce xylooligosaccharides. Like β-glucosidases, β-xylosidases don't carry out inversion of the C1 anomeric configuration and they exhibit strong transferase activity indicating the presence of a stable enzyme-bound intermediate. β-Xylosidases are strongly inhibited by their end product, xylose. *Exo*-β-xylanases, like *exo*-β-glucanases invert the anomeric configuration of the C1 hydroxyl during hydrolysis and do not exhibit transferase activity. Most do not show strong endproduct inhibition, but they are commonly inhibited by divalent cations, particularly Hg^{++} (Matsuo *et al.*, 1977; Takahashi and Kutsumi, 1979), indicating that a free sulfhydryl group might be involved in the catalytic mechanism. Endoxylanases cleave throughout the xylan polymer leaving xylobiosyl and xylotriosyl residues attached to L-arabinofuranose (Takenishi and Tsujisaka, 1973). The substrate specificities and hydrolytic products of *endo*xylanases have been best characterized in a series of enzymes purified from *Aspergillus niger* (Frederick *et al.*, 1981; Frederick *et al.*, 1985; Shei *et al.*, 1985; Fornier *et al.*, 1985). Some appear to require a side branch for maximal attack, but this may simply be related to solubility or accessibility of the substrate.

Biochemically, *endo*-β-(1,4)-D-xylanases can possess much higher turnover numbers than can equivalent *endo*-β-(1,4)-D-glucanases. For example, the V_{max} of an endoxylanase from *Trichoderma harzianum* was found to be 580 U mg^{-1} of protein (Tan *et al.*, 1985), or more than 20 times the equivalent rate observed with the major endoglucanase of *T. reesei* (Table 5). As was pointed out earlier, this difference in rate is attributable at least in part to the relative instability of the β-(1,4) glycosidic bond in xylose.

Xylotriose and xylopentaose have been observed as products of endoxylanases having specificity for long-chain, unsubstituted xylan (Fornier *et al.*, 1985). The presence of xylotriose as a predominant product could be related to the xylotriosyl repeat unit observed in x-ray crystallographic studies of xylan (Marchessault and Liang, 1962). The possible involvement of a DP 5 oligosaccharide in binding is reinforced by the observation that xylopentaose is a strong competitive inhibitor of xylanase from *Trichoderma viride* (Toda *et al.*, 1971). These results can be understood if xylanases having preference for straight-chain, unsubstituted xylan preferentially cleave off a xylotriose repeat unit but will not hydrolyze an oligosaccharide with a DP <8.

Many xylanases are not able to remove glucuronic acid residues from the xylan chain (Biely, 1985) probably because of the 30-fold greater stability of glucuronoside glycolytic linkages noted earlier (Harris, 1975). By comparison, the ester linkages connecting acetyl groups to the xylan are exceedingly unstable and most are removed during isolation of the polymer. Although synergism is observed between xylan esterases and endoxylanases during hydrolysis of partially acetylated xylan, (Biely *et al.*, 1985; Biely *et al.*, in press), it seems unlikely that deacetylation would be a rate-limiting step in the enzymatic hydrolysis of xylan under most practical conditions because of the extreme lability of the ester linkages.

APPLICATIONS

Technical developments, particularly the discoveries of xylose-fermenting yeasts (Jeffries, 1985) and lignin-degrading enzymes have paved the way for specific removal of hemicellulose and lignin in order to recover cellulosic fibers and to produce useful chemicals from the hemicellulose and lignin fractions.

Biological processes might be applied directly to the removal of lignin from whole wood chips by fungi. Lignin removal need not be complete in order to realize a benefit in reduced energy input during thermo-mechanical pulping (Eriksson and Vallanders, 1982). The major problem is in obtaining selective removal of lignin without concomitant cellulose degradation. One approach has been to incorporate glucose into the chips in order to repress cellulase production (Kirk and Yang, 1979, Bar-Lev *et al.*, 1982). Another approach has been to obtain cellulose deficient mutants which will still degrade lignin at the expense of hemicellulose (Eriksson, *et al.*, 1980). The cellulase negative mutants presently available show various ligninase activities (Kirk *et al.*, 1986b). Some native strains of fungi exhibit fairly selective removal of lignin (Blanchette, 1984a,b; Otjen and Blanchette, 1984). Such organisms are presently being evaluated for their abilities to remove lignin from chips. Other applications of lignin degrading fungi in biopulping, biobleaching and waste treatment in the pulping industry have been reviewed recently (Eriksson and Kirk, 1985).

An alternative approach to biological pulping which permits more direct control over the degradation activities is to employ isolated enzymes for lignin or hemicellulose removal. Because the lignin is cross-linked mostly into the hemicellulose, because the hemicellulose is accessible due to its porosity and because hemicellulose is more readily depolymerized than lignin, it has been the initial target of treatment (Paice and Jurasek, 1984). Removing only a small portion of the hemicellulose might be sufficient to open up the polymer and ease solvent removal of the residual lignin. The principal problem in these studies has been in obtaining selective removal of hemicellulose without degrading the cellulose. Selective inhibition of cellulase activity can be realized by the addition of 1 mM $HgCl_2$ (Mora *et al.*, 1986). Xylan was specifically removed from delignified cell walls leading to a decrease in energy demand during beating (Noé *et al.*, 1986). Removing less than 2% of the initial dry pulp weight gave better fiber bonding due to increased flexibility (Barnoud *et al.*, 1986). Therefore, enzymatic treatments of pulp hold the prospect of both decreased costs and improved fiber qualities.

CONCLUSIONS

The physical structure of the tissue and the microstructures of the components determine the accessibility of degradative organisms and enzymes. During the degradation of whole wood, mycelial fungi invade by ray cells and move throughout the tissue by way of the intracellular lumens and pit connections. Degradation of the cells takes place from the inside out with hemicellulose being removed first. Lignified, highly cellulosic tissues are degraded later. The presence of coherent hydrogen bonding in cellulose and the tight covalent cross-linking in lignin prevent access of enzymes to the interiors of these molecules; so erosion occurs principally from the outside. The mechanisms for lignin degradation are determined by the relative chemical stabilities of the bonds, the $C\alpha$ linkage being the weakest. Likewise, cellulose crystallites are more resistant than unbonded chains, so degradation proceeds first through removal of amorphic regions. Degradation of the cellulose crystallite proceeds through synergistic activities of several *endo*- and *exo*glucanases. Substitution of hemicellulose, particularly by uronic acids inhibits degradation, but its open, hydrated structure is both accessible and relatively easy to depolymerize. As porosity increases, diffusion of oxygen and other enzymes into the wood cell structure increases. Removal of hemicellulose can improve the quality of pulps, improve access to and removal of lignin, and may lead to processes in which fuels, chemicals and fibers are co-produced from wood feedstocks.

ACKNOWLEDGEMENTS

The author is grateful to T.K. Kirk, H. Burdsall, M.L. Larsen and P. Kersten for helpful discussions and critical readings of the text, to D. Caulfield for discussions and to H. Schneider for discussions and preprints of work in press.

REFERENCES

Adler, E. (1977) Wood Sci. Technol. 11:169-218.
Ander, P. and Eriksson, K.E. (1976) Arch. Microbiol. 109:1-8.
Bar-Lev, S.S., Chang, H.M. and Kirk, T.K. (1982) Tappi 65:111-113.
Barnoud, F., Comtat, J. Joseleau, J.P., Mora, F. and Ruel, K. In: Proceedings of the Third International Conference on Biotechnology in the Pulp and Paper Industry, Stockholm, June, 1986, pp. 70-72.
Beldman, G., Searle-van Leeuwen, M.F., Rombouts, F.M. and Voragen, F.G.J. (1985) FEBS Lett. 146:301-308.
Biely, P. (1985) Trends Biotechnol. 3:286-290.
Biely, P., MacKenzie, C.R. Puls, J. and Schneider, H. (1985) FEBS Lett. 186:80-84.
Biely, P., MacKenzie, C.R. Puls, J. and Schneider, H. (1986) Bio/Technology (in press).
Blanchette, R.A. (1984a) Phytopathology 74:153-160.
Blanchette, R.A. (1984b) Appl. Environ. Microbiol. 48:647-653.
Browning, B.L. The Chemistry of Wood. John Wiley and Sons, New York, 1963.
Buswell, J.A. and Ericksson, K.E. (1979) FEBS Lett. 104:258-260.
Buswell, J.A., Hamp, S. and Eriksson, K.-E. (1979) FEBS Lett. 108:229-232
Caulfield, D.F. and Moore, W.E. (1974) Wood Sci. 6:375-379.
Caulfield, D.F. (1971) Textile Res. J. 41:267-269.
Chanzy, H. and Henrissat, B. 1983 Carbohyd. Polym. 3:161-173.
Chanzy, H. and Henrissat, B. (1985) FEBS Lett. 184:285-288.
Chanzy, H., Henrissat, B. and Vuong, R. (1984) FEBS Lett. 172:193-197.
Chanzy, H., Henrissat, B., Vuong, R. and Schülein, M. (1983) FEBS Lett. 153:113-118.
Chen, C.-L. and Chang, H.-m. In: Higuchi, T. (ed) Biosynthesis and Biodegradation of Wood Components. Academic Press, New York, 1985 pp. 535-556.
Cowling E.B. (1961) Techn. bull. No 1258. U.S. Dept. Agric., For. Serv. 79 p.
Cowling, E.B. and Brown, W. In: Hajny, G. and Reese, E.T. Cellulases and their applications. American Chemical Society, Washington, D.C. 1969.
Crawford, R.L. Lignin Biodegradation and Transformation. Wiley Interscience, New York, 1981.
Crawford, R.L. and Crawford, D.L. (1984) Enzyme Microb. Technol. 6:434-442.
Dekker, R.F.H. and Richards, G.N. (1976) Adv. Carb. Chem. Biochem. 32:277-352.
Eriksson, K.-E., Grünewald, A., Nilsson, T. and Vallander, L. (1980) Holzforschung 34:207-213.
Eriksson, K.-E. and Kirk, T.K. In: Cooney, C.L. and Humphrey, A.E. (eds) The Principles of Biotechnology, Pergamon Press, New York, 1985, pp. 271-294.
Eriksson, K.-E. and Vallanders, L. (1982) Svensk Papperstig. 85:R33.
Eslyn, W.E. and Highley, T.L. (1976). Phytopathology 66:1010-1017.
Eveleigh, D.E. and Perlin, A.S. (1969) Carbohyd. Res. 10:87-95.
Fägerstam, L.G. and Pettersson, G. (1979) FEBS Lett. 98:363-367.
Fägerstam, L.G. and Pettersson, G. (1980) FEBS Lett. 119:97-100.
Fägerstam, L.G. and Pettersson, G. (1984) FEBS Lett. 167:309-315.
Faison, B.L. and Kirk, T.K. (1983) Appl. Environ. Microbiol. 46:1140-1145.
Fan, L.T. Lee, Y.-H. and Beardmore, D.H. (1980) Biotechnol. Bioeng. 22: 177-199.
Fengel, D. and Grosser, D. (1975) Holz Roh-Werkst. 33:32-34.
Fengel, D. and Wegener, G. Wood. Chemistry, Ultrastructure, Reactions. Walter de Gruyter, Berlin, 1985.
Fournier, R.; Frederick, M.M., Frederick, J.R. and Reilly, P.J. (1985) Biotechnol. Bioeng. 27:539-546.

Frederick, M.M., Frederick, J.R., Fratzke, A.R. and Reilly, P.J. (1981) Carbohydr. Res. 97:87-103.
Frederick, M.M., Kiang, C.-H., Frederick, J.R. and Reilly, P.J. (1985) Biotechnol. Bioeng. 27:525-532.
Freudenberg, K. and Neish, A.A. The Constitution and Biosynthesis of Lignin. Springer-Verlag, New York, 1968.
Gaden, E.L., Mandels, M.H., Reese, E.T. and Spano, L.A. (eds) Enzymatic Conversion of Cellulosic Materials: Technology and Applications. Biotechnol. Bioeng. Symp. 6, John Wiley, Interscience, 1976.
Gardner, K.H. and Blackwell (1974) Biopolymers 13:1975-2001.
Ghose, T.K. (ed) Bioconversion of Cellulosic Substances into Energy Chemicals and Microbial Protein. Indian Institute of Technology, New Delhi. 1978.
Glenn, J.K. and Gold, M.K. (1985) Arch. Biochem. Biophys. 242:329-341.
Grethlein, H.E. (1985) Bio/Technology 3:155-160.
Haider, K., Kern, H. and Ernst, L. (1985) Holzforschung 39:23-32.
Hajny, G.J. and Reese, E.T. (eds) Cellulases and Their Applications. Advances in Chemistry Series, Vol. 95. American Chemical Society, Washington, D.C. 1969.
Hammel, K.E., Tien, M., Kalyanaraman, B. and Kirk, T.K. (1985) J. Biol. Chem. 260:8348-8353.
Hardell, H.-L. and Westermark, U. (1981) In: The Ekman-Days International Symposium on Wood and Pulping Chemistry, Stockholm vol. 1, pp. 32-34.
Harris, J.F. Applied Polymer Symposium No. 28, John Wiley and Sons, Inc, New York, 1975, pp. 131-144.
Hassid, N.Z. (1969) Science 165:137-144.
Henrissat, B., Driguez, H. Viet, C., and Schülein, M. (1985) Bio/Technology 3:722-726.
Highley, T.L. (1978) Material Org. 13:197-206.
Highley, T.L. and Kirk, T.K. (1979) Phytopathology 69:1151-1157.
Highley, T.L., Murmanis, L.L. and Palmer, J.G. (1983a) Holzforschung 37:271-277.
Highley, T.L. and Murmanis, L.L. (1984) Holzforschung 38:73-78.
Highley, T.L., Palmer, J.G. and Murmanis, L.L. (1983b) Holzforschung 37:179-184.
Higuchi, T. (ed) Biosynthesis and Biodegradation of Wood Components. Academic Press, Orlando, 1985.
Higuchi, T. In: Higuchi, T. (ed) Biosynthesis and Biodegradation of Wood Components. Academic Press, Orlando,1985a. pp. 141-160.
Higuchi, T. In: Higuchi, T. (ed) Biosynthesis and Biodegradation of Wood Components. Academic Press, Orlando, 1985b pp. 557-578.
Hintikka, V. and Korhonen, K. (1970) Commun. Inst. For. Fenn. 69:1-29.
Hollaender, A. (ed) Trends in the Biology of Fermentations for Fuels and Chemicals. Plenum Press, New York, 1981.
Hubert, E.E. (1924) J. Agr. Res. 29:523-567.
Huynh, V.-B. and Crawford, R.L. (1985) FEMS Lett. 28:119-123.
Imoto, T., Johnson, L.N., North, A.C.T., Phillips, D.C. and Rupley, J.A. in boyer, P.D. (ed) The Enzymes, 3rd ed. Academic Press, New York, Vol III, 1972, pp. 665-868.
Jeffries, T.W. (1985) Trends Biotechnol. 3:208-212.
Jeffries, T.W. and Macmillan, J.D. (1981) Carbohyd. Res. 95:87-100.
Jeffries, T.W., Choi, S. and Kirk, T.K. (1981) Appl. Environ. Microbiol. 42:290-296.
Jensen, K.F. (1969) For. Sci. 15:246-251.
Kanda, T., Wakabayashi, K. and Kisizana, K. (1980) J. Biochem. 87:1635-1639.
Keyser, P., Kirk, T.K. and Zeikus, J.G. (1978) J. Bacteriol. 135:790-797.
Kersten, P.J., Tien, M., Kalyanaraman, B. and Kirk, T.K. (1985) J. Biol. Chem. 260:2609-2612.
Khan, A.W., Tremblay, D. and LeDuy, A. (1986) Enzyme Microb. Technol. 8:373-377.
Kirk, T.K. (1971) Ann. Rev. Phytopath. 9:185-210.
Kirk, T.K. In: Nicholas, D.D. (ed) Wood Deterioration and its Prevention by Preservative Treatments. Syracuse University Press, Syracuse, 1973 pp. 149-181.
Kirk, T.K. In: Proceedings of a Discussion Meeting, Technology in the 1990's: Utilization of Lignocellulosic Wastes. The Royal Society, London, 1986.
Kirk, T.K., Connors, W.J. and Zeikus, J.G. (1976) Appl. Environ. Microbiol. 32:192-194.
Kirk, T.K., Connors, W.J., Bleam, R.D., Hackett, W.F. and Zeikus, J.G. (1975) Proc. Natl. Acad. Sci. U.S.A. 72:2515-2519.

Kirk, T.K. and Cowling, E.B. In: Rowell, R.M. The Chemistry of Solid Wood. Adv. Chem. Ser. 207. American Chemical Society, Washington, D.C. 1984, pp. 455-487.
Kirk, T.K., Croan, S., Tien, M., Murtagh, K.E. and Farrell, R.E. (1986b) Enzyme Microb. Technol. 8:27-32.
Kirk, T.K. and Fenn, P. In: Frankland, Hedger and Swift (eds) Decomposer Basidiomycetes, British Mycological Society Symposium 4, Cambridge University Press, Cambridge,1982 pp. 67-90.
Kirk, T.K., Harkin, J.M. and Cowling, E.B. (1968) Biochim. Biophys. Acta 165:134-144.
Kirk, T.K., Higuchi, T. and Chang, H.M. Lignin Biodegradation: Microbiology, Chemistry, and Potential Applications. Vol I and II. CRC Press, Inc., Boca Raton, Florida, 1980.
Kirk, T.K., Schultz, E., Connors, W.J., Lorenz, L.F. and Zeikuş, J.G. (1978) Arch. Microbiol. 117:277-285.
Kirk, T.K., Tien, M., Kersten, P.J., Mozych, M.D. and Kalyanaraman, B. (1986a) Biochem. J. 236:279-287.
Kirk, T.K., Tien, M. and Faison, B.D. (1984) Biotech. Advs 2:183-199.
Kirk, T.K., Tien, M., Johnsrud, S.C. and Eriksson, K.E. (1986) Enzyme Microb. Technol. 8:75-79.
Kirk, T.K. and Yang, H.H. (1979) Biotechnol. Lett. 1:347-352.
Kuila, D., Tien, M., Fee, J.A. and Ondrias, M.R. (1985) Biochemistry 24:3394-3397.
Lloyd, C.W. (1984) Int. Rev. Cytol. 86:1-51.
Loewus, F.A. and Runeckles, V.C. (eds) The Structure, Biosynthesis, and Degradation of Wood. Plenum Press, New York, 1977.
Marchessault, R.H. and Liang, C.Y. (1962) J. Polym. Sci. 59:357-378.
Marchessault, R.H. and Sundararajan, P.R. In: Aspinall, G.O. The Polysaccharides, Vol. 2, Academic Press Inc., London, 1983, pp. 12-95.
Matsuo, M., Yasui, T. and Kobayashi, T. (1977) Agr. Biol. Chem. 41:1601-1606
McNeil, M., Darrill, A.G., Fry, S.C. and Albersheim, P. (1984) Ann Rev. Biochem. 53:625-663.
Meier, H. In: Higuchi, T. (ed) Biosynthesis and Biodegradation of Wood Components. Academic Press, Orlando,1985, pp. 43-50.
Miksche, G.E. In: Chemistry of Delignification with Oxygen, Ozone, and Peroxides. Uni Publishers Co. LTD., Tokyo, 1980, pp. 107-120.
Millett, M.A., Baker, A.J. and Satter, L.D. (1975) Biotechnol. Bioeng. Symp. 5:193-219.
Minor, J.L. In: International Symposium on Wood and Pulping Chemistry., Tsukuba Science City, Japan. vol. 1, 1983, pp. 153-158.
Mishra, C., Vaidya, M., Rao, M. and Deshpande, V. (1983) Enzyme Microb. Technol. 5:430-435.
Mora, F., Comtat, J., Barnoud, F., Pla, F. and Noe, P. (1986) J. Wood Sci. Technol. (in press).
Nakatsubo, F., Reid, ID. and Kirk, T.K. (1982) Biochim. Biophys. Acta 719:284-291.
Nicholas, D.D. (ed) Wood Deterioration and its Prevention by Preservative Treatments. Syracuse University Press, Syracuse. 1973.
Noé, P., Chevalier, J., Mora, F. and Comtat, J. (1986) J. Wood Sci. Technol. (in press).
Obst, J.R. (1982) Tappi 65:109-112.
Okada, G. and Nisizawa, K. (1975) J. Biochem. 78:297-306.
Otjen, L. and Blanchette, R.A. (1984) Appl. Microbiol. Biotechnol. 47:670-676.
Paice, M.G., Desrochers, M., Rho, D., Jurasek, L., Roy, C., Rollin., C.F., De Miguel, E. and Yaguchi, M. (1984) Bio/Technology 2:535-539.
Paice, M.G. and Jurasek, L. (1984) J. Wood Chem. Technol. 4:187-198.
Palmer, J.G. , Murmanis, L.L. and Highley, T.L. (1983a) Visualization of hyphal sheath in wood-decay hymenomycetes. I. Brown Rotters. Mycologia 75:995-1004.
Palmer, J.G., Murmanis, L.L. and Highley, T.L. (1983b) Mycologia 75:1005-1010.
Parrish, F.W., and Perlin, A.S. (1960) Nature 187:1110-1111.
Patil, R.V. and Sadana, J.C. (1984) Can J. Biochem Cell Biol. 62:920-926.
Pettersen, R.C. In: Rowell, R.M. (ed) The Chemistry of Solid Wood. Advances in Chemistry Series vol. 207, 1984, pp. 57-126.
Phillips, D.C. (1966) Sci. Am. 215 (5):78-90.
Puri, V.P. (1984) Biotechnol. Bioeng. 26:1219-1222.
Rees, D.A. Polysaccharide Shapes. Halstead Press, John Wiley and Sons, New York, 1977.
Reese, E.T., Siu, R.G.H. and Levinson, (1950) J. Bact. 59:485-497.
Reese, E.T., Parrish, F.W. and Ettinger, M. (1971) Carbohydr. Res. 18:381-388.
Reese, E.T. In Loewus, F.A. and Runeckles, V.C. (eds) The Structure, Biosynthesis and Degradation of Wood. Plenum press, New York, 1974, pp. 311-367.

Reilley, P.J. In: Hollaender, A.E. (ed) Trends in the Biology of Fermentations for Fuels and Chemicals, Plenym Press, New york, 1981, pp. 111-129.

Rowell, R.M. (ed) The Chemistry of Solid Wood. Advances in Chemistry Series, Vol. 207. American Chemical Society, Washington, D.C. 1984.

Russell, J.D., Henderson, M.E.K. and Farmer, V.C. (1961) Biochim. Biophys. Acta 52:565-570.

Rydholm, S.A. Pulping Processes. John Wiley and Sons, Inc. New York, 1965.

Saka, S. and Goring, D.A.I. In T. Higuchi (ed) Biosynthesis and Biodegradation of Wood Components. Academic Press, Orlando,1985 pp. 1-62.

Sarkanen, K.V. and Ludwig, C.H. (eds) (1971) Lignins. Occurrence, Formation, Structure and Reactions. Wiley-Interscience, New York.

Sasaki, T., Tanaka, T., Nanbu, N., Sato, Y. and Kainuma, K. (1979) Biotechnol. Bioeng. 21:1031-1042.

Scheffer, T.C. In: Nicholas, D.D. (ed) Wood Deterioration and Its Prevention by Preservative Treatments. Syracuse University Press, Syracuse, 1973, pp. 31-106.

Shei, J.C., Fratzke, A.R. Frederick, M.M., Frederick, J.R. and Reilly, P.J. (1985) Biotechnol. Bioeng. 27:533-538.

Shoemaker, S., Watt, K., Tsitovsky, G. and Cox, R. (1983) Bio/Technology 1:687-690.

Sjöström, E. Wood Chemistry, Fundamentals and Applications, Academic Press, New York, 1981.

Stephen, A.M. In: Aspinall, G.O. (ed) The Polysaccharides, vol 2. Academic Press, New York, 1983, pp. 97-193.

Stone, J.E. and Scallan, A.M. (1968) Pulp Paper Mag. Canad. 69:69-74.

Stone, J.E., Scallan, A.M. and Aberson, G.M.A. (1966) Pulp Paper Mag. Can. 67:263-268.

Stone, J.E., Scallan, A.M., Donefer , E. and Ahlgren, E. In: Hajny, G. and Reese, E.T. (eds) Cellulases and their Applications, American Chemical Society, Washington D.C., 1969, pp. 219-241.

Takenishi, S. and Tsujisaka, Y. (1973) Agr. Biol. Chem. 39:2315

Takahashi, M. and Kutsumi, S. J. (1979) Ferment Technol. 57:434-439.

Tan, L.U.L., Wong, K.K.Y. and Saddler, J.N. (1985) Enzyme Microb. Technol. 7:431-436.

Tien, M. and Kirk, T.K. (1983) Science221:661-663.

Tien, M. and Kirk, T.K. (1984) Proc. Natl. Acad. Sci. U.S.A. 81:2280-2284.

Tien, M., Kirk, T.K., Bull, C. and Fee, J.A. (1986) J. Biol. Chem. 261:1687-1693.

Timell, T.E. (1961) Tappi 44:88-96.

Timell, T. (1967) Wood Sci. Tech. 1:45-70.

Toda, S., Suzuki, H. and Sisizawa, K. (1971) Hakko Kogaku Zashi 49:499-521.

Towle, G.A. and Whistler, R.L. In: Miller, L.P. (ed) Phytochmistry, Vol 1, Van Nostrand Reinhold Company, New York, 1973. pp. 198-248.

van Tilbeurgh, H., Bhikhabhai, R., Pettersson, L.G. and Claeysens, M., (1984) FEBS Lett. 169:215-218.

van Tilbeurgh, H.V. Pettersson, G., Bhikabhai, R., De Boeck, H. and Claeyssens, M. (1985) Eur. J. Biochem. 148:329-334.

Weimer, P.W. and Weston, W.M. (1985) Biotechnol. Bioeng. 27:1540-1547.

Westermark, U. and Eriksson, K.-E. (1974) Acta Chem. Scand. 28:209-214.

Whistler, R.L. and Richards, E.L. In: Pigman, W. and Horton, D. (eds) The Carbohydrates, vol. IIa, 2nd ed. Academic Press, New York, 1970, pp. 447-469.

Whitaker, D.R., Colvin, J.R. and Cook, W.H. (1954) Arch. Biochem. Biophys. 49:247

White, A.R. and Brown, R.M. (1981) Proc. Natl. Acad. Sci. 78:1047-1051.

Wilcox, W.W. In: Nicholas, D.D. (ed) Wood Deterioration and its Prevention by Preservative Treatments. Syracuse University Press, Syracuse, 1973, pp. 107-148.

Wilkie, C.R. (ed) Cellulose as a Chemical and Energy Resource. John Wiley, Interscience, New York, 1975.

Wood, T.M. and McCrae, S.I. (1979) Adv. Chem. 181:181-209.

Wood, T.M., McCrae, S.S., Hoffman, R.M. and Brown, J.A. (1986) In: Proceedings of the Third International Conference on Biotechnology in the Pulp and Paper Industry, Stockholm. pp. 83-86.

Woodward, J. In: Wiseman, A. (ed) Topics in Enzyme and Fermentation Biotechnology, vol. 8, John Wiley and Sons, New York, 1984, pp. 9-30.

Zeikus, J.G. and Ward, J.C. (1974) Science 184:1181-1183.

Chapter 25

Fermentation of cellulose feedstocks

Matti Linko – VTT Biotechnical Laboratory, Tietotie 2, SF-02150 Espoo, Finland.

SUMMARY

Renewable resources, especially cellulosic materials, came into the limelight after the first oil crisis in the seventies. The present trend is a significant decrease in oil prices. However, the efficient use of cellulosic materials is still a relevant goal.

In terms of potential technical applications there are two categories of cellulosic materials. The long fibres of many trees, especially softwood species, are most valuable for the pulp and paper industry, whereas agricultural residues, other non-wood materials or very fast growing trees are less suitable for this purpose. In the former case, biotechnical applications are closely associated with the chemical processing of wood. In these processes the feedstock for fermentation processes is derived entirely from hemicellulose, whereas both the cellulose and the hemicellulose of agricultural residues are potential feedstocks for fermentation.

Sulphite waste liquor has traditionally been used for production of ethanol or single cell protein (feed yeast or Pekilo). In the sulphite process hemicellulose is solubilized and hydrolyzed primarily to monomeric sugars, which serve as carbon sources for the fermentation. Unfortunately sulphite processes have been largely replaced by sulphate (Kraft) processes, in which the sugars derived from hemicellulose are decomposed to isosaccharinic acids and other compounds. This makes the role of biotechnology extremely difficult.

Other problems encountered are inhibitors formed in the chemical processing of wood and the limitations of pentoses as feedstocks for fermentation compared with hexoses. A billion dollar question is: what bulk product could be produced primarily from pentoses without any substantial loss of the organic material in the form of carbon dioxide?

INTRODUCTION

Lignocellulose gives the strength and rigidity to the plants whereas starch serves as a readily accessible sugar storage. It is only natural that man has traditionally used starch for food (and for production of alcohol) and lignocellulosic materials for shelter (and for production of paper).

Cellulose and starch are both polymers of the same sugar, D-glucose. One detail, however, makes a profound difference. The β-1,4-glycosidic bonds in cellulose make the basis for a very strong fibre structure, which is both physically and chemically resistant. The α-1,4 bonds in starch, however, can easily be hydrolyzed with acids or enzymes.

Starch and cellulose can both be broken down all the way to glucose. From this point on, cellulose and starch offer, of course, exactly the same possibilities for various fermentation processes. In other words, cellulosic materials compete with starch. This is a hard fact, not least in Europe with its surplus of cereals. A self-sufficiency level of 125% means cheap starch and cheap glucose (Thiede, 1985). Furthermore, the EEC's sugar supply surplus is more than 40% corresponding to about four million tons of sugar. On the other hand, the wood balance of the EEC shows a self-sufficiency of less than 60%, which means a net import of about 120 million m^3 (Ollman, 1985).

Moreover, cellulose from high-quality wood is more expensive than glucose. This doesn't leave much space for a feasible process based on hydrolysis of cellulose. It is true that some lignocellulosic materials are cheap, but they are anyway much more difficult feedstocks for fermentation than cereals. So why bother? To replace mineral oil with renewable organic resources? Just take a look at the present discount sale of oil, not exactly a sign of exhaustion.

The present situation is not very tempting in terms of fermentation of cellulose feedstocks, especially not for production of cheap bulk products such as ethanol or single-cell protein, which means wasting half of the raw material in the form of carbon dioxide.

All this is rather depressing. Is there any approach leading to economically feasible processes, provided that the research and development is successful?

My answer anyway is yes. The potential of the lignocellulosic materials came into the limelight during the first oil crisis in the seventies. It was realized that the oil resources are limited after all and the access to these resources is uncertain. The huge amount of lignocellulosic materials synthesized every day made them attractive as potential feedstock for bulk chemicals and fuels. In the long run they will be attractive again, in spite of some recent disappointments and the present cheap oil prices. The long term trend of the oil prices is most certainly an increase, simply because the resources will be more and more difficult to get at, under the deep sea or in the middle of nowhere.

My answer is yes primarily because integration of biotechnology in the cellulose industry may lead to interesting applications. Biotechnology may have a role in pulping, bleaching etc. On the other hand, hemicellulose may be a useful source of sugars for biotechnical processes.

RAW MATERIAL

The raw material in this presentation has been defined to be lignocellulose. However, also within lignocellulose there are various possibilities. A basic scheme has been represented in Fig. 1.

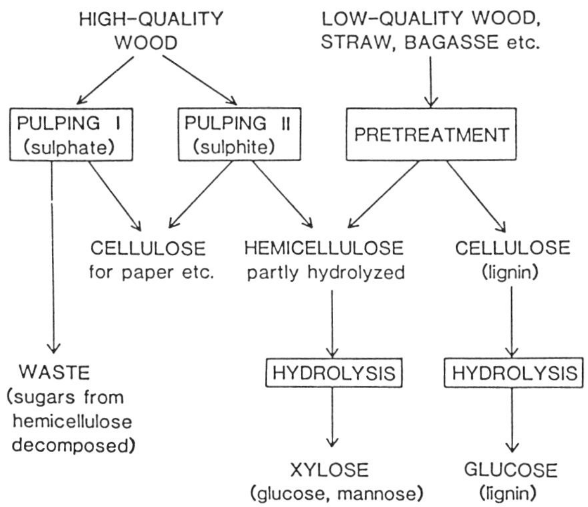

Fig. 1. A simplified scheme of some basic alternatives for utilization of various cellulosic materials (Linko 1983).

Some comments are necessary to clarify and supplement this scheme. To start with, it is more than certain that a high-quality fibre is the most valuable component. This is not necessarily restricted to wood. For example, in Denmark about 30% of the cellulose in paper is derived from cereal straws (Rexen, 1986). We might be approaching the papyrus age again, at least to some extent. However, cellulose of non-wood materials as well as cellulose from some fast growing trees may be potential raw materials for some biotechnical processes whereas cellulose from high quality wood most certainly is not. In fact, at least in Europe there is need for more fibre rather than for more glucose.

At present lignin is primarily removed and burned in the pulp and paper industry. This is useful, because its burning value is high. There is not much to say about fermentation processes utilizing lignin. However, ligninases have been studied quite intensively (Kirk et al., 1986, Eriksson et al., 1986, Leisola et al., 1986). Ligninases could be technically useful only provided that a break-through in their production is achieved by recombinant DNA technology. If ligninases were available with a reasonable price in large quantities, their potential use would be e.g. in biochemical bleaching, i.e. in removal of the lignin residues from the cellulose fibres rather than production of some useful components from lignin (Farrell, 1986, Viikari et al., 1986). Another possible use for ligninases or peroxidases could be oxidation of some harmful waste components (Haemmerli et al., 1986).

In terms of biotechnology integrated with the pulp and paper industry, hemicellulose could be the most important one of the three main components of lignocellulose. Actually this is nothing new. The sulphite processes have been operated this way for quite a while. The sugars derived from hemicellulose have been used for ethanol or SCP, and in fact they are still being used for these purposes in a few remaining sulphite mills, including the Pekilo process in Finland. A new problem at least in the Nordic Countries is the replacement of sulphite by sulphate (Kraft) processes, which destroy the sugars of hemicellulose. The Kraft milling process has been a success in terms of both the cellulose fibre quality and yield. At the time of its development there were no other goals, at least nothing to really pay any attention to. The result is that instead of the monomeric sugars in the sulphite waste liquor there are various organic acids in the black liquor formed in the Kraft process (Alén and Sjöström, 1985). The main component in the black liquor derived from pine is glucoisosaccharinic acid (GISA). As much as 160 kg of GISA per ton of pulp is produced.

The utilization of GISA is a new challenge. The biotechnologists would rather change the pulping process than take this challenge. It is difficult enough to find microorganisms capable of growing on GISA, not to speak of any fermentation process producing a valuable metabolite from GISA (Bailey, 1986). In fact, some people are now ready for a new approach in the pulp and paper industry, taking into account all the main components of wood (Johansson, 1987).

In some cases the whole lignocellulosic material such as cereal straws may be utilized. Enzymes can be used to increase the digestibility of straw and cellulolytic organisms can be cultivated at a relatively low moisture content ("solid state fermentation") to increase the protein content (Tengerdy, 1985, Milstein et al., 1986). The material balance in Europe shows quite significant import of protein for feed. At present SCP is not very tempting, with only a few exceptions. Instead of SCP, protein enrichment of cereal straws is an interesting alternative. Treatment of silage with enzymes also holds promise (Vaisto et al., 1978).

If the starting material is softwood, the sugars derived from hemicellulose are primarily hexoses (glucose, mannose, galactose). However, the major sugar component of many other types of hemicellulose is xylose, which is a much more difficult substrate compared with hexoses, except for a few special products such as xylitol. The enzymatic hydrolysis of xylan is quite complicated because of the side groups in the molecular structure (Biely, 1985, Poutanen et al., 1985, Poutanen et al., 1986, Puls et al., 1986). For cleavage of side groups α-1,2-glucuronidase and acetyl xylan esterase are necessary in addition to endo-1,4-β-xylanase (EC 3.2.1.8) and β-D-xylosidase (EC 3.2.1.37). Moreover, even enzymatic hydrolysis of xylan leads to formation of fermentation inhibitors, e.g. acetic acid. The inhibitory effect of acetic acid is especially strong at a low pH value (Slapack et al., 1983, Faber et al., 1983).

In this presentation I am primarily considering hemicellulose as a raw material source for fermentation processes, because this is relevant in connection with the pulp and paper industry.

PRODUCTS AND PROCESSES

The size of the pulp and paper industry is huge. Really interesting biotechnical products connected with this industry should also have large markets. It may be relatively easy to find out some sophisticated small scale applications, but these would not solve anything essential. For example, there is nothing wrong with xylitol *per se*, it fulfills most of the criteria for a feasible product made of hemicellulose, but its market potential is small compared with the capacity of the cellulose industry.

It should also be kept in mind that it is not always advisable to try to remove all the hemicellulose from the cellulose fibres. Besides adding to the fibre yield hemicellulose may also have an advantageous affect on the strength of the fibres. Oligomers may also offer interesting possibilities. An interesting approach is an "*in situ*" hydrolysis of xylans in pulps in order to modify the fibres (Barnoud *et al.*, 1986). In addition, a challenge could be development of new types of polymers, possibly utilizing a biotechnical process (Esterbauer 1986).

After all these years with intensive research aiming at the biotechnical utilization of "the most abundant renewable organic material" very few products with a large market potential have been suggested. The difficulty in choosing a realistic end product would remain even if cheap hexose sugar could be somehow produced from lignocellulosic materials. An additional problem is to find an end product which would be especially suitable if the feedstock is xylose instead of glucose or mannose. Finally, the real headache comes with the glucoisosaccharinic acid generously offered to the biotechnologists by the Kraft pulping industry.

Studies on production of ethanol from cellulose and hemicellulose have led to a heap of publications - but not to any industrial process, so far. Closest to real life is probably the plant built at the Kyowa Hakko Hofu factory, which produces 200 litres of 95% ethanol per day. The feedstock is rice straw or bagasse, 720 kg/d, saccharified with *Trichoderma* cellulase (Toyama, 1986).

Actually the conventional processes producing ethanol from sulphite waste liquor also mean producing ethanol from hemicellulose. A new possibility in connection with these processes is the fermentation of xylose in addition to hexoses. The organisms capable of fermenting xylose to ethanol include *Pachysolen tannophilus*, *Pichia stipitis*, *Candida shehatae* and *Fusarium oxysporum* (Suihko, 1984, Jeffries, 1985, Slininger *et al.*, 1985, Punnapayak and Emert, 1986, Björling 1986). The tricks tried for improvement of the fermentation of xylose include immobilization (Linko *et al.* 1986), glucose supplementation (Jeffries *et al.*, 1985) and cloning *Schizosaccharomyces pombe* with xylose isomerase gene (Chan *et al.*, 1986).

A major problem in the fermentation of hydrolyzates of lignocellulose is the formation of inhibitors, especially in acid hydrolysis but also in enzymatic hydrolysis, which includes some kind of pretreatment. For example, after steaming at 210°C toxic decomposition products inhibited the fermentation of the hemicellulose fraction of birchwood with *Fusarium oxysporum* completely (Puls et al., 1985).

Mixed bed ion resins have been used to remove inhibitors (Tran and Chambers, 1985). High-level yeast inocula have been investigated to overcome the toxicity of acid hydrolyzate of wood (Chung and Lee, 1985).

European countries are certainly not interested in such gigantic alcohol programmes as Brazil. However, Europe is aiming at leadless petrol, which may mean a role for ethanol as an octane booster. In this respect, higher alcohols such as isopropanol and butanol are in principle even more interesting. The acetone butanol fermentation has a long history, but it has been largely abandoned. Now there seems to be research activity in this field again, *e.g.* in France (Petitdemange & Gay, 1986, Marchal *et al.*, 1986). The organisms involved are various *Clostridium* species.

An interesting product could be 2,3-butanediol (Yu & Saddler, 1985). This compound could also serve as an octane booster. In addition, it could be a chemical feedstock for production of methyl ethyl ketone, butadiene etc. Several advantages to butanediol fermentation by *Klebsiella pneumoniae* have been listed: the organism can use hexoses, pentoses, certain disaccharides such as cellobiose and xylobiose as well as uronic acids and acetate, it grows fast and produces butanediol at near theoretical yields in relatively high final concentrations.

A product with large potential markets is a mixture of calcium and magnesium acetate for deicing roads (Bungay, 1984). This mixture results from neutralizing acetic acid with inexpensive dolomite lime. Calcium chloride commonly used for deicing severely damages not only automobiles but also concrete, asphalt and many other materials such as the coatings of underground cables. The mixture of calcium and magnesium acetates is relatively uncorrosive, so a significantly higher price compared with calcium chloride could be tolerated.

Theoretically, the most advantageous products from xylose in terms of yield are xylitol and xylonic acid. These compounds are also rare examples of products obtained more easily from xylose than from glucose. Several microorganisms are capable of producing xylonic acid by direct oxidation of xylose, catalyzed by xylose dehydrogenase (E.C. 1.1.1.175). For example, Pseudomonas fragi has been shown to produce high yields of xylonic acid, 92% of original xylose (Buchert et al., 1986). It seems hard to find any publications describing the utilization or properties of xylonic acid, but it could be expected that it has similar properties and potential uses as gluconic acid.

While waiting for new pulping processes which would be friendlier to the biotechnologists we have to face the unpleasent black liquor formed in the Kraft process, i.e. we should be able to deal with glucoisosaccharinic acid. At least some microorganisms capable of using this compound for growth have recently been found (Strand et al., 1984, Bailey, 1986). There is a long way to go before developing a feasible process based on the utilization of glucoisosaccharinic acid. However, removal of this compound would increase the combustion value of the black liquor. The microbial biomass produced may also find use e.g. as a nutrient in other biotechnical processes similarly to yeast extracts.

Publications in this field may have promising headlines such as "bioconversion of lignocellulosics into high value products". When looking behind the headline you may find a paper dealing with the breakdown of the polymers to sugars, which can subsequently be fermented to "the desired products". What desired products? The most desired product is anyway a high quality fibre. However, for this also, biotechnology could be really helpful. For example, a specific enzyme capable of breaking down the covalent bonds between lignin and hemicellulose could hit the market (Viikari et al., 1986).

REFERENCES

Alén R and Sjöström E (1985) Degradative conversion of cellulose-containing materials into useful products. In: Nevell and Zeronial (eds), Cellulose chemistry and its applications. Ellis Horwood Ltd, Chichester, UK, p. 531-544.

Bailey MJ (1986) A comparison of bacterial strains growing on glucoisosaccharinic acid. Cellulon Conference, Wales, UK.

Barnoud F, Comtat JP, Joselau P, Mora F and Ruel K (1986) Interest in the enzymatic hydrolysis of xylans for modifying the structure of pulp fibres. Biotechnology in the Pulp and Paper Industry, The Third International Conference, Stockholm, p. 70-73.

Biely P (1985) Microbial xylanolytic systems. Trends in Biotechnology 3, p. 286-290.

Björling T (1986) High yield conversion of pentose sugar to ethanol in spent sulfite liquor. Biotechnology in the Pulp and Paper Inudstry, The Third International Conference, Stockholm, p. 199.

Buchert J, Viikari L, Linko M and Markkanen P (1986) Production of xylonic acid by Pseudomonas fragi. (to be published).

Bungay HR (1984) A timetable for commercializing biomass refining. Third European Congress on Biotechnology, München, Proceedings Vol. IV, p. 93-102.

Chan E-C, Ueng PP and Chen L (1986) D-xylose fermentation to ethanol by Schizosaccharomyces pombe cloned with xylose isomerase gene. Biotechnology Letters 8, No. 4, p. 231-234.

Chung IS and Lee YY (1985) Ethanol fermentation of crude acid hydrolyzate of cellulose using high-level yeast inocula. Biotechnology and Bioengineering 27, p. 308-315.

Eriksson K-E, Ander P and Pettersson B (1986) Regulation of lignin degradation in Phanerochaete chrysosporium. Biotechnology in the Pulp and Paper Industry, The Third International Conference, Stockholm, p. 24-27.

Esterbauer H (1986) New polysaccharide-based microbial polymers. Report on the 2nd Workshop on Lignocellulosic Biomass and its Processing, Stockholm, in press.

Faber MD, Ernst RH and Lefebre PH (1983) Fermentable acid hydrolyzates and fermentation process, Can. Pat. 1 145 277.

Farrell RL (1986) Kraft pulp bleaching with ligninolytic enzymes. Biotechnology in the Pulp and Paper Industry, The Third International Conference, Stockholm, p. 61-63.

Haemmerli SD, Leisola M and Fiechter A (1986) Oxidation of organic pollutants by lignin peroxidases of Phanerochaete chrysosporium. Biotechnology in the Pulp and Paper Industry, The Third International Conference, Stockholm, p. 130-132.

Jeffries TW (1985) Emerging technology for fermenting D-xylose. Trends in Biotechnology 3, p. 208-212.

Jeffries TW, Fady JH and Lightfoot EN (1985) Effect of glucose supplements on the fermentation of xylose by Pachysolen tannophilus. Biotechnology and Bioengineering 27, p. 171-176.

Johansson A (1987) New technologies in pulping. To be presented at the Fourth International Symposium on Wood and Pulping Chemistry, Paris.

Kirk TK, Tien M, Croan SC, McDonagh T and Farrell RL (1986) Production of ligninases by Phanerochaete chrysosporium. Biotechnology in Pulp and Paper Industry, The Third International Conference, Stockholm, p. 5-6.

Leisola M, Troller J, Fiechter A and Linko Y-Y (1986) Production of characterization of ligninolytic enzymes of Phanerochaete chrysosporium. Biotechnology in the Pulp and Paper Industry, The Third International Conference, Stockholm, p. 46-48.

Linko M (1983) Enzymatic hydrolysis - present status and future developments. Kernforschungsanlage Jülich GmbH. Bereitstellung und Verwertung von Lignocellulosen. Status- und Expertenseminar/Workshop, Jülich, p. 317-337.

Linko Y-Y, Kautola H, Uotila S and Linko P (1986) Alcoholic fermentation of D-xylose by immobilized Pichia stipitis yeast. Biotechnology Letters 8, No. 1, p. 47-52.

Marchal R, Ropars M and Vandecasteele JP (1986) Conversion into aceton and butanol of lignocellulosic substrates pretreated by steam explosion. Biotechnology Letters 8, No. 5, p. 365-370.

Milstein O, Vered Y, Sharma A, Gressel J and Flowers MH (1986) Heat and microbial treatments for nutritional upgrading of wheat straw. Biotechnology and Bioengineering 28, p. 381-386.

Ollman H (1985) Wood-balance of the EEC. Report on the 1st Workshop "Agricultural Surpluses", European Federation of Biotechnology, Oberursel, p. 77-84.

Petitdemange H and Gay R (1986) Biochemical mechanism of the aceton butanol fermentation and future prospects. Biotechnology in the Pulp and Paper Industry, The Third International Conference, Stockholm, p. 110-113.

Poutanen K, Puls J and Linko M (1986) The hydrolysis of steamed birchwood hemicellulose by enzymes produced by Trichoderma reesei and Aspergillus awamori. Applied Microbiology and Biotechnology 23, p. 487-490.

Poutanen K, Puls J, Viikari L and Linko M (1985) The role of side-chains of xylan in its biotechnical utilization. International Symposium on Wood and Pulping Chemistry, Vancouver, British Columbia, Canada. Technical papers, p. 205-206.

Puls J, Poutanen K, Körner H-U and Viikari L (1985) Biotechnical utilization of wood carbohydrates after steaming pretreatment. Applied Microbiology and Biotechnology 22, p. 416-423.

Puls J, Poutanen K, Schmidt O and Linko M (1986) Biotechnology in the Pulp and Paper Industry, The Third International Conference, Stockholm, p. 93-95.

Punnapayak H and Emert GH (1986) Use of *Pachysolen tannophilus* in simultaneous saccharification and fermentation of lignocellulosics. Biotechnology Letters 8, No. 1, p. 63-66.

Rexen F (1985) Increased production of straw pulp and paper in the EEC - possibility and constrains. Report on the 1st Workshop "Agricultural Surpluses", European Federation of Biotechnology, Oberursel, p. 212-218.

Slapack G, Edey E, Mahmourides G and Schneider H (1983) Ethanol from hexoses and pentoses in spent sulfite liquor. Proc. Biotechnology in the Pulp and Paper Industy, London, p. 359-381.

Slininger PJ, Bothast RJ, Okos MR and Ladisch MR (1985) Comparative evaluation of ethanol production by xylose-fermenting yeasts presented high xylose concentrations. Biotechnology Letters 7, No. 6, p. 431-436.

Strand SE, Dykes J and Chiang V (1984) Aerobic microbial degradation of glucoisosaccharinic acid. Applied and Environmental Microbiology 47, p. 268-271.

Suihko M-L (1984) D-xylose fermentation by *Fusarium oxysporum* and other fungi. VTT, Technical Research Centre of Finland, publications 17.

Tengerdy RP (1985) Solid substrate fermentation. Trends in Biotechnology 3, No. 4, p. 96-99.

Thiede G (1985) Basic statistic figures for the judgement of the agricultural surpluses within the EEC. Report on the 1st Workshop "Agricultural Surpluses", European Federation of Biotechnology, Oberursel, p. 24-76.

Toyama N (1986) Personal communication.

Tran AV and Chambers RP (1985) Red oak wood derived inhibitors in the ethanol fermentation of xylose by *Pichia stipitis* CBS 5776. Biotechnology Letters 7, No. 11, p. 841-846.

Vaisto T, Heikonen M, Kreula M and Linko M (1978) The use of cellulases for increasing the sugar content of AIV silage. Journal of the Scientific Agricultural Society in Finland 50, p. 392-397.

Viikari L, Ranua M, Kantelinen A, Sundquist J and Linko M (1986) Bleaching with enzymes. Biotechnology in the Pulp and Paper Industry, The Third International Conference, Stockholm, p. 67-69.

Yu EKC and Saddler JN (1985) Biomass concersion to butanediol by simultaneous saccharification and fermentation. Trends in Biotechnology 3, No. 4, p. 100-104.

Chapter 26

Investigations into the production of enzymes from straw by *Aspergillus Fumigatus IMI1255091*

B. Bonakdarpour, S. W. Kim, D. A. John Wase and W. J. McManamey – Department of Chemical Engineering, University of Birmingham, P.O. Box 363, Edgbaston, Birmingham, B15 2TT, U.K.

INTRODUCTION

Of the relatively few organisms producing exocellulase in quantity, *Trichoderma reesei* has been almost exclusively favoured, and subjected to successive mutations, giving higher and higher yields of relatively pure exocellulase [1]. However, many agricultural materials such as hay and straw contain high levels of such materials as hemicellulose and lignin which need to be separated one from another; it is arguable whether pure crystalline cellulase is appropriate for this task. *Trichoderma* is, therefore, possibly not the best source of enzyme for all applications, and we consequently sought an alternative organism with a wide spectrum of activities and whose natural habitat was hay and straw. One isolate, a strain of *Aspergillus fumigatus*, was outstanding; a superior mutant of this is the strain currently used in this study.

Initial work in shake-flasks [2] rapidly reached oxygen limitation, and attention was turned to fermenters. The standard fermenter is a vertical tank with a central shaft to which are attached one or more disc-turbine impellers. These serve both as pumps, to circulate and mix the contents of the vessel, and as shear devices, shearing the bubbles of air (supplied through the sparge) to provide as large an interfacial area for oxygen transfer as possible. Recently, other fermenter configurations, such as the air-lift fermenter, have begun to find application in such areas as production of animal and plant cells, (both of which are shear-sensitive), because the air-lift fermenter is capable of promoting good oxygen transfer and mixing without the local regions of high shear associated with turbine-agitated vessels. It has also been applied in large-scale continuous fermentations, such as single-cell protein or poly-β-hydroxy butyrate.

The recent popularity of the air-lift fermenter is fortunate, for tests showed that conventional disc-turbine agitated fermenters were not very suitable for the A.fumigatus cellulase fermentation. As agitation rate in batch culture was increased, so enzyme yields fell [3]. The enzyme was exceptionally shear-resistant, and it was concluded that the mycelium was damaged, liberating intracellular contents. Similar results have been obtained for Chaetomium cellulolyticum[4]. Use of an air-lift fermenter in our work has avoided these effects, furnishing yields up to 20% higher than conventional vessels; maximum yields were at conditions giving maximum gas hold-up, and hence maximum volumetric oxygen transfer coefficient, (k_La).

Ground straw circulated well in the air-lift fermenter, along with mycelium. However, grinding is energy intensive and time consuming, particularly on scale-up. It was therefore of interest to see what further factors arose through use of chopped rather than ground straw. It was anticipated that there could be problems in circulation, and perhaps in enzymic accessibility, (which grinding is well-known to increase in lignocellulosics). Finally it was expected that air hold-up would be influenced. Some experiments were therefore done to clarify the situation, and are reported in this paper.

EQUIPMENT, MATERIALS AND METHODS

Fermenter The 11-litre air-lift fermenter is described elsewhere by McManamey et al [5]. To aid straw circulation, a wire gauze baffle was suspended at the top of the riser [6].

Organism, medium and growth conditions The cellulolytic organism was A.fumigatus IMI 255091. The medium and growth conditions are given elsewhere [7].

Enzyme assays Culture broth was prepared as described elsewhere [7]. Endoglucanase was measured using the method of Saddler [8]. β-Glucosidase was estimated by a modification by Yap and Wang [9] of Norkrans' method [10] involving p-nitrophenyl β-D-glucopyranoside as substrate. β-Xylosidase was measured by an analogous method which employed p-nitrophenyl β-D-xylopyranoside [11]. Xylanase assays involved adding 0.5ml diluted sample to 1ml of a 2% (w/v) suspension of xylan in 0.05M citrate buffer, pH 4.8 and incubating at 50°C for 20 min before measuring reducing sugar by Nelson's method [12]. Cotton activity was assayed by adding 1.0ml diluted sample and 0.5ml 0.05M citrate buffer, pH 4.8 to 50mg absorbent cotton and incubating at 50°C for 24h. Reducing sugars produced were measured by the dinitrosalicylic acid method [13]. Extracellular protein was estimated by the Folin-Lowry method [14].

Straw preparation Fresh, clean straw, harvested summer 1985 was dried (100°C, 48h) and either passed through a hammer-mill and sieved [7] or chopped into precise lengths with an office paper guillotine equipped with a metal stop. For preparation of aqueous suspensions for investigation of circulation or hold-up, the straw was either soaked for several days or sterilised in a steam autoclave at 15 psig for 20 min, since the wax coating on fresh straw makes it very difficult to wet; unless treated, it floats on the water.

Physical measurements in the air-lift fermenter Volumetric gas hold-up, ε, was defined as the ratio of the volume of gas in an aerated suspension, V, to the total volume of the aerated suspension, V_A; i.e. $\varepsilon = (V_A - V)/V_A$. Superficial gas velocity is the volumetric gas flow rate up the riser, divided by the cross-sectional area of the riser.

EXPERIMENTAL

Circulation of straw, and gas hold-up Initially, 40g batches, each containing a specific length of pretreated straw, were suspended in 11 litres of water in the air-lift fermenter. Although all joints had been carefully aligned and gaskets cut precisely and smoothed, the maximum length which would circulate satisfactorily was only 1cm. Next, the effect of straw concentration was tested. Quantities of 1cm straw up to 50g circulated without problems in 11 litres of water. However, when the quantity was raised to 90g, the fermenter blocked as soon as the air supply was interrupted, and was very difficult to restart. For higher concentrations, it was necessary to add the additional straw whilst the suspension was circulating, and even then the fermenter eventually blocked, usually at the top of the riser or downcomer. Nevertheless, as much as 170g could be circulated for a few hours, even at low aeration rates. The period during which circulation was satisfactory increased with increasing aeration rate until, at the highest convenient air-flow rate, (9 litres min^{-1}), it was 15h. Even 170g was poor in comparison with the 2% ground straw used previously [7], which had supported good growth and enzyme production.

Gas hold-up was measured with water alone and with quantities of straw from 10 to 170g in 11 litres water. Fig. 1 shows the relationship between gas hold-up and superficial gas velocity for the different systems. (In non-agitated gas-sparged equipment, the gas hold-up is the main factor which determines the volumetric mass-transfer coefficient, (k_La), which, in turn, is the principal influence on the maximum enzyme concentration obtainable in the fermenter). At all superficial gas velocities, the addition of straw increased the hold-up above that for water alone, although there is no clear relationship between hold-up and straw concentration because of the scatter of results. However, increases in hold-up up to 25% greater than those in water were found and this improvement in hold-up is likely to produce a significant increase in enzyme titre, provided that the oxygen transfer rate is suboptimal, as is likely.

Influence of straw length on the fermentation Lengths tested ranged from ground straw up to 1.5cm pieces; these are somewhat

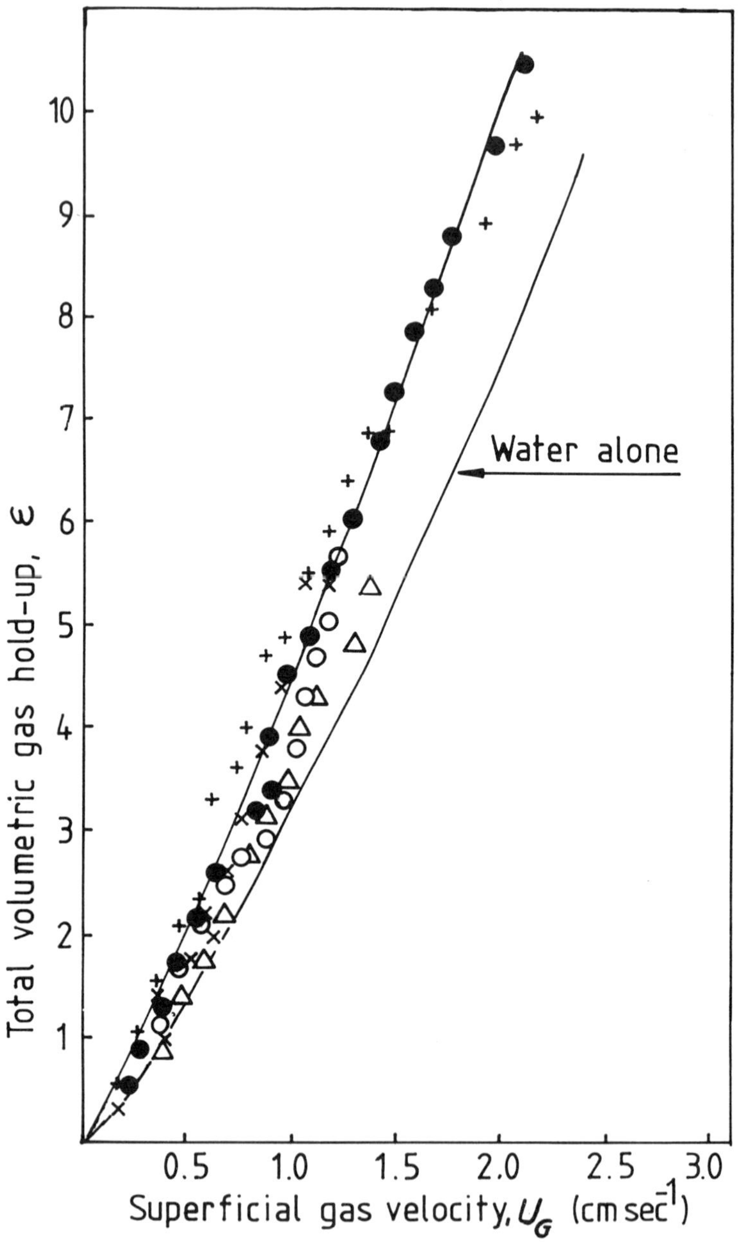

Fig. 1. Total gas hold-up plotted against superficial gas velocity for various concentrations of straw in 11 litres of water.
10g,0: 50g,×; 90g,△; 130g,+; 170g,●.
(Straw length 1 cm)

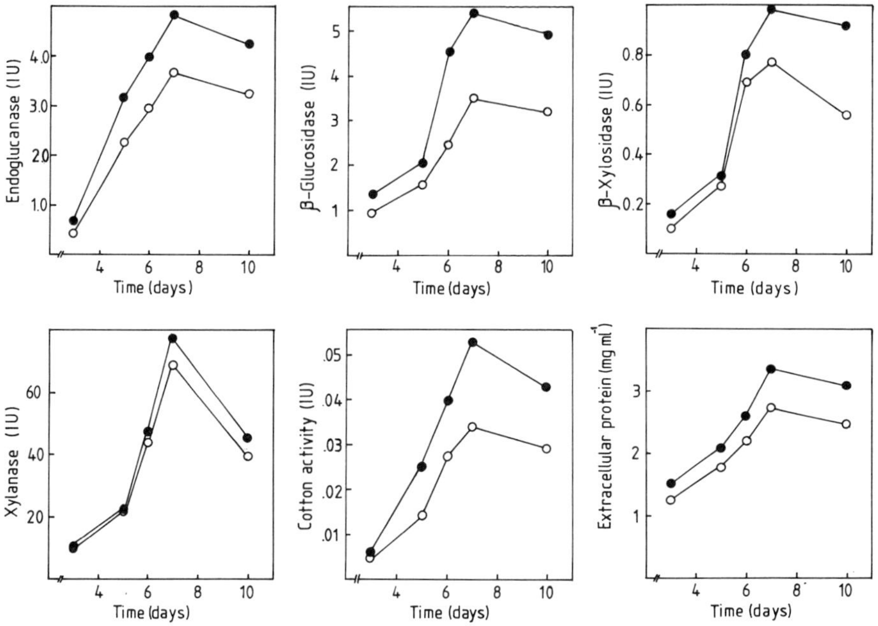

Fig. 2. Enzyme titres produced by *A. fumigatus* on ground straw (•) and on 1.5 cm straw lengths (0). Straw concentration 0.5% (s

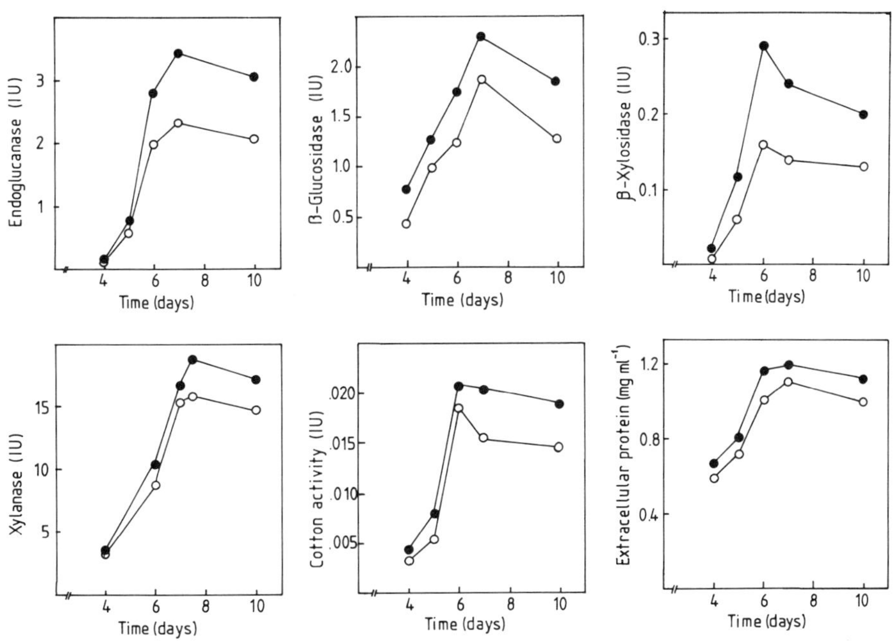

Fig. 3. Enzyme titres produced by *A. fumigatus* on ground straw (•) and on 1.5 cm straw lengths (0). Straw concentration 0.5% (w/v).

smaller than those from a commercial chaff cutter, (c.2-5cm), but anything larger blocks our relatively small air-lift fermenter. Results for 4% w/v straw (Fig.2) show clearly that, in general, production of a range of enzymes peaked at day 7, and that straw length affected the time taken to reach this maximum only marginally. Similar results were obtained for a range of straw concentrations down to 0.5% w/v and even here the differences were only marginal; all the tests peaked at day 7 except for β-xylosidase and cotton activities which peaked on day 6, (Fig.3).

Whilst the time at which maximum activity was found was not seriously affected either by the amount of straw added or the length of the pieces, these parameters exerted considerable influence on the yields. Thus, figure 4 displays β-glucosidase yield on day 7 plotted against straw length, using 4% w/v straw. Around 50% more enzyme was produced for ground straw than for 1.5cm lengths. However, this effect varied from enzyme to enzyme: inspection of figure 5 shows that it was, for instance, much less marked with xylanase.

Finally, it is evident that the yield is also affected by concentration of straw used; as this was increased to 4% w/v, so the amount obtained also increased (Fig.6).

DISCUSSION

Straw surprisingly, is not a cheap substrate. A current price of about £20 per tonne (including dross, unfermentables and up to 15% moisture) can be compared with £100 per tonne for dry, prime sugar outside the EEC and £225 within.+ Nevertheless, there are good reasons for using it: it is plentiful; it can induce a very wide range of enzymes; and sugar will not always be cheap. In our attempts to use straw, we highlighted a number of constraints; for example, the cost of grinding versus the cheaper chopping; the penalties in yield and increase in circulation problems as length increases; the fact that commercial chaff-cutters usually output a range of lengths between 2 and 5 cm which block our vessel.

Rough estimations can be made from our limited data. Commercial fermenters would be much bigger than our 11 litre vessel; if 1.5cm lengths of straw block a 5cm column, 5cm lengths would require a very feasible 20cm diameter. Longer lengths reduce yields. However for 1cm lengths, even the worst cases (e.g. β-glucosidase) give 50% of the yield with ground straw. Other enzymes (e.g. xylanase) are much more marginally affected (Figs. 1, 2 and 5), (probably reflecting differences in substrate accessibility). Moreover the effect of straw length on yield is not linear, (Fig. 4); scale-up involving larger lengths could well carry little further penalty. Thus, for commercial chaff and the worst possible

+Based on Summer 1986 prices; (European Industrial sugar £375 less £150 rebate).

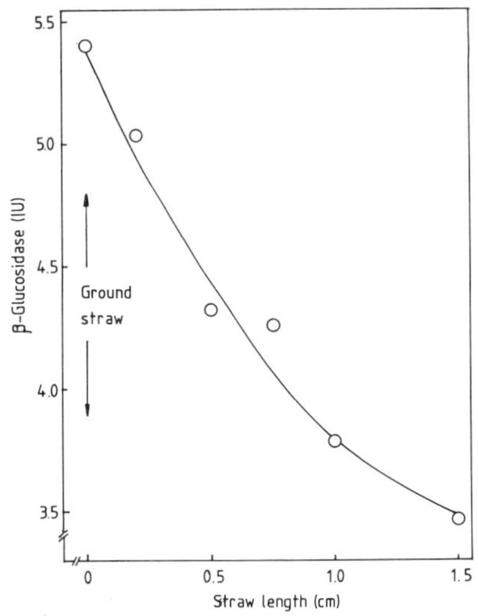

Fig. 4. Effect of straw length on β-glucosidase yield (day 7; 4% (w/v) straw).

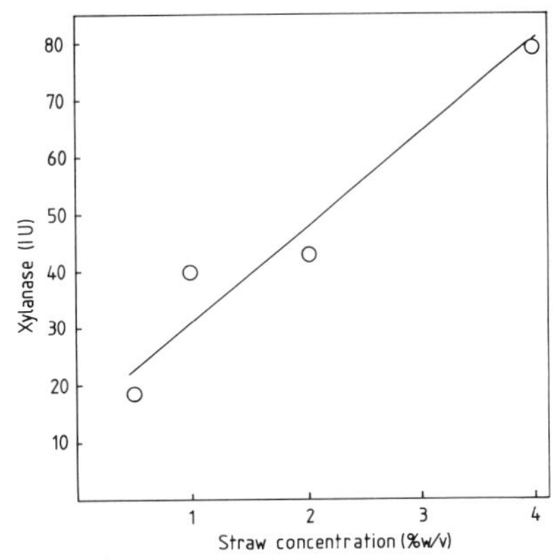

Fig. 6. Yields of xylanase produced by *A. fumigatus* on various concentrations of ground straw after 7 days.

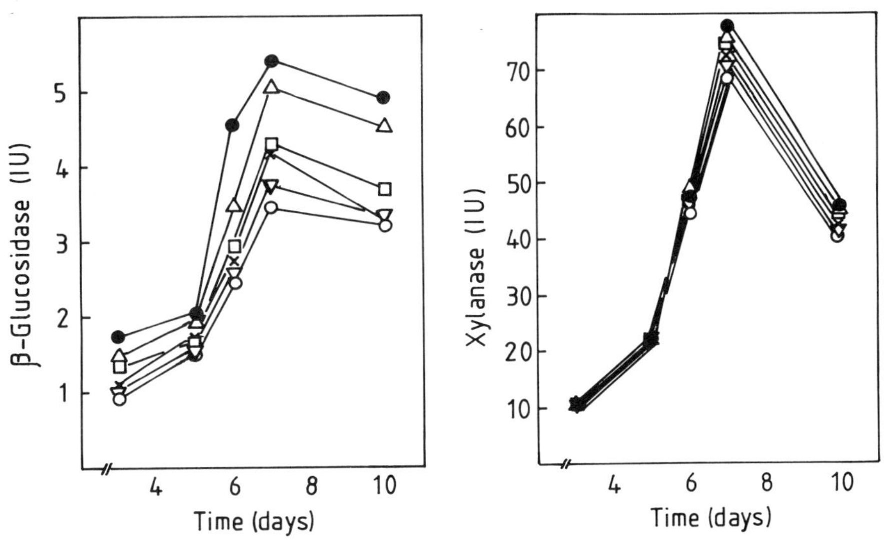

Fig. 5. Comparison of *A. fumigatus* fermentations on various lengths of wheat straw for β-glucosidase and xylanase. Ground straw, ●; 0.2 cm straw, △; 0.5 cm straw, □; 0.75 cm straw, x; 1.0 cm straw, ▽; 1.5 cm straw, 0. Straw concentration 4.0% (w/v).

case, these conditions might well give yields around 30-40% those of ground straw, with the average of all enzymes much better. This figure, moreover, neglects compensating factors. Scale-up is likely to permit higher straw concentrations; at least 4% could well circulate and concentration appears much more important than length; indeed Figure 6 suggests straw availability is the overall limitation. Higher straw concentrations suggest larger increases in hold-up, offering potential improvements in oxygen transfer, and hence yield. (Even for the 11 litre fermenter, hold-up with 1.0cm straw lengths was 20% greater than with ground straw, (Fig. 1)).

It is therefore evident that the outlook for scaling up this process to a pilot scale (at least a 20cm diameter column) offers a fair degree of success. Our crude estimates suggest final yields at least between 50 and 100% of those in shake flasks, and possibly very much higher. Further experiments are now in progress to see if our predictions are justified.

ACKNOWLEDGEMENTS

S.W.K. is grateful for a conference grant from the Society of Chemical Industry.

REFERENCES

1. Wase, D.A.J. and Vaid, A.K. (1983), Process Biochem. 18(6), 35.

2. Wase, D.A.J., Vaid, A.K. and McDermott, C. (1985), Enzyme Microb. Technol. 7, 134.

3. Wase, D.A.J., McManamey, .WJ., Raymahasay, S. and Vaid, A.K. (1985), Biotechnol. Bioeng. 27, 1166.

4. Hecht, V., Schürgerl, K. and Scheiding, W. (1983), J. Chem. Technol. Biotechnol. 33B, 231.

5. McManamey, W.J., Wase, D.A.J., Raymahasay, S. and Thayanithy, K. (1984), J.Chem.Technol.Biotechnol. 34B, 151.

6. Bonakdarpour, B. (1985), M.Sc.project report, Univ.of Birmingham.

7. Wase, D.A.J., Raymahasay, S. and Wang, C.W. (1985), Enzyme Microb. Technol. 7, 225.

8. Saddler, J.N. (1982), Enzyme Microb. Technol. 4, 414.

9. Yap, W.H. and Wang, C.W. Proc. Malaysian Biochem. Soc. Conf. 1981, 7, 16-25.

10. Norkrans, B. (1957), Physiol. Plant, 10, 198.

11. Margaritis, A. and Merchant, R. (1983), Biotechnol. Bioeng. Symp. No.13, 299.

12. Nelson, N.J. Biol.Chem. 1944, 153, 375.

13. Miller, G.L. (1959), Anal. Chem. 31, 426.

14. Lowry, O.H., Rosebrough, N.J., Farr, A.L. and Randall R.J. (1951), J. Biol. Chem. 193, 265.

Chapter 27

Bio-utilization of pineapple waste for ethanol generation

H. K. Tewari, S. S. Marwaha and K. Rupal – Department of Microbiology, Punjab Agricultural University, Ludhiana(Pb.) 141004, India.

J. F. Kennedy – Research Laboratory for the Chemistry of Bioactive Carbohydrates and Proteins, Department of Chemistry, University of Birmingham, P.O. Box 363, Birmingham B15 2TT, England.

ABSTRACT

This paper evaluates the potential of pineapple peels as a substrate for ethanol production. The comparative investigation carried out with sulphuric acid, steam and cellulase enzyme complex (from *Trichoderma reesei* QM 9414) demonstrated sulphuric acid at 2.5% and 15 psi pressure for 15 min to be an efficient saccharifying agent for the hydrolysis of the substrate. Both higher concentrations of sulphuric acid (>2.5% v/v) and treatment durations (>15 min) had inhibitory effects on the hydrolysis of the cellulosic substrate. The maximum saccharification and consequent reducing sugar formation under optimal treatment conditions were 80% and 544 mg/g of pineapple peels respectively. Of the acid and enzymatically hydrolysed mash, the latter supported better ethanol fermentation by *Saccharomyces cerevisiae* var. ellipsoideus. A maximum yield of 57.2 ml ethanol/100 g of reducing sugars was obtained for cellulose hydrolysed substrate.

INTRODUCTION

Biotechnological ethanol production from renewable sources has attracted the attention of biotechnologists in recent years because of sharp increases in the prices and expected shortfalls in the supply of the fossil fuels. A literature survey reveals that during the last decade many substrates including industrial, agricultural lignocellulosic wastes have been evaluated for their potential in ethanol generation. Pineapple peels (constituted mainly of cellulose and starch), a waste product of agrobased industry (pineapple canning industry), appear to be an ideal substrate (following suitable saccharification treatment) for the production of both fuels and potable ethanol via fermentation [1]. Of the total world production of pineapples (8,752 thousand metric tonnes) [2], 40-45% is accounted for by cannary waste (peels, cores and pomace). This waste has high biochemical oxygen demand (BOD) and chemical oxygen demand (COD) values [3]. Unless either properly utilized or disposed of, it may cause serious environmental pollution problems leading to health hazards. Consequently, significant attention has been paid in the past few years for evaluating the potentials of pineapple wastes for the production of valuable fermentative and non-fermentative products.

Most of the previous work in which pineapple cannary waste was utilized as substrate was for the production of bromelain, vinegar, wine, feed yeast, organic acids etc [4]. The present investigation is a continuation of our work on the utilization of industrial and agricultural cellulosics for ethanol generation. The saccharification of the pineapple peels has been carried out comparatively using sulphuric acid, steam and cellulase enzyme (from *Trichoderma reesei* QM 9414) to find out an efficient saccharifying agent for the optimal conversion of the substrate to fermentable sugars. Moreover the saccharified mashes from acid and enzymatic treatment were evaluated for their abilities to support ethanol production by *Saccharomyces cerevisiae* var. ellipsoideus.

MATERIALS AND METHODS

Locally collected pineapple peels from a local market were chopped into small pieces and oven dried at 65°C for 48 h. After powdering to size 40 mesh, the material was packed in polyethylene bags and stored at room temperature.

MAINTENANCE AND CULTIVATION OF ORGANISMS

Trichoderma Reesei: *T.reesei* QM 9414, the fungus for cellulase production, was subcultured regularly on potato dextrose agar (PDA) medium pH 5.0 containing glucose (1%), potato extract filtrate (250 g of chopped peeled potatoes boiled and filtered) and agar (2%) at fortnightly intervals. To prepare fungal inoculum, a medium (pH 5.0) containing glucose (1%), yeast extract (0.2%) and peptone (0.5%), inoculated with macerated mycelium (1% w/v) was incubated at 28 ± 1°C for 3 days.

Saccharomyces Cerevisiae: *S.cerevisiae* var. ellipsoideus , was maintained on glucose yeast extract (GYE) medium pH 5.0 containing glucose (1%) yeast extract (0.5%), peptone (0.5%) and agar (2%) and subcultured at 15 days intervals. The inoculum was prepared in 100 ml of GYE medium, by transferring a loopful of yeast cells from a 24 h actively growing culture. The culture was incubated at 28 ± 1°C for 24 h.

SACCHARIFICATION OF PINEAPPLE PEELS:

Acidic Saccharification: Pineapple peels (1.0 g) were hydrolysed at 10 and 15 psi for 15, 30, 45 and 60 min using sulphuric acid (8 ml) at 2.5–15.0 % v/v. The hydrolysates were cooled to room temperature, diluted to 200 ml, and neutralized with 5% w/v sodium hydroxide.

Enzyme Saccharification: For cellulase production by T.reesei QM 9414, medium [5] (50 ml) containing cellulose (1%) sterilized at 15 psi for 20 min and inoculated with T.reesei culture (2 ml) was incubated at 28 ± 1°C for 8 days. The product was filtered through glass wool and the filtrate freeze dried.

Culture filtrate (15 ml, reconstituted if necessary) and 50 ml 0.1M citrate buffer, pH 4.8 (50 ml) was added to powdered substrate (5 g) and incubated at 50 ± 1°C for 72 h.

Steam Saccharification: Powdered peel (1.0 g) in distilled water (8 ml) was saccharified by steaming for 15, 30, 45 and 60 min in conical flasks.

ETHANOL PRODUCTION

Saccharified and sterilized mash of pineapple peel (150 ml) were inoculated with 3% v/v yeast inoculum and incubated at 30 ± 0.5°C for 48 h.

ANALYSIS

Measurement of Cellulase Activity: Cellulase activities were measured as filter paper activity (FPA) [6], carboxymethyl cellulase activity (CMCase) [6] and cellobiase [7] activity, activities being expressed as international units (IU). A unit of cellulase enzyme was defined as the amount of enzyme required to liberate reducing sugars equivalent to 1 umole/min/g of substrate.

Carboxymethyl Cellulase: Carboxymethyl cellulose (1 g) was dissolved in 0.1M citrate buffer pH 4.8 (90 ml). Enzyme sample (0.5 ml) was added to this solution (0.5 ml) and the mixture was maintained at 50°C for 30 min [6].

Filter Paper Cellulase: Enzyme solution (0.5 ml), 0.1M citrate buffer pH 4.8 (10 ml) and a filter paper strip were incubated together at 50°C for 1 h [6].

Cellobiase: Cellobiose (50 mg) was dissolved in 0.1M acetate buffer pH 5.0 (100 ml). Enzyme solution (0.5 ml) was added to this solution (0.5 ml) and the resultant solution was incubated at 40°C for 2 h [7].

Reducing Sugars: Reducing sugar contents of the powdered peels and saccharified substrate were estimated using 3,5-dinitrosalicylic acid (DNS) [8].

Cellulose: The cellulose content of the substrate was estimated using acetic, nitric and anthrone reagents and 67% v/v sulphuric acid [9].

Starch: The starch content of the substrate was determined using perchloric acid and anthrone reagent [10].

Ethanol: The ethanol content of the fermented mash of the pineapple peels was estimated by using steam distillation of alcohol in a microkjeldahl apparatus [11].

RESULTS AND DISCUSSION

The pineapple peels saccharified using sulphuric acid, cellulase enzyme and steaming contained 41, 162 and 460 mg/g peels of reducing sugars, starch and cellulose respectively.

ACIDIC SACCHARIFICATION

The effect of sulphuric acid (2.5-15.0 % v/v) at 10 and 15 psi pressure on the hydrolysis of substrate is shown in Figs. 1 and 2. In the control samples (samples subjected to autoclaving without the addition of acid) a uniform increase

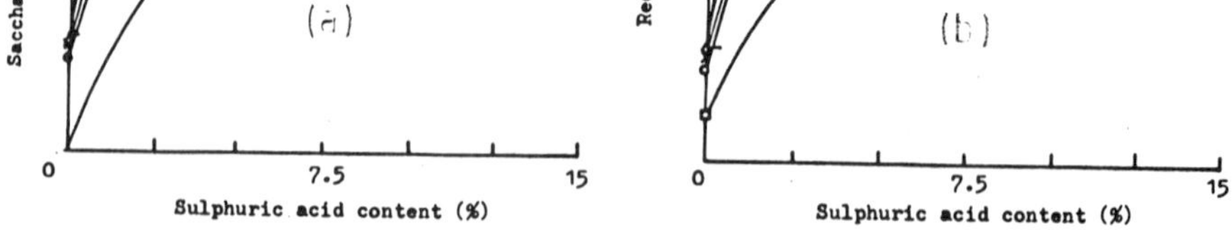

Fig. 1. Influence of sulphuric acid concentration on saccharification of pineapple peels at 10 psi pressure for various time intervals: (a) Saccharification; (b) Reducing Sugars. □, 0 min; ○, 15 min; X 30 min; △, 45 min; ●, 60 min.

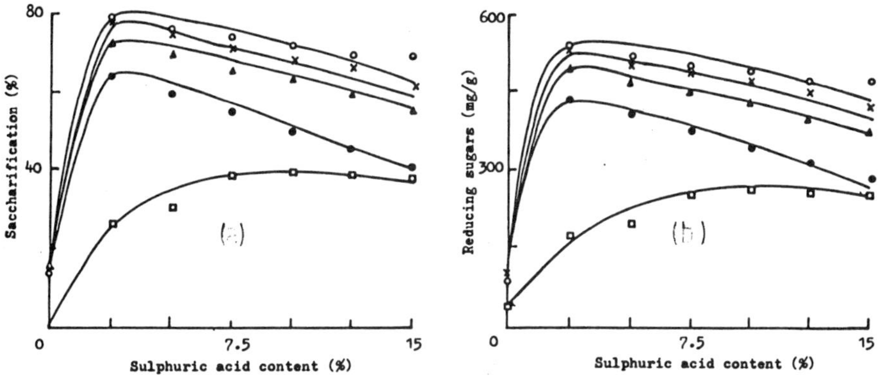

Fig. 2. Influence of sulphuric acid concentration on saccharification of pineapple peels at 15 psi pressure for various time intervals: (a) Saccharification; (b) Reducing Sugars. □, 0 min; ○, 15 min; X, 30 min; △, 45 min; ●, 60 min.

in the saccharification and reducing sugars level was observed with an increase in the treatment duration up to 45 min, both at 10 and 15 psi pressures. Thereafter, a decrease in both functions was recorded. The maximum non-acid control saccharification and reducing sugar values recorded at 10 and 15 psi were 14.7 and 15.8% and 100.0 and 107.5 mg/g peels respectively. Moreover, in zero time acid control saccharification treatment (0 min treatment with sulphuric acid, i.e. addition of acid and immediate neutralization without autoclaving) an increase in the saccharification (26.0-37.5%) and reducing sugars level (41.0-255.0 mg/g) with increasing concntrations of acid (2.5-15.0 % v/v) was recorded.

At 10 psi pressure and various concentrations of sulphuric acid, the saccharification and reducing sugar contents ranged from 26.0-54.8% and 176.0-373 mg/g peels respectively. The maximum hydrolysis (54.8%) was attained at 2.5% v/v sulphuric acid with autoclaving for 15 min. However, similar treatment conditions (2.5% v/v sulphuric acid for 15 min) at psi resulted in a much higher degree of saccharification (80.0%) and consequent reducing sugars liberation (544.0 mg/g) than at 10 psi. Further it was observed that the values of acid >2.5% v/v and treatment duration >15 min higher than the optimal had an adverse effect on the acid hydrolysis of the substrate, both at 10 and 15 psi pressures. This may be attributed to the break down of reducing sugars and subsequent formation of furfurals [12]. These results are similar to our earlier findings on the saccharification of banana peel waste [13]. However variable levels of acid and treatment duration have been reported for the effective hydrolysis of sunflower seed husk [14], rejected

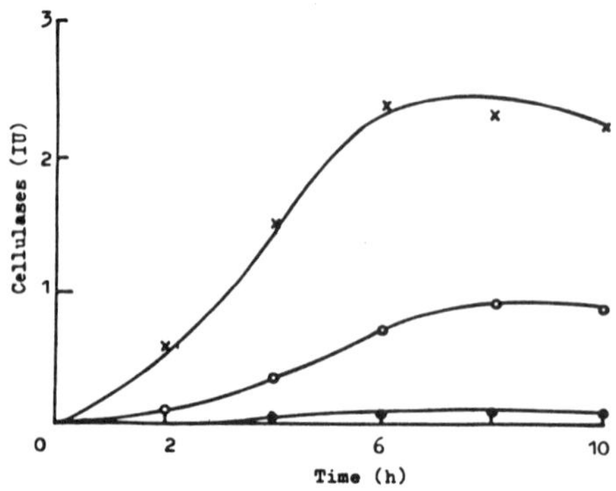

Fig. 3. Time course of cellulases synthesis by *Trichoderma reesei* QM 9414: ×, carboxymethyl cellulase (CMCase); ○, filter paper activity (FPA); ●, cellobiase.

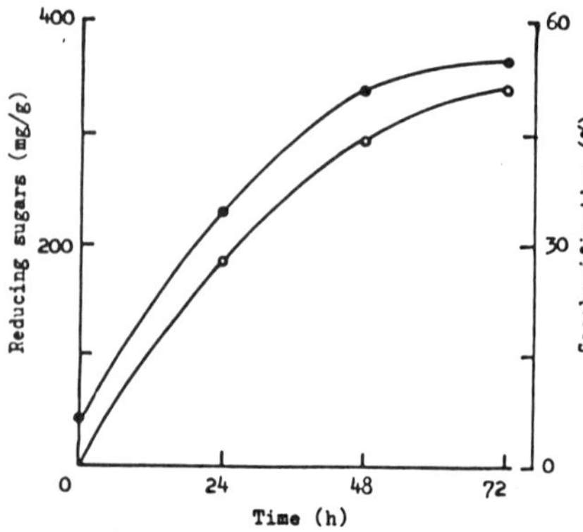

Fig. 4. Time course of enzymic saccharification of pineapple peels with cellulases of *Trichoderma reesei* QM 9414: ○, saccharification (%); ●, reducing sugars (mg/g).

potatoes [15], vegetable wastes [16] and banana peels [17].

ENZYMIC SACCHARIFICATION

The time course studies of cellulase production (Fig 3) revealed that Trichederma reesei QM 9414 synthesized maximum 2.40, 0.96 and 0.11 units/ml substrate of CMCase, FPA and cellobiase at the completion of 6, 8 and 8 days of incubation respectively. Thereafter a decrease in all the cellulase activities was recorded. For the enzymatic saccharification of pineapple peels a uniform increase in the hydrolytic activity with an increase in the treatment period (24-72h) was observed (Fig. 4). The maximum saccharification (57.3%) and reducing sugars (327 mg/g) attained at the completion of 72 h of treatment were higher than that achieved in our earlier investigation on banana peels and vegetable wastes [13,16]. The variability in the susceptibility of the substrates to enzymatic attack may be due to their different cellulose contents and structures [18].

STEAM SACCHARIFICATION

The maximum saccharification (15.7%) with steam was achieved after 45 min of steaming (Table 1) and was similar to the control treatment (autoclaving of the substrate without the addition of acid) at 15 psi pressure (Fig. 2). However, the poorer efficiency of steaming than the control treatment (autoclaving of the substrate without the addition of acid) at 10 psi pressure (Fig. 1) was in line with the observation made

Table 1. Effect of steaming on saccharification of pineapple peels[a].

Time (min)	Reducing[b] sugars (mg/g)	Saccharification[c] (%)	
		Wt. basis	CH_2O basis
0	41	–	–
15	99	9.3	14.3
30	105	9.8	15.4
45	107	10.0	15.7
60	107	10.0	15.7

a Reaction mixture = 1.0 g substrate, 8 ml distilled water
b Reducing sugars (mg/g) = Total reducing sugars - initial reducing sugars.
c Saccharification %: Wt. basis, total weight reducing sugars; CH_2O basis, total starch + cellulose.

Table 2. Production of ethanol from saccharified mash of pineapple peels[a].

Observations	Saccharifying agent	
	2.5% v/v Sulphuric acid	Cellulase[b]
Reducing sugars (g/100 ml)	5.50	1.83
Alcohol (% v/v)		
24 h	2.36	1.05
48 h	2.50	1.00
Alcohol yield (ml/100 g reducing sugars)		
24 h	42.9	57.2
48 h	45.4	54.0

a Fermentation period: 24 and 48 h. Temperature 30°C pH: 4.5
b Cellulase crude filtrate of *T.reesei* QM 9414.

for other cellulosic substrates by our group [13] and others [19].

COMPARISON OF ACID AND ENZYMATIC HYDROLYSATES FOR ALCOHOL PRODUCTION

The comparative suitabilities of acid and enzymatic hydrolysates of pineapple peels for ethanol production by *Saccharomyces cerevisiae* var. ellipsoideus (Table 2) show that the enzymatic hydrolysate supported better ethanol generation (on the basis of reducing sugars conversion), than the acid treated samples. The low alcohol yield in the acid hydrolysed mash may be attributed to the inhibitory effect of the by-products (formed during the hydrolysis reactions) on the activity of yeasts. The formation of furfural, 5-hydroxyfurfural, vanillin and excess of sulphate ions have all been found to suppress the growth of yeast culture and consequently their fermentation ability [20,21].

CONCLUSIONS

The results of the present investigation suggest that the acid hydrolysis of pineapple peels cellulose is better than steam and enzymatic saccharification. However, of the acid enzymatic hydrolysates, the latter supported better ethanol production than the former. The yield of 187 ml alcohol/kg of pineapple peels is 0.25–40 fold higher than the yields of 43, 102, 143 and 150 ml of alcohol/kg of apple powder, cauliflower stalks, pea and banana peels respectively, recorded by our group [13,16] among others [22]. Thus the suitability of pineapple wastes as an alternate substrate for ethanol production is very promising.

REFERENCES

[1] Rupal, K., M.Sc. Thesis Punjab Agricultural University, Ludhiana, India, 1983.

[2] FAO., FAO production year book, 1984, 38, 194.

[3] Burbank, N.C. and Kamagal, J.S., Proc. Ind. Waste Conf. Purde Univ., 1965, (118), 20, 365.

[4] Dev, D.K. and Ingle, U.M., Indian Fd. Packer, 1982, 15–22.

[5] Shewale, J.G. and Sadana, J.C., Can. J. Microbiol., 1978, 24, 1204–1216.

[6] Mandels, M. and Sternberg, D., J. Ferm. Technol., 1976, 54, 267–286.

[7] Toyama, N. and Ogawa, K., Proc. Int. Course Biochem. Eng. Bioconversions (ed. T.K. Ghose), held at II, New Delhi, India, 1977, 120.

[8] Miller, G.L., Anal. Chem., 1959, 31, 426–428.

[9] Updegraff, D.M., Anal. Biochem., 1969, 32, 420–424.

[10] Clegg, K.M., J. Sci. Food Agri., 1956, 7, 40.

[11] Caputi, A, Jr. and Wright, D., J. Assoc. Off. Anal. Chem., 1969, 52, 85–88.

[12] Smith, C.P., Grethlein, H.E. and Converse, A.O., Solar Energy, 1982, 28, 41–82.

[13] Tewari, H.K., Marwaha, S.S. and Rupal, K., Agric. Wastes, 1986, 16, 135–146.

[14] Evacklund, E., Hatrikka, A., Mustrantr, A. and Nybergh, P., Eur. J. Appl. Microbiol., 1976, 2, 143–152.

[15] Tewari, H.K., Singh, L. and Sethi, R.P., J. Res. Punjab Agri. Univ., 1983, 20, 74–80.

[16] Tewari, H.K., Marwaha, S.S., Rupal, K. and Grewal, H.S. In "National Symposium on Microbial Ecology" held at Coimbatore, India from 23–24 Jan., 1986. (In Press)

[17] Pontiveros, C.R., Alcontora, J.A. and Rosodro, E.J., Philippine J. Crop Sci., 1978 3(3), 153-158.

[18] Chahal, D.S. and Overend, R.P., In "Advances in Agriculture Microbiology" (ed. N.S. Subha Rao). Oxford & IBM Publishing Co., New Delhi, 1982. pp. 585-641.

[19] Wayman, M., In "Proceedings of the IVth International Symposium on Alcohol fuels Technol. Brazil, 1980, 79-87.

[20] Banarjee, N. and Vishwanathan, L., Proc. IVth Ann. Conv. Sugar Technologists, Kanpur, 1974, pp. G 1-4.

[21] Vaish, K.N. and Vishwanathan, L., Proc. V. Ann. Conv. Sugar Technologists, Kanpur, 1975, pp. G 85-89.

[22] Hang, Y.D., Lee, C.Y., Woodams, E.F. and Cooley, H.K., Appl. Env. Microbiol., 1981, 42, 1128-29.

Chapter 28

Unusual structural organisation of two cellulases from *Cellumonas fimi*

R. A. J. Warren, N. R. Gilkes, D. G. Kilburn, M. L. Langsford, R. C. Miller Jr., G. P. O'Neill and W. K. R. Wong – Department of Microbiology, University of British Columbia.
C. F. Beck – Institut für Biologie III, Universitat Freiburg.

ABSTRACT

An endoglucanase and an exoglucanase of *Cellulomonas fimi* are each composed of three distinct regions or domains: a short sequence of 20 amino acids containing only proline and threonine (the Pro-Thr box); an irregular domain of about 100 amino acids, which lacks secondary structure; and an ordered domain, of about 300 amino acids, which has secondary structure and contains a potential active site. The Pro-Thr box is conserved almost perfectly in the two enzymes. The irregular domains are 50 per cent conserved. There is very little sequence conservation in the ordered domains, but both potential active sites have the sequence Glu-Xaa_7-Asn-Xaa_6-Thr. These sites occur at different places in the ordered domains. The order of the domains is reversed in the two enzymes: irregular-Pro-Thr-ordered in the endoglucanase; ordered-Pro-Thr-irregular in the exoglucanase. The genes for the enzymes appear to have arisen by shuffling of the two conserved sequences and perhaps other sequences.

INTRODUCTION

Cellulolytic microorganisms produce complex mixtures of cellulases (see Coughlan, 1985 for a review). The enzymes are of two major types: endoglucanases (1,4-ß-D-glucan glucanohydrolases, EC1.3.2.1.4) and exoglucanases (1,4-ß-D-glucan cellobiohydrolases, EC3.2.1.91). These enzymes are of interest for their commercial potential and for those features of their structures which determine their specificities. Although both types of enzyme hydrolyze the 1,4-β-D-glycosidic bond between two glucose residues, endoglucanases hydrolyze bonds in the interior of cellulose molecules to produce oligosaccharides, whereas exoglucanases hydrolyze bonds at the non-reducing ends of cellulose molecules to release the disaccharide cellobiose. Therefore, one expects to see similarities and differences between the enzymes.

A given microorganism may produce more than one of each type of enzyme (Coughlan, 1985). Although this complexity may have physiological origins, the cloning from a given organism of multiple genes for endoglucanases reveals that the complexity, at least in part, certainly has genetic origins (Cornet et al., 1983; Gilkes et al., 1984). The relatedness of the genes encoding these isozymes is also of considerable interest.

CELLULOMONAS FIMI

Cellulomonas fimi is a Gram-positive, non-spore-forming facultative anaerobe which produces a complex mixture of cellulases during growth on cellulosic substrates (Béguin et al., 1977; Béguin and Eisen, 1978; Langsford et al., 1984). We are attempting to characterize all of the cellulases in this mixture, and to determine their relationships Gene cloning, of course, allows the characterization of the genes encoding these enzymes. It has the further great advantage, providing the cloned gene is expressed, of allowing the characterization of a given enzyme absolutely free of related activities. The preparation of a homogeneous enzyme from a complex mixture of cellulases is usually a difficult and time-consuming task.

CLONING AND SEQUENCING OF *C. FIMI* CELLULASE GENES

C. fimi produces an exoglucanase and an endoglucanase which bind very tightly to the substrate in cultures grown with Avicel (Béguin et al., 1977; Béguin and Eisen, 1978; Langsford et al., 1984). These enzymes were purified virtually to homogeneity and their N-terminal amino acid sequences were determined (M.L. Langsford, unpublished observations). Both enzymes hydrolyze carboxymethylcellulose, but only the exoglucanase hydrolyzes 4-nitrophenylcellobioside.

The genes encoding the enzymes were cloned and expressed in *Escherichia coli* (Gilkes et al., 1984) both genes were sequenced (O'Neill et al., 1986; Wong et al., 1986). The reading frames were established by reference to the N-terminal amino acid sequences of the proteins purified from *C. fimi* cultures. The translation stop sites were established by reference to the M_r values of the proteins encoded by the genes using *E. coli* mini-cells (O'Neill et al., 1986; Wong et al., 1986). The enzymes as synthesized by *C. fimi* are glycosylated (Bêguin and Eisen, 1978; Langsford et al., 1984), which makes it impossible to establish realistic M_r values by gel electrophoresis. The enzymes synthesized by *E. coli* are not glycosylated.

STRUCTURES OF THE ENDOGLUCANASE AND OF THE EXOGLUCANASE

The amino acid sequences of the enzymes, predicted from the nucleotide sequences of their genes, were analyzed by computer for their hydropathy profiles, for their potential secondary structures, and for sequence conservation.

Both enzymes contain a short sequence of 20 amino acids, comprised only of proline and threonine, which is almost perfectly conserved. Both contain an irregular domain of about 100 amino acids which is predicted to be free of α-helices and to contain little ß-pleated sheet. The irregular domains are rich in hydroxy amino acids but they have very few charged residues: 6 in the endoglucanase, 4 in the exoglucanase. Some 50 per cent of the amino acids are conserved in these irregular domains.

Fig. 1. Potential active sites in *C. fimi* cellulases. Conserved residues are boxed. The positions of amino acids in the sequences are numbered. ENDO is the endoglucanase, EXO is the exoglucanase.

The irregular domains are separated by the proline-threonine domain from a sequence of about 300 amino acids which is predicted to have significant secondary structure. These ordered domains contain potential active sites of the sequence Glu-Xaa_7-Asn-Xaa_6-Thr (Fig. 1), deduced by analogy with the active site sequences Glu-Xaa_8-Asn-Xaa_6-Thr suggested or proved for lysozymes and a cellobiohydrolase from Trichoderma reesei (Paice et al., 1984). Some 10 per cent of the amino acids are conserved in the ordered domains.

The order of the domains is reversed in the two enzymes (Fig. 2), so that the active sites occur in widely separated loci in them. Each ordered domain appears to be required in its entirety for enzymatic activity (Z.M. Guo and G.P. O'Neill, unpublished observations). Deletion and similar analyses show that the irregular domains, however, are not essential for activity (Wong et al., 1986; G.P. O'Neill, unpublished observations).

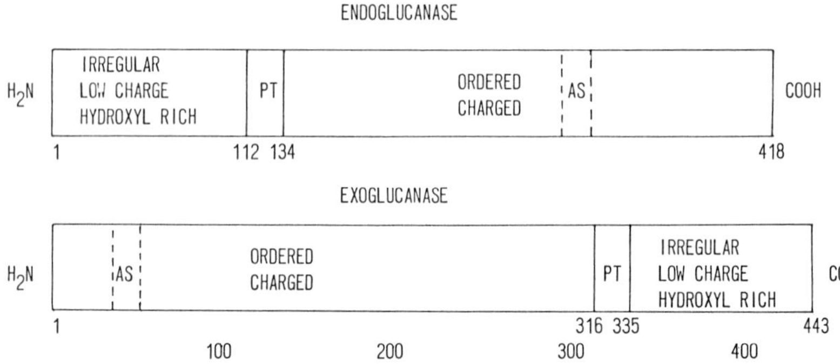

Fig. 2. Overall structures of the endoglucanase and the exoglucanase of *C. fimi*. PT is the Pro–Thr box; AS is the potential active site. Numbers refer to amino acid positions in the sequences.

DISCUSSION

It is clear that the structural genes for the endoglucanase and the exoglucanase arose by the duplication and shuffling of at least three sequences, two of which are highly conserved. The ordered domains of the enzyme are catalytically active and probably determine the specificities of the enzymes. The great similarity of the potential active sites presumably reflects the virtual identity of the reactions catalyzed by the enzymes. The irregular domains are involved in the very tight binding of the enzymes to Avicel. Further cloned genes and their encoded cellulases are being characterized to see if all components of the cellulase system of C. fimi evolved in the same manner.

ACKNOWLEDGEMENTS

This work was supported by Strategic Grant 67-0941 from the Natural Sciences and Engineering Research Council of Canada to DGK, RCM and RAJW; and a grant from the Deutsche Forschungsgemeinschaft to CFB. RAJW thanks the Deutsche Akademische Austauschdienst for support during a Studienaufenthalt in Freiburg.

REFERENCES

Béguin, P. and Eisen, H. (1978) Eur. J. Biochem. 87: 525-531.

Béguin, P., Eisen, H. and Roupas, A. (1977) J. Gen. Microbiol. 101: 191-196.

Cornet, P., Millet, J., Béguin, P. and Aubert, J-P. (1983) Bio/Technology 1: 539-544.

Coughlan, M.P. (1985) Biotechnol. Genet. Eng. Rev. 3: 39-109.

Gilkes, N.R., Kilburn, D.G., Langsford, M.L., Miller, R.C., Jr., Wakarchuk, W.W., Warren, R.A.J., Whittle, D.J. and Wong, W.K.R. (1984a) J. Gen. Microbiol. 130: 1377-1384.

Langsford, M.L., Gilkes, N.R., Wakarchuk, W.W., Kilburn, D.G., Miller, R.C., Jr. and Warren, R.A.J. (1984) J. Gen. Microbiol. 130: 1367-1376.

O'Neill, G.P., Goh, S.H., Warren, R.A.J., Kilburn, D.G. and Miller, R.C. Jr. (1986) Gene, in press.

Paice, M.G., Desrochers, M., Rho, D., Jurasek, L., Roy, C., Rollin, C.F., De Miguel, E. and Yaguchi, M.

(1984) Bio/Technology 2: 535-539.

Wong, W.K.R., Gerhard, B., Guo, Z.M., Kilburn, D.G., Warren, R.A.J. and Miller, R.C., Jr. (1986) Gene, in press.

Chapter 29

Biodegradation of lignin by bacteria and molecular cloning of a gene for arylether cleavage

V. R. Srinivasan and J. W. Cary – Department of Microbiology, Louisiana State University, Baton Rouge, LA 70803, USA.

ABSTRACT

A ligninolytic bacterium identified as Erwinia spp. was isolated by a novel enrichment technique from the sewage. A gene fragment coding for "aryletherase" was cloned into pBR322 and used to transform E. coli host. Since it was found to be partial gene, a larger fragment from Erwinia DNA was reisolated using the original fragment as a probe. The transformants of E. coli containing the larger fragment are stable in their expression of arylether cleavage.

INTRODUCTION

Lignin occurs as one of the major components of woody tissue of plants comprising nearly 20-30% of biomass. It is a very complex and heterogeneous polymer of phenylpropane units bound covalently mainly through three different linkages namely aryl-alkylether, aryl-arylether and diphenyl, of which aryl-alkyl bonds seem to predominate. Degradation of lignin by microorganisms is hindered by the unique structure of the polyphenolic structure of lignin. Because of the recalcitrancy of lignin to microbial degradation, lignin and lignin derived compounds are a major source of pollution in wood pulping and processing industries. Knowledge of the physiology of lignin degrading microorganisms and biochemistry of lignin biodegradation may lead to the development of processes for the use of lignin as a renewable resource of chemical feedstocks for industries. Recently great strides have been made in understanding the growth of ligninolytic while rot fungus Phaenerochete chrysosporium and the mechanism of ligninolysis by enzymes from the organism (1,2,3).

In this report, we discuss a novel method of enrichment for ligninolytic bacteria and our attempts to clone for a gene coding for presumptive arylether cleaving protein from one of the isolates.

ENRICHMENT OF LIGNINOLYTIC BACTERIA

Lignin contains 12-20% methoxy groups. If the organisms have the potential to demethoxylate lignin by ether cleavage, the resulting product would be either methanol or formaldehyde. Hence, such organisms will be capable of tolerating high concentrations of methanol or formaldehyde or utilizing them. Based upon these concepts, we devised a method of enriching ligninolytic bacteria in a continuous culture. A sample of primary sludge mixed with raw sewage obtained from one of the sewage treatment facilities in Baton Rouge, LA was used as the source of ligninolytic organisms. Sewage sludge was introduced into the reactor vessel in a continuous culture apparatus. The apparatus and the method of operation are described previously by Summers et al (4). The nutrient medium with 0.5% methanol as sole carbon source was pumped continuously at a dilution rate of 0.1^{-1}. The pH of the constituents in the reactor was maintained at pH 9.0 by automatic addition of 1M NaOH with the aid of a pH controller. After 10 days when the chemostat culture was in steady state, the carbon source was altered to 0.2% Kraft lignin dissolved in NaOH. The reactor was maintained in steady-state at the same dilution rate for a period of one week. Several single isolates were obtained from the reactor culture by selection on nutrient agar medium with kraft lignin as the only carbon source. One of the isolates which produced a diffusible red pigment was arbitrarily chosen for further studies on lignin degradation. Based on the characteristics of fermentation of carbohydrates and other biochemical tests, the organism was tentatively identified as a saprophytic Erwinia.

DEGRADATION OF LIGNIN AND RELATED COMPOUNDS BY THE ORGANISM

The organisms were grown in liquid cultures, mutated with N-nitro-N'-nitros-N'-methyl guanidine in order to obtain pigmentless mutants since it was found difficult to assay for lignin degradation in the presence of the pigment. The mutant was grown in a nutrient medium with 0.5% sodium succinate and salts. Cells were harvested and resuspended in 0.5M Tris buffer (pH 7.2) such that the suspension contained 1% wet weight of cells. This was incubated with 0.1% lignin at 32°C for 24 hrs. After incubation, the cells were removed by centrifugation and the supernatant liquid was lyophilized. The lyophilized material was extracted with a mixture of ethyl acetate, methanol and glacial acetic acid (4:1:1). The extract was fractionated on thin layer chromatogram developed in a solvent mixture of benzene:ethanol:ammonia (4:1:1) and examined for UV absorbing compounds. The results are presented in Figure 1.

Experiments on the degradation of vanillin vanillic acid, and p-phenoxyphenol were carried out by the original isolate itself. Vanillin is oxidized to vanillic acid in the course of 2 hrs. Both vanillin and vanillic acid were degraded in 8 hrs. to a product which had the same Rf. as catechol on TLC developed with different solvent systems. After 24 hrs., the aromatic nuclei were completely broken down to non-UV absorbing materials. One of the products of degradation of p-phenoxyphenol was identified as p-benzoquinone by NMR spectroscopy.

MOLECULAR CLONING OF ERWINIA GENE FOR "ARYLETHERASE" IN ESCHERICHIA COLI

Our objective in these investigations has been to study the feasibility of developing systems for the production of aromatic chemicals from lignin. Hence experiments were initiated to isolate the gene for "aryletherase" and clone it by recombinant DNA technology into a host so that the gene product can be investigated in the absence of other enzymes involved in ring fission. In order to clone the "aryletherase" gene, Erwinia DNA was digested with the restriction enzyme Bam HI and the fragments were ligated into the Bam HI site of the plasmid pBR322. This ligated plasmid was used to transform E. coli strain Cs412 by the methods of Hanahan (5). The transformants were selected on agar media containing M9 salts, and ampicillin with salicin as the sole carbon source (on the assumption that the glucoside which is formed by a hemiacetal linkage may be considered as a pseudoether). Salicin, an aryl glucoside, supports the growth of Erwinia but not the host strain E. coli Cs412. The recombinant plasmids (pNC1 and pNC2) from two of the transformants conferred on E. coli Cs412, the ability to utilize salicin in subsequent transformation studies. Plasmid pNC1 was shown to contain an 810bp insert by digestion with Bam HI. ^{32}P-nick translated (b) plasmid was used as a probe to demonstrate the sequence homology between the cloned fragment and Erwinia genomic DNA (7). Since no internal Hind III sites were present in the 810bp fragment as determined by restriction enzyme mapping, Erwinia DNA was digested with Hind III and Bam HI before agarose gel electrophoresis to enhance the signal. (Figure 2). Examination of the restriction map of the cloned fragment as well as the hybridization studies point out to the presence of a single "aryletherase" gene copy in the Erwinia spp. chromosome. The transformants of E. coli with the cloned plasmids were able to degrade vanillin, vanillic acid, syringaldehyde syringic acid, and p-phenoxyphenol. The degradation products were detected by TLC as in the experiments with the original Erwinia isolate. (8). These results suggest that the cloned fragment DNA codes for a protein which could initiate the cleavage of aryl-alkyl

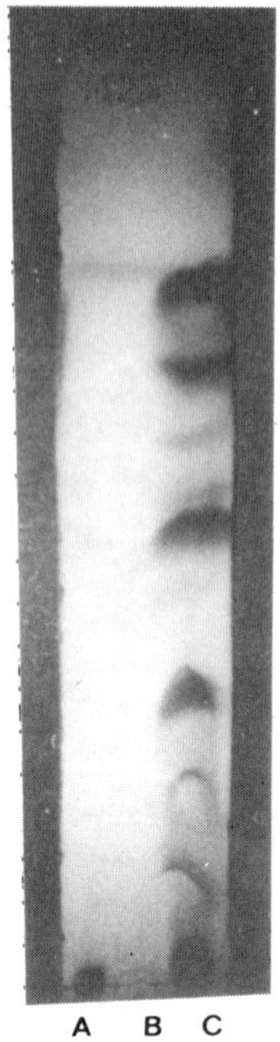

Fig. 1. U.V. absorbing products of degradation of kraft Lignin by Erwinia spp. separated on TLC. A. Lignin incubated in buffer. B. Organisms without Lignin. C. Organisms incubated with Lignin.

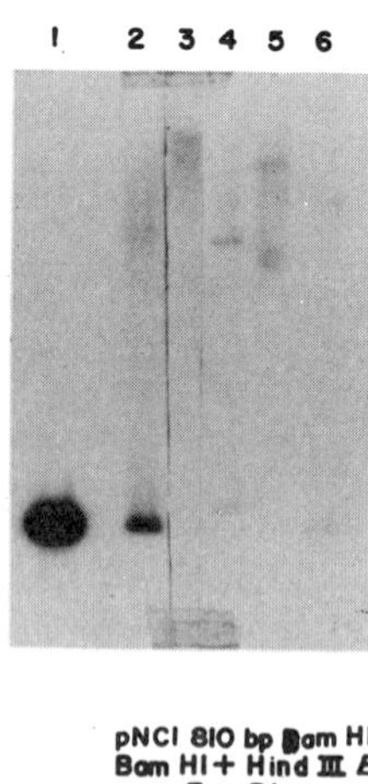

1 pNCI 810 bp Bam HI fragment
2 Bam HI + Hind III *Erwinia* DNA
3 Eco RI " "
4 Sal I " "
5 Hind III " "
6 Pst I " "

Fig. 2.

as well as aryl-arylethers.

CONSTRUCTION OF A GENE LIBRARY FOR ERWINIA CHROMOSOME AND SCREENING FOR A LARGER FRAGMENT FOR ARYLETHERASE

Despite the positive results obtained with the cloned 810bp fragment, it was noticed that the expression was rather erratic and extremely low. A preliminary sequencing of the fragment indicated the absence of a promoter and thus the phenotypic characteristics of the transformants might be due to a read through from the Tet promoter in pBR322. Hence we decided to isolate if possible the complete gene by cloning a larger fragment. In order to accomplish the goal, Erwinia gene library was constructed in bacteriophage lambda. Lambda replacement vector EMBL 3 which contains a 13.7kb non-essential stuffer fragment flanked by polylinkers (having different restriction enzyme sites) was used in these experiments. Foreign DNA of 10-20kb fragments can be packaged with this vector. Erwinia DNA was cleaved partially with restriction enzyme Sau 3A and the fragments of the size between 8 to 23kb were collected. The vector was cleaved with restriction enzymes BAM HI and EcoRI and after annealing the cohesive ends, ligated with Erwinia DNA fragments. The lambda DNA thus constructed was packaged in vitro to form lambda particles which were used to infect E. coli host strain NM538 (9). Approximately $2x10^6$PFV/ug DNA were obtained. The phage was collected from the plaques and amplified on E. coli NM519; (a strain on which the plaques formed are larger in size and which also discriminates against the original stuffer fragment in the vector), 10,000 plaques were screened with ^{32}P-nick translated mp18 RFDNA containing the original 810bp. fragment. Out of 7 positive plaques isolated, DNA was prepared and the fragment were purified. SmaI digests of the fragments were recloned in E. coli plasmid pUC 19. The restriction map of one of the clones is shown in Figure 3. Figure 4 presents the autoradiogram of the proteins obtained by labelling with ^{35}S-methionine under growth conditions which permit preferential transcription and translation of only the plasmids in the cell (10). The arylether cleaving activity of the clones were determine by using the chromogenic glucoside p-nitrophenyl B-D-glucoside (Table 1). The clones are able to break down vanillin into similar products as Erwinia. Figure 5 shows the UV absorbing compounds separated on TLC after incubating vanillin with the respective organisms.

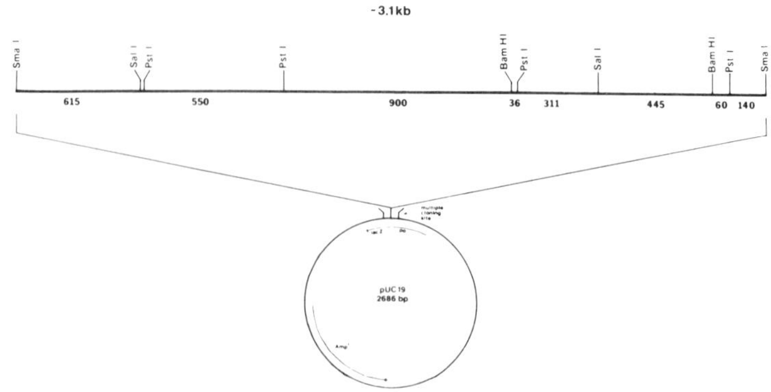

Fig. 3. Restriction Enzyme cleavage sites on the cloned fragment. Distances are given as Base pairs.

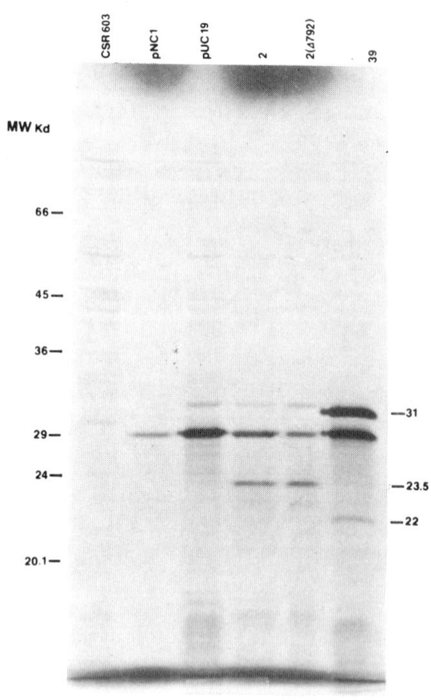

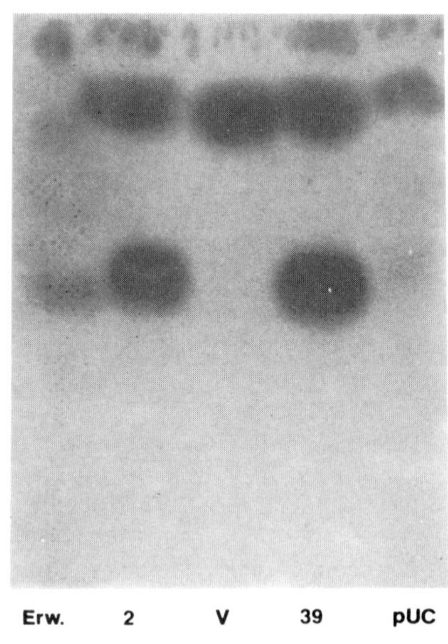

Fig. 4. Autoradiogram of the PAGE analysis of ^{35}S-labelled proteins coded by the plasmid genes.

Fig. 5. TLC separation of degradation of Vanillin by the organisms. Erw-Erwinia 2,39, & pUC are the strains referred on Table 1. V-Vanillin.

Table 1. p-Nitrophenylglucoside cleavage activity of the organism.

	Organisms	*Activity (units/ml)
1.	E. coli RCG 123**	0.037
2.	E. coli RCG 123 with plasmid pUC 19	0.056
3.	E. coli RCG 123 with pUC 19 with insert	0.246
4.	E. coli RCG 123 with pUC 19 with insert in the opp. orientation	0.151
5.	Erwinia spp.	0.386

* Activity is expressed as M moles p-nitrophenol released per mg protein per hour.

** E. coli RCG 123 (lac 74, lon^-, capR, Str^R, Trp^-)

SUMMARY AND CONCLUSIONS

Crawford et at (11) were the first to isolate a bacterium having the potential to cleave the ether bond from arylglycerol-B-arylether. Subsequently, several investigators have shown that different bacterial isolates have the ability to degrade Kraft lignin (12). We have described a novel technique for rapid enrichment of ligninolytic bacteria. Also, a gene fragment coding for a protein which initiates arylether cleavage has been cloned in E. coli by an indirect selection. Although these experiments indicate that the same protein is involved in the cleavage of PNPG and vanillin, further studies are required before we can confirm this observation. Preliminary experiments point out that the bacterial system may be different from fungal ligninase characterized by Kirk and his collaborators since addition of hydrogen peroxide does not enhance the arylether cleavage. Now that we have a stable clone of the gene for the protein, before long, we may be able to isolate the protein in sufficient quantities and gain an insight into the mechanism of arylether cleavage by Erwinia.

REFERENCES

1. Jager, A., S. Croan and T.K. Kirk (1985) Appl. Environ. Microbiol. 50:1274-1278.
2. Tien, M., and T.K. Kirk (1984) Proc. Natl. Acad. Sci. USA 81:2280-2284.
3. Harvey, P.J., H.E. Shoemaker, R.M. Bowen and J.M. Palmer (1985) FEBS Letters. 183:13-16.
4. Summers, R.J., D.P. Boudreaux and V.R. Srinivasan (1979) Appl. Environ. Microbiol. 38:1079-1084.

5. Hanahan, D. (1983) J. Mol. Biol. 166:557-580.
6. Rigby, P.W.J., M. Dickmann, C. Rhodes and P. Berg (1977) J. Mol. Biol. 113:237-251.
7. Southern, E.M. (1975) J. Mol. Biol. 98:503-517.
8. Srinivasan, V.R., K. Narva, J.W. Cary and Y. Chon (1985) Symp. Energy from Biomass and Wastes. 9:1173-1192.
9. Hohn, B. and K. Murray (1977) Proc. Natl. Acad. Sci. USA 74:3259-3263.
10. Sancar, A., R.P. Wharton, S. Seltzer, B.M. Kacinski, N.D. Clark and W.D. Rupp (1981) J. Mol. Biol. 148:45-62.
11. Crawford, R.L., T.K. Kirk, J.M. Harkin and E. McCoy (1973) Appl. Microbiol. 25:322-324.
12. Deschamps, A.M., G. Mahoudeaux and J.M. Lebeault (1980) Env. J. Appl. Microbiol. Biotechnol. 9:45-51.

Chapter 30

Bioconversion of plant xylans

A. Blažej – Faculty of Chemistry and Chemical Technology, Slovak Technical University, Bratislava, Czechoslovakia.

P. Biely – Institute of Chemistry, Slovak Academy of Sciences, Bratislava, Czechoslovakia.

INTRODUCTION

Appreciable quantities of xylan, the main component of plant hemicellulose, are currently not utilized as it would be desired. Released from wood during pulping and pulp processing, xylan is often deposited in water systems causing ecologically harmful effects. Most of the xylan released during pulping is extensively hydrolyzed. However, some wood prehydrolysates and hydrolysates contain significant proportions of xylooligosaccharides which have to be converted to xylose before their use. There are also wastes which contain xylan in non-hydrolyzed form. Examples are the alkaline waste liquors obtained in the process of preparing dissolving pulp from the chemical pulp. The most perspective source of xylan in future could be the water-soluble hemicellulose fraction released from wood and agricultural residues by steam explosion (1,2), if this kind of phytomass pretreatment is adopted industrially. Steady increases in non-wood pulping in the world, especially in wood-deficient developing countries (3), will be accompanied by an additional stream of xylan and xylan fragments the use of which must await the development of an economically beneficial process.

SACCHARIFICATION

Xylose can be converted biologically to single cell protein and to a whole range of fuels and chemicals, and can also be used as a carbon source for production of glucose isomerase (Fig. 1) (4-6). Naturally, to carry out similar conversions with xylan and xylooligosaccharides, their hydrolysis to xylose is required. The hydrolysis can be chemical and enzymic. The chemical hydrolysis accomplished by acids is rapid; however, xylose is partially degraded to furfural which may hinder subsequent microbial fermentations. For these reasons, considerable attention

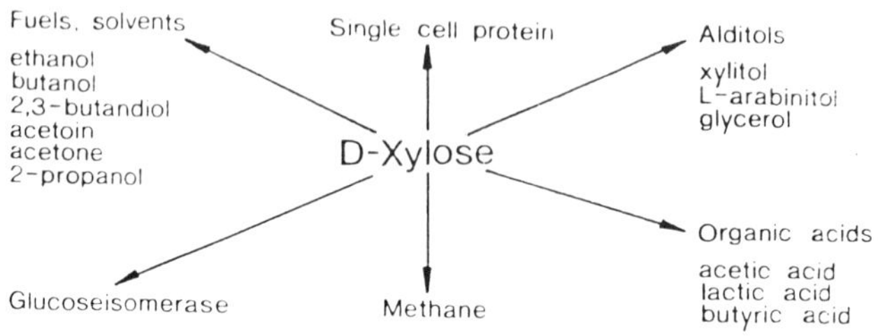

Fig. 1. Products that can be obtained from xylose by microbial fermentations.

is devoted to development of hydrolytic processes under mild conditions, which include also the enzymic hydrolysis (7-9). The complex structure of the substrate and, consequently, the requirement for a special composition of hydrolytic enzymes (Fig. 2), together with the cost for their preparation, may represent a critical holdup of industrial applications of isolated enzymes in xylan conversions in the

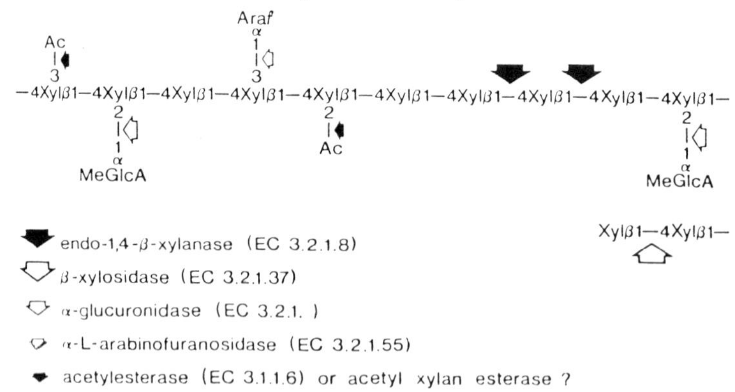

Fig. 2. A hypothetical xylan fragment and the sites of its attack by microbial enzymes. Symbols: Xyl, D-xylosyl residue; MeGlcA, 4-0-methylglucuronosyl residue; Araf, L-arabinofuranosyl residue; Ac, acetyl group.

near future. Processes which would not be dependent on a separate production of saccharifying enzymes would have the virtue of simplicity and low costs.

DIRECT XYLAN CONVERSIONS

Use of microorganisms capable of transforming xylan directly to single cell protein, fuels and chemicals, seems to be the most promising approach to elaboration of economically feasible processes of xylan bioconversions. The literature reports a number of organisms which are able to use xylan as a carbon source and which also possess metabolic pathways to ferment xylose to useful products (Table 1). Of the organisms listed, only anaerobic thermophilic

Table 1. Fermentation abilities of xylanolytic microorganisms.

Microorganisms	Fermentation ability	Ref.
Bacteria		
Clostridium acetobutylicum	xylan - butanol, butyric acid	10
Clostridium thermocellum	xylan - ethanol, acetic and lactic acid	11 12
C. thermohydrosulfuricum	xylan - ethanol	12
C. thermosaccharolyticum	xylan - ethanol	11
Thermoanaerobacter ethanolicus	xylan - ethanol, acetic and lactic acid	12
Thermoanaerobacter brockii	" "	12
Thermoanaerobium sp.	" "	13
Thermoanaerobacter sp.	" "	13
Thermobacteroides sp.	" "	13
Fungi		
Fusarium oxysporum	xylose - ethanol	14
Monilia sp.	xylan - ethanol	15
Neurospora crassa	" "	16
Yeasts		
Aureobasidium pullulans	xylose - ethanol	17
Candida shehatae	xylose - ethanol	17
Cryptococcus albidus	xylan - triglycerides	18
Pichia stipitis	xylan - ethanol	17

bacteria and a few representatives of fungi and yeasts combine the two metabolic pathways necessary for direct xylan bioconversions. Anaerobic thermophilic bacteria are capable of converting xylan to a mixture of ethanol, acetic acid, lactic acid, and ethanol, butanol and butyric acid (10-13). The proportion of the products depends strongly on the concentration of xylan in the medium (12). Scarce information is available on a single-step xylan conversion by fungi and yeasts. Although the xylan-utilizing ability is widely distributed among fungi, only two species, Monilia sp. (15) and Neurospora crassa (16), were reported to ferment xylan directly to ethanol. However, the conditions required by the microorganisms, to grow and produce saccharifying enzymes are different from those required for the fermentation. For these reasons, the fermentation of xylan or cellulose to ethanol by Neurospora crassa was carried out in two steps (19). The fungus was first grown aerobically on an inducing carbon source (e.g. on cellulose), and then a non-aerated fermentation was started after addition of xylan. In this way ethanol was obtained from xylan in 58% yield, which is comparable to that obtained from xylose (65%). It is of interest whether this approach could be applied to another xylanolytic xylose-fermenting fungus Fusarium oxysporum (14).

APPLICATION OF YEASTS

Yeasts can be considered as ideal organisms for biotechnological purposes. In connection with xylan utilization they have not been studied extensively, however. The ability to grow on xylan is restricted among yeasts to genera Aureobasidium, Cryptococcus and Trichosporon (20-22) and to species Pichia stipitis and its anomorph Candida shehatae, and to species Candida ergatensis (17) (Table 1). A strain Aureobasidium pullulans is capable of growing on hemicellulose present in alkaline waste dialysis liquors from the production of viscose fibres and cellophane (23). The liquors, originally containing 1.5% hemicellulose and 1.7% NaOH, were adjusted to pH 4--5 with mineral acids, supplemented with 0.05% yeast autolysate and 0.2% ammonium sulfate and used as growth media. A comparison of the growth yields in such a medium with that in equivalent xylose medium indicated a 70% utilization of hemicellulose. We should note, however, that not all xylanolytic yeasts were able to tolerate high salt concentration and to grow in the media based on the waste liquors.

A number of microorganisms can accumulate considerable amounts of oil, e.g. triglycerides, sterol esters etc., when grown in the presence of high concentrations of sugars under nitrogen limitation (24). Triglycerides can be used as a diesel fuel without major processing (24). The microorganisms also include yeasts, and what is of interest, yeasts that are capable to grow on xylan (18). Table 2 shows the triglyceride production by a series of Cryptococcus albidus strains grown on xylose and xylan. The oil production on the polymer is comparable with that on the monomer.

Table 2. Triglyceride production from xylose and xylan by strains of Cryptococcus albidus (From Ref. 18).

Strain No	Triglycerides from 100 g of xylose	xylan
	g	
FY 6	11.1	8.6
FY 8	15.5	9.8
FY 10	14.8	12.3
FY 47	14.4	8.9
FY 61	15.2	9.9
FY 84	16.0	13.1
FY 152	15.9	13.7

The interest in xylan-utilizing yeasts increased after yeast fermentation of xylose to ethanol had been definitively established (25). Recent screening of xylanolytic yeasts for the ability to ferment xylose and of xylose-fermenting yeasts to utilize xylan as a carbon source, revealed that both metabolic pathways occur among the yeast species Aureobasidium pullulans, Candida ergatensis, Candida shehatae and Pichia stipitis (17). However, only two Pichia stipitis strains were found capable of converting xylan to ethanol directly in a yield of 50--60% of theoretical. The conversion took place in cell recycle experiments, usually starting in the third cycle (17). The recycling was done under oxygen limitation. In the course of such a cycle, ethanol concentration increased with time, reaching a steady-state value after day three or four. In accordance with the ethanol yields, xylan was found to be utilized incompletely (17). The reason for the low efficiency of xylan fermentation in comparison to the efficiency of xylose fermentation (about 80%) may be related to the absence of certain hydrolytic enzymes necessary for complete saccharification of the branched heteropolysaccharide. Further work is needed to learn how to make other xylanolytic xylose-fermenting yeasts to produce ethanol from xylan. It is probable that a two-step process (enzyme production followed by fermentation) outlined with the fungus Neurospora crassa (19) may lead to more positive results. It might be advantageous to use selected yeasts which contain xylanolytic enzymes predominantly in the cell envelope. Such yeasts can actually be constructed from xylose-fermenting yeasts by immobilization of xylanolytic enzymes on their surface (26). Use of mixed cultures with two different microorganisms may be an additional approach.

CONCLUSIONS

The examples of direct xylan conversions by xylanolytic microorganisms suggest that many other useful strains await their discovery. Yeasts deserve special attention in this regard, because their biotechnological potential has not been sufficiently evaluated. Some yeast species produce xylanolytic systems which are free from cellulases, a feature that could be explored in processes where cellulose should remain intact. In order to develop single-step xylan bioconversions it might be useful to direct research to chanelling metabolism of known xylanolytic microorganisms towards useful products. Insertion of genes coding for xylanolytic enzymes into genome of xylose-fermenting organisms is a challenge for genetic engineers.

REFERENCES

1. Bender, F. (1974) US Patent No 4136207.
2. Dietrichs, H. H., Sinner, M., and Puls, J. (1978) Holzforschung 32, 193-199.
3. McGovern, J. N. (1976) TAPPI Nonwood Plant Fiber Program, Alkaline Pulping Conference, pp. 3-8, Dallas, Texas.
4. Flickinger, M. C. (1980) Biotech. Bioeng. Suppl. 1, 27-48.
5. Gong, C. S., Chen, L. F., Flickinger, M. C., and Tsao, G. T. (1981) Adv. Biochem. Engin. 20, 93-118.
6. Natake, M. (1968) Agr. Biol. Chem. 32, 303-312.
7. Blažej, A., Košík, M., and Špilda, I. (1982) Chem. prumysl 32/57, 217-221.
8. Dekker, R. F. H. (1983) Biotechnol. Bioeng. 25, 1127-1146.

9. Yu, E. K. C., Deschatelets, L., Tan, L. U. L., and Saddler, J. N. (1985) Biotechnol. Letters 7, 425-430.
10. Lemmel, S. A., Datta, R., and Frankiewicz (1986) Enz. Microb. Technol. 8, 217-221.
11. Avgerinos, G. C., and Wang, D. I. C. (1980) Ann. Rep. Ferm. Process 4, 165-191.
12. Wiegel, J., Mothershed, C. P., and Puls, J. (1985) Appl. Environ. Microbiol. 49, 656-659.
13. Weimar, P., Wagner, L. W., Knowlton, S., and Ng, T. K. (1984) Arch. Microbiol. 138, 31-36.
14. Linko, M., Viikari, L., and Suihko, M. L. (1983) Proc. Int. Symp. Biotechnol. in Pulp and Paper Industry, pp. 333-358, London, UK.
15. Gong, C. S., Maun, C. M., and Tsao, G. T. (1981) Biotechnol. Letters 3, 77-82.
16. Mishra, C., Keskar, S., and Rao, M. (1984) Appl. Environ. Microbiol. 48, 224-228.
17. Lee, H., Biely, P., Latta, R. K., Barbosa, M. F. S., and Schneider, H. (1986) Appl. Environ. Microbiol. in press.
18. Fall, R., Phelps, P., and Spindler, D. (1984) Appl. Environ. Microbiol. 47, 1130-1134.
19. Deshpande, V., Keskar, S., Mishra, C., and Rao, M. (1986) Enzyme Microbiol. Technol. 8, 149-152.
20. Flannigan, B. (1970) Trans. Brit. Mycol. Soc. 55, 277.
21. Biely, P., Krátký, Z., Kocková-Kratochvílová, A., and Bauer, Š. (1978) Folia Microbiol. 23, 366--371.
22. Leathers, T. D., Kurtzman, C. P., and Detroy, R. W. (1984) Biotechnol. Bioeng. Symp. 14, 225-240.
23. Biely, P., Krátký, Z., Petráková, E., and Bauer, Š. (1979) Folia Microbiol. 24, 328-333.
24. Ratledge, C. (1979) Chem. Soc. Rev. 8, 283-296.
25. Schneider, H., Maleszka, R., Neirinck, L., Veliky, I. A., Wang, P. Y., and Chan, Y. K. (1983) Adv. Biochem. Engin. 27, 57-71.
26. Grüninger, H., and Fiechter, A. (1986) Enz. Microb. Technol. 8, 217-221.

Chapter 31

The role of fungal acetyl xylan esterases in the degradation of acetyl xylan by fungal xylanases

P. Biely – Slovak Academy of Sciences, Bratislava, Czechoslovakia.
C. R. MacKenzie and H. Schneider – National Research Council of Canada, Ottawa, Canada, K1A OR6.
J. Puls – Federal Research Center of Forestry and Forest Products, Hamburg, Federal Republic of Germany.

ABSTRACT

Xylan exists in many plants in an acetylated form. Hydrolysis of such substituted polysaccharides by xylanases was found to be influenced by the presence of deacetylating enzymes in that appreciably more glycosidic bonds were hydrolyzed when both enzyme activities were present simultaneously.

INTRODUCTION

The main chains of some xylans bear substituents such as acetic acid, glucuronic acid, 4-O-methyl glucuronic acid and arabinose [1]. Relatively little is known about the effects of these substituents on enzymic degradation of the main polysaccharide chain. Also, little is known about the effects on degradation of the main chain by enzymes that remove the substituents. The present study showed that the hydrolysis of glycosidic bonds in acetyl xylan occurred cooperatively when both xylanase and acetyl xylan esterase activities were present.

RESULTS

The enzymes used to investigate the role of esterases were obtained by HPLC fractionation of culture supernatants of the cellulolytic fungi *Trichoderma reesei* RUT C-30 and *Schizophyllum commune*

[1]Issued as N.R.C.C. publication number 26016

(Figures 1 and 2). Fractions were obtained that contained esterase or xylanase activity but were not mutually contaminated. Chromatography of T. viride proteins failed to yield suitable fractions, hence were not employed.

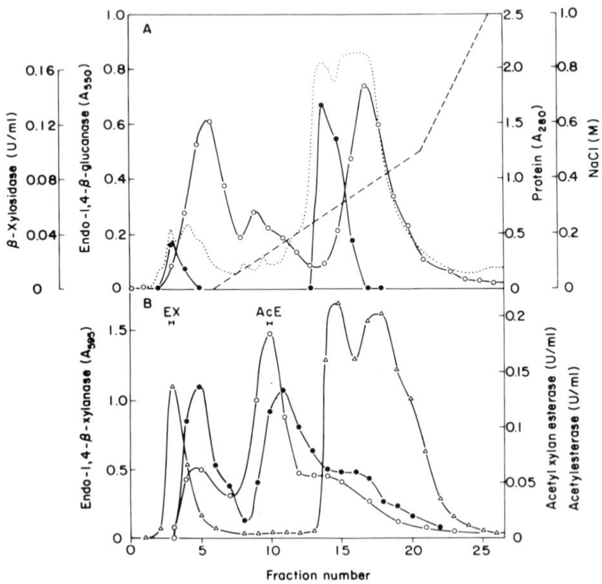

Fig. 1. Fractionation by HPLC of *T. reesei* RUT C-30 Enzymes. Part A: dashed line, NaCl gradient; dotted line, protein; ○, endo-1, 4-β-D-glucanase; ●, β-D-xylosidase. Part B: △ endo-1, 4-β-D-xylanase; ○, acetyl xylan esterase; ●, acetylesterase activity measured using 4-nitrophenyl acetate. The bars in Part B mark the fraction used in further experiments. The enzymes used were obtained as a commercial preparation (Celluclast, NOVO, Denmark). After dialysis 50 mg of protein were applied to the column (TSK-DEAE-3SW, 7.5 × 150 mm) and eluted at a flow rate of 1 ml/min using a 0-1 M linear NaCl gradient in 10 mM phosphate buffer at pH 6. The enzyme assays were carried out as described elsewhere [2].

The fungal esterase fractions were characterized by being relatively more active against acetyl xylan than esterases from other sources; acetic acid release was in the ratio of 64:17:1:1 for enzymes

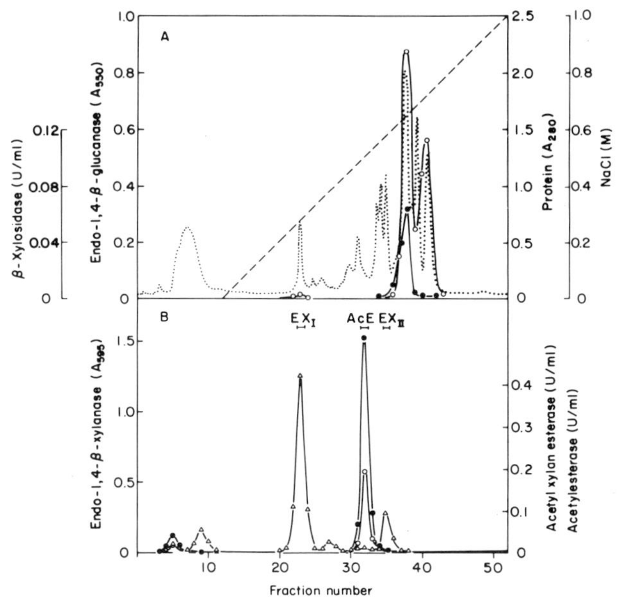

Fig. 2. Fractionation by HPLC of *S. commune* enzymes. Symbols and procedures as in Fig. 1. The proteins used were obtained from the supernatant of a 10-day old culture grown on a cellulose medium [7]. After desalting and five-fold concentration using an Amicon PM 10 membrane followed by lyophilisation, a solution containing 10 mg was applied to the column and eluted as described in Fig. 1.

from, respectively, *T. reesei*, *S. commune*, orange peel and porcine liver [2]. The fungal esterase fractions were also more active in releasing glycerol from triacetin than the enzymes from the other sources tested. The *T. reesei* material was more prominent in this respect than that from *S. commune* (Figure 3).

The *T. reesei* fraction used probably represented only one of several esterases present in the culture supernatant. The position of the two prominent peaks for activity against two different esterase substrates, 4-nitrophenyl acetate and acetyl xylan, were not identical (Figure 1). Also the ratio of activities against the two substrates differed. Additional evidence for the occurrence of multiple esterases was obtained in HPLC fractionation

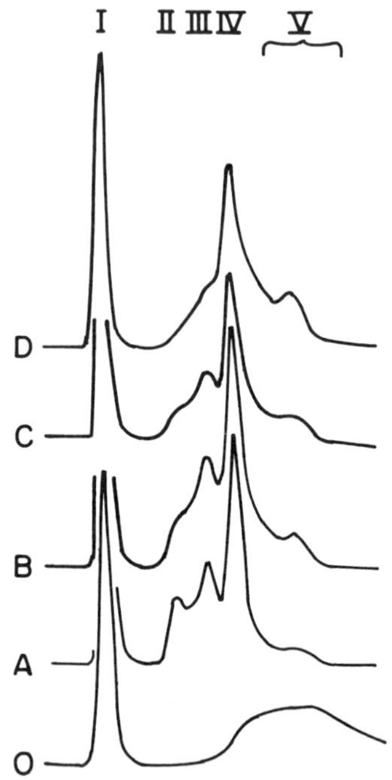

Fig. 3. Products of esterase action on triacetin as determined by HPLC. O, enzyme not present; A, *T. reesei* esterase; B, *S. commune* esterase; C, orange peel esterase; D, porcine liver esterase. Peaks: I, internal standard (xylitol); II, glycerol; III, partially acetylated glycerol; IV, acetic acid; V, triacetin. The reaction mixtures contained 1% triacetin in 0.2 M phosphate buffer (pH 6.5) plus 0.01% sodium azide and were incubated at 30°C. Reaction time for data shown was 50 hours. HPLC analysis was carried out as for acetic acid [8].

experiments with <u>T. viride</u> by the appearance of three distinct peaks which differed in ratio of activity against acetyl xylan and 4-nitrophenyl acetate. Because differences in activity can occur against 4-nitrophenyl acetate and acetyl xylan, the latter is more suitable in screening for esterases that deacetylate acetyl xylan.

The effect of esterases on xylan breakdown was readily demonstrable by TLC of the hydrolyzates [2]. Figure 4 illustrates the results obtained using birchwood acetyl xylan with *S. commune* acetyl esterase and *S. commune* xylanase E_{II}. In the presence of xylanase alone relatively little material was found on the plates and, of that found, an appreciable fraction consisted of acetylated xylooligosaccharides. In contrast, when esterase was present as well, xylose, xylobiose and xylotriose were evident as early as five hours, and their concentration increased with time. Notably, the results with acetyl xylan in the presence of esterase plus xylanase resembled those with chemically deacetylated xylan in the presence of xylanase alone. Similar results were obtained with the other esterase and xylanase fractions, although there were some differences. These included particularly little release of TLC detectable material with beechwood acetyl xylan and *S. commune* xylanase EX_{I}, and differences in the relative amounts of xylose,

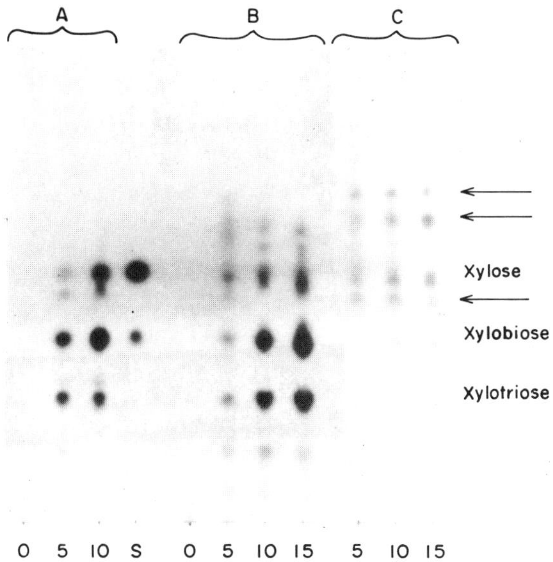

Fig. 4. Thin layer chromatography of hydrolyzates of birchwood acetyl xylan obtained after treatment with *S. commune* xylanase EX_{II} alone or in combination with *S. commune* esterase [2]. Part A; Xylanase alone on chemically deacetylated xylan. Part B; Xylanase plus esterase on acetyl xylan. Part C; Xylanase alone on acetyl xylan. The arrows marked the position of acetylated xylooligosaccharides.

xylobiose and xylotriose when esterase and xylanase were both present.

Measurements of reducing sugar production correlated well with the TLC results. For example, in the presence of S. commune xylanase EX_I alone, relatively little was released from beechwood acetyl xylan (Figure 5). However, appreciably more, corresponding to a three-fold increase, was produced when S. commune esterase was present. Typically, the non-fungal esterases tested were ineffective in enhancing sugar release. With birchwood acetyl xylan in the presence of S. commune acetyl esterase and xylanase EX_{II}, the amount of reducing sugar released was similar to that obtained with the xylanase acting

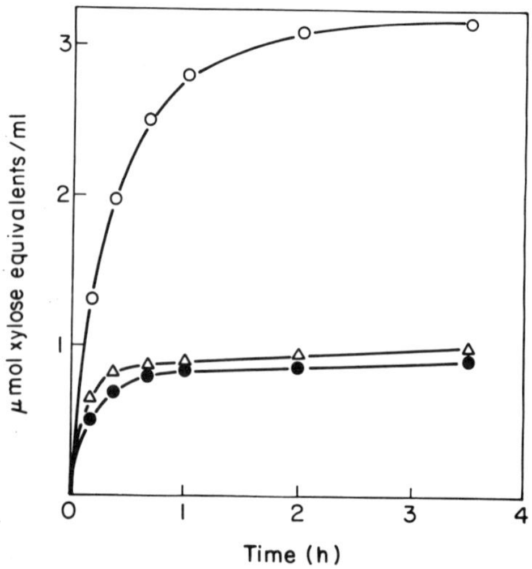

Fig. 5. Release of reducing sugar from beechwood acetyl xylan by *S. commune* xylanase EX_I plus or minus esterase. ○, xylanase plus *S. commune* esterase (0.1 U/ml); △, xylanase alone; ●, xylanase plus orange peel esterase (0.2 U/ml). Esterase activities refer to 4-nitrophenyl acetate as substrate. The level of xylanase activity in all instances was 0.2 U/ml [2].

alone on chemically deacetylated birchwood xylan [2].

Synergism was found also in release of acetyl groups from acetyl xylan by acetyl xylan esterase in the presence of xylanase [2]. The enhancement indicated that the esterase fraction used was more active on shorter fragments of acetyl xylan.

DISCUSSION

The synergism between esterases and xylanases in releasing sugar from acetyl xylan indicated that the role of esterases will require consideration whenever lignocellulosics containing acetyl xylan are degraded by fungal enzymes. The role of enzymes that remove other substituent groups present on xylan will require consideration as well, since their presence can have comparable effects on sugar release [3-6].

REFERENCES

1. Biely, P. (1985). Trends in Biotechnol. 3, 286-290.
2. Biely, P., MacKenzie, C.R., Puls, J., and Schneider, H. (1986). Bio/Technology, in press.
3. Greve, L.C., Labitch, J.M., and Hungate, R.E. (1984). Appl. Environ. Microbiol. 47, 1135-1140.
4. Bailey, R.W., Clarke, R.T.J., and Wright, D.E. (1962). Biochem. J. 83, 517-523.
5. Bailey, R.W. and Gaillard, B.D.E. (1965). Biochem. J. 95, 758-766.
6. Poutanen, K., Puls, J. and Linko, M. (1986). Appl. Microbiol. Biotechnol. 23, 487-490.
7. Desrochers, M., Jurasek, L., and Paice, M.G. (1981). Dev. Ind. Microbiol. 22, 675-684.
8. Biely, P., Puls, J., and Schneider, H. (1985). FEBS Letters, 186, 80-84.

Chapter 32

Solid substrate fermentation of cardboard to protein-enriched fermented fodder

A. S. El Nawawy, M. Allam, M. Musallam, A. Hammoud & R. Al-Daher – Kuwait Institute for Scientific Research, P.O. Box 24885, 13109, Safat, Kuwait.

INTRODUCTION

Lignocellulosic materials can be converted to protein-enriched fermented substrate by direct bioconversion, i.e., without prehydrolysis. This can be achieved by using cellulose degrading bacteria (Han and Callihan, 1975), fungi (Eriksson 1980; Zadrazil 1980; Chahal 1985; and Moo-Young 1982). The inclusion of white rot fungi for direct cultivation on lignocellulose for feed production was investigated by Ek and Eriksson, 1980; and Nilsson 1974; as these fungi produce cellulases besides being able to degrade lignin.

There are two processes for the production of microbial biomass from lignocellulosic wastes by direct bioconversion, i.e., submerged fermentation (SmF) process and solid substrate fermentation (SSF) process. SSF process has a bright future in advanced and developing countries due to its numerous advantages, such as superior productivity, simple technique, low wastewater output, improved product recovery, elimination of foam problems and dispersion with rigorous control of few parameters, over SmF. However, there are certain limitations such as heat build-up, bacterial contamination, problems in scale-up and growth estimation and difficulty in control of the moisture level of the substrate. It seems that the state of art in SSF is still at the stage where its industrial potential has not been fully exploited, due in large part to some intrinsic difficulties of controlling the solid culture process. On the whole, SSF may be an exclusive method for low cost technology.

In Kuwait, cardboard boxes (CB) are the major available source of lignocellulosic wastes, which can be easily collected from the commercial and industrial centres. It was investigated previously for inclusion in feeding sheep in Kuwait, after being supplemented with protein and mineral salts (Razzaque et al., 1982).

In the present work, CB is being investigated for direct bioconversion to protein-enriched fermented fodder in a solid state fermentation process. This was accomplished through incorporating cellulose decomposing mould (T. reesei) with strains able to degrade lignin as well as partially degrade cellulose (Sporotrichum pulverulentum and Pleurotus ostereatus).

EXPERIMENTAL

Samples of CB boxes were cut and ground to particles with one mm diameter. Strains of T. reesei NRRL 6156, Sporotrichum pulverulentum NRRL 6330 and Pleurotus ostereatus NRRL 2366 were used in the study. Separate starter from each strain was propagated on a modified version of Mandels and Weber medium (Hatakka and Pirhonen, 1985). Propagation of starter was at 30°C for three days in flasks put in a rotary shaker with 200 rpm. The resulting mycelium pellets were blended in a stomacher lab blender. The experiments were applied in sterilized polyethylene bags where 50 grams of grounded CB was placed; 150 ml of medium added plus 50 ml of the blended starter.

The experiments were performed in 96 bags to study four different treatments, each in duplicate. These treatments are:

a) The inoculum either T alone, or T + P, or T + S.
b) Carbon nitrogen ratio C:N - 12.5:1 or 25:1.
c) Nitrogen source ammonium sulphate or urea.
d) Period of incubation at 28°C for 2, 4, 6, and 8 weeks.

The bags were sterilized in the autoclave at 121°C for 30 minutes, cooled and then inoculated by starter as mentioned above. So every two weeks, two bags from each treatment were removed, dried and analyzed for their content of moisture, protein, cellulose, lignin and ash. Just after inoculation of each bag as well as every two weeks, filtered air was pumped to each bag through an inlet tube until it was completely filled, followed by mixing the substrate by shaking the bags to allow subjecting of all the particles of cardboard covered by mycelium to aerobic conditions. At the end of the experiments, samples from the bags that lasted for eight weeks were subjected to wet ashing followed by determination of minerals and heavy metals content.

For determination of cellulose, lignin, ash, and ether extract ASTM standard test methods D-1106-56 was applied. Crude protein was determined by the Kjeldahl method using a factor of N x 6.25; minerals and heavy metals by atomic absorption analyses of wet ashed samples.

RESULTS AND DISCUSSIONS

The results of protein, cellulose, lignin and ash in the residual dry weight are given in Table 1. This data indicated the following:

The protein content increased in the protein-enriched fermented fodder (PEFF) gradually, giving the maximum after eight weeks of incubation, especially when C:N was 12.5:1.

Ammonium sulphate seems to be suitable nitrogen source while on the contrary, urea was not suitable for the present case. In the presence of ammonium sulphate, the protein content was much higher than that in the case of urea, which indicated a good growth of moulds in the presence of ammonium sulphate but not in the presence of urea.

The cellulose content in the PEFF was much less than the original CB, which indicates that most of the loss of weight was mainly due to cellulose consumption. This result is more clear when calculating the percentage of cellulose consumed for PEFF production over the original amount of cellulose in CB. These calculations are indicated in Table 2. It was clear that cellulose consumption increased by time, and after eight weeks about 45% of the cellulose in CB was consumed, when using ammonium sulphate as the nitrogen source and whether C:N was 25:1 or 12.5:1, with slight increase in the latter case.

The lignin percentage in PEFF did not change significantly than that in the original CB. This indicate that the rate of lignin degradation was almost the same as that in loss of total weight of CB during conversion to PEFF. So, as in expressing the case with cellulose, the residual amount of lignin in PEFF, when subtracted from the original amount of lignin in CB will give the amount of biodegraded lignin. This amount when compared to the original amount of lignin will give the percentage of lignin biodegradation during PEFF production. This is clear in Table 3, where it was noticed that a range of 24-35% of the lignin was biodegraded after eight weeks incubation.

The highest protein content in PEFF was obtained after 4-6 weeks incubation, followed by a slight increase after eight weeks. The best result in the case of using T + P, followed by T and then T + S, where total protein content was 20.4, 17.6 and 16.9% after eight weeks, was when C:N ratio was 12.5:1. At C:N 25:1, the total protein content was 13.8, 13.9 and 11.8% in the same order.

The PEFF produced contain higher ash content than the original CB. This was ofcourse due to the minerals in the medium, whether that was gained in the microbial cells or attached to the residual PEFF. This led to increasing ash content from 6.80% in CB to 10-13% in the produced PEFF (Table 1). In the case of urea, the apparent low percentage of ash is due mainly to the decrease in loss of CB weight during fermentation period.

Table 1. Changes in the composition of cardboard in solid substrate fermentation over a period of two months*.

C:N Ration	Fungi Used	Ash (Week)				Lignin (Week)				Protein (Week)				Cellulose (Week)			
		2	4	6	8	2	4	6	8	2	4	6	8	2	4	6	8
Ammonium Sulfate (25:1)	T + P	10.9	14.1	12.9	13.7	16.3	16.3	15	17.5	5.5	10.7	12	13.6	53	51.8	43.9	46
	T + S	11.1	13.4	12.8	10.6	15.6	15.2	15.5	16.7	8.7	11.3	11.4	11.8	58.8	51.4	54.2	50
	T	11.0	12.6	12.9	11.6	15	13.3	15.4	15.0	10.4	10.2	11.0	13.9	61.6	54.2	50	48.1
Ammonium Sulfate (12.5:1)	T + P	11.6	14	11.2	11.8	16.8	16.9	16.7	18.7	15.6	16.6	17.1	20.4	48.3	44.6	48.3	45.7
	T + S	10.7	12.5	10.6	10	16.4	15.6	15.9	14.2	16.8	16.0	17.1	16.9	49	50.2	47.8	48.5
	T	10.4	12.2	11.3	10.9	16	15.4	16.3	15.8	11.8	16.7	16.8	17.6	54.1	50.5	45.2	47
Urea (25.0:1)	T + P	9.5	9.3	11	8.5	17.1	16.8	17.1	17.3	5	1.3	5.4	5.8	60.2	58	55.4	57.4
	T + S	9.3	9.8	10.2	8.8	16.6	17.2	16.2	18.2	7.8	8.3	9.9	9.5	60	50.4	58.2	57.7
	T	9.1	9.4	10.1	8.8	15.5	16.2	16.8	17.5	5.8	5.8	6.8	6.4	61.7	60	59.4	59.7
Urea (12.5:1)	T + P	9.3	10.7	10.6	8.6	16.9	17.6	15.9	16.3	6.7	7.7	7.8	11.2	58	62	61	58.3
	T + S	8.8	8.1	9.3	8.6	16.3	15.7	16.9	18.7	13.4	5.7	4.5	9.4	58.6	63.6	61	59
	T	9.2	9.3	9	8.7	16.5	16.3	16.5	17.3	12.3	7.7	8.9	7.4	56.3	63.1	61	57

T = *T. reesei*
S = *Sp. pulverulentum*
P = *Pl. ostereatus*

Initial CB composition %: Cellulose 67; Lignin 18; Ash 6.4; Alcohol extract 3.0; Crude protein 1.2.

*Data expressed as %

Table 2. Percentage of cellulose consumption in solid substrate fermentation over a period of eight weeks.

C:N Ratio	Fungi Used	Weeks 2	4	6	8
Ammonium sulfate (25:1)	T + P	23.87	35.32	47.92	45.89
	T + S	20.40	38.92	37.72	43.83
	T	16.21	35.22	41.86	45.12
Ammonium sulfate (12.5:1)	T + P	29.88	44.63	42.66	47.63
	T + S	29.75	37.49	42.95	47.04
	T	21.93	39.71	48.10	46.52
Urea (25:1)	T + P	10.01	18.41	36.39	34.13
	T + S	8.06	32.10	20.81	34.40
	T	9.44	17.53	21.07	30.57
Urea (12.5:1)	T + P	13.68	13.09	17.52	31.26
	T + S	13.57	9.71	16.96	31.49
	T	16.48	13.18	18.68	34.18

Table 3. Percentage of lignin biodegradation in solid substrate fermentation over a period of eight weeks.

C:N Ratio	Fungi Used	Weeks 2	4	6	8
Ammonium sulphate (25:1)	T + P	15.00	26.16	35.56	27.66
	T + S	26.34	42.12	34.86	28.25
	T	23.53	34.09	35.26	27.52
Ammonium sulphate (12.5:1)	T + P	11.51	24.10	27.94	24.44
	T + S	16.59	33.10	31.88	24.07
	T	14.57	29.51	30.98	25.01
Urea (25:1)	T + P	7.03	14.16	28.63	28.30
	T + S	17.41	19.12	19.10	29.20
	T	7.27	15.78	19.98	25.44
Urea (12.5:1)	T + P	8.58	10.14	22.09	28.16
	T + S	12.73	19.34	19.82	25.48
	T	10.98	18.36	15.68	20.96

Table 4. Minerals content of PEFF from cardboard in solid substrate fermentation by different fungi (g/kg)

C:N Ratio	Fungi Used	K	Na	Ca	P	Mg	Mn	Zn	Fe	Cu	Ni	B	Hg
Ammonium Sulfate (25:1)	T + P	18.8	3.7	6.1	11.9	2.6	1.6	0.10	0.90	0.02	0.001	0.02	0.006
	T + S	21.9	3.8	6.2	11.5	2	2	0.10	0.90	0.02	0.002	0.02	0.008
	T	18.9	3.2	6	12.9	1.7	0.2	0.13	1.0	0.02	-	0.02	0.005
Ammonium Sulfate (12.5:1)	T + P	17.9	3.9	7.9	12.6	1.9	1.5	0.11	0.95	0.02	0.001	0.02	0.005
	T + S	16.1	4.2	7.3	11.7	4.1	1.7	0.11	0.95	0.02	0.003	0.02	0.007
	T	17.4	4.5	6.7	11.7	5.5	1.5	0.13	0.85	0.04	0.004	0.02	0.007
Urea (25:1)	T + P	13.7	3.8	4.5	10	1.5	1.2	0.09	0.55	0.01	-	0.02	0.006
	T + S	20.6	4.4	7.5	12.1	2.8	1.7	0.12	0.95	0.02	0.002	0.02	0.004
	T	17.1	4.2	6.2	10.3	2.5	1.5	0.09	0.95	0.02	0.006	0.02	0.004
Urea (12.5:1)	T + P	14.9	3.3	5.3	8.9	1.4	1.2	0.20	0.56	0.07	-	0.02	0.006
	T + S	14.6	3.3	4.8	9.1	1.6	1.2	0.08	0.21	0.02	-	0.02	0.005
	T	14.6	3	5.6	9.6	1.6	1.4	0.08	0.62	0.02	-	0.02	0.005
Initial CB		0.3	-	4.3	0.07	0.4	0.4	0.07	0.2	0.02	0.02	0.03	-

Cd, Cr, Mo and Pb were also analyzed but found to be absent.

T = *T. reesei*

S = *Sp. pulverulentum*

P = *Pl. ostereatus*

\- = Absent

When comparing the minerals content in PEFF after eight weeks (Table 4) with that in the original CB, it was clear that the product of PEFF became rich in K, Na, Ca, and P. Also the concentration of Mg and Mn had increased more than 4-5 times than that present in the original CB. Also, there was a slight change in the concentration of Zn, Fe and Cu. At the same time, the heavy metals Cd, Cr, Mo and Pb were not detected in PEFF. This data indicates that PEFF, besides being rich in protein, is also rich in minerals required for feeding and free from toxic heavy metals.

However, further investigations should be carried out to optimize the parameters for producing PEFF, followed by scaling up investigations. Such parameters mainly are the efficiency of mixing and aeration, pH and temperature adjustment, and maximizing cellulases activity in solid substrate fermentation. At the same time, the concentration of minerals could be adjusted also to fit feeding investigations.

In conclusion, it was clear that using more than one fungus in SSF had reflected higher cellulose consumption, lignin biodegradation, and protein production. The lignin biodegradation in SSF is higher than that at SmF, due to the deleterious effect of agitation on lignin degradation as indicated by Kirk et al., (1978). Also, it should be noted here that analyses of total protein are based on nitrogen percent x 6.35; and this includes nitrogen of microbial protein soluble nitrogen due mainly to enzymes produced, and the residual inorganic nitrogen given in the media. This is indicated when the PEFF product was washed and filtered. The protein nitrogen in washed PEFF and the soluble true protein in the filtrate gave the actual organic protein in PEFF, while the difference indicates the residual inorganic nitrogen as indicated in Table 5.

Table 5. Nitrogen percentage in washed PEFF.

C:N Ratio	Fungi used	A PEFF Before Washing	B PEFF After Washing	C Nitrogen of True Protein	Inorganic* Nitrogen
Ammonium sulphate (25:1)	T + P	2.072	0.864	0.7424	0.4666
	T + S	1.992	0.712	0.472	0.808
	T	2.224	0.704	0.584	0.936
Ammonium sulfate (12.5:1)	T + P	3.168	1.245	0.816	1.107
	T + S	3.088	0.8496	0.7296	1.5108
	T	2.96	0.904	0.6178	1.439

*Inorganic nitrogen = A-(B+C)

REFERENCES

American National Standard Technical Association of Pulp and Paper Industry (1977). Standard Test Method for Lignin in Wood, ASTM D 1106-56.

Chahal, D.S. (1985). Solid state fermentation with T. reesei for cellulase production. Applied and Environmental Microbiology, January 1985, pp 205-210.

Ek, K. and K.E. Eriksson (1980). Utilization of white-rot fungus Sporotrichum pulverulentum for water purification and protein production on mixed lignocellulosic wastewaters: Biotechnology & Bioengineering, 22: 2273-2284.

Eriksson, K.E. (1978). Enzyme mechanism involved in cellulose hydrolysis by the rot fungus Sporotrichum pulverulentum. Biotechnology & Bioengineering, 20: 317-332.

Han, Y.W. and C.D. Callihan (1974). Cellulose fermentation: effect of substrate pretreatment on microbial growth. Appl. Microbiology, 27: 159-165.

Hattaka, A.J. and T.I. Pirhonen (1985). Cultivation of wood-rotting fungi on agricultural lignocellulosic materials for the production of crude protein. Agricultural Wastes, 12: 81-97.

Kirk, T.K., E. Shultz, W.J. Cannors, L.F. Lorenz, and J.G. Zeikus (1978). Influence of culture parameters on lignin metabolism by Phanerochaete chrysosporium. Archives of Microbiology, 117: 277-285.

Moo-Young, M. (1982). The Waterloo SCP process: direct conversion of cellulosic materials into proteinacious foods. Industry and Environment, 5(2): 30-31.

Nilsson, T. (1974). Comparative study on the cellulytic activity of white-rot and brown-rot fungi. Material and Organismen. 9: 173-198.

Razzaque, M.A., A.R. Suleiman, M.F. Diab and A.J. Salman (1982). Growth performance of sheep feed rations containing corrugated cardboard and poultry manure. KISR Annual Research Report 1982, pp. 9-12.

Zadrazil, F. (1980). Conversion of different plant waste into feed by Basidiomycetes. European Journal of Applied Microbiology and Biotechnology, 9: 243-248.

Chapter 33

Enzymic conversion of cellulose in cardboard for yeast production

A. S. El Nawawy, E. El Rayes and R. Al-Daher – Kuwait Institute for Scientific Research, P.O. Box 24885, 13109, Safat, Kuwait.

INTRODUCTION

The production of enzymes capable of breaking cellulose have been extensively studied during the last 20 years. The studies which cover the development of microbial strains, production media and process control have played a major role in the achievement of high levels of extracellular cellulase production. Trichoderma species, are considered by many workers to be the best producers of a complete system of extracellular cellulolytic enzymes (Mandels and Weber, 1969). Several T. reesei mutants especially QM9414 (Mandels et al., 1971) have been shown to be capable of producing higher amounts of cellulase than wild strain QM6a.

In the present work, T. reesei QM9414 was propagated for enzymic conversion of cellulose in cardboard. The crude cellulase suspension were evaluated under process control. The resulting sugar hydrolyzate was utilized for propagating yeast strains capable to utilize both xylose, glucose and cellobiose.

EXPERIMENTAL

Starter cultures grown on T. viride medium containing 2% glucose, 0.1% Tween 80 were kept on a shaker at 200 rpm; 30°C for T. reesei QM9414 (NRRL 6156). For cellulase production, 60 flasks (capacity 250 ml) each containing 50 ml of T. viride basal medium provided with 0.5% cardboard, 0.05% glucose and 0.1% Tween 80 were inoculated with 5 ml of the above starter mycelium pellets suspension, then incubated as above for eight days. Cellulase activity and protein in the filtrate were assayed. The filtrate of the cellulase production culture was used to hydrolyse the ground cardboard suspended at concentrationsof 1, 2, 3, 4 and 5% in 0.05 M sodium citrate buffer pH 4.8 for 12, 14, 24, 48 and 72 hours at 50°C and with shaking at 150 rpm in a total volume of 100 ml, 50 flasks were used for that. The suspension was filtered under vacuum, washed, and the residue was dried and weighed, while the filtrate was analysed for glucose and reducing sugar determination.

T. reesei QM9414, was propagated in the fermenter (7 litre capacity) on Mandels and Webers medium at 28°C containing 1% cardboard (CB) as the sole carbon source. After eight days, the cellulases in the medium were precipitated with ammonium sulphate (80% saturation), centrifuged in a refrigerated centrifuge at 12,000 g/20 minutes. The precipitates were dissolved in 0.01 M sodium citrate buffer (317.5 ml), desalted by dialysing in collodion sacks against the same buffer. The resulting cellulase activity was 1150 U/ml. giving a total activity of 365125U/317.5 ml. This amount was used to convert 105 cellulose present in 160 g cardboard within 48 hours at 50°C. The enzymes were able to produce 72 g reducing sugar, i.e., 68.6% of the cellulose in CB. The sugar solution was adjusted by evaporation to give 2% reducing sugar, supplemented with ammonium sulfate 0.6% and 0.2% K_2HPO_4 + 0.1% yeast extract and used as a media for yeast propagation (pH 5.0).

Two strains of yeast, i.e., Candida utilis and Candida tropicalis have the ability to assimilate glucose, xylose and cellobiose. They were propagated on the sugar syrup media. The experiment was carried out in the New Brunswick fermenter (7 litre capacity) with two litres working volume. The starter 200 ml was prepared in shake flasks at 28°C for two days. After inoculation, the fermentation started at 28°C with continuous stirring at 400 rpm and aeration 1 vol/vol/minute for 24 hours. At the end, the culture suspension was centrifuged and washed. The precipitate of yeast cells were dried at 85°C overnight, weighed and its protein content was determined by the Kjeldahl method. The filtrate was tested for remaining reducing sugar and pH.

Reducing sugars were estimated by DNS method (Miller 1959; Rosenberg and Oberkotter, 1977) and glucose by Glu-CINET (I.S.C.T. Sclave, SOA, Siena, Italy). The carboxymethyl cellulase (CMC ase) assay (endoglucanase; 1,4-B-D-glucan glucanohydrolase E.C.3.2.1.4), cellobiase assay (B-glucosidase, E.C.3.2.1.21) and filter paper assay (exoglucanase or 1,4-B-D-glucan cellobiohydrolase E.C.3.2.1.91) were estimated by methods described in Rosenberg and Oberkotter (1977), soluble protein was estimated by the Lowry method (Lowry et al., 1951).

RESULTS AND DISCUSSIONS

Susceptibility of Cardboard, Sawdust and Newspaper to Crude Cellulases of *T. reesei:*

The enzyme activity of T. reesei 6156 crude cellulase filtrate (Table 1) was used to hydrolyze CB present in buffer suspension at different concentration (of CB) during 12, 24, 48 and 72 hours. The results shown in Table 2 indicate that cellulose in CB was converted gradually with time by the crude cellulases, giving the highest conversion after 48 hours, with a slight increase in some cases after 72 hours. The crude cellulases present were able to convert 60.8 up to 96.1% of the cellulose in CB according to the concentration of CB in the reaction suspension (Table 2). When the CMC ase was calculated per each gram of cellulose (Fig. 1), it was found that 2000 CMC ase units per gram of substrate was enough to convert 74% of 1.98 g of cellulose within 48 hours, i.e., about 1000 CMC ase unit/g cellulose results in 74% conversion to reducing sugars within 48 hours; when 1000 CMC ase units were produced 8.0 FPU were also produced, which is in accordance with that given by Blanch and Wilke, 1982. However, as CB is considered treated cellulose, so it was decided in this work to concentrate the discussions on CMC ase instead of FPU.

However, it is not the simple relation between enzyme units and substrate concentration that effect the rate of hydrolyses, but there are a lot of interferring factors as was indicated in the present study, i.e., substrate composition (cardboard, newspaper, sawdust). Also Reese and Mandels, 1980 emphasized the sensivity of cellulases to temperature, pH, incubation time, and high substrate and production concentrations, non-sterile environment, presence of foreign molecules and shear stress caused by agitation.

Results on various substrates are clear in Table 3. Analysis indicate that the glucose was only part of the reducing sugars, which includes also xylose, arabinose and cellobiose. It seems that this strain is not so active in β-D-glucosidase. The tested moulds are known to produce hemicellulases as well as cellulases (Linko et al., 1978).

The substrate concentrations used in the present study were 1-5% and 60.8% conversion to sugars was obtained after 48 hours when using 5%. However, this concentration can be reconsidered later and increased by using step-wise addition of substrate as recommended by Blanch and Wilke, 1982.

Utilization of the Sugar Hydrolyzate for Production of Yeast

The results of the two strains given in Table 4 indicate that both C. utilis and C. tropicalis can utilize most of the reducing sugars present in the sugar hydrolyzate of CB. The

Table 1. Cellulases activity of *T. reesei* 6156 after eight days.

Microorganisms Strains	Protein concentration (mg/ml)	Cellulase Activity (u/ml)		
		β-D-glucosidase	CMC ase	F.P. ase
T. reesei	0.24	0.027	65.500	0.528

Table 2. Effect of *T. reesei* 6156 cellulases on the hydrolyses of cellulose in cardboard during 72 hours.

Cardboard (g)	Cellulose (g)	Hydrolyzed Cellulose % After Incubation			
		12	24	48	72
1.0	0.66	86.8	90.2	96.1	96.5
2.0	1.32	59.8	75.0	82.7	86.9
3.0	1.98	57.7	76.7	74.6	72.0
4.0	2.64	45.4	61.0	65.4	64.3
5.0	3.30	47.5	56.1	60.8	60.9

1 ml of crude enzyme contains 65.00 I.U. CMC ase

Table 3. Converted cellulose percentage in respect of substrate types and incubation time using *T. reesei* 6156 cellulase.

Substrate*	Cellulose Conversion (%)				
	Expressed as	Incubation Time (hrs)			
		6	18	24	48
Cardboard (1654)	Loss of weight	48.8	57.0	56.2	62.0
	Reducing sugar	26.9	41.5	40.6	46.3
	Glucose	8.3	8.1	8.4	8.6
Newspaper (2060)	Loss of weight	44.9	53.3	49.1	56.0
	Reducing sugar	20.9	31.6	33.5	40.4
	Glucose	11.4	10.7	10.7	8.8
Sawdust (1915)	Loss of weight	21.8	19.1	12.6	13.0
	Reducing sugar	8.3	7.6	4.7	5.3
	Glucose	0.3	0.6	0.5	0.8

*CMC ase units/gm initial cellulose

Substrate used was 3 g/100 ml

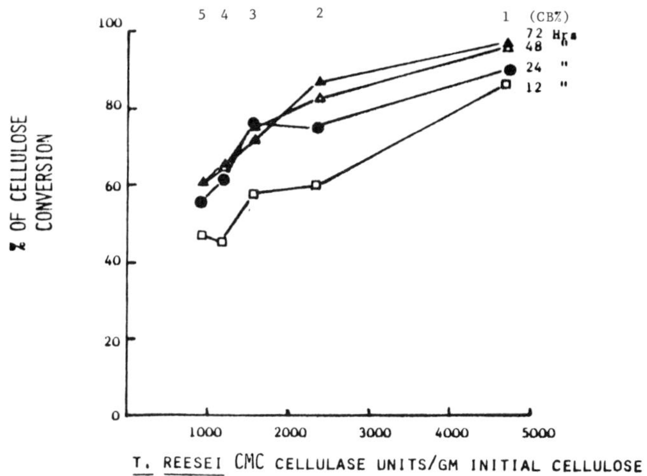

Fig. 1. Effect of *T. reesei* CMC ASE units/substrate ratio on cellulose conversion at different hours.

Table 4. Utilization of sugar in cardboard hydrolyzate for yeast production.

Strains	A Sugar consumed (g)	B Yeast Produced (g)	C B/A	D Protein in Yeast (%)	E (g)	F E/A	G B/X	H E/X
Candida utilis	19.5	10.15	52.1	46.6	4.73	24.2	50.8	23.7
Candida tropicalis	18.5	8.3	44.8	47.5	3.94	21.3	41.5	19.7

X = initial gram sugar (20 gm/1)

percentage of yeast yield compared to consumed sugar (column C) was 52.1, 44.8 for the two yeast strains respectively; but when compared to the initial total reducing sugars (column G), the percentage was 50.8, and 41.5 respectively. Column C indicate the efficiency of yeast growth on consumed sugar, while column G indicatesthat compared to the total sugar in the hydrolyzate. The highest amount of protein was produced by *C. utilis* followed by *C. tropicalis* (column E). When the amount of protein was compared at consumed sugar (column F), and total sugar (column H), the superiority of *C. utilis* was also clear.

Thus, by calculations, each 100 g of reducing sugar in the CB hydrolyzate can give 50.8 g of yeast (column G) with 23.7 g protein (column H), when using *C. utilis*, with lower amounts when using *C. tropicalis*.

Accordingly, *C. utilis* can be selected for further future investigations for producing fodder yeast on CB sugar hydrolyzate, which can be mixed to the residue of CB to form protein enriched fodder. Just an example: if we start with 1 kg of CB to be hydrolyzed chemically or enzymatically, it is possible to have about 500 g reducing sugar in the filtrate, leaving 500 g residue. From the 500 g reducing sugar, *C. utilis* can produce 254 g yeast with 118.5 protein. When the yeast is mixed with CB residue (which contain undegraded parts of cellulose + lignin + minerals) it will result in 754 g product of protein-enriched fermented fodder containing 118.5 g protein i.e., with 15.7% protein.

REFERENCES

Blanch, H.B., C.R. Wilke (1982). Sugars and chemicals from cellulose. Rev. Chem. Eng. 1: 71-119.

Linko, M., P.M. Makkanen, M.J. Bailey and M. Leisola (1978). Production of cellulases and hemicellulases by *T. viride* (In: Bioconversion Symposium, T.K. Ghose (ed.), New Delhi, pp. 329-350).

Lowry, O.H., N.J. Rosenbrough, A. L. Farr and R.J. Randal (1951). Protein measurement with the folin phenol agent. J. Biol. Chem. 193: 263-275.

Mandels, M. and J. Weber (1969). The production of cellulases. In: Cellulases and their application, Adv. Chem. 95: 263-414. Hajny, G.J. and E.T. Reese (eds). American Chemical Society, Washington, D.C.

Mandels, M., J. Weber and R. Parizek (1971). Enhanced cellulase production by mutant of *T. viride*. Applied Microbiology, 21: 152-154.

Miller, G.L. (1959). Use of dinitrosalicyclic acid reagent for determination of reducing sugar. Anal. Chem. 31: 426-428.

Reese, E.T. and Mandels (1980). Stability of the cellulose of *T. reesei* under use conditions: Biotechnol. Bioeng. *22:* 323-335.

Rosenberg, F. A. and L.Y. Oberkotter (1977). Methods for determining microbiol biodegradation in cellulose. In: Biodeterioration Investigation Techniques, pp. 243-259. A.M. Walters (ed.), Applied Publisher Ltd., London.

Chapter 34

Stack-ammoniation of mature cellulosic forages using ammonium salts

R. D. Mills – Brotherton Chemicals Ltd, Calder Vale Road, Wakefield, Yorkshire, WF1 5PH. England.

V. C. Mason – Animal and Grassland Research Institute, Hurley, Maidenhead, SL6 5LR, Berkshire, UK.

J. F. Kennedy – Research Laboratory for the Chemistry of Bioactive Carbohydrates and Proteins, University of Birmingham, P.O. Box 363, Birmingham B15 2TT, England.

C. A. White – Chembiotech Ltd, Institute of Research and Development, Vincent Drive, Birmingham B15 2SQ, England.

SUMMARY

The upgrading of straw for use as an animal feed by low temperature ammoniation in a stack using volatile and nonvolatile ammonium salts has been demonstrated. This method, which requires the minimum of purpose-built equipment, can be performed on the farm by farm workers and provides a beneficial alternative to the unfavourable field burning for the utilization of waste straw. The effect of added water is considered in *in vitro* assessment whilst *in vivo* assessment of the digestibility of upgraded straw, by a number of routes, is made using sheep and cattle.

INTRODUCTION

Much of the cereal straw produced in the UK is currently burned in the field, this being the most economic method for its disposal. However there is pressure, both from within the UK and from outside (particularly from within the European Community) to ban the practise of straw-burning. Since there is a very strong probability of legislation being passed within the next 10 years, alternative methods for using straw are now being actively studied. The simpler methods involve direct soil incorporation (which incurs costs of chopping and of additional nitrogen supplementation, prevents direct drilling and may result in the production of organic acids thereby causing inhibition of germination of subsequent crops) and use as a fuel [1]. The major drawbacks to its use as a fuel are its low calorific value (half that of coal) and low bulk density which makes long-distance haulage uneconomical. Recent research is, however, allowing development of novel, energy-economic compaction techniques which permit more efficient transporation and combustion [2].

Alternative approaches include the utilization of straw for the production of useful compounds such as fine chemicals, industrial chemicals or biofuels. However, the low world price for sugar and the resistance of straw to hydrolysis have made this approach unattractive, although the use of salts (particularly lithium chloride in hydrochloric acid) has improved the bioavailability of straw prior to enzymic or acid hydrolysis to fermentable carbohydrates [3]. Improved production of cellulose- and hemicellulose- degrading enzymes by genetic engineering is also providing an economic route for the biodegradation of straw [4] but utilization of straw in these biotechnological applications requires transportation of the straw to the fermentation plant and can require extensive precautions to be taken to ensure sterility. In an attempt to overcome the problem of contamination of the straw by organisms other than those required to degrade the straw in the fermentation system, a study has been undertaken to improve the production of cellulose-,hemicellulose- and lignin- degrading enzymes in those organisms most commonly encountered in straw [5] namely Penicillium simplicissimum and Aspergillus fumigatus. The use of A.fumigatus is partucularly attractive due to its well-balanced range of cellulases and xylanases [6] and its ability to degrade lignin [7] and the use of such natural organisms for biodegradation to fermentable sugars or production of enzymes is considered a potential method for utilizing waste straw [5].

Alternative, low technology methods for utilization of waste straw which can be performed on the farm, thereby eliminating the high cost of transportation of straw, include its use for mushroom production [8] which currently uses about 5% of the annual wheat straw production or as a feedstuff for livestock. The total energy content of straw is similar to that of hay but much of it is unavailable to animals despite the ability of ruminants to digest most plant materials rich in cellulose. Biological silage additives have been developed which improve the nutritional value of straw at the same time as providing a method for its long term storage [9]. The use of chemicals, in particular alkalis such as sodium hydroxide, ammonia and calcium hydroxide, has been shown to improve the dry matter digestibility in cereal and grass straws by increasing the amount of damaged areas to which bacteria and other organisms, from the rumen, can attach and initiate the digestive processes. Whilst untreated straw or straw treated with calcium or sodium hydroxide requires supplementation with protein, the use of ammonia overcomes this problem [10] and additionally does not require mechanical processing of the straw prior to upgrading.

Whilst specialist, commercially available ovens, permanent buildings or insulated containers can be used for the treatment, it is more economic, by a factor of between 200 and 800, to use treatment within a polythene-clad stack comprising up to 20 tons of straw at any one time. Provided the stack is properly constructed and sealed with polythene sheeting no additional equipment is required. The ammonia is normally applied by specialised contractor using aqueous ammonia.

It is this last method of straw utilization which we have studied using a hitherto unexploited source of ammonia, namely ammonium salts and ammonium fertilizers in preference to anhydrous or aqueous ammonia or the suggested precursor to ammonia, urea. The unavailability of anhydrous or aqueous ammonia in many parts of the world and its volatility limits its widespread adoption for ammoniation whilst the use of urea has often produced disappointing results [11]. The use of stable ammonium compounds is attractive since ammonia can be released, when required, by heat or reaction with a stoichiometric excess of alkalis such as sodium hydroxide or calcium hydroxide in their industrial forms and could obviate the need for specialised contractor applications. By control of the amount of alkali used, the problem of excess urine excretion associated with alkali treated straws is also overcome. This paper examines the feasibility of exploiting this principle of ammoniation to upgrade mature forages by the oven and stack methods.

EXPERIMENTAL

In vitro assessment of dry ammonia sources

Four sources of ammonia were compared under laboratory conditions to assess their acceptability. Samples of chopped barley straw (cv Sonja) were treated, in Kilner jars (1 litre), with ammonia (at the rate 35g/kg straw dry matter) released by volatilization and/or by reaction with a stoichiometric excess of quicklime (70g/kg straw dry weight) or slaked lime (92.5g/kg straw dry weight). The ammonia sources used were anhydrous ammonia, ammonium 'carbonate' (a mixture of ammonium carbamate and ammonium bicarbonate), ammonium bicarbonate (double water and lime rates used) and ammonium sulphate. These were mixed with alkali, where appropriate, and placed in small glass vials within the Kilner jars such that no direct contact with the straw occurred. One set of jars were stored at 20°C for 5 weeks, to simulate stack treatment and another set stored at 90°C for 15 hours, to simulate an oven treatment. The organic matter digestibility was assessed in vitro using rumen fluid taken from sheep to digest the straw samples under laboratory conditions.

In vitro assessment of ammonia sources mixed with water

The same experiment was repeated with the exception that

all ammonia sources were mixed with water (120g/kg straw dry weight).

Production of upgraded straws

Baled, perennial ryegrass hay (cv Cropper) of medium quality (ca 1.5% N) was upgraded either by one of two oven methods or by one of two stack methods. Conventional oven upgrading of 440 kg hay dry weight was performed at 90° for 15 hours using a purpose-built, insulated An-Stra-Verter oven with anhydrous ammonia (35g/kg hay dry weight) being provided from an external pressure tank. Experimental oven upgrading of a similar quantity of ryegrass hay was performed using the same oven, but with in situ generation of ammonia. This was acheived by treating a mixture of ammonium sulphate (fertilizer grade, 58.6 kg) and quicklime (33kg) with water (69kg) in a trough placed in the bottom of the oven prior to heating. The two stack-ammoniation methods were performed using 465 kg hay dry weight treated with ammonia (35g/kg hay dry weight) at ambient temperature over a period of 8 weeks (July/August, 1984). The ammonia was generated by mixing either ammonium sulphate (fertilizer grade, 63 kg), quicklime (33kg) and water (60 kg) or ammonium 'carbonate' (ca. 310g NH_3/kg dry solid, 44kg) and water (50kg) in metal troughs placed close to the stack under the plastic sheeting.

Baled winter barley straw (cv.Igri) was upgraded by one of three upgrading techniques. Conventional oven upgrading, as described above, was performed on 500kg straw dry weight using anhydrous ammonia (35g/kg dry weight) Stack-ammoniation was performed on 500kg straw dry weight in a similar manner to that described for ryegrass hay except that the ammonia was generated by mixing ammonium sulphate (fertilizer grade 66 kg), quicklime (35kg) and water (60kg). In the third method a similar quantity of straw was treated with sodium hydroxide (50g/kg straw dry weight) using a J.F. Straw Processor and then ensiled.

***In vivo* trials with sheep**

Digestibility trials were conducted on Suffolk half-bred wethers (ca. 44 kg live weight), using 6 animals per treatment. Each study consisted of a 16 day preliminary period and a 10 day collection period. Quantities of each of the upgraded ryegrass hays or untreated ryegrass hay was chopped and fed ad libitum, each animal also receiving a daily pelleted supplement providing soya bean meal (1.4g/kg live weight), mineral-vitamin mixture (0.5g/kg) and sodium sulphate (0.1g/kg) with water being available at all times. Organic matter intake, organic matter digestibility, digestible organic matter intake and nitrogen retained were assessed by weighing total input and excrement with analysis of nitrogen by the kjeldahl method.

In vivo trials with cattle

Digestibility trials were conducted on 4 British Freisian steers (ca. 248 kg live weight), tethered in individual standings. Each animal was fed with untreated barley straw or upgraded barley straws (2.8kg/day) according to a randomised Latin Square sequence, together with rolled barley (1.8kg/day) and fishmeal (0.2 kg/day) in diets designed to satisfy approx. 1.3 times the maintenance requirement for energy. The untreated and sodium hydroxide treated straws were further supplemented with urea (ca. 40g/day) to equalise the nitrogen intakes. Water was available at all times. Each trial consisted of a 21 day preliminary period and a 5 day collection period. Organic matter digestibility and dry matter digestibility were assessed by difference between input and excretion weights for each animal.

RESULTS AND DISCUSSION

From the in vitro assessment of dry ammonia sources it has been shown (Table 1) that oven treatment gave improvements in the organic matter digestibility which were slightly inferior or at best equal to those obtained with the positive ammonia control (i.e. anhydrous ammonia treatment). The use of ammonium sulphate gave little or no improvement in the organic matter digestibility. In contrast, the use of the lower temperature

Table 1. Upgrading barley straw (cv. Sonja) using different dry ammonia sources in the absence of water.

	Organic matter digestibility in vitro (g/kg)				
	Control	Source of ammonia			
Alkali added	(no ammonia)	anhydrous ammonia	ammonium 'carbonate'	ammonium bicarbonate	ammonium sulphate
Oven treatment					
None	329	432	376	411	303
Quicklime	364	458	408	401	324
Slaked lime	319	432	435	420	349
Stack treatment					
None	315	433	339	338	322
Quicklime	325	426	349	323	325
Slaked lime	301	455	379	362	350

Standard error difference = 23.5

Table 2. Upgrading barley straw (cv. Sonja) using different ammonia sources in the presence of water.

	Organic matter digestibility *in vitro* (g/kg)				
	Control	Source of ammonia			
Alkali added	(no ammonia)	aqueous ammonia	ammonium 'carbonate'	ammonium bicarbonate	ammonium sulphate
Oven treatment					
None	283	442	467	418	327
Quicklime	294	427	450	453	438
Slaked lime	296	434	449	443	442
Stack treatment					
None	311	440	374	363	344
Quicklime	325	426	349	323	325
Slaked lime	324	466	406	478	473

Standard error difference = 23.5

stack-ammoniation indicated that none of the potential ammonia sources gave any significant increase in the organic matter digestibility and only the positive ammonia control gave any improvement. The effect of lime was inconsistent.

In the presence of added water (Table 2) the situation is much improved with ammonium sulphate giving efficient upgrading when mixed with quicklime or slaked lime in both simulated oven and stack conditions. In addition, for simulated stack conditions, the use of ammonium 'carbonate' and ammonium bicarbonate gave satisfactory improvements when slaked lime was included in the process. The beneficial effect of water was further demonstrated by the improved organic matter digestibility of the straw treated with aqueous ammonia compared with anhydrous ammonia. This therefore indicates that both volatile and non-volatile ammonium salts do have potential for the upgrading of straw as a stockfeed.

For the digestibility studies with sheep (Table 3) the two stack methods improved the digestibility of organic matter almost as efficiently as did the conventional oven ammoniation method. However, organic matter and digestible organic matter intakes for these treatments were inferior to those for the conventional oven method though superior to those for the control. In contrast, the experimental oven technique gave poor results, with only the values for retained nitrogen being greater than those obtained using the untreated straw.

For the studies involving cattle (Table 4) where intakes were standardized, the stack method gave an improvement in organic matter and dry matter digestibility which was very

Table 3. Upgrading perennial ryegrass hay (cv. Cropper) by ammoniation for use as a feed for sheep.

	Untreated	Oven-ammoniation		Stack-ammoniation		Standard error difference
		anhydrous ammonia	ammonium sulphate	ammonium 'carbonate'	ammonium sulphate	
Organic matter digestibility (g/kg)	636	690	619	671	680	10.0
Organic matter intake (g/kg live wt)	18.4	22.4	18.9	21.2	20.2	1.14
Digestible organic matter intake (g/kg live wt)	11.7	15.5	11.7	14.2	13.8	0.69
Nitrogen retained (g/day)	0.93	3.75	1.52	3.67	3.77	0.59

similar to that achieved with the conventional oven ammoniation method. In the case of organic matter this improvement was also close to that obtained with the J.F. Processor ensilage approach. In view of the high content of concentrate feeds in these diets, the improvements achieved were satisfactory.

In general, these experiments demonstrate that ammonium salts, including fertilizers, can be used as a soucre of ammonia for upgrading. The troughs used as reaction vessels in the stack treatments were rather inefficient since they provided no means for mixing the residues subsequent to the addition of water. This meant that ammonia produced in the base of each trough had difficulty in escaping to the surface of the residue. Clearly, the question of the structure of a simple reaction vessel which will permit the free movement of

Table 4. Upgrading winter barley straw (cv. Igri) for use as a feed for cattle.

	Untreated*	Oven - ammoniation (anhydrous ammonia)	Stack - ammoniation (ammonium sulphate)	Silo-J.F.* (sodium hydroxide)	Standard error difference
Organic matter digestibility (g/kg)	668	725	722	730	7.4
Organic matter intake (g/kg)	639	669	670	705	10.9

* supplement included urea

ammonia remains to be solved, as does the optimal ratio of chemical mixture to water.

A promising feature of this approach is that different ammonium salts and alkalis may be used to generate ammonia and they may also be used in complex combinations. We have successfully tested mixtures of ammonium sulphate, di-ammonium hydrogen orthophosphate, quicklime and sodium hydroxide for this purpose in laboratory studies. Thus the efficiency and rate of ammonia release can be altered, and the composition of the residue may be changed so that the latter can ultimately serve as a mineral supplement or lick for animals or as a fertilizer for soils. This may be used to reduce the effective cost of the ammonia treatment.

REFERENCES

[1] A.R. Staniforth, Straw for Fuel, Feed and Ferlilizer, Farming Press, Ipswich, 1982.

[2] T. Dalyell, New Scientist, No 1407 (1984) 28.

[3] T. Shambe and J.F. Kennedy, Enzyme Microb. Technol. 6 (1984) 169.

[4] J.S. Sandhu and J.F. Kennedy, Enzyme Microb. Technol. 6 (1984) 271.

[5] D.A.J. Wase, S. Raymahasay and C.W. Wang, Enzyme Microb. Technol. 7 (1985) 225.

[6] L.A. Harrison, J.B. Parry, J.M. Lynch and J.H. Slater, Proc. Lignocellulosic Biodeg. Conf. Glasshouse Crops Res. Inst., Littlehampton 1983 (Abstract of poster presentation.)

[7] V.S. Bisaria and T.K. Ghose, Enzyme Microb. Technol. 3 (1981) 91.

[8] R. Bisaria and M. Madan, Enzyme Microb. Technol. 5 (1983) 251.

[9] CTB Biotechnology News, 4 (23/24) (1984) 13.

[10] F. Sundstøl and E.M. Coxworth, In Straw and Other Fibrous By-products as Feed (eds F. Sundstøl and E.Owen Elsevier, Amsterdam, 1984, p196.

[11] V.C. Mason and E.Owen ARNAB Workshop, Alexandria, Egypt. 1985.

Chapter 35

Some dietary effects of the ingestion of sodium carboxymethylcellulose by man and rodents

D. M. W. Anderson – Chemistry Department, The University, Edinburgh EH9 3JJ, U.K.

SYNOPSIS

The Joint FAO/WHO Expert Committee on Food Additives (JECFA) recognised in 1984 that toxicological evaluations of viscous hydrocolloids in animals present special problems and called for additional information on caecal enlargement and the role of the gut microflora. The value of data derived from studies in Man was also recognised. This paper summarises (a) variations in caecal enlargements and faecal bulking effects produced in rodents by a range of carboxymethylcellulose (CMC) samples (b) variations in human physiological responses resulting from ingestion of prehydrated CMC. Implications of these variations are discussed.

INTRODUCTION

Since 1981, this laboratory has undertaken studies designed to contribute to demands made by JECFA and the E.E.C.'s Scientific Committee for Food for evidence of safety of hydrocolloids used as food additives. The studies, involving gums arabic (*Acacia senegal*) [1-3], karaya [4,5] and tragacanth [6], in rodents and in Man, have been reviewed [7]; it was concluded that there are no apparent correlations between the chemical compositions and structures of hydrocolloids and their metabolic and physiological effects. Because of their value, there have been requests [8] for the number of studies in Man to be increased. Dietary studies of xanthan [9] and CMC [10] in Man have therefore been made.

Experimental difficulties in administering viscous hydrocolloids to animals at sufficiently high dose levels, combined with the customary 100-fold safety factor [11], led to the award of invidiously low values of Acceptable Daily Intake (ADI) for xanthan (0-10 mg/kg) and CMC (0-25 mg/kg). Unfortunately, some food processors, and regulatory authorities in some countries, consequently regarded these additives as less desirable toxicologically than less viscous substances awarded the classification "ADI not specified". The problems involved in testing hydrocolloids have been recognised [8]; the latest JECFA Report [12] recommends that the gums be reviewed as a group.

Interest continues in the causes and effects of caecal enlargement produced in rodents by some hydrocolloids and the toxicological significance, if any, when extrapolated to Man. Additional evidence was requested and attention was drawn [8] to the role of gut microflora, on which the effect of carrageenan is particularly extensive [15]. A recent study involving one human volunteer [16] has confirmed reports [17-19] that complex polysaccharides induce qualitative and quantitative changes in aerobic and anaerobic gut microflora.

Earlier studies in Man [3,5,6] reported the average changes for five subjects; this concealed, unintentionally, the individual variations in response to ingestion of the Test Article. This was corrected in studies of xanthan [9] and CMC [10]; some conclusions from the latter study are discussed below. Data for faecal bulking efficiencies, and viscosities of some hydrocolloids, are also given. An explanation has also been sought for some differences in the results obtained from studies in Man by different investigators e.g. for xanthan [9, 20], CMC [10,21], gum arabic [3,21] and guar [22,23]. The value of guar in the treatment of hyperlipidemic [24] and diabetic [25] subjects has been attributed to various viscosity effects such as increase in intestinal unstirred layers [26], delay in gastric emptying [27], or inhibition of convective solute movement within the intestine [28]. The effect of simple inorganic salts on the viscosity of hydrocolloids may be important in this context.

MATERIALS AND METHODS

Origin of samples. The food-grade hydrocolloids studied were obtained from European merchants. The ten samples of CMC, differing widely in viscosity, were products of various European, Japanese and American companies.

Animal studies. A powdered breeders' high protein diet was fed to the control group (n=12) and, supplemented with 15% of each hydrocolloid, was fed for 21 days to the test groups. In all cases, water was added to give a thick paste. This diet, and water, were available ad lib. Animals were weighed on alternate days. Body weight gain for seven CMC samples equalled that of the control animals; for one CMC sample body weight gain slightly exceeded that of the controls, and for two CMC samples it was less. Faecal collections, obtained in metabolic

cages, were weighed as quickly as possible. At autopsy, the ends of the caecum were ligated in a standardised procedure; the excised caecum with contents was weighed immediately.
Viscosity measurements. Solutions (1%, w/w) in distilled water, or in a solution containing NaCl (3.5 g/ℓ), KCl (1.5 g/ℓ) and $NaHCO_3$ (1.5 g/ℓ), were measured (Brookfield Model LVF) after 24 hours at 20°C to allow complete hydration.
Human studies. The protocol was identical to that for dietary studies of gums arabic [3], karaya [5], tragacanth [6], and xanthan [9]. Following a 7-day control period, five males consumed 15 g CMC ("Akucell") daily, in three 5 gram prehydrated portions at intervals, for 23 days. The CMC was well-tolerated; there were no adverse effects in any volunteers. This daily intake exceeded, by a factor of 7.5, the current Group ADI for cellulosic products. Publication [10] of the wide range of measurements made before and after the test period has been delayed through replacement of the publisher originally involved in production of a new Journal.

RESULTS

Table 1 shows data for viscosity measurements, average faecal water contents (%), and filled caecal weights (g/kg body weight). Table 2 summarises some of the individual responses of volunteers after ingestion of CMC.

DISCUSSION

The viscosity of anionic electrolytes is reduced in salt solutions; some CMC samples are more salt-resistant than others. A mixture of salts, simulating human physiological conditions, gave (Table 1) decreased viscosities for CMC and gums arabic, karaya and tragacanth, but not for guar, a non-ionic galactomannan. This may provide a simple explanation for observations that the beneficial effects of guar, but not of other viscous hydrocolloids, for diabetic and hyperlipidemic subjects is frequently attributed to viscosity-related effects [24-28]. There are, however, other plausible modes of action, e.g. binding of sodium ions and alterations in gastrointestinal hormones [29]. The wide viscosity variations of guar samples has been noted [29]; close agreement between the results of separate studies cannot be expected unless experimental conditions are standardised and the same sample of guar is used.

CMC is a more efficient bulking agent (Table 1) than gums arabic, karaya, tragacanth, and guar. The CMC samples studied gave a range of faecal water contents. As for the guar samples (Table 1), faecal water content did not always correlate with solution viscosity although there is a tendency, with the exception of sample 7, for CMC of low molecular weight to produce high faecal wet weight. Size-exclusion chromatography

Table 1. Data for filled caecal weights and faecal water content after ingestion of CMC and other hydrocolloids.

Hydrocolloid added to rodent diet		Viscosity (cP) (1% solutions) in water	with salts present	Mol.wt. ($\bar{M}_w$x10^6)	Average faecal water content (%)	Average filled caecal weight (g/kg b.w.)
None, (control animals)		-	-	-	18.5	20.7
CMC, sample	1	35		0.46	73.2	89.4
	2	45	28	n.d.	69.8	87.0
	3	70		n.d.	68.5	75.9
	4	95		0.91	62.9	69.1
	5	140		n.d.	68.8	66.1
	6	540		3.0	62.1	51.0
	7	2,700		2.0	74.7	65.2
	8	3,400		1.6	64.5	51.0
	9	3,900		n.d.	54.3	n.d.
	10	4,500		2.9	57.2	81.5
Guar gum sample	1	4,900	5,000	0.24	35.7	n.d.
	2	3,000		n.d.	36.6	
	3	5,500		n.d.	22.3	
	4	5,000		n.d.	41.5	
Gum arabic		5	4	0.75	31.7	32
Gum karaya		1,470	800	8	28.3	22
Gum tragacanth		1,400	1,000	0.28	31.4	23

(Waters Model 590, 700 p.s.i., "trimodal set" columns, 0.5 mg/ml in 0.2M NaH_2PO_4) indicated that samples 1 and 10 had narrow molecular weight distributions; in contrast, sample 7 had a very broad molecular weight distribution with a large proportion of hemicellulosic low molecular weight material.

The average filled caecal weights (Table 1) also show a tendency to be high for low molecular weight samples, except for samples 7 and 10. Frequently, only the degree of substitution and the solution viscosity are quoted for CMC, but different manufacturers start with different raw materials and employ different reaction processes and kinetics. This leads to differences in regularity of substitution and differences in the position of substitution; as a result, a wide range of CMC products with differing functionalities for particular applications can be marketed. The fine structural differences involved may also differ in their stability to enzymatic and

Table 2. The effect of dietary CMC on some human faecal, serum, and biochemical parameters

Parameter measured	Volunteer A	B	C	D	E	Averaged result
Transit time, h	down	down	up	down	down	down
Faecal wet weight	up	up	up	up	up	up
" dry weight	up	up	up	up	up	up
" fat	up	up	up	up	down	up
" V.F.A.	up	up	down	down	up	up
" bile acids	up	up	up	up	up	up
" neut.sterols	down	down	down	down	down	down
Serum cholesterol	up	same	down	down	up	down
" triglycerides	down	down	up	up	up	up
" phospholipids	up	down	down	up	up	up
Plasma urea	down	same	down	down	down	down
" bilirubin	up	down	up	up	down	up
" AST	down	up	down	up	up	up
" alk.phosphatase	same	up	up	up	up	up
" protein	up	up	same	same	up	up
" albumin	up	down	same	up	up	up

other modes of degradation, and may produce different physiological effects. There can be little surprise that previous studies, in which conditions were not standardised, have led to some different conclusions. Healthy or abnormal subjects were used; prehydrated gels were used in some studies and heat-treated (baked) samples, in which re-hydration properties may have been destroyed or impaired, in others. With the wide range of CMC products available, agreement between studies cannot be expected unless the same Test Article (e.g. a sample of CMC from one production batch) is used.

Study in Man. The mechanism whereby dietary changes influence changes in motility and secretion in the G.I. tract, possibly through alteration of neurotransmitter enzyme activities, remains unclear [7,30]. Further studies of hydrocolloids were therefore desirable in Man, particularly in view of the JECFA recommendations [8,12] and the desirability of clarifying potentially adverse specific comments such as "one report" (no literature reference cited) "suggests that CMC may cause intestinal obstruction, but is generally safe" [31].

The wide range of measurements made before and after the test period showed that ingestion of CMC had no significant effects on plasma biochemistry; haematological indices; urinalysis parameters; glucose tolerance; and serum cholesterol, triglycerides and phospholipids. Faecal wet and dry weights, and bile acids, increased moderately for some volunteers and more extensively for others. Excretion of neutral sterols decreased for all volunteers. Table 2 summarises the variation in volunteers' responses: they are similar to those observed for other hydrocolloids [3,5,6,9,10]. The individual responses may reflect differences in the gut microflora and the previous history of each volunteer; it is becoming clear that individuals may respond in different ways to the ingestion of hydrocolloids that act as forms of dietary fibre. Different conclusions for studies may well be reached if averaged data for small numbers of subjects are considered. Large numbers of single case studies are required; experimental conditions should be standardised if a greater measure of agreement between the results of different investigators is to be expected. Fine structural differences, and the effects of the impurities and microbiological contaminants associated with all complex natural hydrocolloids, may be more extensive and play more significant roles than hitherto realised.

The data obtained [10] do not reflect any adverse toxicological effects resulting from the ingestion of CMC at high dose levels. Indeed, the data support the contention that the hydrocolloids permitted as food additives within the EEC are not only safe, under the conditions specified, and provided they conform in all respects to the established criteria of identity and purity; they are also desirable dietary components. In particular, the data for CMC [10] do not indicate any decreased margin of dietary safety when compared with the data [e.g.3] available for additives awarded the ADI code "not specified". There appears to be no reason for CMC to be restricted invidiously to ADI 0-25 mg/kg as a result of difficulties involved in feeding animals at the dose levels necessary to gain a higher ADI.

Acknowledgements. I thank Dr. M.A. Eastwood for access to animal house facilities, and Mrs. R.G.A. McNab and Miss F.J. McDougall for assistance.

REFERENCES

1. Anderson, D.M.W., Ashby, P., Busuttil, A., Eastwood, M.A., Hobson, B.M., Ross, A.H.M. and Street, C.A., Toxicology Letters, 14 (1982) 221-227.
2. Ross, A.H.M., Eastwood, M.A., Brydon, W.G., Busuttil, A., McKay, L.F. and Anderson, D.M.W., British J.Nutrition, 51 (1984) 47-56.
3. Ross, A.H.M., Eastwood, M.A., Brydon, W.G., Anderson, J.R., and Anderson, D.M.W., Amer.J.Clin.Nutr., 37 (1983) 368-375.
4. Taupin, P.J.Y. and Anderson, D.M.W., Food Chem.Toxicol., 20 (1982) 513-517.
5. Eastwood, M.A., Brydon, W.G. and Anderson, D.M.W., Toxicol. Letters, 17 (1983) 159-166.

6. Eastwood, M.A., Brydon, W.G. and Anderson, D.M.W., Toxicol. Letters, 21 (1984) 73-81.
7. Eastwood, M.A., Brydon, W.G. and Anderson, D.M.W., Amer.J. Clin.Nutr., 44 (1986) 51-55.
8. Evaluation of Food Additives, Technical Report Series No.710, W.H.O. Geneva, 1984, pp.29-33.
9. Eastwood, M.A., Brydon, W.G. and Anderson, D.M.W., Food Additives and Contaminants, 1986, in press.
10. Anderson, D.M.W., Eastwood, M.A. and Brydon, W.G., Food Hydrocolloids, 1986, in press.
11. Vettorazzi, G., Proceeding Int.Colloquium, Luxembourg, 1976 (Pergamon Press, Oxford and New York).
12. Evaluation of Food Additives, Technical Report Series No.733, W.H.O. Geneva, 1986, pp.44-45.
13. Walker, R., in Chemical Toxicology of Food, Eds. Galli, Paoletti and Vetterazzi, Elsevier/North Holland, 1978, pp. 339-348.
14. Leegwater, D.C., de Groot, A.P. and Kalmthout, M., Food Cosmet.Toxicol., 12 (1974) 687-697.
15. Rowland, I.R. et al., Xenobiotica 14 (1983) 251-256.
16. Wyatt, G.M., Bayliss, C.E. and Holcroft, J.D., Brit.J. Nutrition, 55 (1986) 261-266.
17. Saeyers, A.R., O'Brien, M. and Schmetter, B., in A.C.S. Symposium Series No.214 (1983), pp.123-133.
18. Van Soest, P., Horvath, P., McBunney, M., Jeraci, J. and Allen, M.,in A.C.S.Symposium Series No.214 (1983),pp.135-140.
19. Storer, G.B., Illman, R.J., Trimball, R.P., Snoswell, A.M. and Topping, E.L., Nutrition Research, 4 (1984) 701-707.
20. Osilesi, O., Trout, D.C., Glover, F.E., Koh, E.T., Behall, K.M., Dorisis, T.M. and Tartl, J., Amer.J.Clin.Nutr., 42 (1985) 597-603.
21. Behall, K.M., Lee, K.H. and Moser, P.B., Amer.J.Clin.Nutr., 39 (1984) 209-214.
22. Osilesi, O., Trout, D.C., Ryan, R.O. and Knight, E., Nutrition Research, 4 (1984) 259-269.
23. Penagini, R., Velio, P., Vigorelli, R., Bozzani, A., Castagnone, D., Ranzi, T. and Bianchi, P.A., Amer.J. Gastroenterol., 81 (1986) 123-125.
24. Jenkins, D.J.A., Reynolds, D. and Slavin, B., Amer.J.Clin. Nutr., 33 (1980) 575-581.
25. Jenkins, D.J.A., Wolever, T.M.S. and Bacon, S., Amer.J.Clin. Nutr., 33 (1980) 1729-1733.
26. Johnson, I.T. and Gee, J.M., Gut, 22 (1981) 398-403.
27. Holt, S., Heading, R.C. and Carter, D.C., Lancet, 1 (1979) 636-639.
28. Blackburn, N.A., Redfern, J.S. and Jarvis, H., Clin.Sci., 66 (1984) 329-336.
29. O'Conner, N., Tredger, J. and Morgan, L., Diabetologia, 20 (1981) 612-615.
30. Heitkemper, M.M., Nursing Research, 34 (1985) 19-23.
31. Hanssen, M., in "E for Additives", Thorsons Publishers Ltd.

Chapter 36

Highly active α-amylase immobilized on acrylic acid grafted cellulose

M. Shimada, F. Kokubun and Y. Nakamura – Faculty of Technology, Gunma University, Kiryu, Gunma 376, Japan.

INTRODUCTION

Many workers have reported on immobilization of enzymes. But in many cases, the activity of an enzyme decreases by immobilization. The causes of inactivation are thought to be as follows: First, enzyme may lose its active conformation by immobilization, because immobilized enzyme has many reaction sites with the carrier. Second, functional groups close to the active site or inactive site may couple with functional groups on the carrier. Third, substrates with high molecular weight can not approach the active site as enzyme is immobilized on an insoluble carrier with a high molecular weight.

In order to obtain immobilized enzyme with high activity, various causes of inactivation should be avoided. Selection of carrier and protection of the active site of the enzyme are both important factors to improve activity.

Generally speaking, carriers having high specific areas, high hydrophilicity and many functional groups to couple with the enzyme are preferred. As we have carried out γ-ray induced graft copolymerization onto cellulose, we decided to use grafted cellulose as the carrier for immobilization of α-amylase. Grafted cellulose can fill the conditions as mentioned above. Efforts to make conformational changes of the enzyme a minimum and to protect the active site of the enzyme during immobilization were made in this study.

EXPERIMENTAL

Materials

Scoured cotton cellulosic fibers of the Acala variety were purified by extracting with hot ethanol-benzene mixture (1:1 volume ratio). Then the fibers were washed with methanol and distilled water prior to drying.

Crystallized α-amylase from Bacillus subtilis was kindly supported by Nagase Biochemicals, Ltd. Acrylic acid was distilled under reduced pressure before use. Other chemicals used were reagent grade and were used without further purification.

Graft Copolymerization

After cellulose was dried under reduced pressure at 50°C for 20 h, it was irradiated with Co-60 γ-rays at an exposure rate of 1.0 MR/h for 1 h in nitrogen atmosphere at room temperature. Graft copolymerization of acrylic acid onto the irradiated cellulose was carried out in the reaction system containing 10% acrylic acid and 90% water under nitrogen atmosphere at 30°C. After graft copolymerization, the samples were extracted with hot water to remove homopolymer. Extent of grafting was calculated as weight percent increase by grafted polymer to original weight of cellulose.

Immobilization Reaction

1-ethyl-3-(3-dimethylaminopropyl) carbodiimide hydrochloride (water soluble carbodiimide, WSC) was dissolved in 0.2 M acetate buffer solution at pH 4.7. Into the solution, grafted cellulose and α-amylase were added and incubated at 10°C for 12 h. The extent of grafting of the carrier was about 50%. The ratio of each component was as follows:

AA/WSC = 1/0.2 (mol/mol)
AA/α-amylase = 1/0.5 (wt/wt)

where AA stands for the amount of acrylic acid in grafted cellulose.

After the immobilization reaction, samples were washed with water and immersed in 2M acetate buffer solution at pH 4.4 for 6 h to remove α-amylase bonded ionically to the carrier. Some of the reactions were carried out in the reaction system containing molecules which were mimics of substrate (pseudo substrates), i.e. glucose, mannose and xylose. Concentrations of analogue was changed from 0.1 to 0.5 mol/l.

Immobilization reactions were also carried out in the presence of organic solvent, dioxan, methyl cellosolve and DMSO. Another way to increase the activity was by a three step reaction; the first step is the formation of ionic bond between carboxyl group and amino group, the second step is the formation of covalent bond between the residual carboxyl group and amino group, and the third step is the removal of ionic bond. Degree of immobilization was expressed as weight percent increase due to amylase over the original weight of carrier.

Assay of α-Amylase Activity

Free α-amylase and immobilized α-amylase were added to the solution containing 0.1 M acetate buffer solution and 10% soluble starch, and incubated for 5 min at 38°C. Before checking reducing end groups, protein was removed [1]. The amount of reducing end group was determined by the Nerson-Somogyi method [2].

RESULTS AND DISCUSSION

Immobilization of α-Amylase onto Acrylic Acid Grafted Cellulose

The reactions of acrylic acid grafted cellulose (Cell-COOH) with WSC and/or α-amylse were carried out. After the reactions, the samples were washed with distilled water and weighed. The results are shown in Table I. Weight increases by the reactions mean that acrylic acid grafted cellulose reacted with WSC and α-amylase independently. After weighing, the samples were immersed in 2M acetate buffer solution at pH 4.4, washed with distilled water, dried and weighed. The results are shown in Table I. WSC and α-amylase reacted with acrylic acid grafted cellulose independently could be removed easily from the carrier by immersion in 2M acetate buffer solution, but the weight increase of the sample reacted with WSC and α-amylase at the same time did not decrease to the original weight after immersion in buffer solution. It was also clarified that grafted cellulose reacted with α-amylase to yield immobilized enzyme after reaction with WSC and drying.

Table 1. Reaction of acrylic acid grafted cellulose with WSC and/or α-amylase.

Reacted with	W_1/W_0 (%)	W_2/W_0 (%)
WSC	2.4	0
α-amylase	5.4	0
WSC+α-amylase	5.1	3.3

$$W_0 \xrightarrow[\alpha\text{-amylase}]{\text{WSC and/or}} W_1 \xrightarrow[\text{solution, pH 4.4}]{\text{2M acetate buffer}} W_2$$

Therefore the following mechanism is inferred as the immobilization process.

$$\text{Cell-COOH} + R_1N{=}C{=}NR_2 \longrightarrow \text{Cell-COOC}\begin{matrix}\text{NH-}R_1\\ NR_2\end{matrix}$$

$$\text{Cell-COOC}\begin{matrix}\text{NH-}R_1\\ NR_2\end{matrix} + NH_2 - \alpha\text{-amylase}$$

$$\longrightarrow \text{Cell-CONH-}\alpha\text{-amylase (covalent bond)}$$

$$\text{Cell-COOH} + NH_2\text{-}\alpha\text{-amylase}$$

$$\longrightarrow \text{Cell-COO}^{-}\,\overset{+}{N}H_3\text{-}\alpha\text{-amylase (ionic bond)}$$

The relative activities of immobilized α-amylases to free amylase were 24% and 60% for ionically and covalently bonded amylases, respectively. Ionically bonded α-amylase had much higher activity than covalently bonded one. Protein was detected by buiret reaction during hydrolysis of soluble starch with ionically bonded α-amylase, and was not detected with covalently bonded one. So it is assumed that α-amylase ionically bonded is easy to leave during hydrolysis and show high activity. To obtain covalently bonded α-amylase with high activity, efforts to make conformational changes minimum, and to protect active site of enzyme during immobilization reaction were made.

Effects of Analogue

α-amylase was reacted with acrylic acid grafted cellulose in the presence of WSC and pseudo-substrate. The degree of immobilization and relative activity are shown in Figures 1 and 2. The effects of the pseudo-substrate depends on the species and the concentration. The best one was glucose. The activity of immobilized enzyme in the presence of glucose was more than 55% of free enzyme. Mannose also showed positive effects, but xylose did not show positive effects in the concentrations studied. The mechanism of pseudo-substrate effects is shown in Figure 3. As the active site of the enzyme is protected by a pseudo-substrate molecule, the immobilized enzyme showed high activity.

Effects of Organic Solvent

Conformational changes of enzyme are expected when the enzyme is dissolved in aqueous solution of organic solvent, so the immobilization reaction was carried out in the reaction mixture containing organic solvent. The results are shown in Figures 4 and 5. Methyl cellusolv and dimethyl sulfoxide showed positive effects within certain concentrations.

The shifts of fluorescence spectra of α-amylase measured in these solvents supported the results; they explained the differences of solvation among the solvents. The peaks observed both in Figures 4 and 5 are thought to be the reflection of conformational changes of α-amylase in the solvents. The conformational changes during immobilization reaction do not always

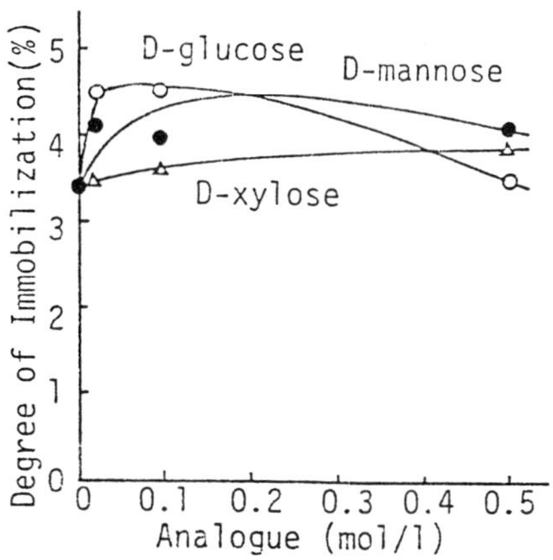

Fig. 1. Effects of analogue on degree of immobilization

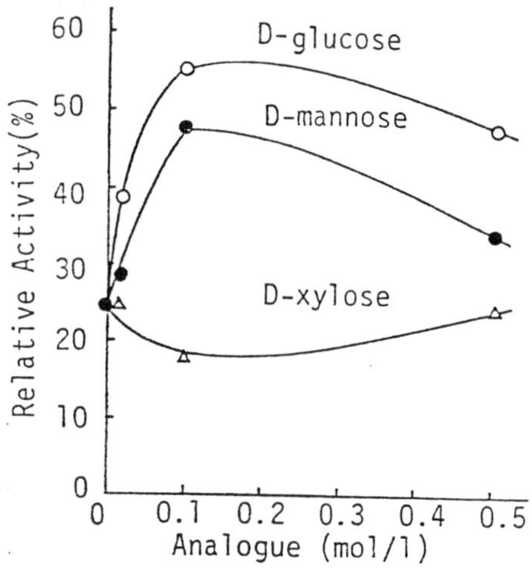

Fig. 2. Effects of analogue on relative activity of immobilized α-amylase.

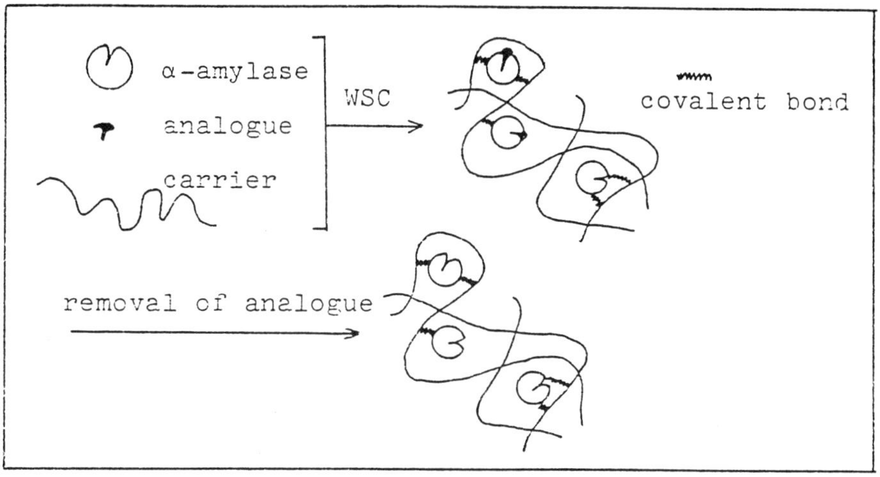

Fig. 3. Mechanism of immobilization of α-amylase in the presence of analogue.

mean the deformation of the active conformation. In some systems the active conformation is recovered in water.

Three Step Immobilization Reaction

Judging from molecular weight of α-amylase [3] and molecular weight per free amino group, there are 10-15 free amino groups in one α-amylase molecule. According to Matsuura et al [4], histidine, lysine, aspartic acid and tyrosine are distributed in the active site of amylase. Therefore many amino groups in

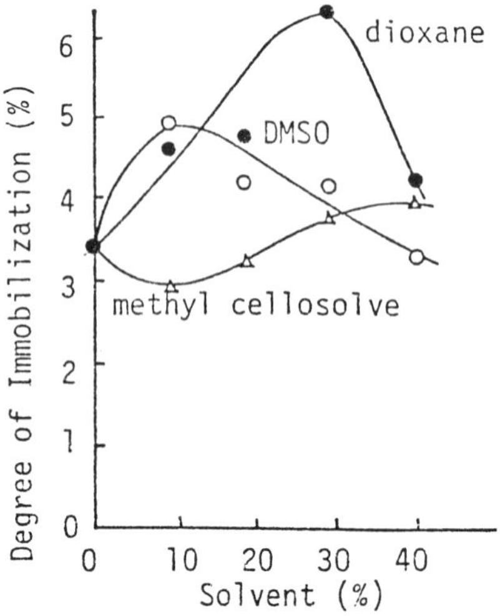

Fig. 4. Effects of organic solvent on degree of immobilization.

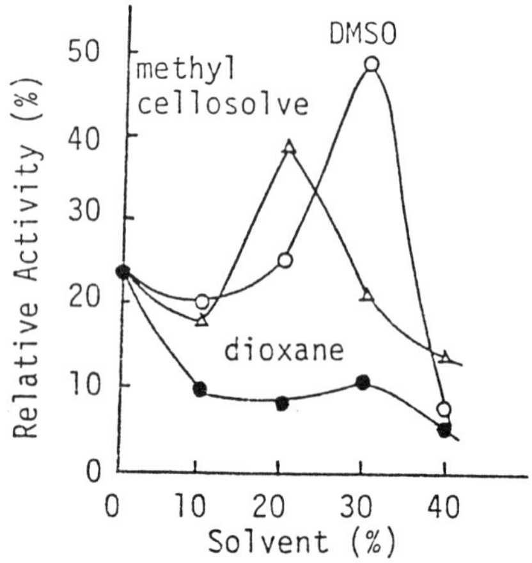

Fig. 5. Effects of organic solvent on relative activity of immobilized α-amylase.

amylase react with carboxyl groups on the carrier. It is inferred that the more the reaction sites the immobilized enzyme has, the less the residual activity is. As shown in Table I, an ionic bond is formed more easily than a covalent bond, and it is easily removed by immersion in acetate buffer solution. Considering these properties, α-amylase was immobilized on acrylic acid grafted cellulose by the three step reaction. The results are shown in Table II. The sample (reaction 1) was immobilized with the ordinary method, and the other two samples were immobilized using three steps. From these results, it is clear that the three step immobilization method is useful to increase the residual activity. The mechanism is shown in Figure 6. The more the ionic bond in the first step, the more the possibility for amino groups in the active site to react with a carboxyl group. That means the amino group in the active site is protected from the formation of covalent bond and few amino groups react with a carboxyl group in the second step. Therefore, conformational changes will be minimized and the immobilized enzyme shows high activity.

Table 2. Immobilization of α-amylase.

Reaction	A_1	Washing	A_2	W_1/W_0(%)	W_2/W_0(%)	Relative Activity (%)
1	-	-	0.50	5.1	3.3	24
2	0.50	-	0.50	7.4	5.0	54
3	0.50	water	0	5.2	3.0	65

$$W_0 \xrightarrow[\text{1st step}]{\alpha\text{-amylase}(A_1)} \xrightarrow[\text{2nd step}]{\text{WSC and } \alpha\text{-amylase}(A_2)} W_1$$

$$\xrightarrow[\text{3rd step}]{\text{2M acetate buffer solution}} W_2$$

A_1, A_2 : weight of α-amylase/weight of acrylic acid

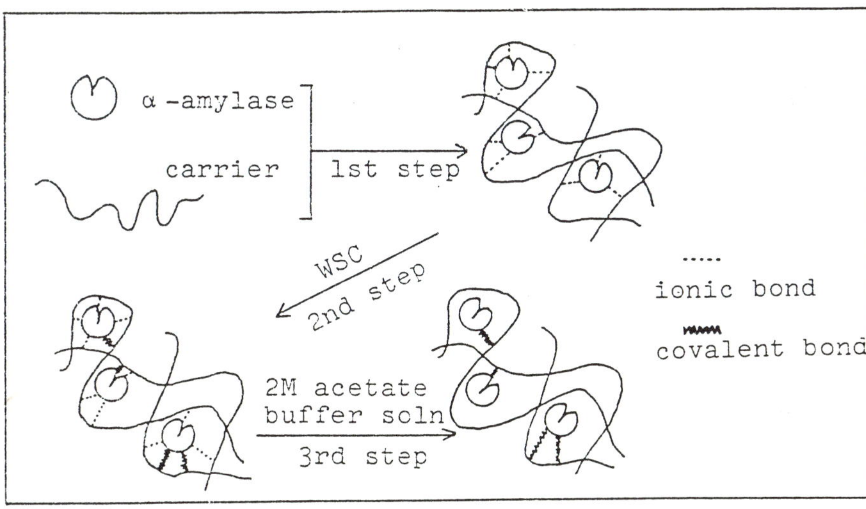

Fig. 6. Mechanism of immobilization of α-amylase by three step reaction.

REFERENCES

[1] Nelson, N., J. Biol. Chem., 1944, 153, 375.
[2] Somogyi, M., J. Biol. Chem., 1952, 195, 19.
[3] Bender, M. L., Mechanisms of Homogeneous Catalysis from Protons to Proteins, Wiley, 1971.
[4] Matsuura, M. et al, Denpun Kagaku, 1980, 27, 165.

Chapter 37

A comparison of bacterial strains growing on glucoisosaccharinic acid

Michael J. Bailey – VTT, Biotechnical Laboratory, Tietotie 2, SF-02150 Espoo, Finland.

INTRODUCTION

Glucoisosaccharinic acid (GISA, see Figure 1) is produced during Kraft pulping under the ambient conditions of high temperature and alkalinity. Glucose moieties are released from the glucomannan-type hemicellulose of spruce and pine pulps in the peeling reaction (Whistler and BeMiller 1958) and undergo a series of reactions resulting in the formation of a mixture of hydroxy acids, of which about 50 % is GISA. The carbohydrate degradation products formed during the Kraft pulping of pine raw material are listed in Table 1, along with their approximate amounts. In Finland, the annual production of GISA is about 500 000 tons (Alén, 1986). On the one hand this represents a major potential reserve of carbon source for the biotechnical industry and on the other the presence of this extensively oxidized component in the black liquor reduces its combustion value.

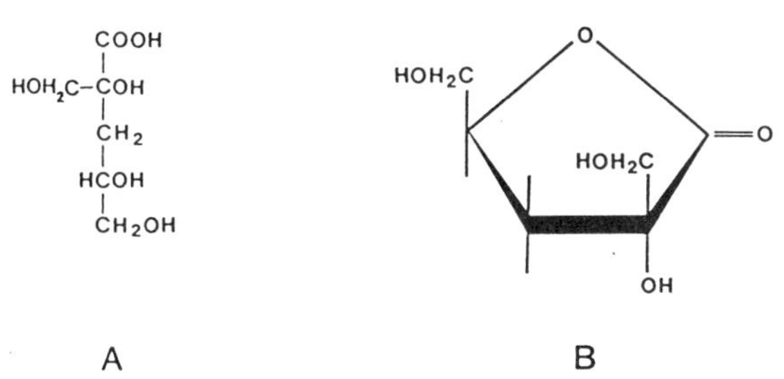

Fig. 1. The molecular structure of α-D-glucoisosaccharinic acid (A) and its lactone (B).

Table 1. Organic acids formed during Kraft pulping of pine (Alen and Sjöström, 1985).

Acid	Amount produced kg per ton of pulp
Formic	70
Acetic	50
Hydroxy acids:	
Glycolic	10
Lactic	45
2-Hydroxybutanoic	15
Xyloisosaccharinic	15
2,5-Dehydroxypentanoic	10
3-Deoxypentonic	5
Glucoisosaccharinic	160

Isosaccharinic acids do not occur in natural, unpolluted environments. Therefore, although these compounds are not toxic (unpublished data), attempts to isolate GISA-degrading micro-organisms from strain collections and unpolluted soil and water samples have been unsuccessful (Strand et al. 1984). Hitherto, the only reported isolates capable of metabolizing GISA as the sole carbon source have been bacterial strains of the genus *Ancylobacter*, all of which were isolated from environments probably contaminated by sulphate pulping wastes (Strand et al. 1984). These strains utilize GISA under neutral (pH 7.2) or alkaline (pH 9.5) conditions.

In this work, bacterial strains capable of utilizing GISA as the sole carbon source were isolated and compared in laboratory experiments.

MATERIALS AND METHODS

Cultivation media

The basic medium used in this work was Bacto Yeast Nitrogen Base without Amino Acids (Difco 0919-15), 6.7 gl^{-1} (YNB w/o AA). Additions to this medium are indicated in the text. GISA, α-D-glucoisosaccharinic acid (calcium salt) was synthesized in the laboratory by the method of Whistler and BeMiller (1963). Other medium components were commercial laboratory reagents. All media were sterilized at 120°C, with separate sterilization of the carbon source.

Isolation of strains

GISA-degrading strains were isolated and maintained as described previously (Bailey 1986) by enrichment cultivation of samples taken from soil, stagnant pools and lake water and mud in the grounds of a Kraft pulping mill in central Finland.

Cultivation methods

Cultivations were carried out either in shake flasks or in a laboratory fermenter with a working volume of 3 l as described previously (Bailey 1986).

Analyses. Glucose was analyzed enzymatically (Werner *et al.* 1970). Analyses of GISA were performed by liquid chromatography as described previously (Bailey 1986). Bacterial growth was estimated qualitatively by direct microscopy and quantitatively by diluting and plating on nutrient agar.

RESULTS

Growth on GISA

A total of 9 bacterial strains capable of growing on the enrichment medium were isolated. Two of the strains, VTT-E-85234 and 85235, have been reported previously (Bailey 1986). The remaining seven isolates (VTT-E-85237...85243) were obtained from the same site on a second occasion. The strains were first tested in shake flask cultivations. The results of this testing are shown in Table 2. All but one of the strains had reached a cell density of > 10^9 CFU ml^{-1} after 50 hours and the GISA was completely utilized from all the flasks within 70 h.

Table 2. Growth and consumption of GISA by 9 bacterial isolates cultivated on a medium containing 2 gl^{-1} GISA. The initial pH was 5.9 in all flasks and the initial cell densities were 5.5...6.0 × 10^6 CFU ml^{-1}.

Strain	After 50 h		After 70 h
VTT-E-	GISA gl^{-1}	cells CFU ml^{-1} [a]	GISA gl^{-1}
85234	0.1	2.7×10^9	
85235	0.1	4.5×10^9	
85237	0.1	3.9×10^9	
85238	0.1	3.7×10^9	
85239	0.1	2.7×10^9	
85240	0.5	3.0×10^9	0.1
85241	1.2	3.4×10^8	0.1
85242	0.1	2.6×10^9	
84243	1.2	1.5×10^9	0.1

[a] CFU = Colony forming units, *i.e.* viable cells

The strains were then compared by cultivation on the same medium in the laboratory fermenter as described previously for strain VTT-E-85235 (Bailey 1986). The time required for complete exhaustion of GISA from the medium was obtained with an accuracy of only a few minutes from the continuous plot of dissolved oxygen in the broth (see Fig. 2) and was later confirmed by HPLC analysis. Under these conditions, the time for complete substrate utilization by the different strains ranged from 22 to > 60 h (Table 3). Curves of GISA consumption for the best and the poorest strains are shown in Figure 3.

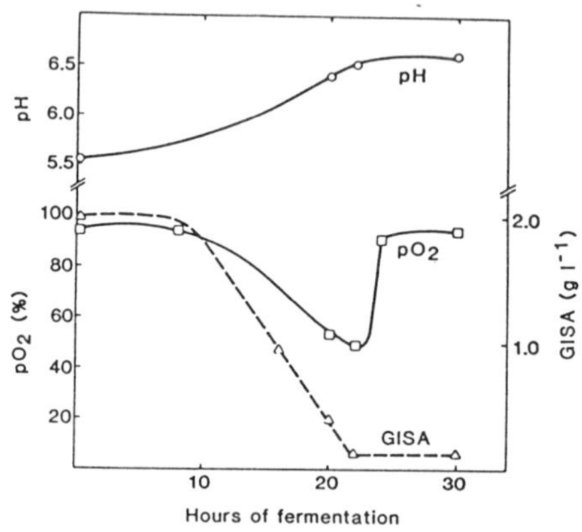

Fig. 2. Growth of strain VTT-E-85235 on 2 gl^{-1} GISA in a laboratory fermenter.

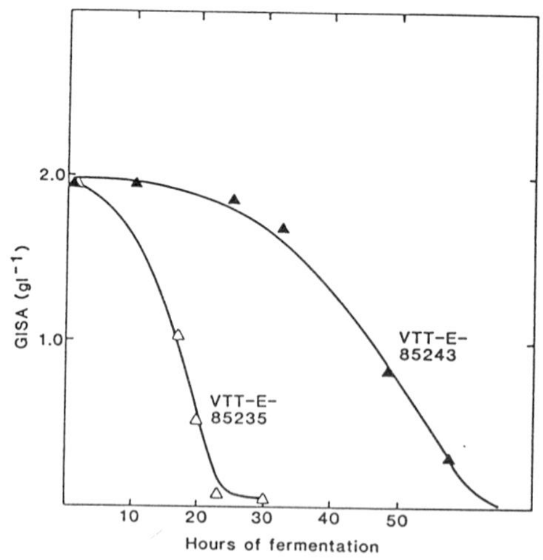

Fig. 3. Consumption of GISA by different isolates in laboratory fermentations.

Table 3. Time required for consumption of 2 gl^{-1} GISA in laboratory fermenter experiments with the 9 GISA-degrading isolates. Fermentation conditions: T = 30°C; initial pH 5.7 ± 0.1.

Strain VTT-E-	Time required for consumption of 2 gl^{-1} GISA h	Maximum cell density CFU ml^{-1}
85234	26	$4x10^9$
85235	22	$2x10^9$
85237	25	$5x10^9$
85238	33	$2x10^9$
85239	55	$2x10^9$
85240	41	$4x10^9$
85241	24	$6x10^8$ [a)]
85242	26	$2x10^9$
85243	> 60	$1x10^9$

[a)] Microscopical observation revealed extensive aggregation of cells

GROWTH ON GLUCOSE

Previous work with strain VTT-E-85235 had shown that this isolate was unable to metabolize glucose (Bailey 1986). The 9 isolates described in this work were tested in shake flasks on a medium containing YNB w/o AA and 2 gl^{-1} glucose instead of GISA. Eight of the strains failed to consume any of the glucose from the medium after a cultivation time of 6 d. The strain VTT-E-85243, however, consumed 25 % of the glucose after 3 days, after which no consumption was observed, almost certainly due to the formation of acid (final pH 4.8). When this strain was cultivated at a higher initial pH (7.0) the glucose was consumed steadily to exhaustion after about 80 h. All the other strains were unable to consume any glucose at the higher pH, with or without the addition of peptone and yeast extract.

GROWTH ON DELIGNIFIED BLACK LIQUOR

Partially concentrated black liquor (dry weight ≃ 30 %) was obtained from a Finnish Kraft mill after the stage of removal of tall oils. This liquor was coarsely delignified by the procedure outlined in Table 4. The pH of the delignified liquor was then adjusted with ammonium hydroxide before being used as carbon source for cultivation experiments at a dilution of 1:10 or 1:20 with YNB w/o AA.

Table 4. Delignification of black liquor.

Stage no.	Procedure
1	Dilute to ~ 15 % d.w. with water
2	pH → 4.0 (H_3PO_4)
3	Heat to 80°C, cool to room temperature
4	Separate the liquor from the crust of coagulated lignin
5	Clarify by centrifugation

In the first experiment, the pH of the black liquor medium was adjusted to reach 5.8 ± 0.1 after sterilization. No significant growth was observed in any of the flasks at either dilution. Addition of 2 gl^{-1} GISA to the same media did not induce growth in a second experiment (results not shown). However, when the pH of the black liquor medium was increased to 6.5, the strains were able to grow (Table 5).

Table 5. Growth of the isolates VTT-E-85235 and 85243 on delignified black liquor medium. Initial pH was 6.5 and initial cell densities were ~ 10^7 CFU ml^{-1}.

Strain VTT-E-	Medium strength	2 d -sample		6 d -sample	
		pH	cells, CFU ml^{-1}	pH	cells, CFU ml^{-1}
85235	1:10	6.8	$2x10^7$	7.6	$2x10^9$
	1:20	7.4	$1x10^9$	7.4	$1x10^9$
85243	1:10	7.4	$4x10^9$	7.9	$< 10^8$
	1:20	7.6	$2x10^9$	7.6	$< 10^8$

DISCUSSION

The strains isolated in this work exhibited some similarities but also major differences. Although it is by no means certain that all the isolates are representatives of different species, at least some of them are non-identical on the basis of colony colouration and morphology in agar plating. All 9 strains have been despatched to Deutsche Sammlung von Mikroorganismen, Göttingen, FRG for identification. This taxonomical information will be of significance in the estimation of their potential for the biotechnical utilization of Kraft pulping wastes.

Three of the salient properties of these GISA-metabolizing strains, apart from growth on GISA, were:

- Growth at low pH values (< 6)
- Non-utilization of glucose (with one exception)
- Non-requirement of organic nitrogen.

The strains are thus very different in nature from the Ancylobacter strains isolated by Strand et al. (1984). Despite repeated efforts, no alkalophilic GISA-degrading strains were isolated from the field samples taken in this study using an enrichment medium containing peptone and yeast extract as well as GISA at pH 7.5 or 9.0.

The results reported in this work must be regarded as preliminary. However, the rather efficient consumption of GISA by some of the strains isolated and the demonstration of the growth of at least some of these strains on very coarsely pretreated black liquor media indicate that there exists a real potential for the biotechnical utilization of Kraft pulping wastes. Further investigations are in progress.

REFERENCES

Alén R (1986) VTT, Laboratory of Fuel Processing and Lubrication Technology. Personal communication.

Alén R and Sjöström E (1985) Degradative conversion of cellulose-containing materials into useful products. In: Nevell and Zeronial (eds) Cellulose chemistry and its applications. Ellis Horwood Ltd, Chichester, UK, pp 531-544.

Bailey MJ (1986) Utilization of glucoisosaccharinic acid by a bacterial isolate unable to utilize glucose. Appl Microbiol Biotechnol, in press.

Strand SE, Dykes J, Chiang V (1984) Aerobic microbial degradation of glucoisosaccharinic acid. Appl Environ Microbiol 47: 268-271.

Werner W, Rey H-G, Wielinger H (1970) Uber die Eigenschaften eines neuen Chromogens für die Blutzuckerbestimmung nach der GOD-POD-Methode. Z Anal Chem 252: 224-228.

Whistler RL, BeMiller JN (1958) Alkaline degradation of polysaccharides. Adv Carbohydrate Chemistry 13: 289-329.

Whistler RL, BeMiller JN (1963) "α"-D-Isosaccharino-1,4-lactone. In: Whistler RL, Wolfram ML (eds) Methods in carbohydrate chemistry. Academic Press, New York and London, pp 477-479.

Chapter 38

Cellulose fibres treated by dimethylolmelamine (DMM) as reinforcement in unsaturated polyester composites

Li Hua and Per Flodin – Department of Polymer Technology, Chalmers University of Technology, S-412 96 Göteborg, Sweden.

ABSTRACT

Dimethylolmelamine (DMM) was used as a surface modifying agent of cellulose fibre and a coupling agent of cellulose composites. The unsaturated polyester composites were prepared with treated cellulose as reinforced material. The hydrophilicity of composites can be improved. The mechanical properties of composites in wet condition have been enhanced.

INTRODUCTION

Plastic composites reinforced with natural cellulose fibre demonstrate observable advantages: low cost, light mass, and greater strength and stiffness. Many studies have reported the work on the modification of cellulose fibre and the selection of matrices. (1), (2), (3), (4), (5).

Cellulose is a hydroscopic glucan polymer consisting of linear chain of 1,4-β-bonded anhydro -D- glucose units. (6). There are three OH groups in the single repeat unit of cellulose. OH groups on the surface of the crystallites or in the amorphous regions may be available for bonding water.

The cell wall of cellulose fibre can adsorb water until saturated. Water beyond the saturation point is free water in the void structure (7). In addition, the cellulose molecules also contain crystal water which is hardly removed. So bonding water, free water, and crystal water bring about the high water uptake of cellulose. The dimensional stability, mechanical , electrical properties of composites reinforced by cellulose are affected by the moisture content of cellulose.

It has been reported that triazine-based coupling agents were used for surface treatments of cellulose fibres. (8). The aim of this work is to search for cheaper and more easily available chemicals as surface modifying agents and coupling agents.

Melamine products have been used extensively in a variety of fields including plastics, surface coatings, tanning agents, petroleum and rubber chemicals, and surface modifying agents (9).

We worked on the surface modification of cellulose by the reaction between methylolmelamines and OH groups of cellulose. The reduction of OH groups, and the formation of crosslinking between the cellulose molecules and covalent bonds in the interface between treated cellulose and polyester matrix result in lower water uptake and higher mechanical properties of composite. Melamine compounds contribute the rot-proof and reinforcement for cellulose fibre as well.

EXPERIMENTAL

Synthesis of dimethylolmelamine

Melamine reacts with formaldehyde under these conditions: pH:8 – 10, reaction temperature: 75 - 85°C, time 5 - 10 min., melamine versus formaldehyde molecular ratio: 1:1:5 - 2.1, catalyst: sodium-bicarbonate. The reaction product was diluted with water as soon as the reaction finished.

Treatment of cellulose

The cellulose fibres used in this study were bleached softwood kraft paper and obtained from Klippans Finpappersbruk AB, Sweden. The untreated cellulose was impregnated in DMM dispersion with 5% solid content for 10 min. Lewis acid mix was used as curing agent for methylol compounds (10). The treated paper was dried and cured in an oven at 90°C for 3 hrs, then washed with water and dried.

Composite and specimen preparation

The polyester used was a commercial unsaturated polyester resin, Norpol P1 2614 made by Jotun A/S, Norway.

100 mm x 100 mm sheets of paper were impregnated with liquid polyester resin containing 2% benzoyl peroxide (BPO) as the curing agent. Air was removed under 15 mm Hg vacuum for 10 min. Stacks of five impregnated sheets paper aligned in the machine direction were placed between two aluminium plates and restrained by a steel picture frame mould. The laminates were cured at a pressure of 10 kg/cm² for 1 hr at 80°C and postcured at 90°C for 23 hrs. 10 mm wide composites in the orientation of the machine direction were cut for tensile tests and water absorption studies (8).

Water absorption

The samples with unsealed edges were dried at 80°C to constant weight and then immersed in water at room temperature. At chosen time intervals the free water on the surface of the samples was removed and weighed using an analytical balance. The mean value of 3 samples was considered as the test result.

Mechanical properties of composites

The tensile strength and tensile modulus of composite test specimens were determined by the INSTRON 1122 model test machine in wet and dry conditions. The test method followed ASTM D638. At least 5 specimens were tested for each condition, and the mean values and standard deviations were calculated (8).

RESULTS AND DISCUSSION

Synthesis of DMM

Various reaction conditions effect on the molecular size of DMM, which effects the penetrating power to cellulose. We are sorry to say that a definite molecular size cannot be offered quantitatively due to the instability of the molecular weight of DMM, but it has been proved that the molecular size of DMM increases with reduced pH, increased temperature, prolonged time, and increased mole ratio of formaldehyde to melamine. It was evident that the viscosity varied with the reaction conditions.

Cellulose treatment

Mono- and di- methylolmelamine resin is the desirable surface modifying agent of cellulose (11). Molecular size of DMM effects the penetrating power to cellulose. The retention of MF resin on cellulose increases with an increased dispersion concentration. After a certain interval, immersing time has little influence on retention. Over age resin could cause the cellulose to be brittle and the formation of a weak boundary layer between matrix and the surface of treated cellulose due to the lower crosslinking extent of DMM.

Water absorption (Fig. 1)

Water absorption was calculated using the formula:

$$W = \frac{Ww - Wo}{Wo} \times 100\%$$

Ww = the weight of wet composite specimen
Wo = the weight of dry composite specimen

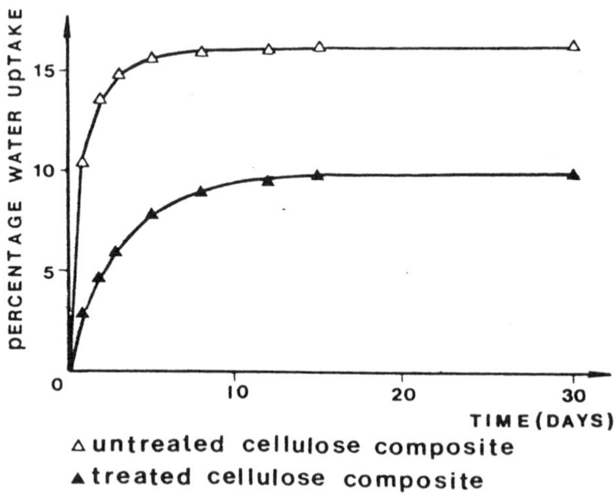

Fig. 1. The plot of water uptake versus immersion time for various cellulose composites.

The treated cellulose composite had a lower water uptake. DMM dispersion with suitable concentration and molecular size can penetrate the amorphous region or enter the lumina of cellulose. The OH groups in cellulose can react with DMM. The reduction groups and formation of crosslinking between cellulose molecules result in the reduction of the amount of water bonded by hydrogen bonding. On the other hand, void structure is covered with a layer of DMM or filled with polyester matrix resin in composite. The change in volume caused by swelling and shrivelling can be decreased and the amount of free water existing in void structure can be reduced.

Mechanical properties (Table 1)

The mechanical strength of the composite is contributed by 55% (wt) polyester matrix (and melamine resin), 45% (wt) treated or untreated cellulose and interface adhesion between cellulose and matrix. The improvement of wet strength means better interface adhesion against moisture. On the other hand, the value of the modulus reflects the situation of interface adhesion as well. The untreated cellulose composite demonstrates the lower modulus and that fibres are pulled out from the matrix and bigger elongation at break after the composite undergoes tensile force.

Table 1. Mechanical properties of composites.

reinforcing fibres	composites	strength Mpa	s*	modulus Gpa	s*
untreated	dry	129.8	2.7	4.8	0.5
	wet **	56.1	1.6	1.7	0.9
treated by 5% DMM	dry	139.6	2.9	5.1	0.5
	wet **	86.3	1.8	3.2	0.4

* S: standard deviation. $S=\sqrt{(nx^2-n\bar{x}^2)/(n-1)}$.
** Samples were immersed in water for one month.

CONCLUSION

(1) DMM is a desirable surface modifying agent of cellulose. It is important to control the molecular size of DMM and the concentration of dispersion for cellulose treatment. The reduction of OH groups and the formation of crosslinking between cellulose molecules result in the reduction of the water sensibility of composites.

(2) DMM is used as a coupling agent as well in cellulose-unsaturated polyester composites. The interface adhesion was improved by physical and chemical crosslikings. The free $-NH_2$ groups in melamine molecules can react with the -OH and -COOH

end groups of polyester. The wet mechanical properties were improved considerably.

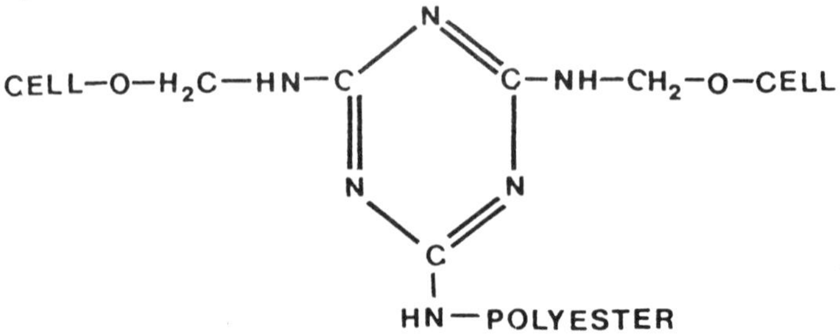

Fig. 2. The functions of DMM as surface modifying agent and coupling agent in cellulose composites.

V. References

1. Anthony J. Mitchell, Janet E. Vaughan, and Donald Willis, J. Polymer Sci., No 55, 143 - 154, (1976).
2. O. Ovolabi, T. Czuikouszky, and J. Kovaks, J. Applied Polymer Sci.,Vol. 30, 1827 - 1836, (1985).
3. Matsushita Electric Works, Ltd, Japan Patent 57-111323 (82-111323).
4. National Electric Works, Ltd, Japan Patent 57-197152 (82-197150).
5. National Electric Works, Ltd, Japan Patent 57-197152 (82-197152).
6. Roger C. Pettersen, The Chemistry of Solid Wood, edited by Roger Rowell, American Chemical Society, Washington, D.C., (1984), p 58.
7. Roger M. Rowell, The Chemistry of Solid Wood, American Chemical Society, Washington, D.C., (1984), p 176.
8. Pawel Zadorecki and Per Flodin, J. Applied Polymer Sci., Vol 30, No 6, 2419, (1985)
9. W. F. Herbes, S. J. O'Brien, and R. G. Weyker, Chemical after Treatment of Textiles, edited by H. Mark ect., John Wiley & Sons Inc, N.Y.,(1971), p 319-320.
10. Reference (9), P 328.
11. Richard Steel and Lorrain E. Giddings, Industrial and Engineering Chemistry, Vol 48, P 111-114, (1956).

PART 3

INDUSTRIAL UTILISATION

Chapter 39

Chemical aspects of wood processing

E. J. Gibson – Department of Forestry and Wood Science, University College of North Wales.

INTRODUCTION

This review will be concerned with chemical aspects of processing for wood products; it will not consider pulping for paper or board manufacture. This is certainly a major topic, but I consider it to be an entirely separate one.

Generally speaking, chemical processing is carried out for one of three purposes; to bond two pieces of wood together, to modify its properties so that it becomes less affected by water, or to protect it against degradation by biological agencies (insects or fungi) or by elements of climate (mainly light and water).

The properties of wood are determined by its structure, both anatomical and chemical elements being of prime importance, and it will be necessary to touch upon a few aspects of structure before reviewing the state of the art in chemical processing.

The basic anatomical structure of timber from coniferous species is relatively simple (fig. 1). However differences in genetics and growth conditions, together with differences in sizes, numbers and distribution of the cell types between species mean that there are very large differences in properies. Chemically the key components are crystalline cellulose, non crystalline hemicelluloses and lignin.

The cellulose is largely crystalline, and consists of linear chains of 1,4-bonded anhydroglucose units with a degree of polymerisation between 10,000 and 15,000 (Fig.2). It is believed that the cellulose molecules in any one sample have the same D.P. The chains of the anhydroglucose units pack in layers held together by Van der Waals forces and there is intramolecular hydrogen bonding between atoms of adjacent glucose residues.

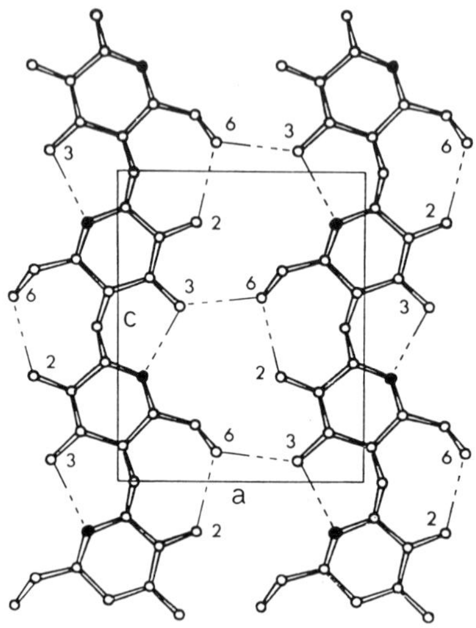

Fig. 2. Arrangements of glucose molecules in chain of cellulose 1.

Molecular chains may pass from one crystallite to another through non-crystalline zones giving what has become known as a fringed micellar structure. The crystallites are collected together in microfibrils of considerable length, orientated generally at a low angle to the axis of the cell (Fig 3).

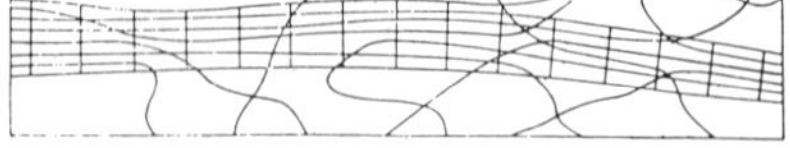

Fig. 3. Model of a fringed fibril structure.

Hemi-celluloses in wood are mixtures of polysaccharides containing glucose, mannose, galactose, xylose and arabinose together with some uronic acids. They have a much lower molecular weight than cellulose and are often branched rather than linear. They are intimately associated with the cellulose chains throughout the cell walls and appear to contribute to their strength. There is extensive hydrogen bonding between polymer chains. The composition of the hemicelluloses varies with species of tree, and there are, generally, substantial differences between those occurring in conifers and deciduous species. For example glucomannans occur predominantly in conifers and xylans in the deciduous species. Because they are largely non-crystalline they are accessible to water molecules and this makes it possible for water and other polar compounds to enter the cell wall by disturbing the hydrogen bonding. Non polar solvents cannot readily penetrate wood cell walls.

Lignin consists of an irregular 3-dimensional array of variously bonded hydroxy - and methoxy substituted phenyl propane units. Molecular weights of at least 15,000 have been determined. It contains both phenolic and alkyl hydroxyl groups available for reaction. Its concentration is highest in the middle lamella and cell corners but it does occur throughout the walls of many cells, and the greatest quantity occurs in the secondary walls of tracheids.

There are some structural features which are common to most species and important in the context of this review. The orientation of the main cells, and of the chains of crystalline cellulose within those cells, along the grain is the reason for wood's high tensile strength parallel to the grain and its marked anisotropy in physical and mechanical properties.

The non-crystalline polysaccharides are generally aligned parallel to the cellulose chains; they are accessible to water and so the swelling which results from absorption of water and the creation of new hydrogen bond systems takes place largely in the transverse direction; transverse swelling can be 10 to 100 times longitudinal swelling.

The other property which results from a common anatomical feature is permeability. Longitudinal permeability is in many species much higher than transverse permeability. This is a major influence on the uptake of water and processing fluids. Fluid flow within softwoods is often controlled by the small openings called pits, between adjacent cells, at which there are membranes. In some species, e.g. the spruces, the membranes move to block the pits during drying with the result that lateral penetration of treating fluids is limited to 1 or 2 mm and that longitudinal penetration may only be a few millimetres. In Scots pine by contrast, sapwood can be completely penetrated.

PRESERVATION OF WOOD

The technology of preserving wood against biological degradation has developed steadily over the past 20 years. There have been few changes in the chemicals being used; chlorinated hydrocarbons are now little used as insecticides as they have become unacceptable for environmental reasons, and there are growing pressures on the use of arsenic and chlorinated phenols as fungicides for the same reasons. Alternative insecticides have been found, particularly synthetic pyrethroids, but a search continues for replacement fungicides.

It is rather in the technology of processing that advances have been made, and in which further improvement is still being sought. Once it became evident and accepted that penetration in depth was essential to achieve long life from wood in the ground or in exposed exterior situations, commercial techniques involving combinations of vacuum and pressure were successfully developed and are in widespread use. By varying the parameters of pressure (positive or negative), time and solution strength, the desired loadings and penetration can usually be obtained, the limitations being set by the permeability of the species, and by costs.

The solvents used are generally either water or white spirit; the former is cheap and takes the preservative into the cell wall, but causes grain-raising - which can be important with planed wood, used, for example, in joinery. It also often necessitates long periods of storage for drying. Oil-based systems, being non polar, generally leave the preservative coating the cell lumen; they are widely used, especially for building timbers. Because of the high cost of petroleum products a water-dispersible petroleum based fluid has recently been developed for treating poles; this was designed to replace PCP in heavy petroleum oil. It appears to be giving promising results [1] in terms of the gross penetrations obtainable and the biological effectiveness. The spruces, which are resistant to treatment, are beginning to dominate softwood supplies in the U.K. The preservative treatment of dry sawn spruce remains a difficult and important problem, research on which is being funded by some EEC money in several countries. However it should be noted that Sitka spruce, for example, has a low rate of water absorption. This must reduce the risk of its moisture content reaching the 20% required for fungal decay, and may mean that in some situtations where the hazard is due to intermittant wetting, a shallow treatment with preservative will prove to be reasonably effective.

Never-dried spruce sapwood is very permeable and the treatment of spruce in the form of poles for use by telegraph or power transmission industries can be successfully achieved using a sap displacement process. A pressure difference is applied along the length of a never-dried pole and the preservative solution moves under the pressure gradient and replaces the natural sap. This process avoids long storage periods while poles dry before treatment, and it thus has some attractions. It has recently been used more widely and even with the more easily treated pines.

DIFFUSION TREATMENTS

Diffusion treatments in which a high concentration is applied to the outer surface of wet wood, and penetration takes place by ionic diffusion, have been used and have obvious attractions with impermeable species. Most work has been done with borates, which diffuse satisfactorily provided the moisture content in the wood is over 60%. The salts remain soluble and therefore are subject toleaching. They are widely used in Australia and New Zealand. There have been several attempts to develop double diffusion treatments. This involves applying a second solution to the wood after the first has diffused, and as diffusion of the second salt takes place interaction between the two solutes within the wood results in an insoluble product, toxic to fungi. No really successful process has yet been developed largely because of the time required for the treatment and problems of obtaining a good distribution of the deposited preservative in the wood.

Diffusion treatments require little equipment and are very attractive in developing countries, particularly as they can effectively double what would otherwise by a service life of perhaps only 3 years.

WATER REPELLENTS

These are materials which can reduce the rate of water uptake by wood without affecting its equilibrium moisture content. They can be valuable in situations where wood is subject to only intermittent wetting. Such treatments could become more attractive in future as they could avoid or reduce the use of fungi-toxic materials and hence be environmentally preferable. There is good evidence from field trials [2] that impregnation of joinery-type samples with waxes or resins reduces the rate of water uptake, the moisture content remaining below the 20% required for decay (Fig.4).

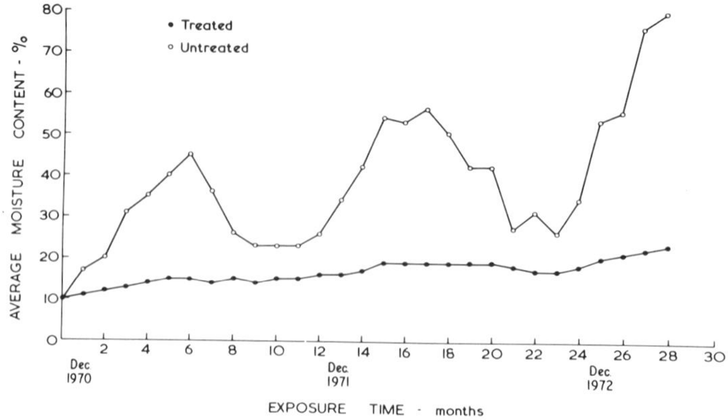

Fig. 4. Moisture content of Scots pine sapwood specimens during exposure.

However the effect lasts for only a few years; this decline in performance is also well established from laboratory tests. The reasons for this are likely to be complex but some clues are available from the work of Voulgaridis and Banks[3] (Fig.5).

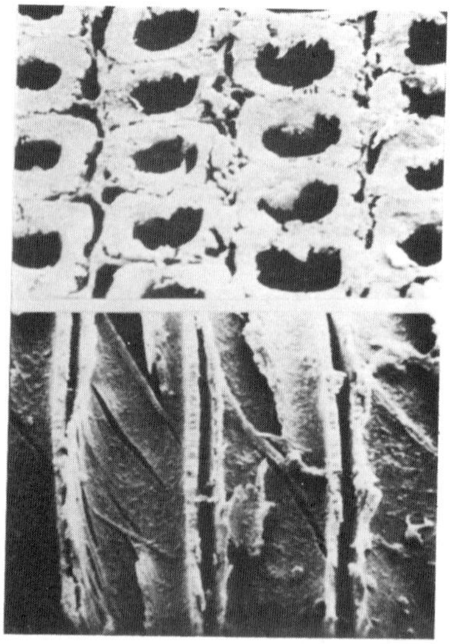

Fig. 5. Surfaces of pine treated with water repellent and exposed out of doors for 1 year.

They obtained evidence of severe degradation of the wood substrate within the outer one millimetre after weathering for one year, some degradation still being observed at 2-3 mm depth. Within the heavily degraded zone water repellent efficiency appears to be reduced because of the loss of any coherent cell-wall support for the resin/wax deposit.

These wax and resin deposits act by blocking pores and/or presenting hydrophobic surfaces; they are held to the wood by "Van der Waals" forces, which are easily broken when water is present. Attempts are therefore being made to achieve much stronger bonding of water repellents to wood surfaces. Treatments which result in covalent bonding very often involve reaction conditions which cause weakening of the wood itself [e.g. 4]. For example, treatment with chromium (VI) salts in acid conditions imparts significant water repellency. It is thought that the water-proofing effect is associated with the formation of chromium-wood complexes, probably mainly through reactions with lignin [5,6]. The chromium is reduced to the trivalent state.

However recent research using trivalent chromium soaps under reaction conditions which do not degrade the wood has shown that it is possible to achieve deposits which are resistant to leaching and which retain their water repellency for much longer than do conventional treatments [7](Fig. 6). Further developments in this field can be expected.

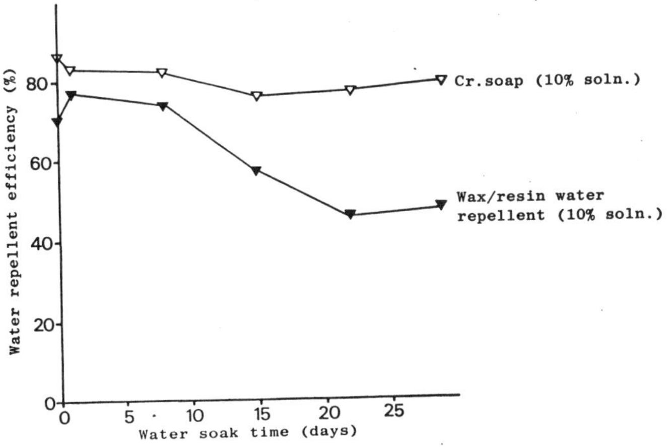

Fig. 6. Persistence of water repellency after treatment with chromium soap.

PENETRATION AND DISTRIBUTION

The distribution of a chemical in a piece of wood after a processing treatment is always uneven. On a gross scale the concentration is normally higher on the outside than in the centre, an effect which is accentuated by movement of the solvent during drying [8] and reduced over longer periods of time because of loss by volatilisation [9]. To some extent these distribution profiles can be improved by adjustments to the processing conditions, especially by applying vacuum as the last stage of the treatment cycle. Penetration is relatively low when the wood is just immersed in the treating fluid but can be greatly increased by use of an initial vacuum to remove the air and subsequently flooding with treating fluid and applying pressure [10].

Because both water and decay fungi can enter wood through small untreated areas, and so by-pass substantial areas which have been properly treated, the distribution of preservatives at cellular and micro-level is now regarded as a matter of some significance. At the cellular level there is good evidence that preservatives and water repellents tend to concentrate in certain types of cell.

Early work with pine showed penetration first through ray cells, then into tracheids, with the latewood cells being more easily accessible than springwood cells (Fig. 7).

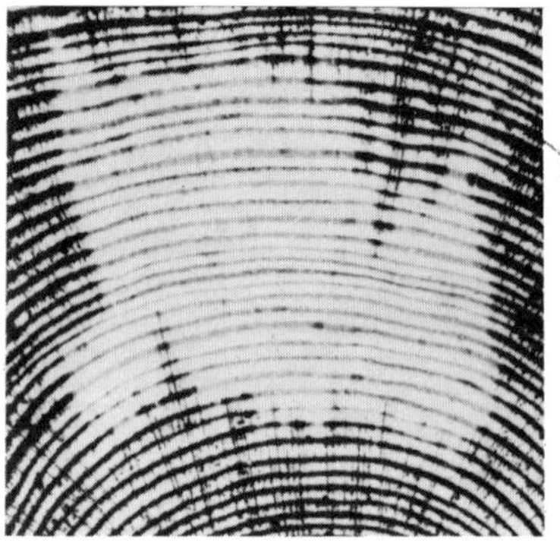

Fig. 7. Penetration pattern of liquids into lateral surfaces of Scots pine.

More detailed studies of distribution have been made using scanning electron microscopy, X-ray micro analysis and auto radiography of wood treated with radio-actively labelled materials.

Levy and Greaves for example [11] examined the distribution of copper, chromium and arsenic in eleven hardwood species treated with a standard preservative. The distribution was more even in those species for which the treatment was most effective. There was clear evidence for disproportionation of the elements in the preservative within the cell walls of some species, elemental ratios varying considerably. Retallick has used PCP and waxes labelled with carbon-14 to study the distribution in pine and spruce. He has obtained evidence [12] that whereas PCP enters predominantly into ray parenchyma, wax is deposited in bordered pits, resin ducts and ray tracheids (Fig. 8).

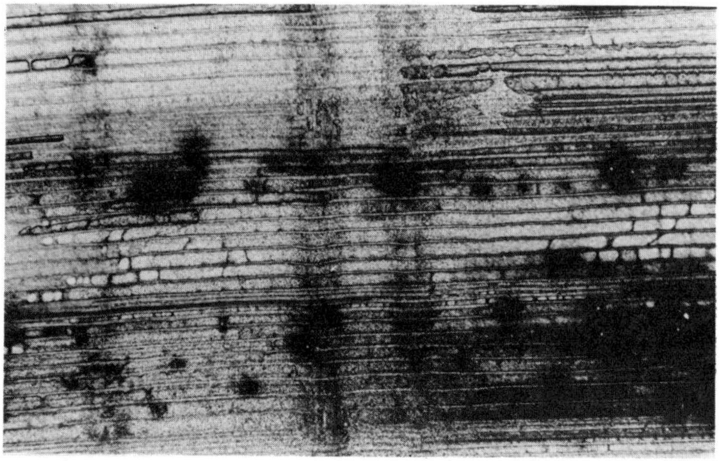

Fig. 8. Penetration pattern of 14C labelled PCP in softwood showing deposits in ray parenchyma but not in ray tracheids.

Both materials are deposited in latewood tracheids. We do not yet have explanations for the partitioning of the constituents in the fluid within the wood. There is no doubt from what we now know about the way in which fungal hyphae attack wood, that we need to achieve better control of preservative and water repellent distribution.

EMULSIONS

Over the past 10 years, emulsions have been developed for wood treatment primarily for remedial treatment against insect attack. The penetration of preservative from such emulsions is of some interest. Evidence so far [13] suggests that using oil:water ratios in the range of 1:2 to 1:20, penetration from surface applications into pine lateral surfaces is only 6 mm compared to 10 mm using white spirit alone. It was calculated however that not all cells were penetrated throughout these distances. Higher concentrations were found in the outer 0.5 mm when using the higher oil:water ratio emulsions. The short term rate of loss of insecticide by evaporation was slightly reduced by use of emulsions.

The physics of the penetration of wood by emulsions is complicated; water will tend to enter the cell wall but will not carry oil-soluble materials with it. The viscosity of the remaining emulsion will increase, thus reducing ease of penetration and eventually leaving pockets of oil containing a high concentration of solute. However emulsion technology is highly developed, and it would be surprising if systems are not developed which will make it possible to obtain a desired distribution of all kinds of chemicals within wood.

PERMANENCE AND LOSS

The essence of good preservation is long term effectiveness and so possible loss of toxic material with time is an important topic. Copper- chrome- arsenic preservatives in wood are not soluble in water and present no problems of loss. Nor do the metallic soaps such as copper naphthenate. PCP, chlorinated hydrocarbon insecticides, and tri n-butyl tin oxide are subject to loss by evaporation. Research has shown that loss of PCP, Dieldrin and HCH from treated pine takes place primarily from the outer 2 mm in which the original concentrations were comparatively high, and that even after a period of 8 years the concentration remaining in the depleted zone is still sufficient to provide protection [9]. The use of c-solvents reduces the rate of loss (Fig. 9).

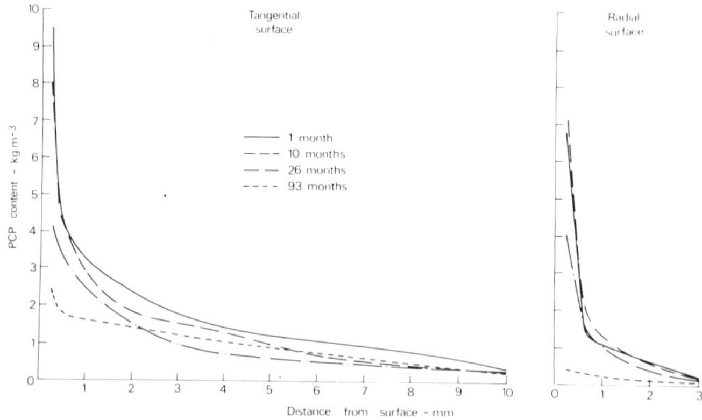

Fig. 9. Distribution profiles of PCP in exposed painted and unpainted Scots pine treated by immersion in 5% solution containing co-solvent.

When wood treated with these materials is painted, the paint film dissolves material from the outer 2mm, resulting in a more even distribution profile in the wood [9]. Loss by evaporation continues through the paint film.

In the last few years it has become evident that TnBTO is subject to loss by decomposition and that this could be more significant than evaporation. The tri-butyl compound breaks down to the less fungi-toxic di- and mono-butyl compounds, a process which seems to be due to free radicals, perhaps originating from extractives or other components of the wood itself [14]. There is also evidence that TnBTO is subject to degradation by enzymes from wood rotting fungi [15]. Nevertheless, evidence so far from field and service trials indicates that this compound can give protection for a period of about ten years.

SURFACE FINISHES

The science and technology of surface coatings for wood was a completely neglected subject until the mid 1960's. Changes in paint composition, building design and wood quality combined to bring about an increase in the rate of failure of paints on cladding and joinery, and an alarming increase in maintenance costs,and brought into prominence the need to find improvements.

Largely as a result of Miller's work over the past 10 years we now have a much better understanding of the factors which control the performance of paints, stains and varnishes on wood [16,17]. He has shown that while painting problems may stem from deficiences in coating materials, a general factor which is often of overriding importance is deterioration in the condition of the wood substrate during service. He has shown for example [16] that when clear finishes are used the wood surface suffers photochemical degradation not only from UV light (which we have known about for a long time) but also from visible light, resulting in considerable mechanical weakening and increase in water solubility. These changes are due to depolymerisation of both lignin and carbohydrate components of the cell wall. It follows that there is a limitation in principle on the performance of clear and semi-transparent finishes, and that either opaque finishes, or pretreatments which stabilise the wood against photodegradation, will be required to obtain long life. Treatment with solutions of chromic acid or chromates are claimed to be beneficial [18].

Recent American research has demonstrated that penetrating chemical agents based on polyethylene glycol PEG G-400 with 1-octadecanol, and clear films containing homo- and co-polymers of substituted hydroxy benzophenones are very effective in inhibiting discolouration and surface deterioration during ultra violet irradiation [33]. The authors suggest that these latter compounds are able to absorb energy from wood through hydrogen bonds and dissipate it through the keto-enol tautomerism and within the benzo-phenone moiety.

The incorporation of extenders in the form of minerals with a refractive index similar to the binder improves the performance of clear finishes quite markedly [19]. Material such as mica, pumice, foundry slag and pulverised fly ash have been used; these materials have a relatively high absorbance in the UV and it was originally thought that this property would be a major benefit. Use of PFA and mica certainly resulted in improved adhesion, implying less degradation of the wood surfaces; with some binders, they also brought about a change in the mode of failure and this is just as significant. The finishes containing the minerals failed by loss of gloss and chalking rather than cracking and flaking.

Miller has also established that it is vital to prevent the entry of water through end-grain in order to ensure even reasonable paint performance on softwood joinery (Fig. 10).

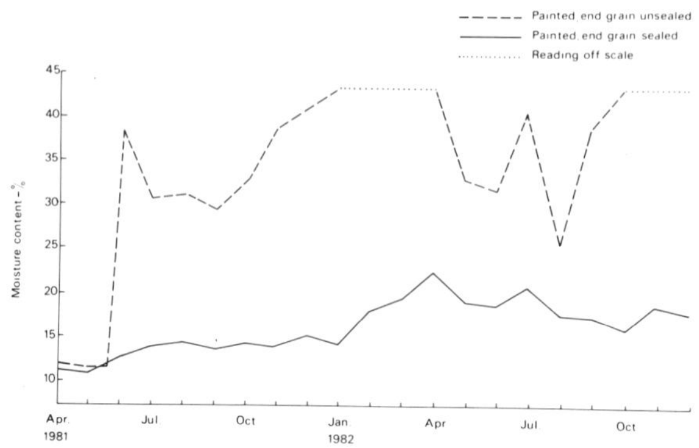

Fig. 10. Moisture contents in L joints exposed to the weather, showing effect of sealing end-grain.

Sealing end-grain has always been regarded as a difficult problem, for the sealant must not only close the open grain but must remain flexible enough to accommodate, without cracking, the large strains caused by swelling of the wood. Glues, primers and various resin treatments have been evaluated as end-seals [17], solvent thinned primers being the most promising to date.

More recently Banks and Evans [20] have shown that even de-ionised water, in the absence of light and oxygen, can bring about significant reductions in the tensile strength and toughness of 100 micron wood sections at temperatures of 50 to 65°C (Fig 11).

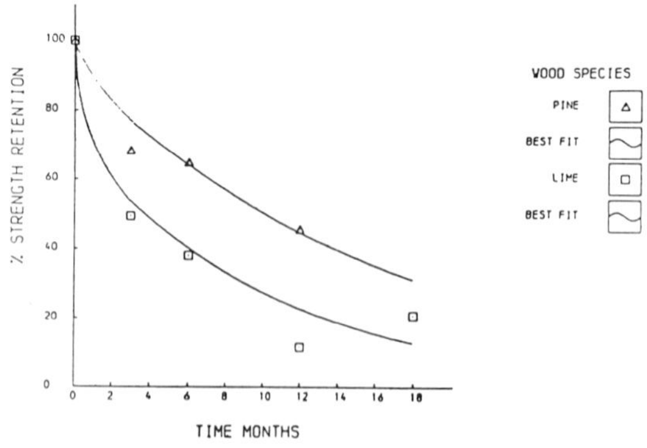

Fig. 11. Loss in tensile strength of 100μ strips of wood due to action of de-ionised water.

Such temperatures are frequently attained in exposed painted wood surfaces. The evidence suggests that degradation occurs primarily in the lignin and hemi-cellulose rather than in cell-wall cellulose, and is more pronounced in a hardwood like lime than in pine.

There is plenty of evidence for paint films splitting and peeling from the wood with wood fibres still attached, indicating failure in the wood surface, and it is clear there is a need to think about stabilising treatments for surfaces.

Because of the cellular and heterogeneous nature of the wood surface the strains imposed by swelling on a film adhering to the surface will not be the same throughout, but will be very high in localised areas. Thus one requirement for a good surface finish is sufficient elasticity to accommodate these strains, and more importantly but more difficult to achieve, the retention of elacticity on ageing. Another requirement for a good finish is an appropriate permeability to water vapour. Too low a permeability will inhibit evaporation of water which may have entered through cracks or end-grain, while too high a permeability will lead to high and frequent movement in the surface and the formation of surface cracks. The desirable level of permeability is now being recognised [21] and can be obtained using either oil or water based paints.

Our improved understanding of the science behind the behaviour of finishes on wood has been accompanied by developments in paint technology. There are now available products which provide a continuous range from lightly pigmented stains through highly resinous stains to fully pigmented gloss paints. Appearance, application and maintenance costs, and situation of use are all factors which will determine choice in any particular case. Water-based acrylics have good colour retention and retain their elasticity on ageing better than the newer oil-based alkyds, but they are less good from an operational point of view, being less tolerant of weather conditions and having poor compatibility with materials such as glazing compounds and previously painted surfaces.

DIMENSIONAL STABILISATION

Three main approaches have been explored; the removal of reactive groups by reaction with esterifying agents, bulking wood in the swollen state by depositing polymers within the cell wall, and blocking the pores. To date the effectiveness and practical utility of such treatments has been limited.

A measure of the degree of stabilisation obtained is the anti-shrink efficiency (ASE) defined as:

$$\frac{S_1 - S_2}{S_1} \times 100$$

where S_1 is the volumetric swelling coefficient of the untreated wood and S2 is the corresponding coefficient of the treated wood. The maximum ASE obtained is 86%, more usual values being in the range 40-60%.

Most work has been on acetylated wood; the product is dense, with good mechanical properties but it can be corrosive. Also, it has not been possible to bring about the reaction with hydroxyls without using conditions of low pH, or the generation of free acid which it is impossible to remove completely. This causes degrading hydrolysis of the wood substance in the longer term[22].

Reaction of wood with simple epoxides and with methyl isocyanate has been shown to produce material with improved dimensional stability, ASE up to 70% being recorded [22]. The reactions take place readily without the formation of undesirable byproducts. Reaction is best carried out on dry wood to minimise the formation of homo-polymers not bonded to the wood. Dimensional stability is at a maximum at 25-30% weight gain; beyond this level the cell wall becomes excessively swollen and split by the polymer and the wood can then swell through water absorption beyond its original swollen volume. Infra-red evidence suggests bonding of isocyanate both to carbohydrate and to lignin.

Mechanical properties are generally reduced by these treatments. The effect of treatment varies considerably with the species of wood being used.

It is quite possible that treatments of this type may provide a useful route to stabilising surfaces.

In the mid 1960's, attempts were made to stabilise wood by impregnating with vinyl monomers and subsequently polymerising using either a heat sensitive catalyst or gamma radiation. Products with improved strength and stability were obtained but the process has had limited practical success [23]. Polymer is usually deposited in the pores but can be made to form in the swollen cell wall by using polar solvents. Polymerisation using radiation did not bring about any grafting of vinyl polymers to carbohydrate but there was some evidence of grafting to lignin[24].

Bulking, by treating with solutions of polyethylene glycol and allowing the polymer to diffuse into the wood, has been widely investigated and has been successfully used-particularly for preserving archaeological artefacts, raised wooden ships and old wooden carvings. The polymer (which can have a molecular weight of several thousand) diffuses into the cell wall which becomes swollen but stable. ASE up to 80% can be obtained[23].

It should be said that for industrial purposes, it will often be cheaper to use a more stable, but more expensive species, than to treat a cheaper but readily available species by such processes as have been discussed. Specialised treatments for special needs may be attractive; for example beech veneers heavily impregnated with PF resin and heated under pressure give a product which has been very successfully used for propellor blades and high voltage insulators.

ADHESIVES

Glueing of wood is important for furniture, building products such as laminated beams, doors and windows, and for board materials such as plywood and chipboard. Most current adhesives are based upon resins involving reaction of phenols or amines with formaldehyde, or vinyl polymers such as polyvinyl acetate. Because these are largely derived from petroleum there is an incentive to find new adhesives which are durable but derived from natural products. Adhesives based on the liquor from sulphite pulping, and from wattle tannin have been developed, the latter particularly receiving attention in Australia and South Africa. While they can provide reasonable quality products they are most likely to be used in conjunction with synthetic resins. PVA based adhesives are regarded as having relatively low durability and being subject to creep. Newer cross-linked PVA products are thought to be an improvement but experience of their use is limited at present.

The outstanding technical problem is durability. The durability of a glued product depends upon three factors; the integrity of the adhesive itself, the integrity of the substrate and the maintenance of the bond between the two. The first two of these are dependent on chemical stability, and in particular resistance to hydrolysis. But there are other factors which are at least as important. The maintenance of the bond will also depend on whether the various materials can resist the stresses applied, not only those from an applied load but also those resulting from changes in the environment.

In a glued laminated beam, the pieces of wood are aligned with the grain parallel, but in many situations, such as a joint in a window frame or within a piece of plywood or chipboard, the two pieces of wood which are glued have their grain aligned at up to 90°. Given the large anisotropy in swelling of wood, any change in moisture content will in these circumstances impose a large differential stress across the adhesive, and there is a growing body of evidence and opinion that this is a major cause of failure in glued products. By adding thiourea to UF resin and assessing the location of the sulphur atom using X-ray analysis, Beele, Bolton and Dinwoodie [25] have shown that in spruce chipboard there is little or no resin in the cell lumena but good evidence for penetration of resin into the cell wall. Resin penetrates up to 3 cells deep into the spruce chips.

Dinwoodie [26,27] has shown that under conditions of alternating humidity which weaken chipboard, there is no change in the properties of UF resin per se. However, applying quite small mechanical strains to glued wood joints under constant conditions results in substantial weakening. He concludes that mechanical fatigue resulting from changing moisture content is more likely to be the reason for loss in strength than change in resin itself. Canadian work [28] showed that when joints are submitted to cycling moisture contents, the shear strength of those made with UF resin weakened more rapidly than those using PF or PVAc, despite having the highest initial strength.

Results of long term tests on glued joints in the UK showed that whereas initially failure was normally in the wood, after a period of years only joints made with PF or RF resins failed in this way, other resins failing in the glue-line or at the glue/wood interface [29].

The stress/strain behaviour of PF and UF resins has been examined in work at Bangor [30]. PF resins tend to be more hygroscopic, have a lower MOE and a faster rate of stress relaxation. They are thus better able to withstand stresses imposed by differential expansion of the adherends. These observations seem to explain the differences in behaviour of UF and PF bonded boards and joints at room temperature. They also imply failure for mechanical rather than chemical reasons. At higher temperatures, resin hydrolysis will probably play a larger role in joint failure, particularly with UF resins.

The nature of the adhesive to wood bond is still subject to much uncertainty despite a good deal of research. The ideas on the mechanism of failure emerging from the above discussion do not seem to imply the presence of a covalent bond. Canadian workers have studied the development of cohesive and adhesive strengths of wood/glue bonds during hot pressing using UF resin. They conclude that the cohesive strength of the resin developes faster than adhesive strength [28]. They also report an activation energy for the polymerisation of PF resin in the presence of various cellulosic materials of 6 to 7.8 Kcal/mol, compared with 11.2 Kcal/mol for the resin alone [31]. Humphrey and Bolton [32] report an activation energy of 11.4 Kcal/mol for bond formation in particle board using UF resin. Canadian data on rates and activation energies for the hydrolysis of resin/wood combinations are difficult to interpret as they are based only on measurement of formaldehyde production. Thus although there is some circumstantial evidence for a covalent bond between wood and resin, the issue of the significance of this type of bonding remains unresolved.

FINAL COMMENTS

In the 1960's, chemical treatment for wood was a popular topic for research, but the new processes that emerged, such as the production of wood-plastic composites, have had but limited success in application

The present emphasis on improving the durability of glued wood products, preservative treatments, and paints and other finishes on wood is giving a new impetus to research and some new lines of investigation are opening up. I think there is now hope that some useful new processes will soon be developed.

REFERENCES

1. Kirchner R.P. and Taylor J.M. (1983). Record of the Annual Convention of the British Wood Preserving Ass. pp 26-37.

2. Purslow D.F.(1982) The effect of water-repellent preservative treatment on moisture levels in window joinery. BRE Information Paper IP 20/82. Dept. of the Environment.

3. Voulgaridis E.V. and Banks W.B. (1981) Degradation of wood during weathering in relation to water-repellent long-term effectiveness. J.Inst.Wood Science,9,No2, pp72-83.

4. Feist W.C. (1979) Protection of wood surfaces with chromium trioxide. USA Forest Products Lab.Report No AD-A077432

5. Feist W.C. and Ellis N.D. (1978) Fixation of hexavalent chromium onto wood surfaces. Wood Sci.11,No2.pp76-81.

6. Pizzi A. (1980) Waterproofing and lignin cross-linking by means of chromium trioxide/guaiacyl unit complexes. J.Appl.Polymer Sci. 25, No11.pp 2547-2553.

7. Wright J.K.,Banks W.B. and Eilbeck W.J. (1985) A substantively bonded water-repellent treatment based on chromium carboxylates. Paper to International Research Group on Wood Preservation,Brazil,March 1985.

8. Banks W.B. and Carragher J.G. (1984) Record of the Annual Convention of the British Wood Preserving Ass. pp3-12.

9. Orsler R.J. and Stone M.W.S. (1979) The permanence of pentachlorophenol in preserved joinery timbers. International Journal of Wood Preservation, 1 No.4, pp177-184.

10. Purslow D.F. (1974) Methods of Applying Wood Preservatives. BRE Report. H.M.S.O.

11. Levy J.F. and Greaves H. (1978) Penetration and distribution of copper-chrome-arsenic preservative in selected wood species. Holzforschung, 32, No6 pp209-213.

12. Retallick S. Personal Communication. Dept. Forestry and Wood Science, U.C.N.W.

13. Orsler R.J. and Stone M.W.S. (1982) Remedial spray treatments used against insect attack pt.1. Permanence and distribution. Material und Organism 17 No3 pp161-180.

14. Orsler R.J. and Holland G.E. (1984) Experiments on the degradation of tri-butyl tin oxide: a progress report. Paper to International Research Group on Wood Preservation, Sweden. March 1984.

15. Orsler R.J. and Holland G.E. (1982) Degradation of tri-butyl tin oxide by fungal culture filtrates. Int. Biodet. Bull. 18, No4 pp 95-98.

16. Derbyshire H. and Miller E.R. (1981) The photodegradation of wood during solar irradiation. Holz als Roh-und Werkstoff, 39, pp 341-350.

17. Miller E.R. and Boxall J. (1984) The effectiveness of end-grain sealers in improving paint performance on softwood joinery. Holz als Roh-und Werkstoff, 42 pp27-34.

18. Feist W.C. and Hon D.N (1984) Chemistry of Weathering and Protection; The Chemistry of Solid Wood, Ed.R.M.Rowell American Chemical Society.

19. Boxall J., Hayes G.F., Laidlaw R.A., and Miller E.R. (1984) The performance of extender modified clear finishes on exterior timber. JOCCA, No 9 pp227-233.

20. Banks W.B. and Evans P.D. (1985) Hydrolytic degradation of wood surfaces. Symposium on Weathering and Photo-degradation of Wood and Cellulose Fibres. A.C.S. Autumn Meeting, Chicago.

21. Graystone J.(1984) Architects Journal 20th June p 75.

22. Rowell R.(1984) Penetration and Reactivity of Cell Wall Components; The Chemistry of Solid Wood. Ed.R.M.Rowell American Chemical Society.

23. Meyer J.A.(1984) Wood-Polymer materials: The Chemistry of Solid Wood. Ed.R.M.Rowell. American Chemical Society.

24. Gibson E.J.,Laidlaw R.A. and Smith G.A.(1966) Dimensional stabilisation of wood. J.Appl.Chem.16 pp 58-64.

25. Bolton A.J.,Dinwoodie J.M. and Beele P.M.(1985) The micro-distribution of UF resins in particleboard. Proceedings of Conference "Forest Products Research International: Achievements and the Future". Pretoria.

26. Dinwoodie J.M.(1981) Characterising the performance of chipboard in the UK. Proceedings of 15th International Particleboard Symposium, Washington State University.

27. Dinwoodie J.M.(1977) Causes of deterioration of UF chip-board under cyclic humidity conditions. 1. Performance of UF adhesive films. Holzforschung 31 No.2 pp50-55.

28. Chow S.and Steiner P.R.(1975) Catalytic exothermic reactions of urea formaldehyde resins. Holzforschung 29 No.1 pp5-10.

29. Rodwell D.F.G.(1984) Ageing of wood adhesives - loss of strength with time. BRE Information Paper IP8/84. Dept. of the Environment.

30. Irle M.(1985) Personal Communication. Dept. Forestry and Wood Science, U.C.N.W.

31. Chow S.(1969) A kinetic study of the polymerisation of phenol formaldehyde resin in the presence of cellulosic materials. Wood Science 1 No.4 pp 215-221.

32. Humphrey P.E. and Bolton A.J.(1979) Urea formaldehyde resin bond strength development with reference to wood particleboard manufacture. Holzforschung 33 No. 4 pp 129-133.

33. Hon D.N.,Chang S.T. and Feist W.C.(1985) Protection of Wood Surfaces against photo-oxidation. J.Appl.Polymer Sci.30 pp 1429-1448.

Chapter 40

Ester pulping of wood

Raymond A. Young, James L. Davis, Eva-Barabara Wiesman, Timothy R. Young – Department of Forestry, University of Wisconsin, Madison, Wisconsin 53706.

Kenneth W. Baierl – Biodyne Chemicals, Inc., 100 W. Lawerence Street, Appleton, Wisconsin 54911.

INTRODUCTION

A variety of alternate pulping methods have been explored over the past century but most have never reached commercial operation. This is due to a variety of reasons which include poor pulp quality, chemical losses and recovery and environmental problems. Even in those cases where potential exists for new process development, sufficient funding was not forthcoming to properly evaluate the process technology at the pilot plant scale.

With increasing environmental restrictions such as reductions in total reduced sulfur for kraft pulps and water pollution limitations on sulfite mills, the climate is steadily improving for commitments to new pulping technology. A new process termed ester pulping will be described in this paper. This innovative method could revolutionize the pulp and paper industry. Ester pulping provides the following innovations:

1. A system based on organic solvents rather than aqueous sulfur compounds.
2. A system which utilizes very low energy phase separation rather than expensive furnace systems or heavy reliance on distillation for recovery of the pulping chemicals.
3. A system which practically eliminates air and water pollution by recycling all chemicals.
4. A system which increases yields of pulp fiber from wood for more efficient utilization of our renewable resources.

5. A system which has a much lower capital expenditure necessary for installation. It is estimated that the capital requirements would be one-half those of conventional processes.

Earlier studies on organic acid pulping of wood provide important background information and the rationale for new pulping developments in organic media.

ACETIC ACID PULPING

Wiltshire [1] demonstrated in 1944 that acetic acid would effectively delignify wood at low temperatures. Later DeHaas and Lang [2] demonstrated that strong pulps could be obtained from both hardwoods and softwoods by cooking at high acetic acid concentrations and high temperatures. They suggested that a water content in excess of 15% of the cooking liquor reduced the pulp yield as a result of polysaccharide hydrolysis, slowed the delignification process, and caused precipitation of the lignin.

Great potential exists for organic acid pulping because of the good pulp strength possible and the development of new methods for recovery of organic acids. A variety of methods are available for recovery of organic acids and include: direct and azeotropic distillation, osmosis, liquid-liquid extraction, esterification and adsorption. Solvent extraction methods appear to be the most desirable for recovery of organic acids from organic acid pulping [3-5].

Pulping With Acetic Acid. A further evaluation of acetic acid pulping was carried out in our laboratories and these results and pulp properties are summarized in Table 1 (kraft pulp properties are included for comparison). Generally the highest yield at each cooking time was obtained with the highest acetic acid concentration (87.5%) [6-8].

The conditions for acetic acid pulping which gave the highest strength properties at comparable lignin contents are noted by asterisks. In all cases, a high concentration of acetic acid was necessary to obtain low Kappa numbers and high strengths [7, 8].

The results obtained for acetic acid pulping present a dilemma. It is necessary to have a high acetic acid concentration for pulp strength and delignification but it is not possible to economically recover high acid concentrations by liquid-liquid extraction methods (or any other method). A substitute solvent which was compatible with this system was therefore sought.

Solvent Substitution

Schuerch [9] tested the ability of more than 30 organic solvents and water to dissolve a variety of lignin

Table 1. Acetic acid pulp strength properties.

Temp °C	HAC CONC %	Time hr	Yield %	Kappa No.	B.L.[a] M	Burst Factor	Tear Factor
ASPEN							
160	50	1.5	59.1	48	2429	4.73	81.4
		3	52.3	33	5054	9.1	92.5
160	75	1	64.7	60	2654	2.94	99.5
		3	52.5	11	6308	25.87	78.3
160	87.5	1	70.2	60	1437	0.56	43.1
		3*	55.6	14	8128	24.17	109.7
170	50	1	57	43	2982	4.82	74
	75	1	57.5	28	4926	12	65.7
	87.5	1	62.9	40	4048	11.8	96.9
185	50	1	50.9	30	7780	9.09	57.1
	75	1	51.3	9	6563	17.77	64.3
	87.5	1*	55.1	10	9101	28.3	98.8
180 KRAFT		3	48.5	14	6515	25.27	103.7
SPRUCE							
170	75	2	68.5	96	1175	0.6	53.3
		4	54.6	85	3863	12.9	78.2
185	75	2	49	45	6506	26.1	101.8
		3	51.6	80	3602	16.3	69.7
185	50	2	55.5	132	924	0.5	58
	75	2	48.2	40	5815	21.5	77
	87.5*	2	55.3	46	6370	24.6	126.5
225	50	0.75	48.8	125	2883	7.2	91.2
	75	0.75	45.6	26	5897	23.6	50.9
	87.5*	0.75	46.2	16	8321	37	141.1
180 KRAFT		3	47.1	31	6072	31.7	200

[a] B.L. = Breaking Length

preparations. He concluded that those solvents with a Hildebrand solubility parameter (δ) of approximately 11.0 were the solvents best able to dissolve large lignin fragments. If the value of the solubility parameter for a solvent lay far from the optimum value, Schuerch asserted that this solvent would only be able to dissolve small lignin fragments. Although these data were obtained with isolated lignin preparations, it servės as a guideline for native lignin solvation in pulping reactions.

Since a solvent with a value of δ close to 11.0 would be able to dissolve large lignin fragments, use of a better lignin solvent in the pulping liquor would reduce the degree of lignin degradation necessary in order for the lignin fragments to become soluble in the cooking liquor. The high value of δ for water (23.4) is consistent with the observation that lignin is practically insoluble in water, unless the lignin is chemically modified to make it hydrophilic. In order for the effective solubility parameter of an aqueous acetic acid solution to approach 11 to 12, the concentration of acetic acid would have to be quite high (87–95%). The experimental results are consistent with the solubility parameter prediction. It is proposed that ethyl acetate (δ = 9.1) serve as the substitute solvent for water in acetic acid pulping to give a new solvent based pulping process comprised of ethyl acetate, acetic acid and water.

ESTER PULPING SYSTEM

That ethyl acetate can be directly substituted for water in the system is shown in the ternary phase diagram in Figure 1

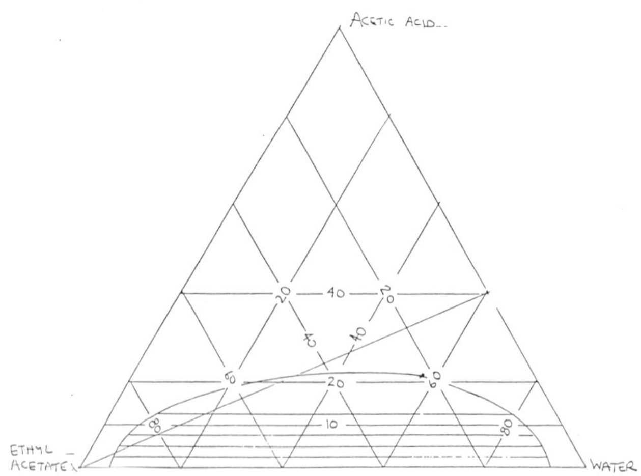

Fig. 1. Ternary phase diagram for acetic acid-ethyl acetate-water.

[10]. The composition of each phase is given by the interaction of the tie-lines in the diagram. Outside the phase envelope at the bottom of the diagram no phase separation occurs and the system is one homogeneous phase of ethyl acetate/acetic acid/water. However, by proper adjustment of the solvent ratios so that the liquor composition falls within the phase envelope, a two-phase system results; an organic phase composed of ethyl acetate and acetic acid, and an aqueous phase. This facile separation (decantation) drastically reduces energy requirements for chemical recovery. Sugars are obtained from the aqueous phase for fermentation and a pristine lignin can be easily precipitated from the organic phase. The lignin properties are characteristic of other organosolv lignins which are uncondensed and contain no sulfur. Thus, possible byproduct utilization of lignin is also a distinct possibility.

The positive implications of the addition of ethyl acetate to the pulping medium are profound and include:

1. A reduction in the amount of water required in the system.
2. Enhanced solubility of lignin in the pulping solvent.
3. Reduced cooking time.
4. Lower pulping temperatures for given cooking pressures.
5. Higher strength pulps.
6. Higher yields.
7. Simplified recovery of organic chemicals.

Ester pulping offers great flexibility for production of a wide variety of pulps, from high yield chemimechanical pulps for container board to dissolving grades. The pulp grade is determined by the ratio of ethyl acetate:acetic acid:water in the liquor, the temperature of the cook (160-200°C) and the time of the cook (15-120 minutes). A typical ester pulping reaction for aspen chips is shown in Figure 2. In just 30 minutes a Kappa number of 30 is attained, with an outstanding yield of 60%.

Selectivity curves for ester pulping of aspen and spruce compared to acetic acid and kraft pulping are shown in Figure 3 and 4. The selectivity curve for aspen dissolution (Figure 3) is superior to even 87.2% acetic acid, the optimum acid concentration suggested by DeHaas and Lang [2] for maximum pulp strength. Figure 4 shows that the ester pulping selectivity for spruce lignin removal is superior to the kraft pulping process. Optimization for specific pulp properties is currently in progress.

A total pulp mill including chemical recovery has been designed for aqueous acetic acid-ethyl acetate pulping. The system involves liquid-liquid phase separation, liquid-liquid extraction, distillation, centrifugation, evaporation and burning to give the most economical, commercially viable process [11].

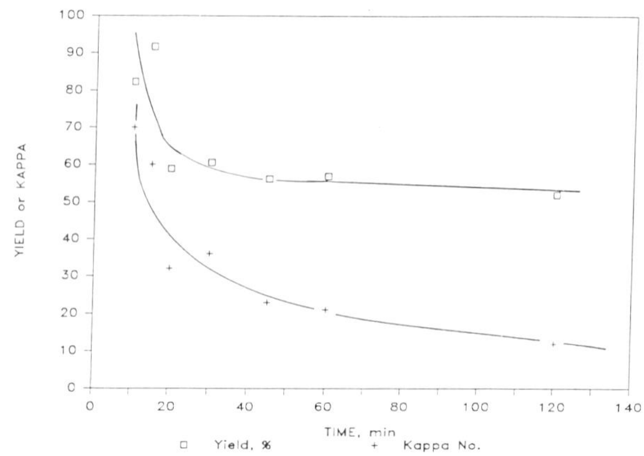

Fig. 2. Percent yield and Kappa number versus reaction time for ester pulping of aspen chips.

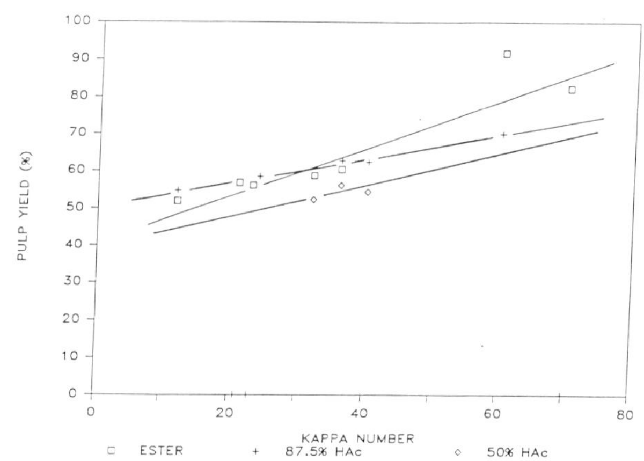

Fig. 3. Selectivity curve for aspen delignification in aqueous acetic acid compared to ester pulping.

SUMMARY AND CONCLUSIONS

The ester pulping process improves wood utilization since higher pulp yields are obtained and no expensive inorganic chemical recovery system is required. The organic chemicals can be readily produced from wood so that it would probably not be necessary to purchase large quantities of chemicals. The process should be environmentally safe since no air or water pollution sources are part of the system. Biodyne Chemicals,

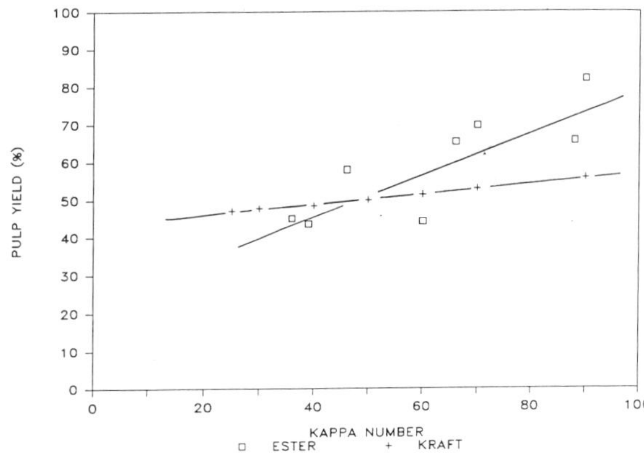

Fig. 4. Selectivity curve for spruce delignification in kraft and ester pulping.

Inc. is currently licensing the process for use on a worldwide basis.

REFERENCES

1. Wiltshire, W. A., Proc. Tech. Sect. Paper Makers' Assoc. of Great Brit. & Ireland. 24 (1944) 347.
2. DeHaas, G. and C. Lang, U.S. Patent 3,553,076, 1971.
3. Young, R. A. and S. Achmadi, In: Biomass Utilization. W. A. Cote, ed., Plenum Press, N.Y., p. 585-610, 1983.
4. Data, R., Paper presented at the Third Symposium on Biotechnology in Energy Production and Conservation, Gatlinburg, Tennessee, May 12-15, 1981.
5. Biggs, W. A., J. T. Wise, W. R. Cook, W. H. Baxley, J. D. Robertson and J. E. Copenhaven, Tappi 44 (1961) 385.
6. Young, R. A., J. L. Davis, E.-B. Wiesmann and K. Baierl, Organic Acid Pulping of Wood. Part I. An Overview of Applications. Proceedings of the Intn'l Symp. Wood Pulping Chem., Vancouver, Canada, August, 1985.
7. Young, R. A., J. L. Davis and E.-B. Wiesmann, Organic Acid Pulping of Wood. Part II. Acetic Acid Pulping of Aspen, Holzforschung 40 (1986) 99.
8. Davis, J. L., R. A. Young and S. S. Deodhar. Organic Acid Pulping of Wood. Part III. Acetic Acid Pulping of Spruce, Submitted to J. Japan Wood Res. Soc. (1986).
9. Schuerch, C., J. Am. Chem. Soc. 74 (1952) 5061.
10. Sahoni, V. R. and V. R. Warhadpande, Ind. Eng. Chem. 44 (1952) 1428.
11. Young, R. A. and K. W. Baierl, Southern Pulp Paper 48 (1985) 15.

Chapter 41

Increasing wood yields by silviculture and tree breeding

Christine Cahalan – Department of Forestry and Wood Science, University College of North Wales, Bangor, Gwynedd.

INTRODUCTION

Foresters and wood users have different understandings of what is meant by 'yield'. Timber growers usually estimate the volume production of their crops, but wood users may be more concerned with the total dry matter production of a forest plantation or the yields of sawn timber or pulp that can be expected from it. This paper attempts to consider the ways in which yield in the broadest sense could be increased by silviculture and tree breeding, mainly in the British context.

SILVICULTURE

Silviculture is the tending of forest crops throughout their rotation (the time taken for them to reach maturity). In upland Britain the commonest silvicultural system is one of clearfelling of the whole crop followed by replanting, usually with a single species. Site preparation, weeding and fertilization are done in Britain to ensure the satisfactory establishment and early growth of the crop. More intensive ground preparation and higher rates of fertilization might increase the growth rates of young crops on some sites, but where soil fertility and drainage are adequate and tree growth is limited by climatic factors beyond the forester's control there is little scope for further improvement.

The density of forest plantations is controlled by the initial spacing of trees at planting and by thinning during the rotation. Over a wide range of densities these have little effect on total yield, but trees grown in closely spaced, unthinned plantations are smaller than those grown at wide spacing and/or in plantations which have been thinned.

Windthrow (uprooting of trees), which is the major constraint to forestry in the uplands, is exacerbated by thinning, and as a result many plantations in upland Britain are not thinned at all ("no thin" regimes). These plantations can stand and grow for longer than thinned ones, and therefore produce higher total yields in a single rotation than thinned crops on the same site (Rollinson 1985). However, the yield of sawlogs from no thin crops is often very low.

The need for protecting forests against pests and diseases is obvious when the damage they do proves fatal, can be identified easily and occurs regularly. Serious growth reductions can also result from diseases and pest attacks which lack obvious symptoms, and the yield of utilisable wood may be affected by fungal rots which attack the trees late in the rotation. These two categories of pests and diseases have received less attention than their economic importance would seem to justify, and important yield increases could probably be achieved by reducing levels of such chronic pest and disease damage.

There are alternatives to the single species plantations of conifers grown on a 45-65 year rotation and then clearfelled before replanting. Four such alternatives are discussed here: high input regimes; mixtures; "biomass" plantations; whole tree harvesting.

HIGH INPUT REGIMES

These are intensive systems of tree growing which depend on high inputs of fertilisers and regular heavy thinning to produce high value (sawlog) crops in a short rotation. They are practised in various parts of the world where a premium is paid for high quality timber, for example in New Zealand (Manley and Knowles 1980). In Britain the main advocate of high input systems is Davies (1979), who has proposed a regime for growing Sitka spruce (*Picea sitchensis*) in this way on a 38 year rotation. The economics of the regime are debatable given that there is no premium for quality Sitka spruce logs, and the combined effect of rapid growth and heavy fertilisation on the wood quality of the final crop, mainly as a result of reduced density (Brazier 1977), must be a cause for concern.

MIXTURES

The literature on yields of mixtures and pure stands of the same species is conflicting, and most of the work has been done on herbaceous crops. There are at least some instances of higher yields from a mixture of two species than from either species grown alone, but the more usual finding is that the mixture produces less dry matter than a pure stand of the higher yielding species (Harper 1977).

Mixtures of conifers and broadleaves and of two coniferous species are both planted in Britain. The main reasons for growing conifer-broadleaved mixtures is the early return on investment produced by thinning of the conifer component and, for some broadleaved species, the benefits of sheltering (nursing) by the conifers. Recent interest in broadleaves (Forestry Commission 1985) seems certain to lead to increased planting of broadleaved species, and if mixtures are used there will be an associated increase in conifer production. The prospects for increased total yields from mixed plantations are best on poor sites. Mixtures of Sitka spruce and lodgepole pine (Pinus contorta) or Japanese larch (Larix kaempferi) are used on infertile soils because they grow well without the inputs of nitrogen required by pure spruce (Taylor 1985). Alnus spp. have potential on soils low in nitrogen because of the ability of bacteria associated with their roots to utilise (fix) atmospheric nitrogen (Tarrant and Trappe 1971).

"BIOMASS PLANTATIONS"

Following the oil crisis of the 1970s interest in wood as an energy source increased in many parts of the developed world (Anderson et al 1983). Work has been done on single stemmed trees grown in very dense plantations and on coppice crops, which regenerate from shoots produced from the cut stump and which can be harvested on a 3-10 year cycle. Most broadleaved species can be coppiced, and trials in Europe and North America have shown that yields of 15-20 tonnes dry matter. hectare^{-1}. year^{-1} can be achieved under the best conditions. Single stemmed trees grown on rotations of 12-20 years can produce 8-12 tonnes. hectare^{-1}. year^{-1} (King 1981).

However, such yields are only realised on fertile lowland sites and/or with high inputs of herbicide and fertilisers. It is important to recognise that so-called "energy plantations" do not increase absolute yields from the same area of land (Cannell and Smith 1980). Closely spaced crops simply capture the site earlier and reach their maximum productivity earlier than widely spaced ones, which has economic consequences but does not alter the underlying biological limit to production of biomass on a single site.

One possible use of coppice in Britain has not been explored very thoroughly. The two hardwood pulpmills in the south of the country have to haul their raw material long distances, and will accept small sized thinnings for which there may be no other market. Coppice grown on longer rotations (10 to 25 years) can produce stems of similar dimensions to those taken out in early thinnings, and yield per hectare is much higher than from a conventional first or second thinning. Is such material suitable for pulping?

WHOLE TREE HARVESTING

In most conventional harvesting systems the branches and tips of felled trees are cut off the merchantable stems before these are removed from the site. Whole tree harvesting is the harvesting and removal from the forest site of all above ground parts of a tree crop (Anderson 1985).

It is clearly possible to raise total biomass yield from a forest simply by switching from the former method of harvesting to the latter. However, there are potential problems in doing so and the advantages may not be as great as they first appear. It is true that in young crops a substantial proportion of the total biomass is contained in the upper stem and branches, but this percentage decreases with age. More important, at least for foresters concerned with long term productivity of forest sites, is the fact that most of the nutrients held in trees are found in the foliage and small branches, and their complete removal could have serious effects on nutrient levels in soils of inherently low fertility (Weetman and Algar 1983).

Moreover, it is not obvious (at least not to foresters) what the extra biomass would be used for. Branches and foliage can be chipped and used as a source of fuel (King 1981), but the market for green chipboard is probably limited, and green material would presumably create problems in the pulping process. Given the potentially serious long term consequences of this form of forestry, it should not be attempted on any scale, and certainly not on marginal sites.

SPECIES

Correct choice of species is obviously vital to the long term success of any forestry enterprise, but it is not always appreciated that correct choice of seed source (provenance) may be equally important. Nearly all commercially grown tree species exhibit considerable geographic variation in traits of interest to foresters and wood users, including growth rate, form, disease resistance and wood properties (Callaham 1964). Provenance trials, in which trees from a number of seed sources are grown together in the same place, have shown that some of this variation is due to heritable, genetic differences between the provenances (IUFRO 1978).

PROVENANCES

Forest geneticists have often recommended the use of local seed sources when growing native species (Campbell 1974). For exotic species a provenance which shows good growth on a range of planting sites is usually required, and in Britain Queen Charlotte Island (QCI) origins of Sitka spruce are recommended for general planting (Kennedy 1974). Both approaches place a premium on reliable performance rather than rapid growth, and there is therefore scope for increasing productivity by more precise matching of provenance to site. The Forestry Commission intends to use different seed origins to develop three populations of Sitka spruce, one of which will be used on mild sites (Faulkner 1985). Here trees which go on growing well into the autumn can be used without incurring the risk of frost damage, and the extended period of growth can be expected to result in yield increases.

SELECTION AND BREEDING

There is usually as much genetic variation within provenances (i.e. between individual trees) as there is between provenances (Perry 1979), and this can be exploited by tree breeders in improvement programmes. Direct selection for volume production is usually done by identifying trees with fast growth rates and using them as the parents of the next generation. Less frequently, indirect selection for yield is done by screening for tree form, disease resistance and wood properties.

There is also genetic variation in characteristics which are rarely included in breeding programmes including nutritional characteristics and traits which enable trees to perform well in plantation conditions rather than as individuals. Genetic improvement of nutrient uptake and utilisation could increase productivity on poor sites (Nambiar 1985). Cannell (1982) has suggested that conventional selection procedures, which evaluate field tests at a relatively early age, may select for performance of trees growing as individuals free from competition rather than for yield per unit area in plantation conditions.

Genetic gain is an estimate of the percentage improvement that can be expected from a programme of selection and breeding. In the first round of selection (i.e. starting from a natural, unmanaged population) genetic gains for volume production are often in excess of 10% (Gill 1983a; Zobel and Talbert 1984). Estimates for wood properties are lower, around 4-6% (Nicholls et al 1980; Jett and Talbert 1982). Percentage gains decrease as breeding programmes move into the second and third generations of selection, partly because there is a progressive narrowing of the genetic base of the breeding population and therefore less genetic variation from which to select (Kleinschmit 1979).

CLONES

Breeding programmes are usually based on reproduction by seed. The nature of the cell divisions which define sexual reproduction (meiosis) means that the exact genetic makeup (genotype) of an individual cannot be passed on in seed; most commercial species are also outcrossing, and it is rarely possible to identify the pollen (male) parents of seed collected from a single tree. Vegetative propagation is the production of plants from non-reproductive parts of the parent. The most common method in forestry is the rooting of stem cuttings, though propagation is also possible using root cuttings, grafting and culturing of tissues or organs. In such methods the genotype is passed on intact to the next generation.

The identical individuals which are produced by vegetative propagation from a single tree constitute a clone, and there has been much talk in recent years of the advent of "clonal forestry". The advantages of using clones in commercial forest plantations are considerable; not only can greater genetic gains be realised, but trees are more uniform, which makes both management and utilisation easier. The best example of the use of clones in forestry comes from Brazil, where the Aracruz company is growing *Eucalyptus* species, mainly *E. grandis*, for pulp and paper manufacture (Zobel et al 1983). Trees are selected for a number of characteristics and clones are produced by taking cuttings from coppice shoots. In the early 1980s there was a planting programme of 2000 $ha.yr^{-1}$, using 25 clones in 10-20ha clonal blocks (i.e. of a single clone); the number of clones is expected to rise to 100. Volume yields since the start of the clonal selection programme have risen from 36 to 70$m^3.ha^{-1}.yr^{-1}$, and this figure is expected to rise to 100$m^3.ha^{-1}.yr^{-1}$. Not only are yields high, but the wood from clonal blocks is expected to be extremely uniform and to produce a higher quality, more uniform paper (Anon 1985).

Methods for the vegetative propagation of Sitka spruce have now been perfected (Mason 1985). However, the use of clones is not without its problems. The production of planting stock from cuttings is usually much more expensive than production from seed, and this is often enough to make vegetative propagation uneconomic unless anticipated gains are extremely high (Gill 1983b). The risk of losing entire plantations as a result of insect or fungal attack is greater if plantations consist of a single clone than if it contains a large number of different genotypes (Libby 1980), but there are no empirical data with which to attempt some kind of analysis of the risks involved.

BIOTECHNOLOGY

The use of "biotechnology" in forestry is a subject which is currently being debated in Britain. Micropropagation and tissue culture are both being investigated at the research level (Dodds 1983) and considerable progress is being made, but at the moment costs are too high to make either a commercial proposition for the large scale production of planting stock. Should genetic engineering in tree species ever become a reality, major breakthroughs could be achieved by the introduction of genes for nitrogen fixation into species which lack this ability and the creation of hybrids between species which do not normally cross.

SUMMARY

Growth of trees in the British uplands is frequently limited by site conditions which cannot be altered by foresters, and the greatest increases in yields from conventional conifer plantations will probably be achieved by reducing levels of chronic pest and disease damage. Of the various alternative silvicultural regimes which have been discussed, the most promising are short rotation coppice for the production of hardwood pulp and a more imaginative use of mixed species plantations; whole tree harvesting may be possible on some sites, but the long term effects of the practice on soil fertility are unknown.

The genetic variation which exists in most commercial tree species can be exploited in a variety of ways to increase yield. Conventional selection and breeding techniques have already resulted in substantial increases, and more precise matching of provenance to site, selection for traits other than volume production, the use of vegetative propagation to produce clonal planting stock and (at some time in the future) genetic engineering all have the potential to raise them still further. As long as forestry remains a low input practice carried out on land of poor quality it is through genetic manipulation rather than intensive management that productivity will be increased; in the long term this is true of all commercial forestry.

REFERENCES

ANDERSON, M. 1985. The impact of whole tree harvesting in British forests. Quarterly Journal of Forestry 79, 33-39.

ANDERSON, H.W.,C.S. PAPADOPOL and L. ZSUFFA. 1983. Wood energy plantations in temperate climates. Forest Ecology and Management 6, 281-306.

ANON. 1985 Brazil's new forest soars to success. World Wood 26(3), 40-41.

BRAZIER, J.D. 1977. The effect of forest practices on quality of the harvested crop. Forestry 50, 49-66.

CALLAHAM, R.Z. 1964. Provenance research; investigation of genetic diversity associated with geography. Unasylva 18, 40-50.

CAMPBELL, R.K. A provenance-transfer model for boreal regions. Report of the Norwegian Forest Research Inst. 31, 541-566.

CANNELL, M.G.R. 1982. "Crop" and "isolation" ideotypes; evidence for progeny differences in nursery grown Picea sitchensis. Silvae Genetica 31, 60-66.

CANNELL, M.G.R. and R.I. SMITH. 1980. Yields of minirotation closely spaced hardwoods in temperate regions: review and appraisal. Forest Science 26, 415-428.

DAVIES, E.J.M. 1979. High input prospects. Scottish Forestry 39, 120-132.

DODDS, J.H. (Ed) 1983. Tissue Culture of Trees. Croom Helm.

FAULKNER, R. 1985. Sitka spruce breeding programme. Forestry Commission Genetics Branch Information Sheet No. 15.

FORESTRY COMMISSION. 1985. Broadleaves in Britain. Review of Policy. A Discussion Paper. Forestry Commission.

GILL, J.G.S. 1983a. Genetic improvement in some forestry practices - with special reference to natural regeneration. Scottish Forestry 37, 250-258.

GILL, J.G.S. 1983b. Comparisons of production costs and genetic benefits of transplants and rooted cuttings of Picea sitchensis. Forestry 56, 61-73.

HARPER, J.L. 1977. Population Biology of Plants. Academic Press

IUFRO. 1978. Proceedings of the IUFRO Joint Meeting of Working Parties, Vancouver, Canada, 1978. (2 volumes). British Columbia, Ministry of Forests.

JETT, J.B. and J.T. TALBERT, 1982. Place of wood specific gravity in the development of advanced-generation seed orchards and breeding programs. Southern Journal of Applied Forestry 6, 177-180.

KENNEDY, J.N. 1974. Selection of conifer seed for British forestry. Forestry Commission Leaflet 60.

KING, G.H. 1981. An assessment of how the forest production which is currently not used would be obtained for energy purposes, and from this an evaluation of the management methods most appropriate to the production of wood for energy purposes. Commission of the European Communities, Luxembourg.

KLEINSCHMIT, J. 1979. Limitations for restriction of the genetic variation. Silvae Genetica 28, 61-67.

LIBBY, W.J. 1980. What is a safe number of clones per plantation? Proceedings of the Third International Workshop on the Genetics of Host Parasite Interactions in Forestry. Wageningen, Netherlands, September 1980. pp 342-360.

MANLEY, B.R. and R.L. KNOWLES. 1980. Growth of radiata pine under the direct sawlog regime. New Zealand Journal of Forestry 25, 15-34.

MASON, W. 1985. Clones to replace forest seedlings? Forestry and British Timber 14(1), 24-26.

NAMBIAR, E.K.S. 1985. Increasing forest productivity through genetic improvement of nutritional characteristics. In Forest Potentials, Productivity and Value. Weyerhauser Science Symposium 4, 191-215.

NICHOLLS, J.W.P., J.D. MORRIS and L.A. PEDERICK. 1980. Heritability estimates of density characteristics in juvenile Pinus radiata wood. Silvae Genetica 29, 54-61.

PERRY, D.A. 1979. Variation between and within tree species. In E.D. Ford, D.C. Malcolm and J. Atterson (Eds.) The Ecology of Even-Aged Forest Plantations. ITE, Cambridge pp 71-98.

ROLLINSON, T. 1985. Windthrow and price are main thinning practice factors. Forestry and British Timber 14(10),22-24.

TARRANT, R.F. and J.M. TRAPPE. 1971. The role of Alnus in improving the forest environment. Plant and Soil (Special volume), 335-348.

TAYLOR, C.M.A. 1985. The return of nursing mixtures. Forestry and British Timber 14(5), 18-19.

WEETMAN, G.F. and D. ALGAR. 1983. Low-site class black spruce and jack pine nutrient removals after full-tree and tree-length logging. Canadian Journal of Forest Research 13, 1030-1036.

ZOBEL, B.J. and J.T. TALBERT. 1984. Applied Forest Tree Improvement. John Wiley and Sons.

ZOBEL, B.,E. CAMPINHOS AND Y. IKEMORI. 1983. Selecting and breeding for desirable wood. Tappi 66(1), 70-73.

Chapter 42

The use of lignins in polymeric controlled release systems

A. Chanse and R. M. Wilkins – Department of Agricultural Biology, University of Newcastle upon Tyne. U.K.

ABSTRACT

Lignins are a largely unused polymer source with protective properties under environmental conditions. They can be used for formulation with agrochemicals to provide controlled release of the bioactive agent with benefits in certain applications. The release characteristics of physical incorporation of pesticides with kraft lignin and other polymers is described.

INTRODUCTION

Controlled release is used for a variety of applications from drug delivery to antifouling paints for ships. An important use is in agriculture and other external environment applications [1]

Controlled release can be defined as the permeation-moderated transfer of an active material from a resevoir to a target at a predetermined concentration for a specified time [2]. When used in external environmental conditions, benefits over conventional formulations can be obtained such as:

(a) the targeting of the active agent is improved, thus reducing impact on other non-target organisms,
(b) losses due to degradation or leaching, etc, are reduced enabling short-lived agents to be used with increased efficiency,
(c) less active agent is required.

A number of controlled release systems are available, including reservoir devices with and without rate limiting membranes and monolithic (matrix) methods. Others release by erosion or retrograde chemical reactions [3]. For applications in agriculture, in addition to the environmental requirements, the low cost for formulations is most important.

Lignins are available as a natural waste or byproduct and represent an underused resource. They occur in most land plants and constitute the largest natural source of aromatic compounds [4]. Biologically, they are fairly inert, serving as structural support materials in the plant and protecting it from chemical or physical damage. Thus, lignins are an ideal choice for agricultural controlled release formulations. Further, they form a glassy matrix when physically incorporated into certain pesticides [5]. Lignins can also be esterified with active agents possessing suitable functional groups or can be cross-linked to form a gel into which the active agent can be incorporated by swelling in an aqueous medium and then drying [6]. Lignin can also be chemically modified to serve as a nutrient source or as a pesticide in its own right.

Lignins are available as a byproduct of many pulping processes but alkali (kraft) lignin is most appropriate for pesticide formulation as it is insoluble at most soil pH's. The preparation, properties and release chracteristics of lignin-based physical combinations of a soil-applied insecticide and herbicide are described here.

MATERIALS AND METHODS

Indulin AT, a commercially available pine kraft lignin (Westvaco Corp). provided a standardized source of the polymer. Indulin AT is a free flowing brown powder soluble in polar organic solvents and aqueous base. Carbofuran is a widely used systemic insecticide, acaricide and nematicide applied to the soil. Diuron is a pre-emergence herbicide used for general weed control.

Formulation was carried out in an electrically heated open vessel. The active agent was melted first and any plasticizer or modifiers added followed by the lignin. The melt was constantly stirred using a high torque low speed mixer to ensure homogeneity.

On cooling a brittle glassy matrix is formed which can be granulated or cast into sheets. Granulation was either by hand or by hammer mill. The granule size range used throughout was 500-1000 μm.

Initial assessment of active agent release was by immersion of granules containing 17.5 mg active agent in 250 ml static distilled water at 23°C. The release rate was monitored by measurement of the concentration of active agent released into the aqueous phase, based on HPLC. A Gilson isocratic HPLC system was used with a Spherisorb ODS column. Detection was by a variable wavelength Holochrome ultra-violet detector.

RESULTS AND DISCUSSION

Diuron-lignin controlled release matrix. Diuron was chosen on the basis of its wide usage as a soil-applied herbicide. Its compatibility for formulating with lignin was assessed over a range of diuron-lignin proportions, and these formulations were then used to investigate this relationship on release rate.

Table 1. Diuron-lignin formulations evaluated.

Formulation	% diuron	% kraft lignin
DL 20	20	80
DL 35	35	65
DL 50	50	50

The release profiles for DL 35 and DL 50 (Fig.1) suggested a similar rate determining step operating in both formulations. When plotted against $t^{\frac{1}{2}}$ (square root of time), as in Fig. 4, a partial linear dependence was noted, characteristic of a diffusion-dependent release kinetic.

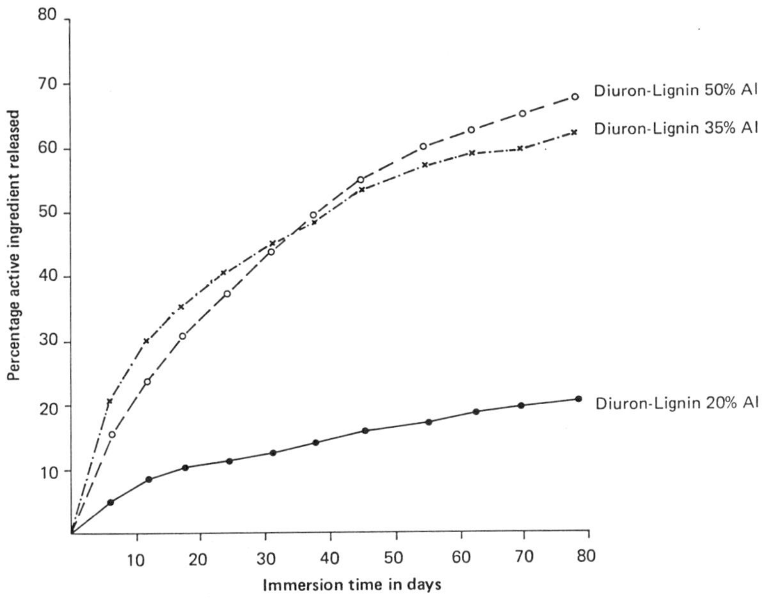

Fig. 1. Release of diuron into static water at 23°C from a lignin matrix.

Release from DL 20 was much slower and also did not show the linear relationship for $t^{\frac{1}{2}}$. In this case, a different release mechanism was rate determining.

The release characteristics can be altered by the inclusion of other polymers. Using the 35% diuron-lignin matrix (DL 35) the following modifications were made (Table 2).

Table 2. Polymer-modified lignin matrix.

Formulation	modifier	% diuron	% lignin
D PVC L 35	polyvinyl chloride	35	55
D PS L 35	polystyrene	35	55
D PVAc L 35	polyvinylacetate	35	55

Inclusion of polymers usually reduced the release rate. The compatibility of the active agent, lignin and modifier can be determined from the solubility parameters of the components, which need to be similar.

Table 3. Solubility parameters of some components $(cal/cm^2)^{1/2}$.

carbofuran	11.8	polyvinyl chloride	11.0
diuron	16.3	polystyrene	9.1
kraft lignin	12.8.	polyvinyl acetate	11.3

However, the release profile of the diuron is not dependent on compatibility and the effects of adding 10% of polystyrene or polyvinyl-chloride are similar (Fig.2).

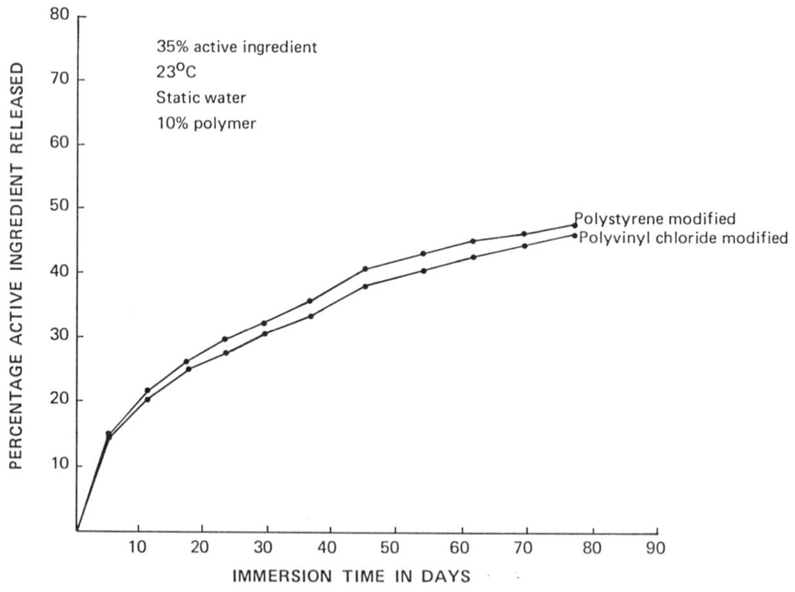

Fig. 2. Release of diuron from polymer modifed lignin.

This could be the result of the uniform dispersion of the polystyrene or polyvinylchloride throughout the matrix, thus filling void space and reducing the size or rate of pore formation. Polyvinyl acetate gave a marked reduction in release rate (Fig. 3). This could be attributed to the greater adhesion between the polymer and the lignin further reducing void character.

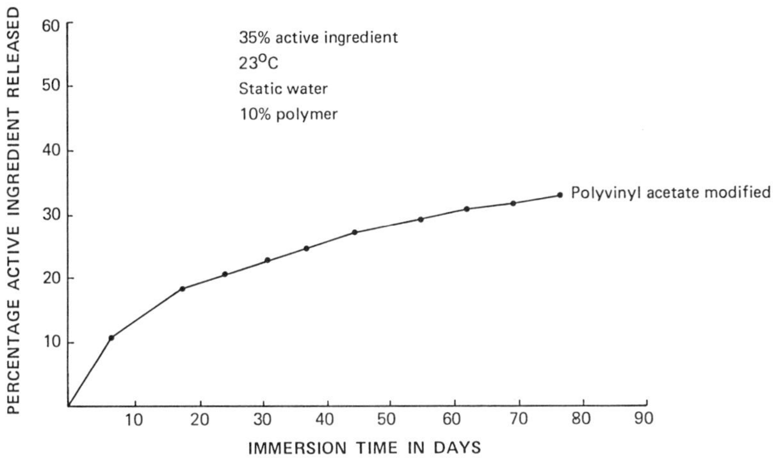

Fig. 3. Release of diuron from a hydrophilic polymer modified lignin.

The dependence of release from both modified and unmodified diuron-lignin matrices on square root time is shown in Fig. 4. The linear relationship is extended for the polymer modified formulations for greater periods, possibly indicating the dominance of diffusion-determined release.

Another widely used, soil applied agrochemical is the insecticide and nematicide, carbofuran. Its solubility parameter (Table 3) is close to kraft lignin and in fact, it is a good plasticizer. The two components form a glassy matrix over a wide range of proportions (Table 4).

Table 4. Carbofuran-lignin matrics.

Formulation	% carbofuran	% lignin
CL 25	25	75
CL 35	35	65
CL 45	45	55

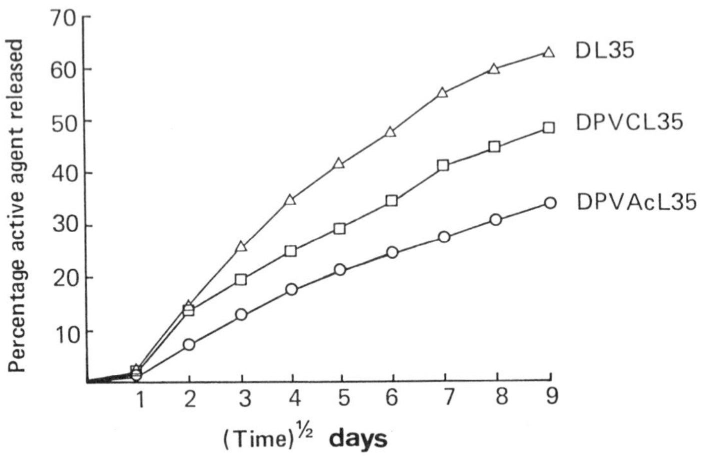

Fig. 4. Comparison of release from diuron-lignin formulations.

The release profiles for CL 35 and CL 45 are similar (Fig. 5), which is in line with the results for release from diuron-lignin formulations (Fig. 1). Similarly, the CL 25 gave a lower release rate paralleling that of DL 20.

GENERAL

For the active agents used here, the lignin matrices showed release dependent on active agent concentration, and most probably based on a diffusion-dissolution mechanism. Below 30% initial active agent or during the later stages of release other factors may be more important. Polymeric modifiers help to maintain the diffusion mechanism over a longer period.

The ideal release rate is zero order; that is, it is constant with time and independent of concentration. Although, these lignin-based matrix systems do not achieve this, they can be more economical and the delivery rate to the target organism provides a rapid build-up followed by maintenance of the effective dosage. Such formulations have shown promise in field trials and use, for example, for weed control in conifer forests, using the herbicide 2,4-D [7], and for insect and virus control in tropical flooded rice, using carbofuran [8].

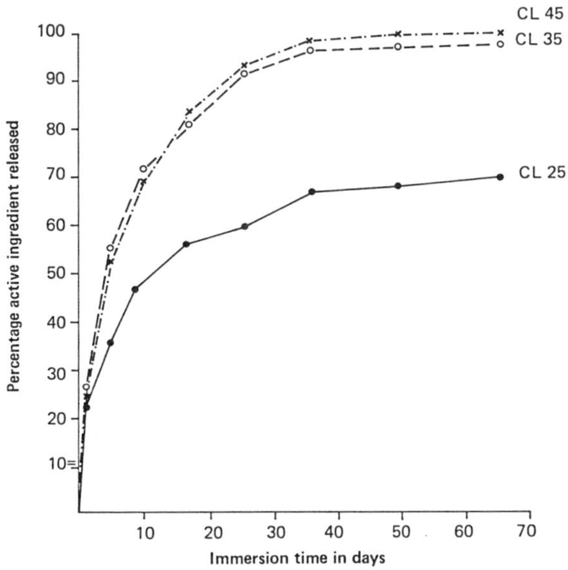

Fig. 5. Release of carbofuran into static water at 23°.

REFERENCES

1. Wilkins, R. M., Controlled release - present and future, Proc. Int. Con. Plant Protection, Vol. II, 554-559, 1983.

2. Lewis, D. H. and Cowsar, D. R., Principles of controlled release pesticides, in Scher, H. B. (ed.), Controlled Release Pesticides, Amer. Chem. Soc., Washington D.C., 1977.

3. Kydonieus, A. F. (ed.), Controlled Release Technologies: Methods, Theory, and Applications, 2 vols., CRC, Boca Raton, FL, 1980.

4. Glasser, W. G., Potential role of lignin in tomorrows wood utilization technologies, For. Prod. J., 33(3), 24-29, 1981.

5. Wilkins, R. M., Lignins as formulating agents for controlled release in agriculture, Brit. Polymer J., 15, 177-78, 1983.

6. Dellicolli, H. T., Pine Kraft lignin as a pesticide delivery system, in Controlled Release Technologies, Kydonieus, A. F. (ed.), 225-34, 1980.

7. Wilkins, R. M., Use of controlled release herbicides in the establishment of conifer plantations, Proc. Crop Protection Northern Britain, 261-67, 1981.

8. Wilkins, R. M., Batterby, S., Heinrichs, E. A., Aquino, G. B., Valencia, S. L., Management of the rice tungro virus vector _Nephotettix virescens_ with controlled release carbofuran, J. Econ. Entomol., 77, 495-99, 1984.

Chapter 43

Problems associated with storage of waterlogged archaeological wood

R. Mouzouras – School of Biological Sciences, Portsmouth Polytechnic.

INTRODUCTION

In recent years there has been world wide interest in the recovery of wooden ships from aquatic environments [1, 2, 3, 4] with the result that a large number of waterlogged wooden objects require to be conserved. Such objects are liable to dehydration leading to shrinkage and distortion if their storage or conservation is inadequate [5]. In many cases, wooden objects are recovered and stored until a suitable consolidation method is selected or funds for conservation are raised. Ten different storage methods have been used in the past [6]; enclosure in polythene or immersion in water (with or without biocide) are most commonly used, on a temporary basis or for extended storage periods [6, 7, 8]. Recently [9], plans for the extended storage of timbers (from the Kollerup wreck) in nature were reported. This involved the burial of wood in a sandy soil and comparing this to more traditional storage methods. Unfortunately no results are available since it is a long term study.

So far, there are only a few reports on the microflora of waterlogged archaeological wood. Several workers report the presence of fungal propagules on timbers [8, 10, 11] and some record the species of fungi found [12, 13] but a comprehensive list of species found has never been reported. In October 1982 the Mary Rose was recovered following 437 years on the seabed [4]. Many of the structural timbers were removed from the hull structure and are presently in storage.

METHODS

About 3000 timbers from the Mary Rose have been in storage for a minimum of 3 years. The majority of these are wrapped in polythene sheeting (Figure 1) in order to maintain their waterlogged state and kept in bays below ground. The environment within the storage site is humid with relative humidity around the 95% level and temperatures ranging between 9°C and 20°C depending on the time of year.

Structural timbers were assessed for decay using a non-destructive method. A Pilodyn with a spring load of 2J and a blunt steel pin 40mm in length and 2mm in diameter was used to take readings of depth of penetration at regular intervals. Timbers were surveyed on two occasions; the second survey was carried out 16 or 20 months after the initial survey.

Timbers in storage were examined for the presence of fungal fruiting structures and wood specimens were also examined with the scanning electron microscope (S.E.M.) using the method of Moss and Jones [17].

RESULTS

Figure 2 illustrates an oak partition support, showing all four faces (A, B, C, and D) and the Pilodyn survey of this sample is illustrated diagrammatically by Figure 3. Numbers represent depths of penetration (in mm) of the steel pin and values within broken lines are those which exceed 12 mm. The depth of penetration in sound waterlogged oak being between 10 and 12mm [13]. Figure 6 indicates that decay exists in localised pockets, rather than extensive decay of whole timbers. This picture is true for the majority of timbers in storage. The second Pilodyn survey showed little change in penetration values (Table 1).

Table 1. Mean Pilodyn values for initial survey (x) and second survey (x^1).

Timber code	x (mm)	x^1 (mm)
81/T302	18.8	18.5
81/T332	20.5	19.8
81/T757	12.4	12.4
81/T1004	17.5	18.1
81/T1005	16.8	16.3
81/T1006	18.1	17.8
81/T1010	13.8	12.8
81/T1077	15.0	14.3
81/T1082	15.0	14.6

Fig. 1. Polythene wrapped timbers within the storage site.

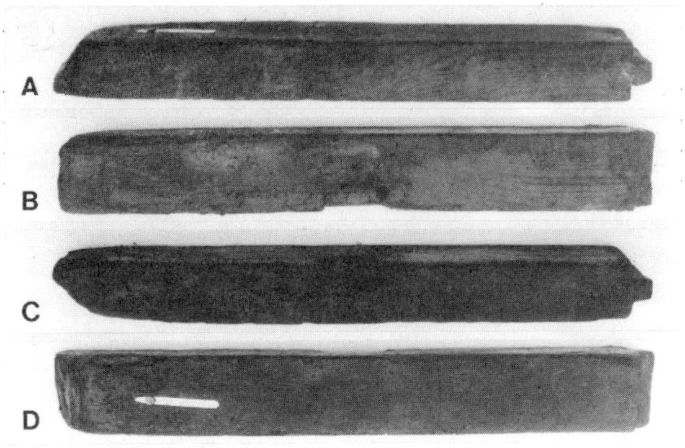

Fig. 2. Oak partition support showing all four faces (A, B, C, D).

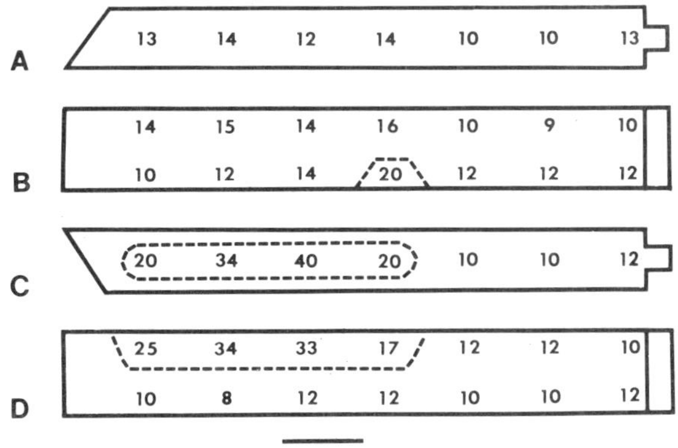

Fig. 3. Pilodyn survey of partition support seen in Fig. 5. Bar = 0.1 m.

Table 1. continued.

Timber code	x (mm)	x^1 (mm)
81/T1109	14.4	13.2
81/T1186	12.6	12.2
81/T1352	19.9	19.5
81/T1374	14.4	13.4
81/T1429	14.8	13.8
81/T1434	11.2	12.4
82/T190	15.7	14.8
82/T251	14.7	14.5
82/T272	13.8	13.9

Table 2. Frequency of microorganisms recorded on Mary Rose timbers in storage.

Microorganism	Occurrence
ACTINOMYCETES	
White mycelial type	70
White spherical type	51
DEUTEROMYCOTINA	
Cirrenalia macrocephala (Kohlm.) Meyers & Moore	2
Dictyosporium toruloides (Corda) Guegen	1
Gliomastix murorum (Corda) Hughes	1
Humicola alopallonella Meyers & Moore	2
Monodictys pelagica (Johnson) Jones	31
Stachybotrys atra Corda	6
Trichocladium achrasporum (Meyers & Moore) Dixon	11
Zalerion maritimum (Linder) Anastasiou	2
Z. varium Anastasiou	1
ASCOMYCOTINA	
Ceriosporopsis halima Linder	1
Corollospora maritima Werdermann	1
Lulworthia sp. Sutherland	17
Nereiospora cristata (Kohlm.) Jones, Johnson & Moss	10
Zopfiella-like sp.	1
BASIDIOMYCOTINA	
Digitatispora marina Doguet	3
Nia vibrissa Moore & Meyers	3

Table 2 lists the species of microorganisms observed on the surface of Mary Rose timbers.

Sixteen fungi were recorded to fruit on the wood surface, from 134 samples examined. Fourteen of these are marine and two are terrestrial fungi (*Stachybotrys* and *Gliomastix murorum*). All the fungi recorded sporulated within the storage site, with the exception of the marine basidiomycete *Digitatispora marina*. The white, waxy basidiocarps (Figure 4) developed when the wood samples were kept in cold storage (4°C) for several months.

In many cases slime is present along with surface microbial growth. However, despite surface growth, very little recent microbial attack has been observed. The majority of timbers have a decayed outer layer with a sound inner core. In the decayed zone hyphae can be found (Figure 5) but do not seem to be degrading the wood cell wall components. Around the decayed/sound interface of the wood on some occasions recent bacterial attack was observed (Figure 6).

DISCUSSION

The initial problem for the conservator is how to maintain the wood, recovered from the aquatic environment, in a waterlogged state. Most wood which has spent hundreds of years on the seabed, would have decayed to various degrees by the combined action of biological and chemical factors. When drying, archaeological waterlogged wood shows shrinkage, cracking and warping, due to considerable loss of cell wall material, leading to cell collapse. Slightly decayed oak heartwood (120-145% water content) from the Bremmer Kogge showed 9-10% radial and 20% tangential shrinkage, which is twice the value of new waterlogged wood and is mostly irreversible [5]. The way in which such wood is stored is therefore important. Recovery of old shipwrecks results in the accumulation of a large number of timbers which need to be stored so that their condition can be assessed. Analysis of old wood is necessary, since its physical, chemical and morphological condition will dictate the conservation process to be adopted [14].

During the storage period, it is vital that the wood is not further decayed by microorganisms. The method of storage used for the Mary Rose is sufficient for maintaining their waterlogged state, providing the wrapping is adequate. However, a number of microorganisms did appear on the wood surface (Table 2). Nine of the fourteen marine fungi recorded cause soft rot and two cause white rot [15], while *G. murorum* and *S. atra* can also degrade cellulose [16]. Despite the presence of these organisms microscopic examination did not reveal recent decay and this is reflected in the results of the second Pilodyn survey.

Evidently, archaeological waterlogged wood presents problems to the conservator and several considerations must be taken

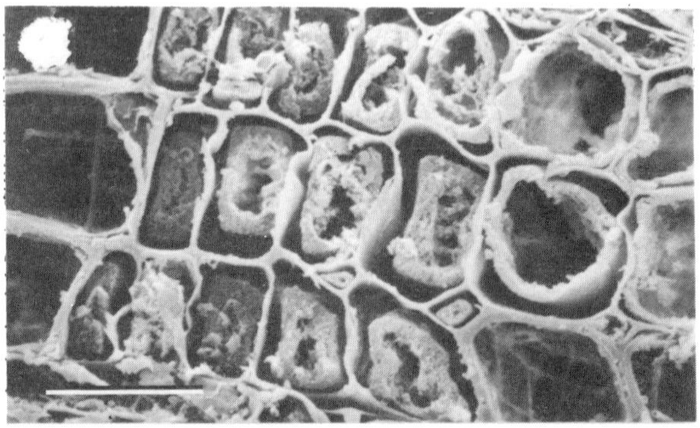

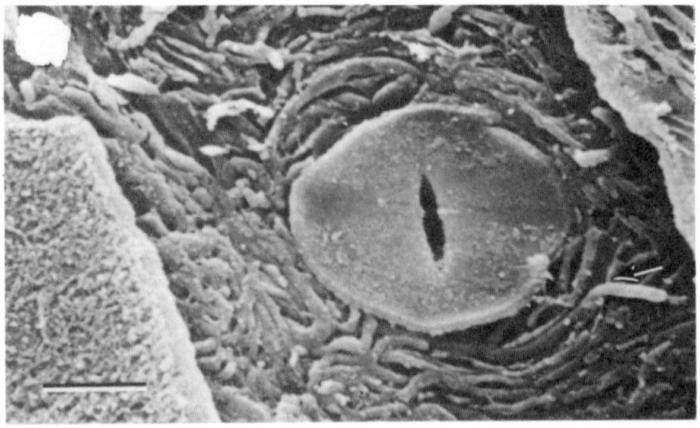

Fig. 4. Basidiocarps of D. Marina on an oak partition plank. Bar = 0.1 m.

Fig. 5. S.E.M. of a transverse section of Mary Rose pine. Bar = 50 μm.

Fig. 6. S.E.M. of Mary Rose oak showing bacterial attack (arrow). Bar = 2 μm.

into account, regarding storage methods. They must be (i) inexpensive, as usually when wooden finds are recovered funds for restoration are unavailable; (ii) easy to maintain; (iii) compatible with subsequent conservation treatments; (iv) non-toxic to the workforce and public and (v) adequate in preventing or minimising further biological decay.

ACKNOWLEDGEMENTS

The author would like to thank the S.E.R.C. and the Mary Rose Trust for financial support, and C. Derrick for photographic assistance.

REFERENCES

1. BARKMAN, L. (1967). On resurrecting a wreck. Wasastudier, 6, Statens Sjohistoriska Museum, Stockholm.
2. KATZEV, M.L. (1970). Resurrecting the oldest known Greek ship. National Geographic, 137: 841-857.
3. GREEN, J.N. (1975). The Voc ship Batavia wrecked in 1929 on the Houtman Abrolhos, Western Australia. International J. of Nautical Archeol., 4: 43-63.
4. RULE, M.H. (1983). The Mary Rose: The Excavation and raising of Henry VIII's Flagship, Windward: Leicester. 240p.
5. JONG, J. de. (1977). Conservation techniques for old waterlogged wood from shipwrecks found in the Netherlands. In Biodeterioration Investigation Techniques, ed. A.H. Walters. pp.295-338. Applied Science: London.
6. DAWSON, J.E., RAVINDRA, R. and LAFONTAINE, R.H. (1982). A review for storage methods for waterlogged wood. In: Proc. ICOM Waterlogged Wood Working Group Conf., Ottawa 1981, eds. D.W. Grattan and J.C. McCawley, pp. 227-235. ICOM: Ottawa.
7. LUCAS, D.A. (1982). On site packing and protection for wet and waterlogged wood. In: Proc. ICOM Waterlogged Wood Working Group Conf., Ottawa 1981, eds. D.W. Grattan and J.C. McCawley, pp. 51-55. ICOM: Ottawa.
8. JOHNSON, R. (1985). Large waterlogged timber structures excavated on the Billingsgate excavation, London 1982. In: Waterlogged Wood, Study and Conservation, eds. R. Ramiere and M. Colardelle, pp. 63-69. CETBGE-CENG ORIS: Grenoble.
9. JESPERSEN, K. (1985). Extended storage of waterlogged wood in nature. In: Waterlogged Wood, Study and Conservation, eds. R. Ramiere and M. Colardelle. pp. 39-54. CETBGE-CENG ORIS: Grenoble.

10. WATSON, J. (1982). The application of freeze-drying on British hardwoods from archaeological excavations. In: Proc. ICOM Waterlogged Wood Working Group Conf., Ottawa 1981, eds. D.W. Grattan and J.C. McCawley, pp. 237-242. ICOM: Ottawa.
11. SEIFERT, B. and JAGELS, R. (1985). Conservation and wood analysis of the Ronson ship timbers. In: Waterlogged Wood, Study and Conservation, eds. R. Ramiere and M. Colardelle, pp. 269-297. CETBGE-CENG ORIS: Grenoble.
12. BOUTELJE, J.B. and KIESSLING, H. (1964). On water-stored oak timber and its decay by fungi and bacteria. Archiv. f. Mikrobiologie, 49: 305-314.
13. JONES, A.M., RULE, M.H. and JONES, E.B.G. (1986). Conservation of timbers of the Tudor ship Mary Rose. In: Proc. 6th. Int. Biodet. Symp., Washington, August 1984. In press.
14. FLORIAN, M.L.E., SECCOMBE-HETT, C.E. and McCAWLEY, J.C. (1978). The physical, chemical and morphological condition of marine archaeological wood should dictate the conservation process. In: Papers from the 1st. Southern Hemisphere Conference on Maritime Archaeology, pp. 128-144. Oceans Society of Australia: Melbourne.
15. MOUZOURAS, R., JONES, E.B.G., and VENKATASAMY, R. (1986). Decay of wood by micro-organisms in aquatic habitats. Rec. Ann. Conv. B.W.P.A.
16. NILSSON, T. (1973). Studies on wood degradation and cellulotic activity of microfungi. Studia Forestalia Suecica No. 104: 5-41.
17. MOSS, S.T. and JONES, E.B.G. (1977). Ascospore appendages of Marine Ascomycetes: *Halosphaeria mediosetigera*. Trans. Brit. Mycol. Soc., 69: 313-315.

Chapter 44

The dynamic mechanical analysis of wood and wood products

C. Birkinshaw, M. Buggy and G. G. Henn – National Institute for Higher Education
Limerick, Ireland.

INTRODUCTION

Dynamic mechanical analysis (DMA) is a powerful technique for the characterisation of viscoelastic materials. Application of a simple harmonic stress results in a phase difference between stress and strain. The tangent of the phase angle (tan δ) measures the internal friction or damping, $\Delta E/E$ where ΔE is the energy dissipated in taking the sample through a stress cycle, and E is the energy stored in the sample when the strain is a maximum. Tan δ goes through a maximum when the dynamic modulus of the material is changing due to a physical transition. Thus the temperature at which these tan δ maxima occur can be used to characterise physical changes in viscoelastic materials on the molecular level and consequently the technique finds extensive use in the structure-property analysis of polymeric materials.

Wood is a naturally occurring composite material made up of polymeric components, and therefore can be considered a suitable system for the application of the DMA technique. A limited amount of work has been reported on the application of DMA to wood and wood products, but this work has generally been restricted to examination of specific property variation in a single species or type (1,2,3). The results considered in this paper come from a more general and comparative study of the applicability of DMA.

EXPERIMENTAL

Materials

Whilst it is recognised that density, microstructure and chemical constitution all contribute to the mechanical behaviour of wood, the exact relationship between these variables is not apparent. Therefore species were chosen to represent a wide range of microstructural types. Details are given in Table I together with the moisture content and measured density at time of testing.

Table 1.

SPECIES	DENSITY ($Kg\ m^{-3}$)	MOISTURE (%)
Ash (Fraxinus excelsior)	613	10.7
Brazilian Mahogany (Swietenia macrophyla)	503	11.2
Iroko (Chlorophora excelsa)	589	9.0
Keruing (Dipterocarpus spp.)	681	8.0
Kingwood (Dalbergia cearensis)	1049	9.0
Obeche (Triplochiton scleroxylan)	395	11.5
Parana Pine (Araucaria augustifolia)	481	9.2
Pine (Pinus sylvestris)	482	9.8
Sapele (Entandophragma cylindricum)	607	10.6
Spruce (Picea abies)	340	10.0

All solid timbers were from conventional commercial sources and had been subject to a further 2 years air drying prior to use. DMA test pieces were milled from the bulk material using a vertical milling machine to give approximate dimensions 60 x 12 x 4.2mm. These dimensions were determined from experimental optimization trials. The long dimension of the specimen was cut parallel to the grain whilst the width was in the direction of radial growth. Experimental difficulty was initially encountered in obtaining reproducible results from unconditioned specimens. This was ascribed to shrinkage associated with dehydration during testing. Using an air circulated oven a heating programme of one hour at 40°C, followed by a rise of 10°C per hour to 100°C, plus a further 2 hours at 100°C was employed to condition DMA specimens. Mechanical restraint was applied to the specimens to prevent distortion during this conditioning period. Final adjustment to finished dimensions was then carried out as required.

The fibreboard material was commercially prepared from milled spruce of average particle dimensions length 2.67mm and width 0.19mm, using a urea-formaldehyde binder. The resin content of the final material was 9% of total dry weight. Variation in density was achieved by alteration of the compression factor used during curing(4). Specimen dimensions were as those used for solid wood and the samples were tested without pre-conditioning.

Testing

A Du Pont 982 DMA interfaced to a 9900 programmer and data analysis station was used to obtain the shear storage modulus and loss tangent between -100° and 150°C. A heating rate of 5°C per minute was used, and during the test the samples were blanketed in nitrogen to eliminate oxidative degradation. Clamp length was 43.6 mm and oscillation amplitude 0.1mm for the solid wood and 0.2mm for the fibreboard. The configuration of the test system was such as to produce shear deformation perpendicular to the grain and tangential to the radial growth direction in the solid material. Shear storage modulus and loss tangent (tan δ) were calculated using Du Pont programme DMA Modulus 2.0F (5). Duplicate tests were conducted on each specimen type to confirm reproducibility.

A small number of materials were also examined by differential scanning calorimetry, a Du Pont 990 system being used. The samples were prepared as fine powder and weighed approximately 2mg. A heating rate of 10°C per minute was used between -100° and 150°C.

RESULTS AND DISCUSSION

Figures 1 and 2 show the variation in shear storage modulus and relative damping with temperature for the solid wood. Although there is variation in the absolute values observed in storage modulus it is apparent that the general form of the curves obtained is similar, and independent of species. The fall off in the value of storage modulus is constant and can be explained by the thermoplasticity of the proto-lignin (6). Tan δ curves show a low temperature (-50°C) transition and a progressive increase in thermoplasticity from 50°C upwards. The low temperature transition is considered to be associated with the structural polymers of the wood, and its magnitude and position suggest that it involves side chain conformational changes.

Figure 3 shows the calorimetry response of keruing, obeche, fibreboard and powdered cellulose. Again the general form of the curves is consistent, the broad endothermic event at low temperatures is thought to correspond with the low temperature event observed on the DMA, suggesting that it is a function of the cellulose polymer. The larger endotherm on the DSC curve

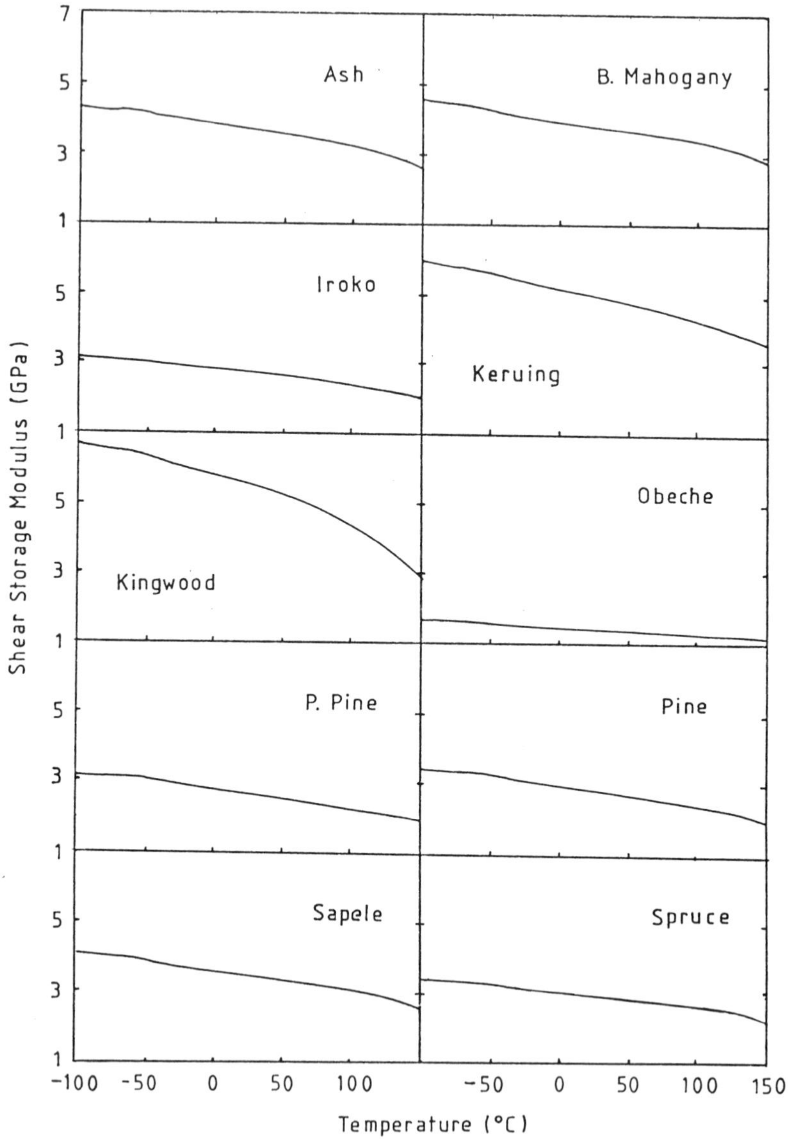

Fig. 1. Variation of shear storage modulus with temperature.

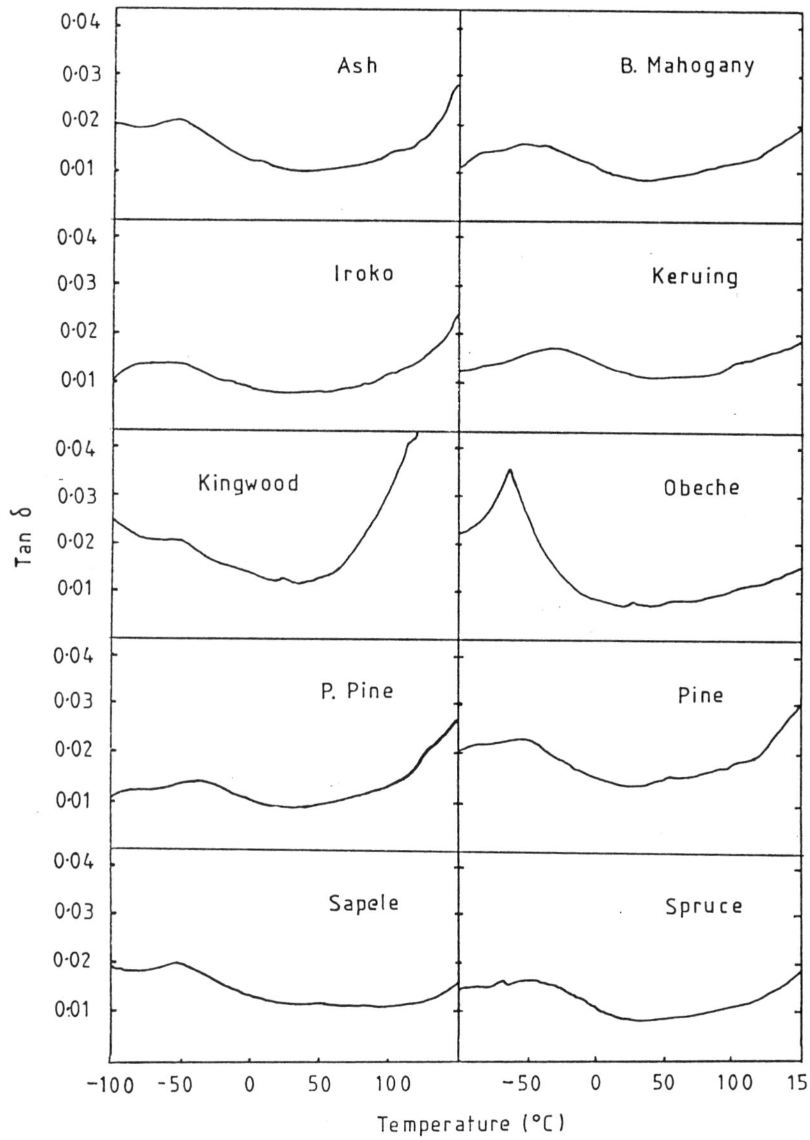

Fig. 2. Variation of relative damping (tan δ) with temperature.

Fig. 3. Calorimetric responses.

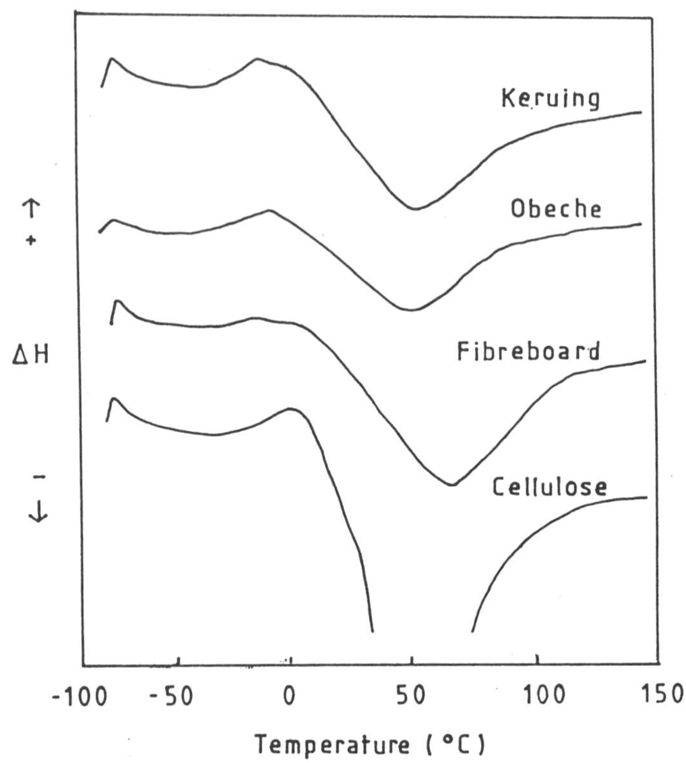

is considered to be recording the dehydration of the sample. The enthalpy associated with this event was calculated from peak areas and found to agree with the heat requirement for evaporation of the measured moisture contents of the samples.

The DMA response of the fibreboard is displayed in Figures 4. A relationship is apparent between shear storage modulus and density. As expected the fibre board moduli are lower than those of solid timber of equivalent density. This arises from the substantial differences in the fibre length and distribution, in the two types of material.

The tan δ curves show the low temperature transition previously observed in the solid wood specimens and this is therefore associated with the spruce fibres. A further maxima is

Fig. 4. DMA response of fibreboard.

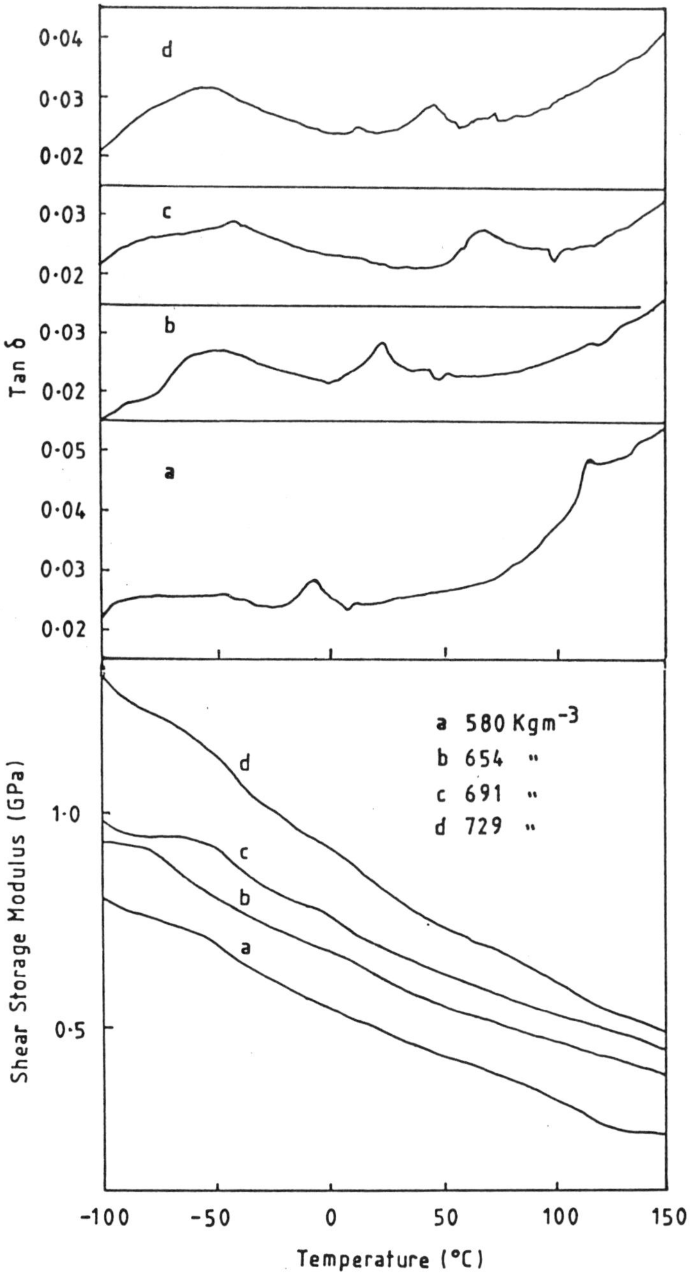

apparent in all samples, in the temperature range -10° to $+60^\circ$C, the actual temperature showing a general relationship to density. Further work indicates that this peak is not associated with loss of entrained moisture from the system. Neither can it be ascribed to a concentration effect of the matrix material as such an effect would be to change the magnitude of the damping peak, not its position on the temperature axis. It is thought that this transition arises from fibre-matrix shear slippage. In the higher density materials this process is hindered by more effective packing which ensures more uniform stress distribution. Additional energy is therefore required to induce slippage and the event occurs at higher temperatures.

CONCLUSIONS

These initial results indicate that the DMA technique is potentially useful in the characterisation of wood and wood products. It offers a route by which the microstructural variation within these materials may be related to macroscopic mechanical behaviour. The composition of the polymer systems constituting wood may be studied without resorting to structure changing chemical techniques. A better understanding of the structure-property relationship of wood may lead to significant improvements in the construction and performance of synthetic composites.

REFERENCES

1. R.N. Subramanian and R. Hofmann. J. Polym. Sci, Letters 21 (1983) 105.

2. T. Handa, S. Yoshizawa, M. Suzuki and M. Fukuoka. Proc. 22nd Jap. Constr. Mater. Res. Conf. Kyoto 1978.

3. D. Atack. Philosophical Magazine 43A (1981) 619.

4. Private Communications. Medite of Europe Ltd. Clonmel, Ireland.

5. Du Pont DMA Calibration Programme PN996227-001 December 1984.

6. A.J. Panshin and C. de Zeeuw. "Textbook of Wood Technology" 4th. Ed. (McGraw-Hill 1980).

Chapter 45

Wettability and water repellency of wood: A faster, more convenient method

Martins A. Kalnins, Research Chemist and Clairemarie Katzenberger, Chemist – USDA Forest Service, Forest Products Laboratory Madison, Wisconsin 53705-2398.

INTRODUCTION

Water repellency of wood is important in outdoor uses of wood and wood-based products. Use of effective water repellents on wood has been linked to improved performance of wood outdoors (Banks and Voulgaridis 1980; Voulgaridis and Banks 1981; Purslow 1982; Miniutti, et al. 1961). Conversely, good wetting is required in operations involving coatings, adhesives, and some types of protective treatments that require penetration of wood by liquids. Efficient wetting of wood by adhesives has been correlated with strength of adhesive bond (Collett 1972; Huntsberger 1978; Bodig 1962; Freeman 1959; Subramanian, et al. 1983).

Wettability is usually defined as the contact angle that a liquid makes when in contact with a solid surface. From contact angle data, surface free energy of a solid can be estimated. Good water repellency is the special case of poor wettability (large contact angle). Wood is both hygroscopic and briefly water repellent. Direct contact angle measurements (Gray 1962; Nguyen, et al. 1978), tilted plate (Freeman 1959; Bodig 1962), and capillary rise techniques (Freeman 1959) have been used to study forces acting on liquids in contact with wood surfaces, but all involve experimental complications, and are relatively slow. A rapid, reliable procedure for measuring wettability and related properties on wood and on wood products was clearly needed, and we devised one. A modified Wilhelmy method had been used to determine the forces acting on immersed fiber samples (Young 1976; Klungness 1981). Casilla, et al. 1981; 1984 used a larger, conical wood sample to obtain a "wettability index". We used that procedure as a starting point, and combined an automatic surface tension apparatus with a microcomputer for control, data collection, and processing, for the study of wood surfaces.

EXPERIMENTAL

Wood materials

Green logs of various species were obtained immediately after harvesting and were stored at 2°C prior to sawing into edge-grain boards about 1.3 centimeters (cm) thick. The wood was kiln dried below 70°C. The boards were planed, and sawn into 3 x 0.64 cm sticks.

Automated surface tension analyzer

Surface tension measurements of liquids and determinations of forces on wood samples during immersion in liquids were done with a Fisher Model 215 Autotensiomat Surface Tension Analyzer that was interfaced with an IBM-PC-XT microcomputer using Metrabyte Corporation Dash 16 multifunction high-speed analog to digital expansion board.

The analyzer consists of a strain sensitive wire affixed at one end to the balance beam and the other to a transducer. The transducer's signal is proportional to the load. The output of the analyzer is 0 to 1 mv which is calibrated to equal 0 to 100 dynes/cm. This signal is amplified and filtered; noise above 2 HZ is attenuated before entering the Metrabyte Dash-16 Analog to Digital interface board installed in the IBM-PC-XT.

The elevator in the surface tension analyzer is controlled by the digital output port of the Dash-16.

Computer program: wood wettability study*

The BASIC program automates experimental procedures, and is divided into three major modules: Surface Tension, Run New Sample, and Analyze Data, and one major submodule, Graph.

The Surface Tension module has two main features: electronic calibration and automatic apparatus control. Surface tension of a liquid is determined, and the value is filed. Run New Sample routine calibrates and controls the apparatus. The collected data is filed under a name based on sample statistics. Analyze Sample Data module retrieves file to be analyzed and reads it into computer memory. The main module has sub-modules: Graph, Calculation of Contact Angle from Attractive Force, from Work of Adhesion, Estimate of Surface Free Energy, Calculation of Interaction Factor, and Data Summary.

GRAPH

The Graph submodule features variable axis scaling, and rescaling that enlarges sections of the graph. Figure 1 is the graph of a typical wood sample run.

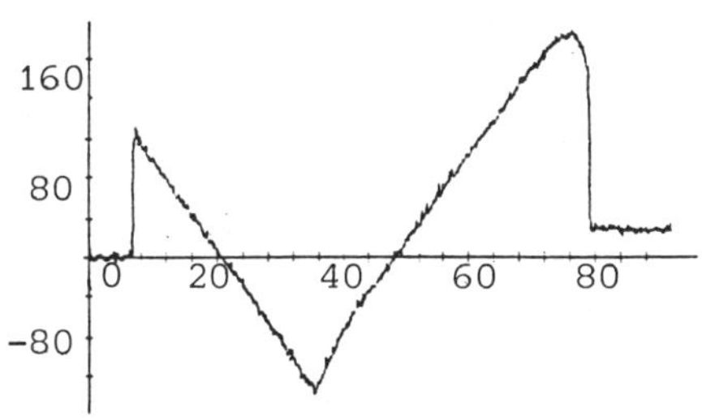

Fig. 1. Force (dynes) – vs – time (seconds).

Contact angle from attractive force

The contact angle is calculated by use of the Wilhelmy equation:

$$\text{Attractive force} = \gamma_L \, P \cos \theta$$

*A more detailed description of the program is available upon request.

where

γ_L = surface tension of the liquid,

P = the perimeter of the wood sample, cm, and

θ = the contact angle.

The point of initial contact of the wood with the immersion liquid is located, and correction for buoyancy is made. This gives the attractive force at zero immersion from which the contact angle, θ, is determined.

Contact angle from work of adhesion

The contact angle, θ, is calculated with the Young-Dupré equation:

$$\text{Work of adhesion} = \gamma_L\ (1 + \cos\theta)$$

The work of adhesion is obtained from the area under immersion curve (Fig. 1).

Surface free energy estimation

The well-known Zisman (1963, 1964) critical surface tension γ_c, was determined by plotting the cosine of the contact angle on a wood surface versus the surface tension of a series of liquids. The point at which cos θ = 1 (zero contact angle) line is intercepted, determines γ_c. Mixtures of water with ethylene glycol and with glycerol provided a range of surface tensions. End grain of wood samples was sealed with a small amount of beeswax to prevent rapid capillary flow of liquids into wood samples.

Interaction parameter calculation

The interaction parameter, Φ, was calculated according to the procedure used by Good (1979) and Becher (1977):

$$\cos\theta = 2\,\Phi \left(\frac{\gamma_s}{\gamma_L}\right)^{1/2} - 1$$

γ_s = surface tension of solid

Aging effect

Wood samples, cut at the same time, were stored under constant 27°C, 30% relative humidity conditions for several weeks; elapsed time shown denotes length of exposure to laboratory illumination and atmosphere. Each point represents a wood sample cut from the same board.

RESULTS AND DISCUSSION

Aging effect

The sensitivity of this experimental procedure in measuring wettability of wood samples is shown by the decreasing wettability of aspen (Fig. 2) with time of exposure to laboratory conditions. This effect has been noted before by Gray (1962), who attributed it to surface contamination and by Nguyen and Johns (1979) who suggested environmental rather than wood factors. Subtle effects of light and oxygen should also be considered.

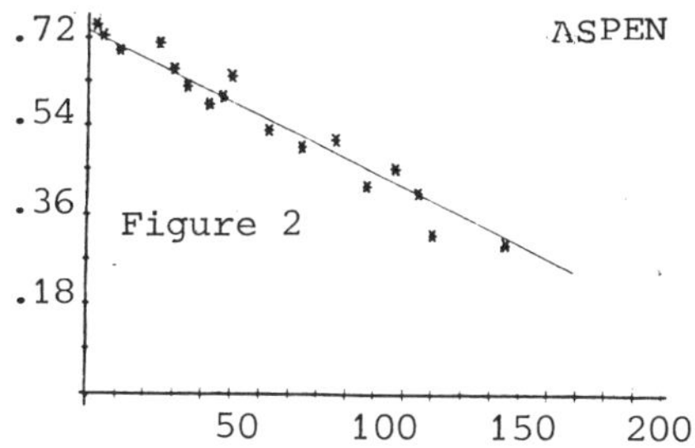

Fig. 2. Cosine of contact angle – vs – time (minutes).

Surface free energy estimates

The critical surface tension γ_c, (an approximation of the surface free energy) of birch was obtained with the Zisman plot (Fig. 3) from measured surface tension values and advancing contact angles calculated from attractive forces. The value, 49.5 dynes/cm for the birch wood surface free energy is in the range of values reported for wood previously by others (Marian 1962; Herczeg 1965; Nguyen and Johns 1979).

Interaction parameter calculation

If γ_c (critical surface tension) is taken for an approximation of the surface free energy of wood, γ_s, it becomes possible to calculate the interaction

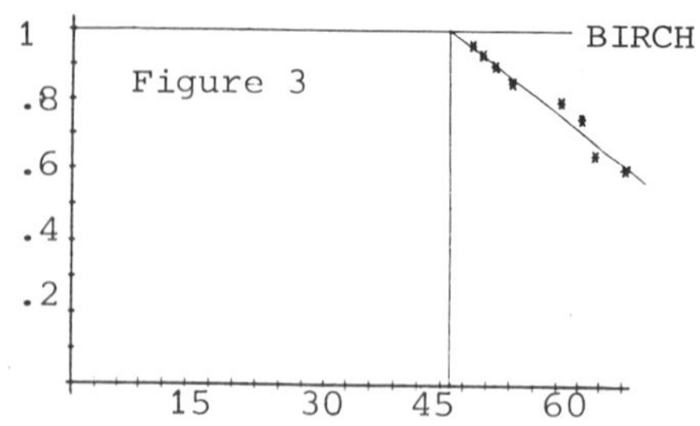

Fig. 3. Cosine of contact angle – vs – liquid surface tension

parameter Φ. Φ is a function of the properties of the solid and the liquid on a molecular level (Good 1979; Becher 1977). Its value is near unity for nonpolar liquids and solids, and generally from 0.5 to 1.0.

$$\cos \theta = 2\,\Phi \left(\frac{\gamma_s}{\gamma_L}\right)^{1/2} - 1$$ (Good 1979;

Becher 1977). For birch, $\gamma_c \simeq \gamma_s = 49.5$ dynes/cm, $\cos \theta = 0.608$ with H_2O, $\gamma_L = 72.4$ dynes/cm; then $\Phi = 0.97$. With ethylene glycol $\cos \theta = 0.866$, $\gamma_L = 52.3$ dynes/cm; then $\Phi = 0.96$. With glycerol $\cos \theta = 0.653$, $\gamma_L = 66.3$ dynes/cm; then $\Phi = 0.96$.

Further work is planned on the interaction parameter for wood and liquids and on the contributions of dispersion and polar forces to the free energy values. These data are presented to illustrate the information that is readily obtainable with this apparatus. The rapid determination of wettability and of liquid absorption by wood lends itself to practical applications that are related to improved solid wood products and wood-derived materials.

ACKNOWLEDGEMENTS

The authors thank Richard E. Kinney for the electronics work.

REFERENCES

Banks, W.; Voulgaridis, E. 1980. Brit. Wood Preservers Ann. Conv. Record: 43-53.

Becher, P. 1977. J. Colloid and Interface Sci. 59, (3) 429-432.

Bodig, J. 1962. For. Prod. J. 12, (6) 265-275.

Casilla, R.C., Chow, S. and Steiner, P. R. 1981. Wood Sci. Technol. 15, 31-45.

Casilla, R.C., Chow, S., Steiner, P.R. and Warren, S.R. 1984. Wood Sci. Technol. 18, 87-96.

Collett, B. 1972. Wood Sci. Technol. 6, 1-42.

Freeman, A. 1959. For. Prod. J. 9, (12) 451-458.

Gray, V. R. 1962. For Prod. J. 12, (9) 452-461.

Good, R. J. 1979. In: Surface and Colloid Science, Good, R. J., Stromberg, R. R., eds. Vol. 11, Plenum Press, New York and London, 1979, 1.

Herczeg, A. 1965. For. Prod. J. 15, (11) 499-505.

Huntsberger, J. 1978. Adhes. Age. 23-27 (Dec.)

Klungness, J. 1981. Tappi 64, (12) 65-66.

Marian, J. R. 1962. In: Symp. on Prop. of Surf., ASTM Mater. Sci. Ser. 4: Spec. Tech. Publ. 340, Phila.:ASTM 122-149.

Miniutti, V. P., Mraz, E.A. and Black, J.M. 1961. For. Prod. J. 11, (10) 453-462.

Nguyen, T.; Johns, W. E. 1978. Wood Sci. Technol. 12, 63-74.

Nguyen, T.; Johns, W. E. 1979. Wood Sci. Technol. 13, 29-40.

Purslow, D. F. 1982. Bldg. Res. Establ., Info. Paper 20/82, Nov. 1982, 3 p. (Garston, Watford, WD2 7JR, UK).

Subramanian, et al. 1983. Holzforschung 37, 117-120.

Voulgaridis, E.; Banks, W. 1981. J. Inst. Wood Sci. 9, (2) 50, 72-83.

Young, R. A. 1976. Wood and Fiber 8, (2) 120-128.

Zisman, W. A. 1963. Ind. Eng. Chem. 55, (10) 19-38.

Zisman, W. A. 1964. In: Contact angle, wettability, and adhesion. Adv. Chem. Ser. 43. Washington, D.C.: Am. Chem. Soc. 1.

Chapter 46

Formaldehyde release from cellulosic textiles treated with formaldehyde and its amide derivatives for smooth drying properties

B. A. Kottes Andrew and Robert M. Reinhardt – Southern Regional Research Center, ARS, USDA New Orleans, Louisiana 70179, U.S.A.

HISTORICAL PERSPECTIVE

There can be no consideration of formaldehyde release from textiles without a historical perspective of the theory of cellulose crosslinking and the development of stable finishing agent systems. Progress in the two areas are parallel. Creaseproofing of rayon and other cellulosics emerged in the last half of this century as a response to consumer demand for easy care textiles. The creaseproofing process involved impregnating the textile with a solution of finishing agent and catalyst, drying the wetted textile, and heating the textile at a temperature high enough to "set" or cure the finish.

The early finishing agents to find commercial success were those which could polymerize, such as urea formaldehyde and melamine formaldehyde[1,2,]. Although the initial patent on use of urea-formaldehyde in production of anti-crease textiles was granted to Tootal, Broadhurst, Lee Co., Ltd. in 1928 [3], crosslinking as the operative mechanism for the phenomenon was first proposed much later[1,4,5].

The establishment of covalent bond formation as the mechanism for "permanent" creaseproofing explained some of the problems associated with the first generation creasproofing systems and the basis for development of improved systems. The performance problems with the early polymer-former finishes (urea- and melamine-based reactants) concerned their durability, both to laundering and to bleaching[6,7,8]. Replacement of the second hydrogen on the urea nitrogens by ring closure, as in 2-imidazolidinone, produced a more chlorine-resistant finish[9]. This discovery led to the proliferation of the methylol derivatives of other cyclic amides for wash-and-wear finishing of cotton and cellulosics. In the early 1960's, in addition to urea formaldehydes and melamine formaldehydes, the reactants included methylolated propyleneureas, triazones, urons and dicarbamates[10]. These agents are not greatly unlike the urea-formaldehyde and phenol-formaldehydes used in many wood products, such as particle board, plywood and other bonded woods. A recent book provides a discussion of the similarities and differences between formaldehyde release in textiles and wood[11].

AN UNDERSTANDING OF THE CELLULOSE METHYLOLAMIDE CROSSLINK

With so many agents in use, and with such a range of stabilities to the varied hydrolyzing conditions of launderings represented among the finishes, there was a need for research to a) compare the hydrolysis resistance of the finishes, and b) establish a general operative mechanism for the differing stabilities. The reactions between the amidomethylol compounds and cellulose to form the amidomethyl ether of cellulose are reversible. Depending on the electronic configuration around the amido nitrogen, the covalently bonded ethers are susceptible in varying degrees to hydrolytic cleavage. Mechanisms for hydrolysis of the amidomethyl ethers of cellulose under acidic and basic conditions, were proposed by Reeves, et al [12], and Andrews, et al [10] respectively. Acidic hydrolysis proceeds by a carbocation mechanism; the more electropositive the groups surrounding the nitrogen are, the more susceptible the N-C bond is to cleavage under acid conditions. During alkaline hydrolysis, electron attracting groups cause decreased electron density on the amido nitrogens which favors the attack of the negative ion. Vail, et al, proposed a reactivity-hydrolysis mechanism for cellulose crosslinking [13]. Finishes that require strong acidic conditions for application require strong acidic conditions for hydrolysis and, therefore, are the most stable to acid hydrolysis.

Added to the lack of durability of certain finishes to laundering was a strong formaldehyde odor associated with processing these fabrics. The first tests to measure formaldehyde release, reported in the late 1950's and early 1960's, were designed to quantify subjective odor (sniff) tests [14,15]. Based on this research, the American Association of Textile Chemists and Colorists established two tests for formaldehyde release from textiles: the sealed jar method (AATCC Test Method 112-1965) and the steam tube method (AATCC Test Method 113-1965) [16]. AATCC Test Method 112 has persisted as the test of commerce for formaldehyde release in the United States. Fabrics finished with the first generation creaseproofing systems had formaldehyde release levels greater than 4000 μg/g of fabric in the AATCC Sealed Jar test [17]; current fabric levels are $<$ 500 μg/g.

REDUCTION OF FORMALDEHYDE LEVELS IN FABRICS

Formaldehyde release of fabrics finished for durable press gradually decreased over the years as performance demands led the industry to development of more stable finishes. Dimethyloldihydroxyethyleneurea (DMDHEU) emerged in response to the need for a finish stable enough on post-cure fabric applications (the reactant is set or cured after garment fabrication) to remain uncured during garment fabrication [17]. From the reactivity-hydrolysis relationship, the reaction characteristics of DMDHEU also proved an advantage in pre-cure fabric applications wherein garment making followed fabric treatment through the curing step. DMDHEU became the workhorse finishing agent of the textile industry.

Methylolated monocarbamates also found applications in flat goods finishing [18]. The seemingly anomolous formaldehyde-release characteristics of carbamate-finished fabric, i.e., high release of uncured, or cured-unwashed fabric and low release of cured-washed fabrics, focused attention on the different types of liberated formaldehyde. The equilibrium dissociation of methylolamides to the amide and formaldehyde was recognized as a contributing factor to formaldehyde released from non-washed fabrics finished for durable press [6]. It is important to maintain reaction conditions in the preparation of the finishing agents that promote the equilibrium distribution of methylolamide:amide:formaldehyde.

Formaldehyde liberation from textiles is attributed to several sources. Free, or loosely bound formaldehyde from partial- or non-reaction of the agent with cellulose is present in uncured (sensitized) fabrics and in cured, but unwashed fabrics [19]. Formaldehyde that is excess to the reactant equilibrium requirements is also present in an uncombined state in these fabrics. Formaldehyde from finish hydrolysis by storage conditions and by the incubation conditions of the test methods themselves is liberated from both unwashed and washed fabrics. To a lesser extent, formaldehdye liberated from washed fabrics is evolved from a finish's attempt to reestablish its equilibrium position. Sources of potential formaldehyde release within a finish include:

$$\mathrm{HCHO,\ HOCH_2NC({=}O)NCH_2OH,\ Cell{-}OCH_2OH,}$$

$$\mathrm{Cell{-}OCH_2NC({=}O)NCH_2OH,\ {>}NC({=}O)NCH_2OCH_2NC({=}O)N{<},}$$

$$\mathrm{Cell{-}OCH_2NC({=}O)NCH_2O{-}Cell,\ {>}NC({=}O)NCH_2NC({=}O)N{<},}$$

$$\text{and } \mathrm{Cell{-}OCH_2O{-}Cell}$$

These residues only reflect sources from within the finishes themselves, not contributions from chemicals that have been added purposely, such as buffering agents or hand modifiers, or materials added inadvertently.

CONTROL OPTIONS AVAILABLE FOR REDUCTION OF FORMALDEHYDE RELEASE

From the hydrolysis-reactivity relationship, agent choice should be foremost among the processing variables that influence formaldehyde release. Research has shown that formaldehyde release under different humidity conditions from washed fabrics finished with methylolated agents in some use today follows the decreasing order: dimethylolurea > dimethylolethyleneurea > dimethylolpropyleneurea > dimethyloldihydroxyethyleneurea > dimethylolcarbamates (Figure 1). Within each class it is possible to modify the formaldehyde release characteristics further by replacing the OH of the methylol group with certain alkoxy groups [6,20,21]. Formaldehyde release from methylated DMDHEU finishes is far less than that of the DMDHEU finish itself. Although lower boiling alcohols must be prereacted with the methylolamide to form a precondensate, higher boiling, water soluble polyols are as effective in formaldehyde reduction when they are merely added to the methylolated agent (blended) and allowed to react *in situ* at the time of cure [22]. The levels obtained by the two methods of alkylation are seen in Table I. The use of methoxy precondensates and blended polyols for reduction of formaldehyde release is also effective with carbamate-based agents [24,25].

Catalyst type and concentration used to promote crosslinking between the agent and cellulose also heavily influence formaldehyde release. Table II shows the adverse effects on formaldehyde release of reducing the catalyst to agent ratio beyond the minimum necessary to produce optimum crosslinking.

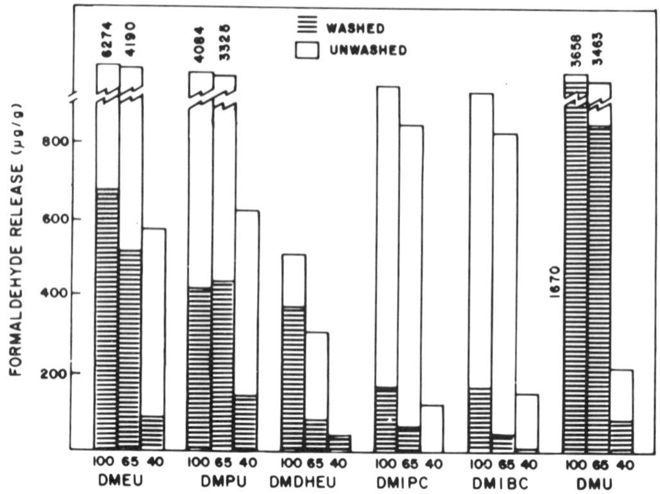

Fig. 1. Comparison of formaldehyde released at 49°C and 100, 65, and 40% relative humidity from unwashed and afterwashed printcloth finished with the dimethylol derivatives of: ethyleneurea, propyleneurea, dihydroxyethyleneurea, and isopropyl and isobutyl carbamates [20].

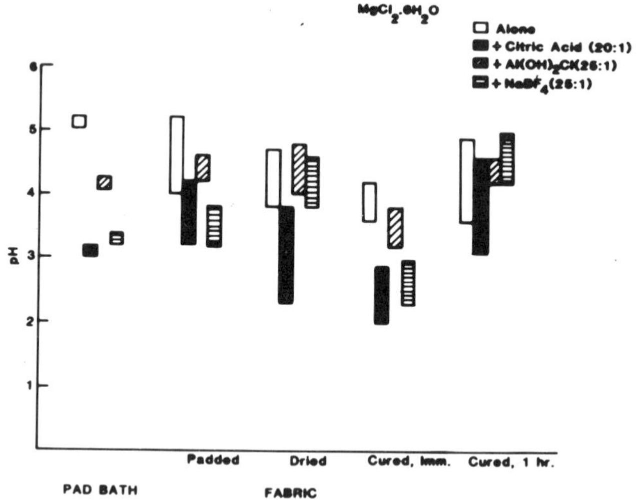

Fig. 2. The pH of pod bath and fabric after various stages of processing: immediately after drying, immediately after curing and one hour equilibration following curing in treatments of cotton printcloth with 9% DMDHEU and 2.7% $MgCl_2$ or its catalyst mixtures. (When used with N_aBF_4, concentration of $MgCl_2.6H_2O$ was 1.43%) [23].

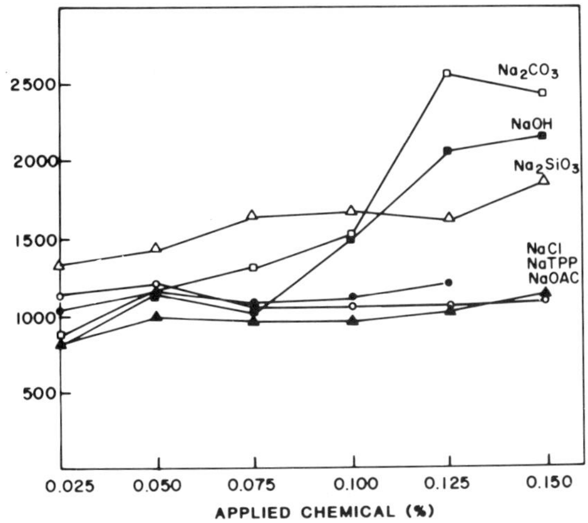

Fig. 3. Formaldehyde release from fabrics crosslinked with DMDHEU following pretreatments with increasing concentrations of (●) sodium chloride; (○) sodium acetate; (▲) sodium tripolyphosphate; (△) sodium silicate; (□) sodium carbonate and (■) sodium hydroxide [27].

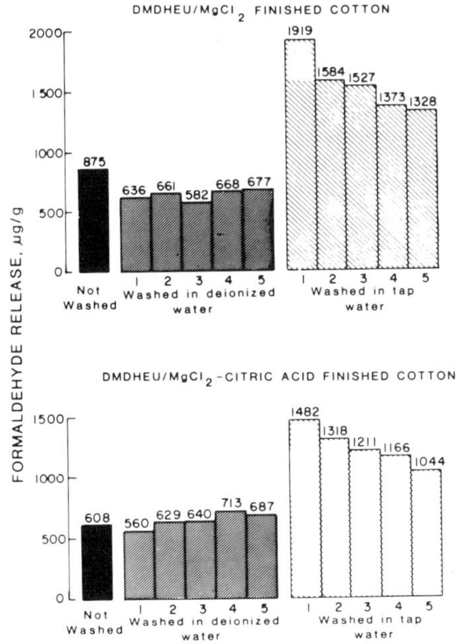

Fig. 4. Formaldehyde release after multiple wash cycles in deionized or tap water [3u].

Table 1. Comparison of effectiveness of DMDEU systems [23].

Catalyst	Durable press rating	N (%)	AATCC-112 HCHO release (μg/g) unwashed	washed
		5% DMDHEU		
0.277% $Zn(NO_3)_2.6H_2O$	3.5	0.64	646	282
1.5% $MgCl_2.6H_2O$	3.7	0.71	592	233
+ citric acid (20:1)	4.0	0.71	528	265
		5% DMDHEU, 2% propylene glycol		
0.277% $Zn(NO_3)_2.6H_2O$	3.5	0.64	371	197
1.5% $MgCl_2.6H_2O$	3.9	0.69	449	179
+ citric acid (20:1)	3.9	0.68	260	124
		5% Partially methylated DMDHEU		
0.277% $Zn(NO_3)_2.6H_2O$	3.3	0.56	237	52
1.5% $MgCl_2.6H_2O$	3.5	0.60	118	34
+ citric acid (20:1)	3.6	0.61	94	28

a/ Curing conditions were 160°C for 3 min.

Table 2. Effect of $Zn(NO_3)_2.6H_2O$ catalyst concentration on properties of printcloth finished with 9% solids DMDHEU [20].

Catalyst (%)	Durable press rating	HCHO Released AATCC test, (μg/g)	HCHO Extracted Japanese test, (μg/g)
	9% solids unbuffered DMDHEU		
0.25	3.3	1787	460
0.50	5.0	897	413
1.00	4.7	818	454
1.50	4.3	909	487
	9% solids buffered DMDHEU		
0.50	2.3	2734	1440
1.00	3.2	1436	909
1.50	4.3	952	351

Table 3. Variation of formaldehyde release with cure conditions [23].

Catalyst (gfw ratio)	Formaldehyde release, AATCC-112 (μg/g) Cure Conditions[°C/time(min)] 120/3	140/3	160/3	180/1	200/20 s
2.7% $MgCl_2.6H_2O$	1951	1389	1363	1368	1460
2.8% $MgCl_2.6H_2O$: citric acid (20:1)	1015	1013	854	876	998

Predominantly Lewis acid catalysts produce fabrics with lower formaldehyde release than do those of predominantly Bronsted activity. When proton donors are added in sufficient proportion to catalysts with very little Bronsted activity, improvements can be realized in appearance-related properties while low formaldehyde release is maintained. Figure 2 shows the pH of finishing bath and fabric after various stages of processing with 9% DMDHEU and catalyst mixtures with $MgCl_2$. Catalyst concentraton must be sufficiently high to overcome the effects of buffering if a buffered agent is used, but not so high that finish hydrolysis occurs.

Curing conditions are another set of variables influencing formaldehyde release. Both acidic medium and heat energy are necessary to optimize crosslinking. This requirement is shown in formaldehdye release levels from treatments with 9% DMDHEU and two $MgCl_2$-based catalysts with a range of cure times and temperatures (Table III). The necessity for optimizing both catalysis and curing conditions can be seen in the two distinct levels of formaldehyde release from the two catalyst systems.

EXOGENOUS CONTRIBUTORS TO FORMALDEHYDE RELEASE

The previous sections discussed influences on formaldehyde evolution from within the crosslinking system itself, reactant, catalyst, and/or finish residues. Another contributor to formaldehdye release is the state of the fabric before finishing. Residual salts and chemicals from pretreatments to remove the natural impurities and processing lubricants on cotton fabrics can increase the pH and alkalinity of fabrics to an extent that interferes with crosslinking. The crosslinking interference is essentially interference with or neutralization of the catalysts [26,27]. Figure 3 summarizes the effect on formaldehyde release from impurities likely to be present on the fabric from poor preparation.

Certain chemicals, formaldehyde scavengers, are added to finishing systems to react with formaldehyde to reduce the levels of free formaldehyde in the pad bath. The scavengers, however, often interfere with the reactant equilibrium and produce fabrics with less than optimum appearance properties. Examples of scavengers include urea, ethyleneurea (2-imidazolidinone), dicyandiamide, and carbohydrazide [28,29].

After-handling can either increase or decrease the formaldehyde release of textiles finished for durable press. Response to pH, water type, water hardness, detergent type, presence of buffers is a function of the hydrolysis resistance of the finish itself and its extent of cure [30]. In Figure 4 is shown the response of finishes from two DMDHEU-based finishing systems to pH differences during laundering. More acidic catalysts in processing increase the finish's alkali resistance.

Response to heat may proceed by a different mechanism than response to hydrolysis in an aqueous medium. Formaldehyde release always decreases in fabrics subjected to heat stress [31].

TESTS FOR FORMALDEHYDE LIBERATED FROM TEXTILES

Tests have been developed to differentiate among the sources of formaldehyde within a finish [32]. Formaldehyde release usually refers to the combination of formaldehyde both hydrolyzable and loosely bound. The levels from the incubation conditions of AATCC Test Method 112 reflect this type of formaldehyde. Other tests for liberated formaldehyde with somewhat stringent incubation conditions, but not as severe as the AATCC Test Method 112, are the Japanese Law 112 Test [33] and the Shirley Institute Test [34]. These tests attempt to assess formaldehyde liberated from garments in direct contact with a perspiring human. The terms "free", "surface", "residual", and "steady-state" formaldehyde reflect a need to characterize the formaldehyde which is not intimately bound in the finish. This formaldehyde can be unbound, sorbed, present as a cellulose hemiformal; it can be excess to reaction, or from liberation to satisfy equilibrium ratios within a finish [35,36]. Transport formaldehyde is that formaldehyde which can be transferred to or from a textile under static conditions. It is inherently free, regardless of from which of the above defined sources it comes [37].

SUMMARY

Basic to a reduction of formaldehyde release from cellulosic products treated with formaldehyde derivatives of ureas is an understanding of the variables contributing to the release. The abundance of recent research conducted by the U. S. textile industry, in general, and the Southern Regional Research Center, in particular, has provided the textile finisher with the information to exercize control options for lowering formaldehyde release to de minimus levels. By analogy, mechanisms and methods established for cotton cellulose modifications should be valid for application to other cellulosics. The wood products industry can benefit also from this pioneering research to minimize liberation of formaldehyde from modified wood cellulose.

REFERENCES

[1] Cameron, W. G., and Morton, T. H. J. Soc. Dyers Colour. (1948), 64(10), 329-336.

[2] Nickerson, R. F. Am. Dyest. Rep. (1950), 39(1), P46-50.

[3] Foulds, R. P., Marsh, J. T., Wood, F. C., Boffy, J., and Tankard, J. British Patent 291,473.

[4] Gagliardi, D. D., and Nuessle, A. C. Am. Dyest. Rep. (1950), 39(1), P12-19.

[5] Cooke, T. F., Dusenbury, J. H., Kienle, R. H., and Linekin, E. E. Text. Res. J. (1954), 24, 1015-1035.

[6] Peterson, H., presented in part at the American Association of Textile Chemists and Colorists National Technical Conference, Philadelphia, PA, USA, Sept. 1972.

[7] Enders, H., and Pusch, G. Am. Dyest. Rep. (1960) 49(1), 25-28.

[8] Piedmont Section, AATCC Am. Dyest. Rep. (1960) 49(24), P843-855.

[9] Frick, J. G., Jr., Kottes, B. A., and Reid, J. D. Text. Res. J. (1960) 30, 495-504.

[10] Andrews, B. A. Kottes, Arceneaux, R. L., Frick, J. G., Jr., and Reid, J. D. Text. Res. J. (1962) 32, 489-496.

[11] "Formaldehyde Release from Wood Products", Andrews, B. A. K., Meyer, C. B., and Reinhardt, R. M., eds., Symposium Series No. 316, ACS, Washington, D. C. (1986).

[12] Reeves, W. A., Vail, S. L., and Frick, J. G., Jr. Text. Res. J. (1962) 32, 305-312.
[13] Vail, S. L. Text. Res. J. (1969) 39, 774-780.
[14] Bacon, O. C., Parker, M. F., Horn, L. F. Am. Dyest. Rep. (1957) 46, P933-935.
[15] Reid, J. D., Arceneaux, R. L., Reinhardt, R. M., and Harris, J. A. Am. Dyest. Rep. (1960) 49(14), 29-34.
[16] Nuessle, A. C. Am. Dyest. Rep. (1966) 55(17), P646-648.
[17] Goldstein, H. B. Text. Chem. Color. (1973) 5(10), 209-214.
[18] Arceneaux, R. L., Frick, J. G., Jr., Reid, J. D., and Gautreaux, G. A. Am. Dyest. Rep. (1961) 50(22), P849-853.
[19] Kullman, R. M. H., Pepperman, A. B., and Vail, S. L. Text. Chem. Color. (1978) 10(9), 195-199.
[20] Andrews, B. A. K., Harper, R. J., Jr., and Vail, S. L. Text. Res. J. (1980) 50, 315-321.
[21] Vail, S. L., Kullman, R. M. H., Reeves, W. A., and Barker, R. H. Text. Res. J. (1970) 40, 355-362.
[22] Andrews, B. A. K., Harper, R. J., Jr., Smith, R. D., and Reed, J. W. Text. Chem. Color. (1980) 12(11), 287-291.
[23] Andrews, B. A. K., and Reinhardt, R. M. Text. Res. J. (1982) 52, 123-132.
[24] Andrews, B. A. K., and Reinhardt, R. M. U. S. Patent 4,488,878.
[25] Andrews, B. A. K., and Reinhardt, R. M. U. S. Patent 4,539,008.
[26] Palmetto Section, AATCC Text. Chem. Color. (1982) 14(12), 249-256.
[27] Andrews, B. A. K., and Reinhardt, R. M. Text. Chem. Color. (1984) 16(5), 100-105.
[28] Cooke, T. F., and Weigmann, H.-D. Text. Chem. Color. (1982) 14(6), 136-144.
[29] Cashen, N. A. Text. Res. J. (1982) 49, 480-484.
[30] Reinhardt, R. M., and Andrews, B. A. K. Text. Chem. Color. (1984) 16(11), 235-240.
[31] Reinhardt, R. M., and Andrews, B. A. K. Book of Abstracts, 190th National ACS Meeting, Chicago, IL, USA CELL 67 September 8-13, 1985.
[32] Andrews, B. A. K., and Reinhardt, R. M., Advances in Chemistry Series No. 210 Formaldehyde: Analytical Chemistry and Toxicology, V. Turoski, Ed., ACS, Wahsington, D. C. (1985), p 83.
[33] "Law for Control of Household Goods Containing Harmful Substances," Law No. 112; Welfare Ministry Ordinance: Japan, 1973; No. 34-1974.
[34] Shirley Inst. Publ. (1975) 48, 17.
[35] Andrews, B. A. K., Simoneaux, J. M., and Harper, R. J., Jr. Die Angew. Makromol. Chem. (1983) 115(1807), 115-129.
[36] Andrews, B. A. K., and Reinhardt, R. M. Text. Res. J. (1986) 56, 115-120.
[37] Reinhardt, R. M., and Andrews, B. A. K. Text. Res. J. (1986) 56, 144-150.

Chapter 47

Cellulose ethers- properties and applications

Lennart Dahlgren, Ph.D. – Berol Kemi AB, Stenungsund, Sweden.

ABSTRACT

The industrial production of cellulose ethers started in the 1920's. To a large extent, these products were used as adhesives and binders. Today, the cellulose ethers are used in a wide range of applications, e.g. as additives in latex paints, mortars and plasters, detergents, shampoos, hair conditioners and ... ice-cream. The total production capacity in the world is around 450,000 tons per year.

The cellulose ethers have gained their position on the market due to their multifunctional effects. They are used for flow (rheology) control, as stabilizers (protective colloids), as water retention aids, as film formers or adhesives. These combined effects are not generally obtainable by the use of fully synthetic polymers.

This paper will mainly deal with non-ionic cellulose ethers, their chemical, physical and application properties as well as some recent development trends. The apparent similarities and distinctions between the cellulose ethers will be discussed.

THE MARKET

Let us start this review of the cellulose ethers on the market. What are the needs of the customer?

The customer is asking for specific effects. He is not usually interested in buying a particular chemical product, or in the chemical structure of such products. What he wants is to obtain a maximum performance with his own product.

The customer will very probably ask for one or several of the following important performance characteristics:

Thickening or flow control
Stabilization
Long open time/workability
Easy handling
Adhesion/strength development
Film formation

How can these needs be satisfied? Which alternatives exist? The customer has indeed many alternatives, which will meet his needs more or less completely.

The following products should be mentioned:

Natural polymers

guar gum
alginates
starch
and many more

Synthetic and derivatised natural polymers

cellulose ethers
polyacrylates
polyacrylamides
polyurethanes
and others

Inorganic thickeners

clay
other minerals

The sales value of the cellulose ethers represent an important part of these products.

What makes cellulose ethers attractive?

* They offer a combination of properties

  The thickening function itself can very often be fulfilled by, for example, starch or fully synthetic thickeners. The achievement of a specific consistency or flow is however a very demanding task.

  Water retention, absolutely vital in the formulation of cement or gypsum based mortars or plasters, can in practice only be achieved by the use of cellulose ethers.

  The cellulose ethers are excellent protective colloids and can stabilise dispersions by means of steric repulsions.
  Other functions are adhesion, film formation or coating.

* The cellulose ethers offer a consistent quality and are easily available

  Availability can be a problem when using natural polymers, where the supply depends on the weather conditions and on many small producers.

* Price/Performance

  The cellulose ethers usually perform at low addition levels, typically 0.1 to 1 per cent. In some cases however they are used as the main component, e.g. in wall paper paste.

  The ethers can be designed for specific purposes. There is a great number of possible variations.

* Health aspects

  Many cellulose ethers have been proven to be toxicologically harmless. Some of them are used as food additives.

THE PROCESS

The main raw materials for the production of cellulose ethers are cellulose and alkylene oxides or alkyl chlorides.

The building stones in cellulose are anhydroglucose units. The chains are linked together by strong inter- and intramolecular bonds. Consequently cellulose is not water soluble, in spite of its apparent hydrophilic character.

In order to make a soluble polymer, using the cellulose chain as a backbone, the bonds between the chains must first be broken. This is usually done for commercial purposes by treatment with sodium hydroxide, which can penetrate into the cellulose. This process is called activation or mercerisation. The resulting alkali cellulose can be reacted with the alkylating agents (figure 1).

The manufacture of cellulose ethers is complicated and demands a high technological competence. The production process involves the handling of solids, liquids and gases.The performance of the products is determined not only by their chemical structure, but also by their physical form and other means of modification. The tailoring of the polymers into their final physical form therefore involves substantial know-how.

THE PRODUCTS

The total world capacity of cellulose ethers is close to 450,000 tons. The anionic ethers (almost all CMC) represent about 250,000 tons and the nonionics 150,000 tons. The cationic types are produced in small volumes only.

The prices vary considerably. The list prices were for example in USA June 1986 1.40 USD per kg for technical grade CMC and 10.80 USD for printing ink quality of EHEC.

The commercially important cellulose ethers are listed below. Each type is produced in many different qualities. Most products are water soluble (WS), some are soluble in organic solvents (OS).

ACTIVATION

$$\mathrm{CELL-OH + Na^{+}OH^{-} \rightleftharpoons CELL-O^{-}Na^{+} + H_2O}$$

HYDROXYALKYLATION

$$\mathrm{CELL-O^{-} + H_2C\overset{O}{-}CH_2 \rightarrow CELL-O-CH_2CH_2O^{-}}$$

$$\mathrm{CELL-O^{-} + H_2C\overset{O}{-}CH-CH_3 \rightarrow CELL-O-CH_2-\underset{\displaystyle CH_3}{\underset{|}{CH}}-O^{-}}$$

ALKYLATION

$$\mathrm{CELL-O^{-} + R\,Cl \rightarrow CELL-O-R + Cl^{-}}$$

CARBOXYMETHYLATION

$$\mathrm{CELL-O^{-} + Cl-CH_2-COO^{-}Na^{+} \rightarrow CELL-O-CH_2-COO^{-}Na^{+}}$$

BY–PRODUCT FORMATION

GLYCOLS
GLYCOL ETHERS
SODIUM CHLORIDE
ETC.

NONIONIC CELLULOSE ETHERS

Methyl hydroxyethyl cellulose	MHEC	WS	OS
Methyl hydroxypropyl cellulose	MHPC	WS	OS
Hydroxyethyl cellulose	HEC	WS	
Ethyl hydroxyethyl cellulose	EHEC	WS	OS
Hydroxypropyl cellulose	HPC	WS	OS
Methyl hydroxybutyl cellulose	MHBC	WS	OS
Ethyl cellulose	EC		OS

IONIC CELLULOSE ETHERS

Carboxymethyl cellulose	CMC	WS
Carboxymethyl hydroxyethyl cellulose	CMHEC	WS
Cationic hydroxyethyl cellulose		WS

The cellulose ethers are characterized by their chemical structure as well as their physical properties:

- The average degree of polymerization (DP) or average number of anhydroglucose units. The viscosity of the cellulose ether in solution depends on the DP.

- The average degree of substitution (DS). There are three hydroxyls available for substitution per anhydroglucose unit. DS can therefore reach the maximum value of three.

- The average molar substitution (MS). When reacting cellulose with alkylene oxides, new reactive hydroxyls are formed. MS can therefore exceed three.

- The properties of the products are influenced by the distribution of the substituents along the chains. The detailed structure can be determined by the use of modern analytical techniques like nuclear magnetic resonance (NMR) spectroscopy and chromatographical methods, but the analysis can for the more complex structures still represent a major challenge.

- The hydrophilic/lipophilic balance is influenced by the choice of substituents. The hydrophilic ones are represented by hydroxyethyl and carboxymethyl. Lipophilic substituents are methyl, ethyl, hydroxybutyl and also hydroxypropyl. One of many possible structures of ethyl hydroxyethyl cellulose is shown in figure 2.

- Dissolving characteristics. The dissolution speed is regulated by the variation of the particle size and crosslinking of the ethers, as well as by various additives. The incorporation of the cellulose ethers into the end product or in the actual application, without lump formation but with the necessary speed, is one of the key elements in the cellulose ether technology (figure 3).

The physical properties of the important cellulose ethers in water solution are shown below

	CMC	CMHEC	HEC	EHEC	MHPC MHEC	HPC
Thickening	+	(+)	+	+	+	+
Surface activity	-	-	-	++	+	+
Hot water solubility	+	+	+	-	-	-
Electrolyte tolerance	-	++	++	+	+	+
Polyvalent ion tolerance	--	-	++	+	+	+

The best recognized function of the cellulose ethers is the thickening capacity. The commercially available CMHEC is however usually optimized for other properties, e.g. stabilization of slurries. In order to properly judge the performance of the cellulose ethers as flow control additives it is necessary to measure the characteristics of the formulated end-use product, taking into consideration the interaction with other ingredients like fillers and binders.

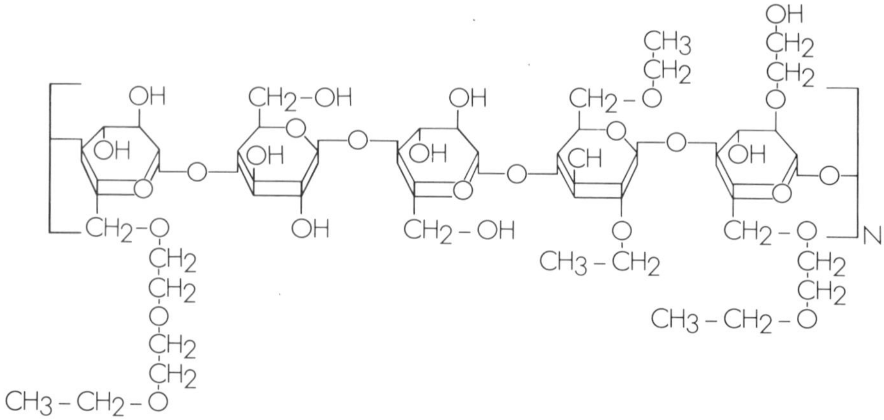

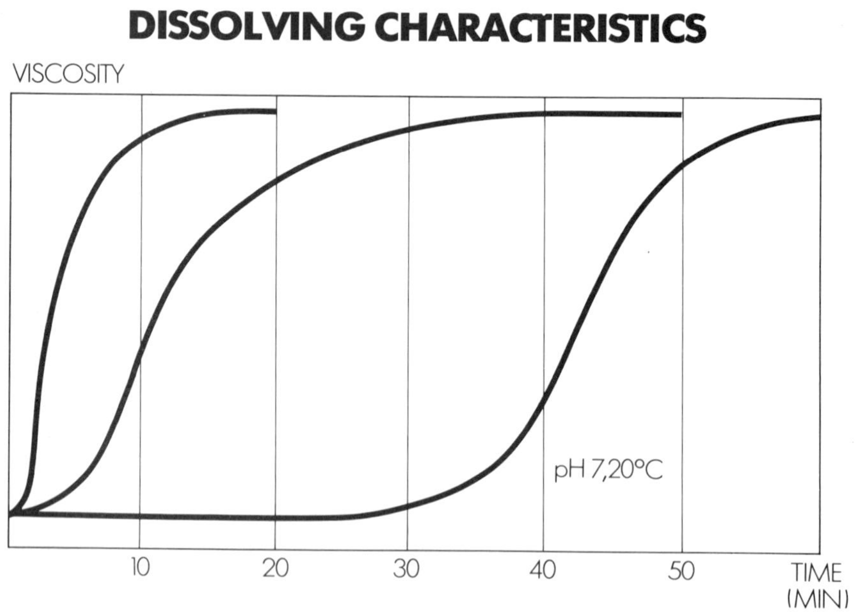
DISSOLVING CHARACTERISTICS
VISCOSITY
pH 7,20°C
10
20
30
40
50
TIME (MIN)

All cellulose ethers reduce surface or interfacial tension to some extent, but the most pronounced effect is achieved when a lipophilic substituent like methyl or ethyl is introduced. Some of the cellulose ethers can thus be regarded as polymeric surfactants, which have interesting properties regarding e.g. dispersion and emulsification.

The more lipophilic cellulose ethers (containing methyl, ethyl, hydroxypropyl or hydroxybutyl substituents) are soluble in cold water but not in hot water. This is a reversible reaction.

Depending on their hydrophilic/lipophilic character, the products are more or less sensitive to electrolytes (e.g. sodium chloride). See figure 4.

CMC is crosslinked by polyvalent metal ions like Ca^{2+}, Mg^{2+}, Fe^{3+}, and is therefore particularly sensitive to hard water.

APPLICATIONS

In most applications a <u>combination</u> of functional properties is utilized (figure 5).

The methyl celluloses have their strongest position in building materials, wallpaper adhesives, detergent powders, suspension polymerization, paints and tobacco products.

The hydroxyethyl celluloses with their more hydrophilic character have their best position in paints, emulsion polymerization and oil exploration.

Water soluble EHEC has its largest markets in paints, building materials, wallpaper adhesives and detergents.

CMC is mostly used in detergents, paper and textile industry, oil exploration and the food industry.

One particular application area will be discussed in some detail: Paint.

ELECTROLYTE TOLERANCE

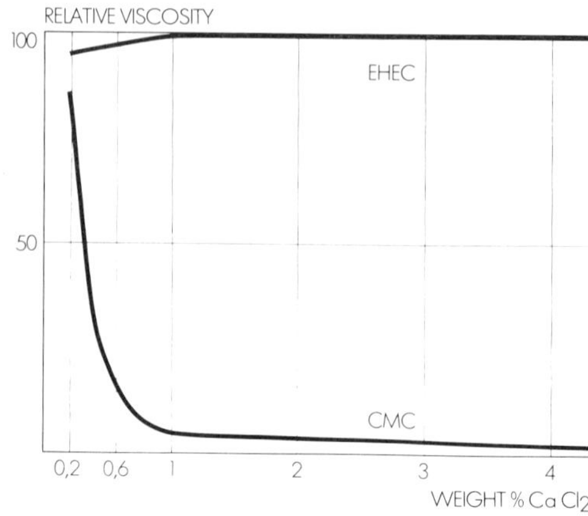

ELECTROLYTE TOLERANCE

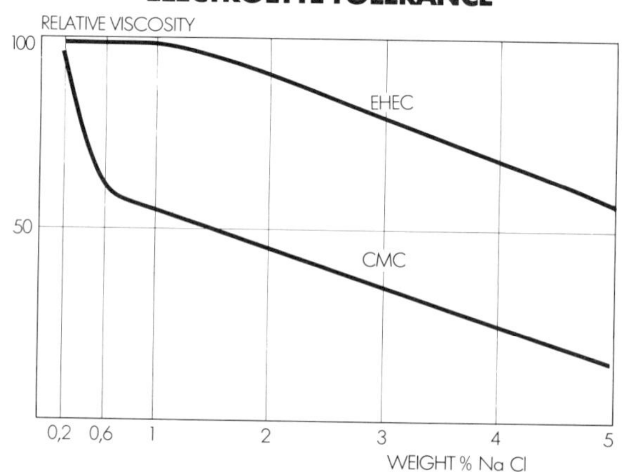

THE IMPORTANCE OF FUNCTIONAL PROPERTIES IN VARIOUS APPLICATIONS

	WATER-BORNE PAINTS	BUILDING MATERIALS	WALLPAPER ADHESIVES	PAPER COATING	DETERGENTS	POLYMER DISPERSIONS	TOBACCO	COSMETICS & PHARMACEUTICALS	CERAMICS
DISPERSANT	■	■		■				■	■
PROTECTIVE COLLOID	■			■	■	■		■	■
WATER RETAINER	■	■						■	■
THICKENER	■	■	■	■	■	■		■	■
FILM FORMER	■		■	■			■		■

The cellulose ethers play an important role in the formulation of water based (latex) paints. Key functions are flow control, dispersion/stabilization and water retention.

Rheological (flow) characteristics of three cellulose ethers of high, medium and low viscosity in water solution are shown in figure 6. The high viscosity (high DP) type shows a more pronounced shear thinning than the low viscosity type.

The cellulose ethers thicken the water phase in the paint, but can also interact with other components. The structure is broken down as the paint is being applied and it is gradually restored when the paint is at rest. This effect is called thixotropy.

The tendency of the paint to spatter when applied is linked to its elasticity. Paints formulated to the same viscosity (at medium shear rate) with a high viscosity or low viscosity cellulose ether behave differently. They show different flow characteristics; the paint formulated with the low viscosity cellulose ether is less prone to spatter when applied.

In the manufacture of paints, infections by microorganisms can degrade the products. The sensitivity to degradation by the attack of enzymes produced by fungi or bacteria is demonstrated in figure 7. The different stability of the cellulose ethers is related to the distribution of the substituents along the cellulose chain.

PRESENT DEVELOPMENT TRENDS

This paper has hopefully demonstrated the wide versatility of the cellulose ethers. We have seen that they usually provide a combination of interesting performance characteristics, which are of value in a great number of applications. Besides, it has been pointed out that the production of cellulose ethers requires a high technological competence.

FLOW CONTROL

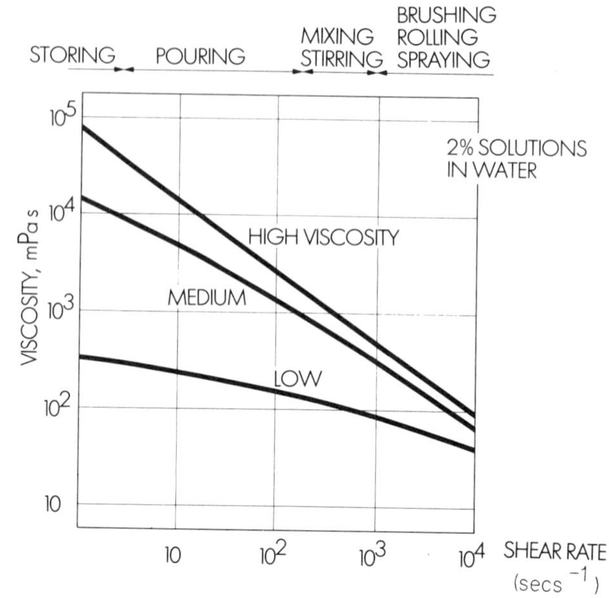

ENZYME RESISTANCE

1% AQ. SOLUTION, 12.5ppm CELLASE 1000, 20°C, pH 7

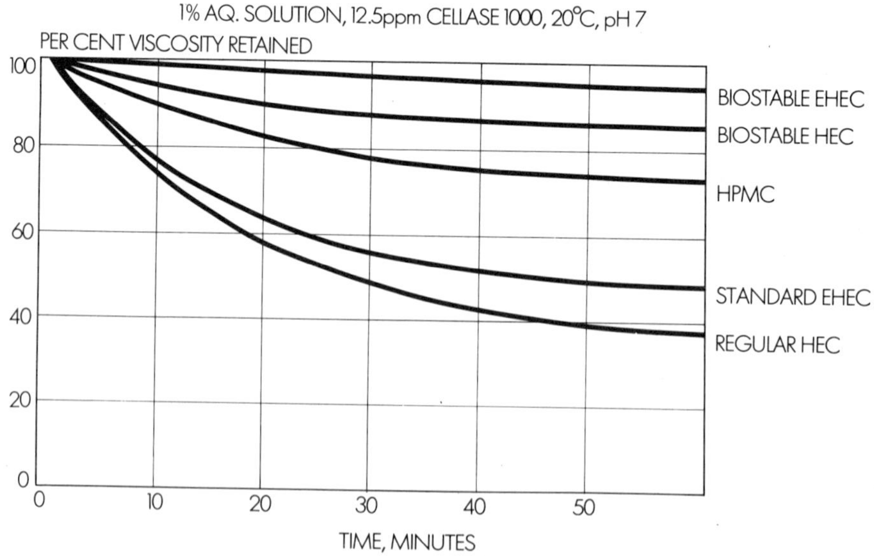

Finally, some development trends can be pointed out:

- A move towards more specialized products, designed to meet the customers' specific needs, for example associative thickeners and biostable cellulose ethers.
- Continued process development to meet increased demands of flexibility and efficiency.
- The development of a rational understanding of process- structure- activity relationships.

Chapter 48

The use of hydroxyethylcellulose in completion/workover fluids

H. J. G. Beckers – Hercules BV, Zwijndrecht, The Netherlands.

INTRODUCTION

The application of water-soluble polymers in petroleum recovery operations has shown a significant increase during the last few decades.

Already in the 1940's starch was used as a filtration control agent in salt water drilling muds [1]. Somewhat later CMC was introduced as a more effective and more versatile mud additive than starch [1]. Besides a strong reduction in filtrate loss, CMC also provided for rheology control and, to some extent, inhibition of swelling clays. A product, especially suited for stabilizing reactive shales, was polyanionic cellulose (PAC) [2]. Other products which are nowadays used in drilling fluids are xanthan gum, a biopolymer, and synthetics like acrylic acid/acrylamide (co)polymers [3].

Guar and its derivatives are mainly used in fracturing fluids [4]. The fact that these products can be easily crosslinked, makes them pre-eminently suitable to carry the proppant into the fractured zones.

Since the first oil crisis, enhanced oil recovery has also become an outlet for polymers. The products which are mostly used for mobility control in flooding operations [5], are polyacrylamide and xanthan gum.

This treatise covers several aspects of the application of hydroxyethylcellulose in nondamaging completion/workover fluids. Subjects treated comprise a.o. thickening efficiency, brine solubility, formation damaging character and temperature stability.

Furthermore, comparisons are made between HEC and other products used in completion/workover fluids.

COMPLETION/WORKOVER

A completion/workover fluid is a fluid which is placed against the oil-bearing formation during operations such as well cleaning (removal of cuttings, mud filter cake, cement etc.), perforation, sand control or repair of subsurface equipment. The basic functions of such a fluid are to remove solids from the well bore and to control formation pressures.

In the past, it was common practice to carry out a completion/workover operation with normal drilling mud. However, as one began to realize that the use of drilling mud caused irrepairable formation damage, the industry started the development of "clean", nondamaging fluids. At the moment, fluids are used which, basically, consist of water, water-soluble salts, acid-soluble solids and polymers for viscosity and fluid loss control.

GENERAL REQUIREMENTS OF COMPLETION/WORKOVER FLUIDS

- **Fluid density**
  In order to prevent uncontrolled flow of formation fluids into the well bore, the completion/workover fluid should have a density high enough to counterbalance the formation pressure.
  Higher densities are not desirable since they may cause penetration of fine solids into the formation pores, which can lead to a reduction in permeability.

- **Solids removal**
  During circulation, the fluid should have enough carrying capacity to remove undesired solids from the well.
  Since the carrying capacity of the circulating fluid is directly related to its annular viscosity, it may be obvious that the cleaning efficiency of the fluid can be improved by addition of a viscosifier. Furthermore, when circulation of the fluid is interrupted, a certain gel strength is needed to keep the solids in suspension.

- **Fluid loss controll**
  Fluid loss to the formation should be as low as possible, because the filtrate can cause considerable swelling of formation clays thereby reducing permeability. Moreover, very fine solids present in the filtrate cause plugging effects.

- **Formation damage**
  As mentioned earlier, invasion of fine, solid particles into the formation pores can cause a decrease in permeability.

In order to avoid this phenomenon, the completion/workover fluid should not contain solids other than those which can be removed from the formation by acid, enzymes, oxidizing agents or by dissolution in the crude oil itself. This means that materials like bentonite (viscosifier/suspending agent) or barite (weighting agent) are not suitable for a nondamaging completion/workover fluid.

- **Temperature stability**
  The compounds present in the completion/workover fluid should be stable at bottom hole temperatures. If this is not the case, circulation of the fluid can cause considerable change in rheological behaviour and fluid loss control.

- Further items which should also be considered in selecting a completion/workover fluid are corrosion inhibition, available mechanical equipment and, of course, economics.

HEC IN COMPLETION/WORKOVER FLUIDS

HEC is a water-soluble polymer which is prepared by reacting cellulose and ethylene oxide in alkaline medium. The chemical structure of the product is given in Fig. 1.

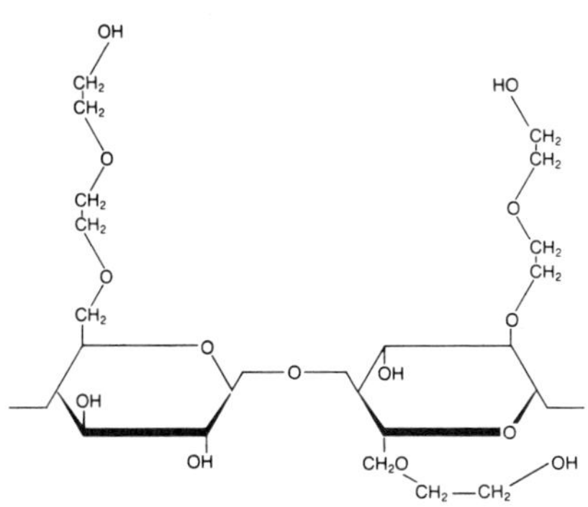

Fig. 1. Chemical structure of hydroxyethylcellulose.

One of the main functions of HEC in a completion/ workover fluid is to provide viscosity and carrying capacity to the fluid. Table 1 lists some data which express the high thickening efficiency of HEC in fresh and salt (NaCl) water.

The thickening efficiency of a viscosifier in fresh or salt (NaCl) water systems can also be expressed by means of the so-called OCMA Yield [6], i.e. the number of cubic meters of fluid with an apparent viscosity of 15 mPa.s that can be prepared from one metric ton (MT) of polymer (OCMA = Oil Companies Materials Association). Table 2 shows that, both in fresh and in salt water, HEC has a

Table 1. Rheological data of HEC in fresh and salt (NaCl) water solutions.

Concentration lb/bbl	Marsh funnel viscosity sec.	Apparent viscosity mPa.s	Plastic viscosity mPa.s	Yield point lb/100 ft²
Fresh water				
0,5	30	7	6	2
1	37	15	10	10
2	83	47	20	54
Salt water (4%)				
0,5	30	7	6	2
1	37	16	11	10
2	85	48	21	54
Saturated salt water				
0,5	34	10	9	2
1	42	21	14	14
2	122	63	25	76

superior thickening efficiency in comparison with other products used in the oil field.

Table 2. Typical OCMA yields of HEC and other water-soluble polymers.

	OCMA Yield (m^3/ton)		
	DW*	SW**	SSW***
HEC	320	320	400
CMC	300	180	220
Guar Gum	290	290	330
Xanthan Gum	150	190	220

* DW = distilled water
** SW = 4% salt (NaCl) water
*** SSW = saturated salt (NaCl) water

The density of a nondamaging completion/workover fluid is normally adjusted by water-soluble salts and/or acid-soluble solids. The most currently used water-soluble salts are sodium chloride (NaCl), potassium chloride (KCl), calcium chloride ($CaCl_2$), calcium bromide ($CaBr_2$) and zinc bromide ($ZnBr_2$). Acid-soluble solids used as weighting agents are calcite ($CaCO_3$), dolomite ($CaCO_3.MgCO_3$), siderite ($FeCO_3$) and iron oxide (Fe_2O_3).
A viscosifier which is used in fluids with high salt concentrations should have a good salt compatibility. HEC is a polymer which can pre-eminently meet this requirement. Because of its nonionic character, HEC can tolerate salt concentrations as high as 70%, e.g. a 16,7 lb/gal brine containing 17% $CaBr_2$ and 52% $ZnBr_2$). The thickening efficiency of HEC in various brine systems is given in Fig. 2.
A comparison of HEC and other products currently used in completion/workover fluids (xanthan, modified starch) points out that HEC is clearly superior in viscosifying high density brines (Fig. 3).

When added as a dry powder to a heavy brine (density > 11 lb/gal), HEC might require some time before reaching full viscosity. This can be avoided by "activating" the polymer before use. Polymer dissolution by means of

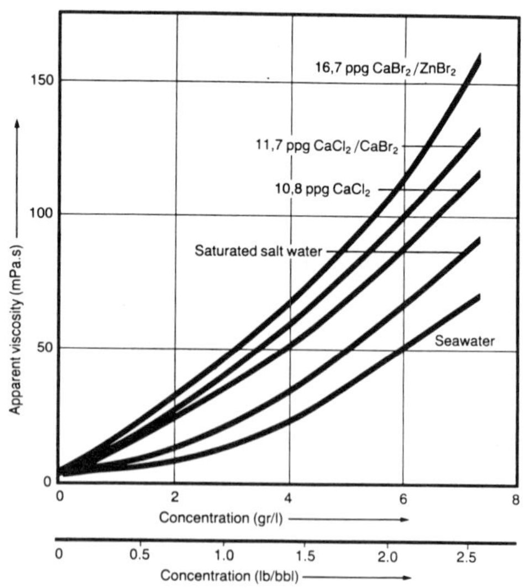

Fig. 2. Apparent viscosity of HEC in different brine systems.

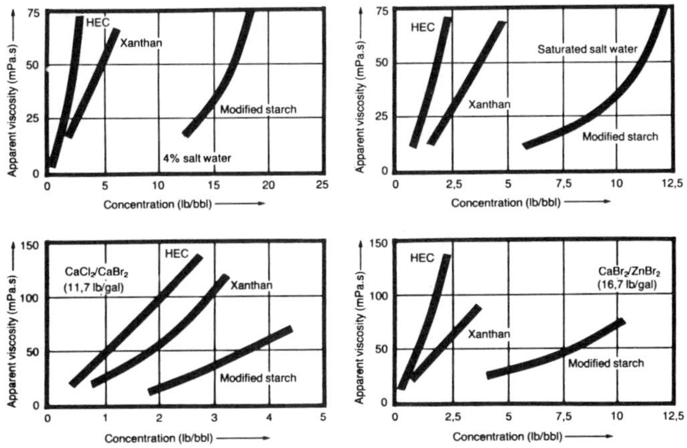

Fig. 3. Thickening efficiency of HEC, xanthan and modified starch in several brine systems.

activation comprises the following steps:

- the required amount of HEC, e.g. 2 lbs, is dispersed under mild stirring into 4 lbs of isopropanol
- to this mixture 4 lbs of water is added and stirring is continued for 10-20 minutes. After this period, the polymer is in a swollen (activated) state
- finally, the complete mixture is added, under stirring, to 1 bbl of brine.

This method results in a rapid viscosity development (Fig. 4) without formation of "fish-eyes".

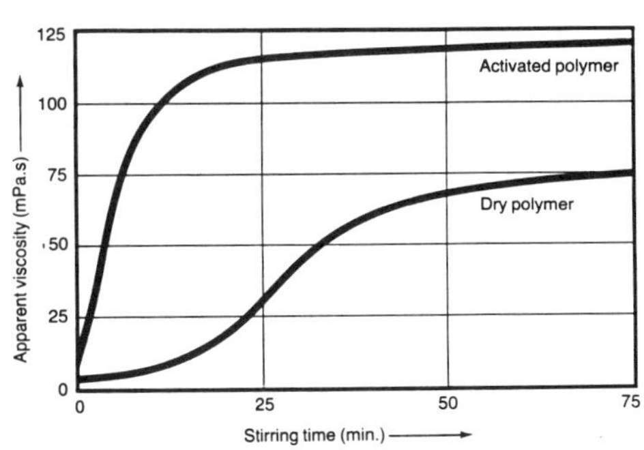

Fig. 4. Viscosity development of HEC in $CaBr_2/ZnBr_2$ brine (17.9 lb/gal).

During a completion/workover operation the viscosifier can cause a significant permeability reduction by penetration into the formation pores. Removal of such damaging materials is usually done by means of 15% hydrochloric acid. In contrast with xanthan, HEC can be degraded completely by 15% hydrochloric acid. At 60°C, for example, HEC-breakdown occurs within 1 hour (Fig. 5). Xanthan is not broken completely, not even after one week at 60°C (Table 3).

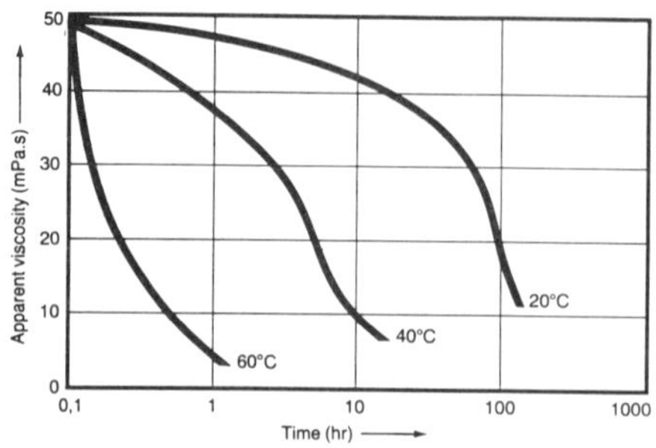

Fig. 5. Breakdown (with 15% HCl) of 2 lb/bbl HEC solutions at different temperatures.

Only by raising the HCl concentration and/or the temperature, the biopolymer can be degraded completely. After acid breakdown, HEC leaves practically no residue (Table 3), whereas xanthan yields a relatively large amount of insolubles, which can be highly formation damaging. Other polymers like CMC, guar and modified starch leave also higher residues than HEC.

Table 3. Polymer residue after breakdown with 15% HCl.

	Breakdown		Acid-insoluble residue (%)
	Temperature (°C)	Time (hr)	
HEC	60	1	0,1
CMC	60	1	3
Xanthan	60	144	8
	80	1,25	2,5
Guar	60	1	2,5 - 5
Modified starch	60	1	1

Fluid loss control in a completion/workover system is normally achieved by means of a combination of a bridging solid and a polymeric viscosifier. The bridging particles prevent flow into the formation by forming a stable, low-permeability bridge at the formation face. The polymer is required to keep the solid particles in suspension and to improve the effectiveness of the sealing bridge by a plastering action. After the completion/ workover operation, the bridge should be removable either by an acid treatment or by the action of the crude oil itself.

Table 4. Fluid loss control of HEC thickened $CaCO_3$ dispersions.

Concentration (lb/bbl)		API Fluid loss (ml)		
HEC	$CaCO_3$	SW[1]	NaSSW[2]	CaSSW[3]
0	6	250	250	250
2	0	250	250	250
2	6	16	11	15

1. SW = seawater
2. NaSSW = seawater saturated with NaCl
3. CaSSW = seawater saturated with $CaCl_2$

A combination of HEC and graded $CaCO_3$ appears to give a good fluid loss control in various salt solutions (Table 4). Again, HEC is preferable over other polymers because of its low acid-insoluble residue.

However, if the $CaCO_3$ particles have to be kept in suspension for extended periods of time, small additions of xanthan gum will enhance the carrying capacity of the fluid. Instead of $CaCO_3$, one can also make use of an oil-soluble resin as a bridging agent. Just like $CaCO_3$, the resin particles will plug the formation pores and prevent the fluid from going into the formation. After the operation, the oil-soluble resin particles can be removed by dissolution in the crude oil. Table 5 shows some figures of the fluid loss control of such a resin in 9,9 lb/gal $CaCl_2$ brine thickened with HEC. Because the resin has a lower specific gravity (approx. 1,0 g/cc) than that of the brine, it may show a tendency to float to the surface of the fluid once circulation has stopped. Raising the yield point of the fluid by means of xanthan can reduce this floating tendency and thereby improve the fluid loss control (Table 5).

Table 5. Fluid loss control of an oil-soluble resin in a 9.9 lb/gal $CaCl_2$ brine thickened with HEC or HEC/Xanthan.

HEC conc. (lb/bbl)	Xanthan conc. (lb/bbl)	Resin conc. (lb/bbl)	Resin particle size (micron)	Marsh funnel viscosity (sec/qt)	API fluid loss (ml)
0,96	–	6	>150	47,5	100
0,96	–	6	75 < d < 150	48	68
0,96	–	6	<75	48	40
0,96	–	10	<75	52	27
0,79	0,26	10	<75	50	16
0,61	0,61	10	<75	50	11

Nowadays, oil wells are drilled to much greater depths than some decades ago. This implies that the compounds which are presently used in completion/workover fluids should be able to endure increasingly high bottom hole temperatures. Just like other polysaccharides, HEC has a limited thermal stability in aqueous solution. However, the stability can be improved significantly by the addition of antioxidants like magnesium oxide (MgO), magnesium chloride ($MgCl_2$), sodium thiosulfate ($Na_2S_2O_3$), thioglycerol ($C_3H_8O_3S$) or combinations thereof. This can be illustrated by means of the following example.

An HEC thickened packer fluid (Table 6) retained only 10% of its initial viscosity after storage for 24 hrs at 210°F. After addition of MgO (0,3 lb/bbl) and thioglycerol (1,14 lb/bbl), no loss of viscosity was observed.

Table 6. Thermal stabilty of HEC in a packer fluid.

Thermal stabilizer	Viscosity (%) retained after 24 hrs at 210° F
–	10
MgO (0,3 lb/bbl)	23
MgO (0,3 lb/bbl) + Thioglycerol (1,14 lb/bbl)	100
$MgCl_2$ (2 lb/bbl) + Thioglycerol (0,38 lb/bbl)	100

Packer fluid formulation: fresh water
$CaCl_2$ (10 lb/bbl)
NaCl (84 lb/bbl)
HEC(1,1 lb/bbl)
NaOH (pH = 9,5)

Excellent results were also obtained when the above mentioned combination of antioxidants was replaced by $MgCl_2$ (2 lb/bbl) and thioglycerol (0,38 lb/bbl).

HEC IN OTHER PETROLEUM APPLICATIONS

In addition to its application in completion/workover fluids, HEC can also be used as thickener in so-called nondamaging drilling fluids. These fluids which are used to drill the payzone, are very similar to completion/workover fluids, i.e. they are not allowed to contain formation damaging solids like bentonite and barite. This means that viscosity, fluid loss control and density of such a fluid should also be achieved by means of acid-degradable materials. An example of a nondamaging drilling fluid is given in Table 7.

Table 7. Example of a nondamaging drilling fluid.

Fresh water	:	1 bbl
HEC	:	1,5 lb/bbl
Xanthan	:	0,25 lb/bbl
Starch	:	2,5 lb/bbl
$CaCO_3$	:	15 lb/bbl
MgO	:	0,3 lb/bbl
Preservative	:	0,5 lb/bbl

Furthermore, HEC can also be applied as fluid loss reducer and rheological modifier in cementing operations [7]. In this application only the low viscosity types are usually applied. The high viscosity types may give a too high consistency and, hence, a poor pumpability of the cement slurry.

Finally, it should be mentioned that high viscosity HEC can also be used as a mobility control agent in flooding operations (E.O.R.).
Usually, polyacrylamides and xanthan gum are preferred in this application. However, the effectiveness of these polymers is significantly impaired by the presence of polyvalent cations [8]. HEC, which is practically insensitive to such ions, is therefore a good alternative when dealing with highly saline formation waters.

REFERENCES

1. G.R. Gray, H.C.H. Darley, "Composition and Properties of Oil Well Drilling Muds" (Fourth Edition) Gulf Publishing Company, Book Division, Houston (1980)

2. J.J. Denny, J.L. Shannon, "New Way to Inhibit Troublesome Shales", World Oil (July 1968), 111-117

3. R.K. Clark, R.F. Scheuerman, H. Raath, H. v. Laar, "Polyacrylamide-Potassium Chloride Mud for Drilling Water Sensitive Shales", paper SPE 5514 presented at the 50th SPE Annual Fall Meeting, Dallas, TX, Sept. 28 - Oct. 1, 1975

4. C.J. Githens, J.W. Burnham, "Chemically Modified Natural Gum for Use in Well Stimulation", Soc. Petr. Eng. J., (Febr. 1977), 5-10

5. V. Martin, J. Klein, "Tertiärförderung von Erdöl - ein neues Einsatzgebiet für Polymere", Erdöl-Erdgas-Zeitschrift, 95 Jg., (Mai 1979), 164-173

6. Oil Companies Materials Association, "Specification No. DFCP-7, Drillling Fluid Materials, "High Viscosity" Carboxymethylcellulose (Revised October 1973)"

7. D.K. Smith, "Cementing, Monograph Vol. 4 of Henry L. Doherty Series" (Second Printing) Society of Petroleum Engineers of AIME, Dallas (1976)

8. E. Unsal, J.L. Duda, E. Klaus, "Comparison of Solution Properties of Mobility Control Polymers", in Chemistry of Oil Recovery (Ed.: R.T. Johansen, R.L. Berg), ACS Symposium Series 91, ACS, Washington, D.C. 1979

Chapter 49

Cellulose derivatives: Post-harvest control of fresh produce

A. Reveley – Courtaulds Chemicals, Spondon, Derby.

ABSTRACT

The recognition that the film forming properties of water soluble cellulose derivatives, combined with suitable surfactants, could give biologically significant surface membranes has given rise to an enormous potential for improving the ripening control of a wide variety of produce, especially fruit and vegetables.

This paper gives an overview of the effect of the membranes, their potential and some suggestions to account for their function. Major benefits are in waste reduction and quality improvement of a range of produce.

The likely areas of interest both large and small crop growing and harvesting organisations particularly in countries with ideal growing climates, for both international and domestic produce trading.

INTRODUCTION

One of the many fascinating aspects of cellulose derivatives is the wide range of uses of which they are placed. This applies especially to the water soluble cellulose derivatives where, over the years, the number of applications has grown considerably.

The film forming properties of cellulose derivatives are well known and are the basis of a number of well developed applications including, for example, pharmaceuticals, tobacco sheet production, pencil lead extrusion, refractories and metal sheet manufacture from powders instead of billets.

Use is also made commercially of these properties in the food industry in general and in such particular applications as fried batters and coatings. Some work was carried out on methods for controlling the maturing rate of stored cheeses and it is from this work that the potential for controlling the atmosphere of individual fruits during the post-harvest ripening period was developed.

The establishment of these ideas, measurements of efficacy and approval for use as food additives, for example by the F&DA in the United States has taken time but a commercial venture is now emerging.

BACKGROUND

Once produce, eg fruit, is harvested its respiration rate is largely governed by the rate at which oxygen is taken in through its permeable surface. The nature of the surfaces varies in type and the effect, for example, of natural coatings like wax. Other factors include the species type, place and regime of growing and harvesting.

Large amounts of produce are lost from point of harvest to consumption due to a combination of circumstances including high temperatures and hence rapid respiration leading to loss of taste and texture, and the effects of disease and damage.

Considerable efforts have been made since the early work in the 1920's at Cambridge University to develop suitable gas atmospheres, in which to store and transport produce to reduce this waste. One of the early workers in this field, Dr West, who died only recently aged 98, played a significant part.

A major contribution to the present knowledge of semi-permeable membrane ripening systems was made by Danny Cutts, especially while head of Research at Geest, a major UK produce marketing company.

PRODUCE MARKETING REQUIREMENTS

1 Approximately 40% of the world's harvest of fresh fruit is wasted between the point of harvest and

consumption and reducing this waste is therefore of considerable interest. Cooling fruit is one way of reducing its respiration rate and retaining a reasonable fresh quality but this poses both cost and organisational restraints. A partner for temperature control was therefore needed.

2 There is an increasing demand for more fresh produce. This is related to the need for users in developed countries to maintain a healthier diet and in developing countries to establish an improved diet.

3 Particularly in developed countries there is a greater emphasis on fruit of better quality, and

4 Wide choice including the more exotic varieties grown in tropical countries and therefore generally requiring long transportation distances and times to the consumption point.

DISCOVERY AND INITIAL DEVELOPMENT

The discovery that films of cellulose derivatives and sucrose esters could affect the chill damage of bananas started a chain of investigation. The easy, year round availability of the banana made this an ideal research tool although commercially the highly refined cropping, distribution and marketing chain was less likely to see the major benefits subsequently found with other fruits.

After initial work on the banana, the system was tried on a wide range of fruits all of which showed some effect and many of which were clearly subjects for further development.

With the help of Cambridge University, the Wolfson Foundation, Geest and the Science Research Council, a specialised laboratory was set up in Cambridge with the object of investigating the structure and function of surface coatings for fruit and vegetables particularly in respect of oxygen access control.

It soon became clear that the original liquid composition used to make the application solutions were not transportable especially to countries with hot climates and suitable dry products were sought. Solutions could then be made easily at the point of application. A number of commercial trials were conducted to see if the laboratory tests could be confirmed on a larger scale and this was shown to be so on a number of produce lines.

PRODUCE CONCEPTS

Because the information accumulated to date was very encouraging, it became important to define more precisely what the products should consist of and what they were intended to do.

The concept was summarised as follows:

- Edible: and approved for food use
- Non-toxic
- Odourless
- Tasteless
- Economic, on both large and small scale
- Give shelf life extension
- Support refrigeration
- Work on a wide variety of produce

In order to establish the appropriate approvals, petitions were made to various regulatory bodies and after a time-consuming procedure they were accepted. These bodies include F&DA in the United States, the EEC, the Japanese Food Organisation, WHO, amongst others.

TECHNICAL

Although the effects on fruit in particular could be quantified and the benefits identified there was little understanding of the bio-physical mechanisms accounting for these effects. A good deal of work has been carried out in this area and useful contributions made to the understanding of the gas exchange routes at fruit surfaces. As a broad description, products can be considered to form semi-permeable membranes but it is important they reduce but do not stop the intake of oxygen and do not damage the intrinsic surface cell structure.

Novel techniques have also been devised for analysing the gas atmospheres generated inside individual fruits.

Another practical conclusion found from larger scale trials is that products inhibit the development of some of the diseases which affect texture, appearance and low yield of some important produce lines. This is particularly significant in the case of pineapples where internal brown spotting can be a major problem with fruit from some growing areas. There is also evidence of this feature on pears and mangoes. Explanations of these characteristics are as yet incomplete, but they probably relate to oxygen availability and delaying the onset of sugar concentrations of a sufficient level to trigger and support growth of attacking organisms.

APPLICATION

Preparation techniques have been devised for a range of different produce lines and the wide variety of mechancial arrangements currently used in their preparation. The dispersions of the products can be applied by drenching and dipping and the dipping procedure can be either direct or through a flotation system, the latter being particularly useful in transporting some fruit in order to reduce damage during preparation. Combinations with other active agents, eg anti scalds and fungicides can be used.

BENEFITS

The value of the benefits which the preparation system offers varies according to the position one takes in the chain 'harvest to consumption'. With some fruits the benefit is mainly to the grower in avoiding waste at the harvest point. In other cases the increase in either transportation time and/or distance can be a major feature in opening up wider volumes or new kinds of produce from well established growing areas. Particularly beneficial is the increased proportion of grade A quality in both taste and appearance which can be offered at the point of consumption and this does fit in with the greater emphasis being laid in many world markets on increased fresh fruit consumption.

The availability of a cheap, approved, low capital system has great possibilites for increasing the output of fresh produce, even locally grown, to improve the diet of the population in low income countries with little distribution and storage infra structure.

CONCLUSION

The contribution made by the film forming properties of edible grades of water soluble cellulose derivatives to this concept for post harvest control and the right discipline can yield very significant benefits in the complete chain from harvest to consumer and therefore has the potential for developing into a major new preparation system for not only fresh fruit but vegetables and other lines for example cut flowers offered dry. New markets and produce lines can be opened up, waste can be reduced and produce quality improved.

A greater understanding of the function of the products will be of great value and interest to academic and applied biologists.

Chapter 50

The rheological properties of plasticised nitro-cellulose as a function of molecular weight

F. S. Baker and G. J. Privett – Royal Ordnance Research & Development Centre, Waltham Abbey, Essex.

M. J. Healey – RARDE, Waltham Abbey, Essex.

INTRODUCTION

The manufacture of nitrocellulose (NC) based propellants involves the mixing of the polymer with solvent and other additives until a pasty mass is formed. This is then extruded to the required shape and dried. Folklore gathered over the years has suggested that the molecular weight of nitrocellulose is an important factor in determining the processing properties of a propellant paste.

This paper presents for the first time, data which shows that molecular weight not only affects the rheological properties of the pasty mass, but influences the mechanical properties of the finished propellant in a subtle manner.

EXPERIMENTAL METHOD

Two samples of NC were used in this investigation. These were made from a cotton precursor and had molecular weights of 167,000 and 550,000.

The molecular weights were calculated from intrinsic viscosity measurements of the NC dissolved in tetrahydrofuran (THF), (ref 1).

The samples were processed to give similar fibre lengths, and fibre length distribution. These were measured on a prototype machine developed by the Paper Packing Research Association.

To each sample was added an equal mass of nitroglycerine and 1% of diethyl diphenyl urea as a stabiliser. The resultant material although fibrous is called paste in the propellants trade.

The paste was incorporated in a highly instrumented internal mixer using contra-rotating blades (speed ratio 1:1.95) with a separation (nip) of 1mm between the blades and the mixer wall (ref 2). The dry ingredients were mixed with acetone to form a gelatinised dough-like material. The dough temperature was maintained at 20°C using Churchill heaters and a thermocouple. The energy input per kilogram (work unit) to the dough was monitored using a torque transducer mounted on the drive shaft of the mixer and integrating under the resultant torque versus time curve using a microcomputer.

Extensive preliminary trials in which solvent levels were varied between 50 pph and 15 pph (pph is the parts per hundred of dry weight of material) were undertaken with both NC types. These showed that doughs containing negligible quantities of ungelatinised material could be made at a solvent concentration of 25 pph. To avoid build-up of dry or partly gelatinised ingredients on the following edge of the blades, it was necessary to reverse the blade direction for a short period twice during incorporation. The work input to the dough was controlled to 2 MJ/kg. In all cases this took approximately 2 hours.

The mixer loading density used (proportion of mixer free volume occupied by ingredients) was 0.8, a value shown to produce highly efficient mixing in mixers of this kind (ref 3,4).

After mixing, the resultant pasty mass was transferred to a capillary extrusion rheometer specially designed for handling hazardous materials (ref 5). Measurements suggested that the transfer operation did not result in any significant loss in solvent from the dough.

The dough was extruded through a 2.02 x 20 mm die over a range of volumetric flow rates. The extrusion pressure, extrudate die swell and extrudate surface temperature were measured. The shear stress and strain rates observed during extrusion were calculated and used to give an apparent flow curve. From this the apparent rheological properties of the solvent-wet material were determined (ref 6,7).

During extrusion trials it was found that seemingly identical incorporations did not lead to identical extrusion properties. In an effort to identify the source of the variation the NC was dried over silica gel for 2 weeks. This was found to change the mean value of the observed rheological parameters. Similarly, changing from an industrial grade acetone to analar acetone (dried using a molecular sieve) altered the values derived from the data. It must be assumed that during extrusion the water present within the ingredients influences the flow of extrudate within the die, possibly by contributing to the wall slip phenomena (ref 8). This would distort the flow curve derived. Unless stated otherwise, the results herein are those found using dried ingredients.

The extrudate was cut into 150 mm lengths and dried in an air oven for two weeks at a temperature of 45°C. The strain rate at which each stick was extruded was noted for later reference.

The sticks were conditioned at 20°C and 55% relative humidity for a period of 2 weeks prior to the tensiometer measurements. These were carried out at ambient temperature (approx 23°C) using an Instron tensiometer operating at a jaw separation rate of 50 mm/min.

Measurements of the dynamic mechanical properties of the sticks were undertaken using a Polymer Laboratories DMTA system.

RESULTS

Dough Rheology

Doughs made from both NC types were successfully extruded. It was found that the cords had smooth surfaces for all apparent shear rates lower than 500 s^{-1}. At higher shear rates it was found that the surface of some cords was rough and uneven.

Analysis of the apparent shear rate/shear stress curves (fig 1) showed that the doughs all behaved as Herschel-Bulkley fluids ie they obeyed the equation:

$$\sigma - \sigma_o = k\,\dot{\gamma}^n$$

Where σ_o is the yield stress, σ is the applied stress, k is the consistency index, n the flow behaviour index and $\dot{\gamma}$ the shear rate.

It was found that the σ_o and n values of the doughs were significantly different from each other (Table 1). This was true for both the dried and undried ingredients. The k values were not found to differ for different types of NC but were found to differ significantly between dried and undried ingredients.

Table 1. Extrusion parameters.

Mol weight	dry/wet	σ_o	k	n
167,000	wet	55.1	14.5	0.46
554,000	wet	59.7	15.6	0.66
167,000	dry	62.5	24.3	0.43
554,000	dry	47.2	35.7	0.60

Scale of Error

	σ_o	k	n
Error	8	7	.05

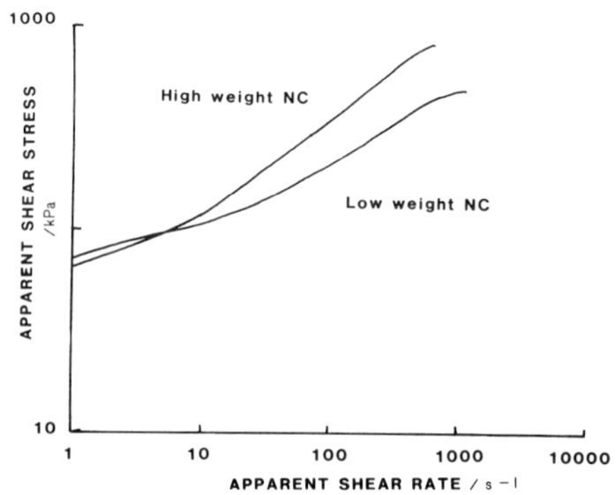

Fig. 1.

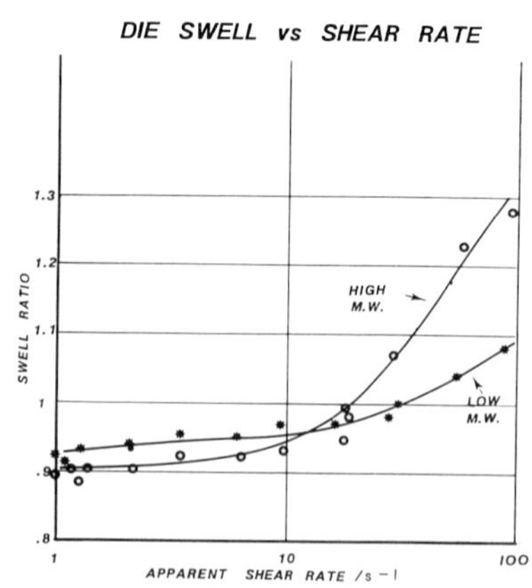

Fig. 2.

The k values for dried ingredients were twice those measured with "wet" ingredients.

At high extrusion rates (> 500 s^{-1}) the rate of increase of the shear stress with extrusion rate decreased. The infra red pyrometer showed that this was due to shear heating effects occurring within the die (ref 9). These reduce the effective viscosity of the dough, thereby lowering the pressure required to extrude the dough at a given flow rate. To ensure that this did not influence the analysis of the rheological parameters, data points at shear rates greater than 15 s^{-1} were excluded.

Some of the shear rate, shear stress curves were slightly s-shaped. This suggested that at moderate shear rates there was some wall-slip taking place, even for dried ingredients. This has been observed in a number of propellant formulations, as has shear heating. The presence of wall slip made accurate analysis difficult, and may account for much of the variation in the parameters determined during this work.

Die swell

The line scan camera showed that all doughs exhibited the Barus effect and thereby displayed evidence of elasticity. The amount of swell was found to be related to the NC molecular weight and the shear rate (fig 2). It was seen that swell increased most rapidly at shear rates above 8 s^{-1}.

By measuring the diameter of the dried cords, it could be seen that the dieswell measured during extrusion was reflected in the diameters of the finished dried propellant (fig 3).

Shear heating

The infra-red pyrometer showed that the surface temperature of the extrudate was related to the rate of extrusion (fig 4). At very low extrusion rates (< 1 s^{-1}) the surface temperature was found to be lower than that of the unextruded dough. This probably results from the evaporation of solvent from the surface. At extrusion rates in the range 1-8 s^{-1} there was no appreciable rise in surface temperature. Above this the skin temperature rose sharply in all cases. At high extrusion rates, approaching 1000 s^{-1}, comparable to those found in many production facilities, the skin temperature was sometimes 60°C or higher. At such a temperature the solvent would boil off from the extrudate surface. This may account for the rough surface found for some high speed extrusions.

The ultimate tensile strength of the extruded cords were all around 32 MPa. It was found that the strength of both types of cord dropped as the extrusion rate increased (fig 5). The drop is of the order of 15% at a shear rate of $100s^{-1}$.

Dynamic mechanical analysis

The dynamic mechanical spectra showed that two processes were present within the temperature range investigated. These correspond with the alpha and beta peaks described elsewhere (ref 10). The alpha process is known to be related to processing conditions, whereas the beta process is not affected. For this reason only the alpha process was studied. It was found that it's position and amplitude were influenced by the molecular weight of the NC present and the rate of extrusion.

As the rate of extrusion was increased the peak temperature was seen to drop (fig 6). This may result from differing degrees of fibre alignment caused by the passage of the dough through the die.

DISCUSSION

It is clear from the data that the molecular weight of the nitrocellulose has a strong influence on the rheological properties of the solvent wet dough.

Cellulose morphology is complex, and this complexity is transmitted to the nitrate ester as with other esters (ref 11). Evidence is gathering that there is much order and structure in the NC and the NC/NG system (ref 12, 13).

The greater the molecular weight, the longer the chains and fibrils which are known to exist in cellulose. It is not unexpected therefore to note that the higher molecular weight material exhibited more die swell at higher shear rates, since there would be considerably more elastic energy stored in the system as longer fibres are forced to align in the die.

CONCLUSION

It has been shown that the molecular weight of NC effects the processing of a propellant. In general the higher molecular weight gives a stiffer dough at production level extrusion rates. The molecular weight of the NC influences the physical properties of the dried propellant markedly.

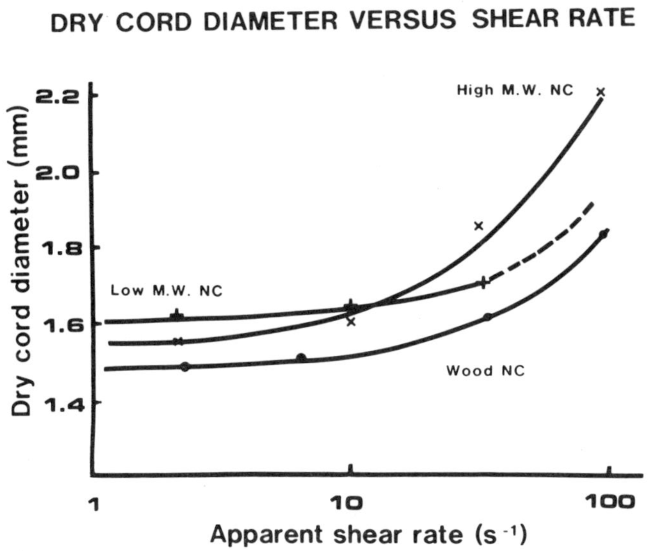
DRY CORD DIAMETER VERSUS SHEAR RATE
High M.W. NC
Low M.W. NC
Wood NC
Dry cord diameter (mm)
2.2
2.0
1.8
1.6
1.4
1
10
100
Apparent shear rate (s⁻¹)

Fig. 3.

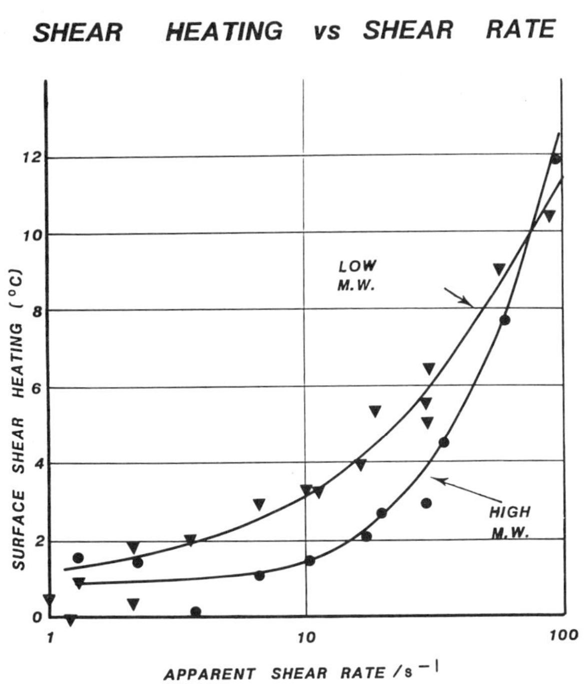
SHEAR HEATING vs SHEAR RATE
LOW M.W.
HIGH M.W.
SURFACE SHEAR HEATING (°C)
12
10
8
6
4
2
0
1
10
100
APPARENT SHEAR RATE /s⁻¹

Fig. 4.

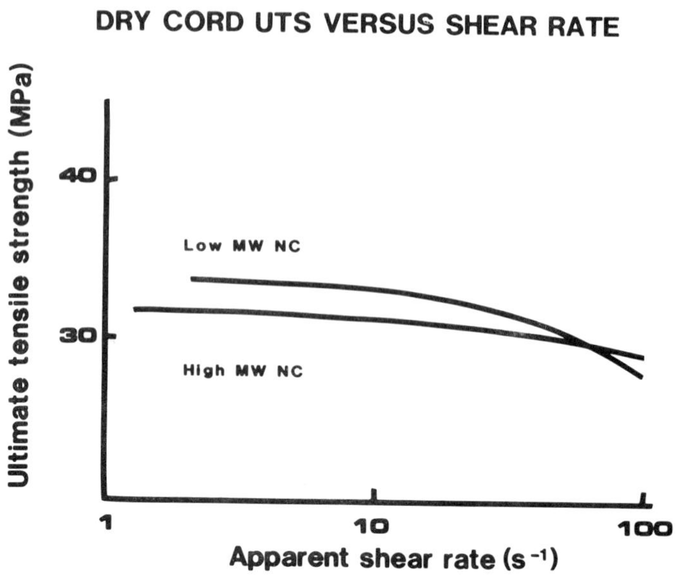
DRY CORD UTS VERSUS SHEAR RATE
Ultimate tensile strength (MPa)
40
30
Low MW NC
High MW NC
1
10
100
Apparent shear rate (s^{-1})

Fig. 5.

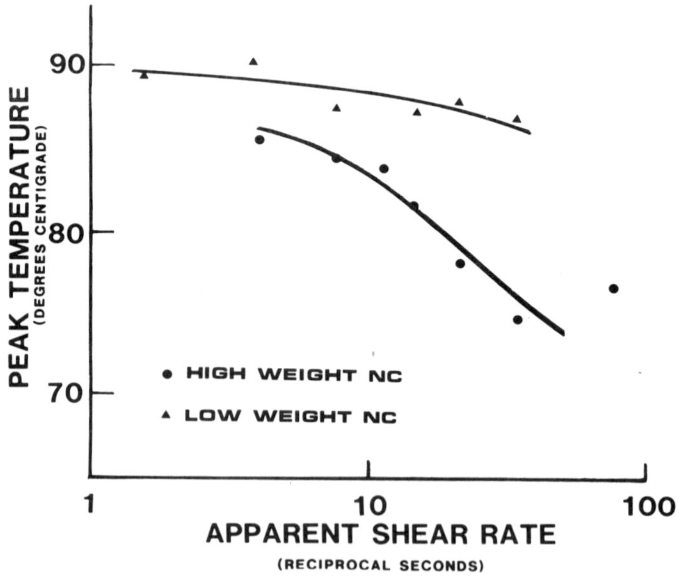
PEAK TEMPERATURE
(DEGREES CENTIGRADE)
90
80
70
• HIGH WEIGHT NC
▲ LOW WEIGHT NC
1
10
100
APPARENT SHEAR RATE
(RECIPROCAL SECONDS)

Fig. 6.

ACKNOWLEDGEMENTS

The authors wish to thank the Procurement Executive, M.O.D. for supporting this work.

REFERENCES

1. Cosgrove J D, Head B C, Lewis T J, Graham S G, Warwicker J O, A GPC Study of Cellulose Degradation, Cellulose and its Derivatives, chpt 11, Ellis Horwood, London, 1985, pp143-152.
2. Baker F S, The Processing Properties of Double- and Triple-Base Propellant Solvent Doughs Part 1, PERME TR 119, HMSO 1979.
3. Freakley P, Van Idris, Rubber Chem Techn, 52, 134-145 1979.
4. Baker F S, Carter R E, Processing Characteristics of Gun Propellants, Prop Exp, Pyro, 7, 139-147, 1982.
5. Carter R E, The Processing Properties of Double- and Triple-Base Propellant Solvent Doughs Part 2, PERME TR 137, HMSO 1979.
6. Baker F S, Griffiths T, Hall P, Privett G J, The Rheology of Pyrotechnic Mixtures, Proceedings 16th Int Pyro Con, Karlsruhe, 1985, ICT, W Germany.
7. Tadmor Z, Gogos C G, Principles of Polymer Processing, J Wiley & Sons, New York 1979.
8. Brydson J A, Flow Properties of Polymer Melts, Chpt 3, George Godwin, UK, 1984.
9. Carter R E, Warren R C, Extrusion Stresses, Extrudate Swell, Viscous Heating effects in Double-Base Propellants, to be published.
10. Baker F S, Gledhill R A, Jones M, Lewis T J, Privett G and Warren R C, Relaxation Behaviour of Nitrocellulose-Nitroglycerine Systems, Cellulose and its Derivatives, Chpt 22, Ellis Horwood, London 1985, 263-271.
11. Doyle S E J, The Physical Characteristics and Dielectric Behaviour of Cellulose Acetate Polymers, thesis, Univ of Strathclyde, UK, 1985.
12. Isles M, Jennings B R, B Polym J, March 1976.
13. Warren R C, The Effect of Liquid Crystal Structure on the Rheological Properties of Nitrocellulose-Dimethylacetamide Solutions, Rheol Acta 23, 541-547, 1984.

Chapter 51

Progress in physical chemistry of cellulose and its application to biomedical membranes

K. Kamide – Fundamental Research Laboratory of Fibers and Fiberforming Polymers, Asahi Chemical Ind. Co. Ltd. 11-7 Hacchonawate, Takatsuki, Osaka 569, Japan.

INTRODUCTION

Studies on the molecular characterization of cellulose and cellulose derivatives (CD) until recently remained at a very primitive stage due to experimental difficulty. The experimental data, unsystematically obtained for some CD solutions, were inconsistent with solution theories applicable to many synthetic polymers. For this reason, as late as the 1970s, the molecular characterization of cellulose and CD was one of the unsolved problems in macromolecular science. This has motivated our research in this area over the last twenty years (1-31).

DISTRIBUTION OF DEGREE OF SUBSTITUTION

Information on the distribution of the average degree of substitution of j-th polymer in the sample ($g(\langle F^j \rangle)$) and the weight-average degree of the substitution of overall glucopyranose units of all molecules $\langle\langle F \rangle\rangle$ were obtained for cellulose acetate (CA) and cellulose nitrate (CN) (2). The average probability of substitution of hydroxyl groups attached to the C_k position $\langle\langle f_k \rangle\rangle$ of CA with $\langle\langle F \rangle\rangle = 0.49 - 2.92$, sodium cellulose sulfate (NaCS) and sodium carboxymethyl cellulose (NaCMC) was determined precisely by NMR (1). The sum $\langle\langle f_2 \rangle\rangle + \langle\langle f_3 \rangle\rangle$ is a predominant factor for the anti-coagulant activity χ of NaCS (3) and the absorbency A' of NaCMC for water is unambigously determined by $\langle\langle f_6 \rangle\rangle$ (4) (See, Fig. 1 and 2, respectively). For NaCMC, detailed investigation of the NMR spectrum of the C_6 carbon peak region revealed that 1) the hydroxyl group at the

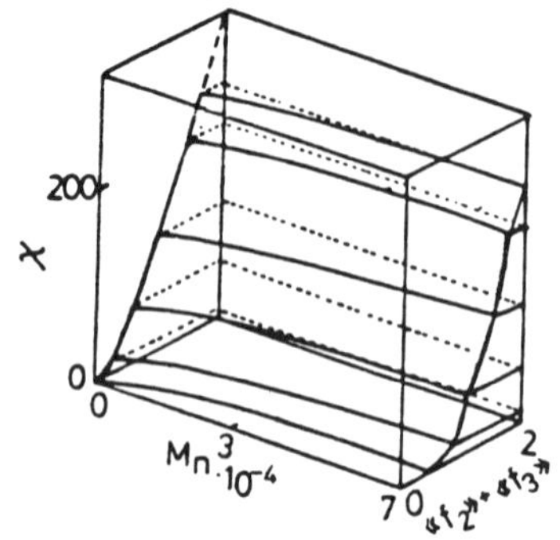

Fig. 1. Anticoagulant activity X against number-average molecular weight M_n and the sum of the substitution degree of the substitution degrees, $\langle\langle f_2 \rangle\rangle + \langle\langle f_3 \rangle\rangle$.

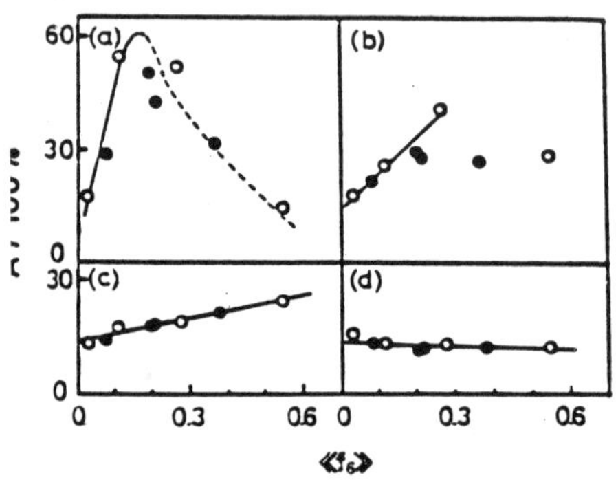

Fig. 2. Absorbency A' of Na CMC towards various liquids as a function of $\langle\langle f_6 \rangle\rangle$; a, H_2O; b, 0.9wt%NaCl; c, 0.9%$CaCl_2$; d, 0.9%aq.$AlCl_3$.

C_6 position for cellulose II participate in two types of intramolecular hydrogen bonds (O_2-H···O_6' and O_6-H···O_2', the latter being much more acidic (Fig.3)),

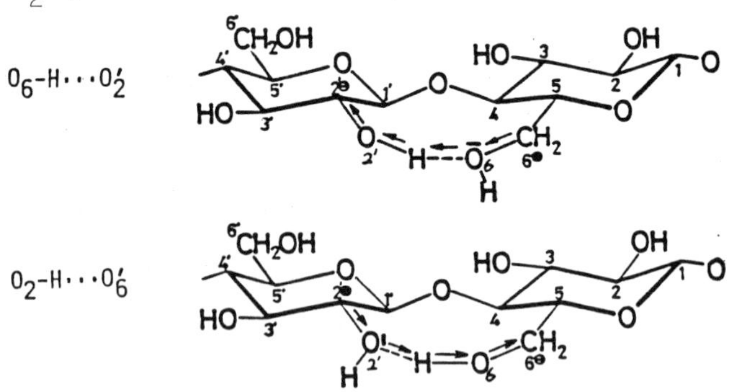

Fig. 3. Schematic representation of intramolecular hydrogen bonds in cellulose molecules. → indicates the direction of movement of electron. +, NMR peak shifts to lower magnetic field; –, NMR peak shifts to higher magnetic field.

2) for cellulose I O_6-H...O_2' type intramolecular hydrogen bonds do not occur and 3) electron density on C_6 carbon for cellulose II in both ordered (crystalline) and amorphous regions is higher than that for cellulose I, and the reverse is true for the electron density on C_2 carbon. These facts suggest that the reactivity of the hydroxyl group at the C_6 position of cellulose II is relatively higher than that of cellulose I and in turn that at C_2 position is reversed.

FRACTIONATION AND DETERMINATION OF MOLECULAR WEIGHTS

Cellulose derivatives are extremely difficult to prepare with reasonably narrow molecular weight distribution. For example, the numerous attempts made to fractionate CA(2.92) (numbers in parentheses denotes the <<F>>) so far have met with very limited success. The successive solution fractionation (SSF) method, originally advocated by us (8), was applied, and a series of fractions with relatively narrow molecular weight distribution were isolated successfully (8, 12) (the ratio of the weight- to the number-average molecular weight M_w/M_n of CA(2.92-0.49) is represented in Fig 4).

Experimental difficulty was encountered

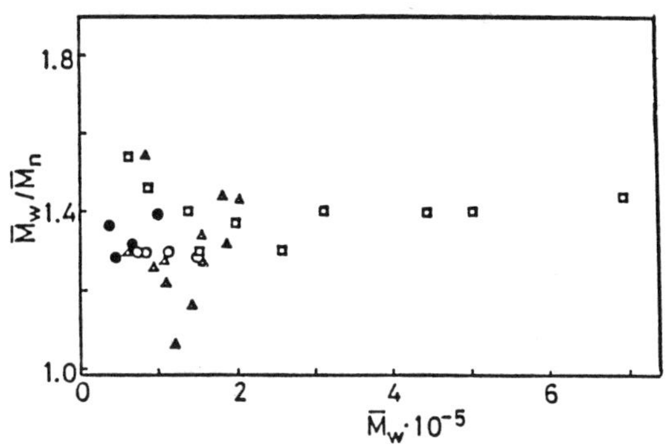

Fig. 4. Plot of M_w/M_n against M_w for CA fractions with <<F>> = 0.49 (○); 1.75 (○); 2.46 (△); and 2.92 (□), isolated by SSF method.

with light scattering (LS) measurements of CD even in a good solvent, owing to contamination by gel-like materials in the solutions. By manipulation of the sample preparation and refinement of the solutions, LS measurements on CD were made possible (5,7). In many solvents, CA molecules dissolve molecularly, and the same values of M_w for a given CA sample were obtained in various solvents (8-10).

CRITICAL PHENOMENA

Lower and upper critical solution temperatures (θ_L, θ_U) were determined (14,15); θ_U=310, θ_L=433 K for CA (2.46)/2-butanone (Fig.5).

Fig.6 shows a schematic representation of the mechanism of pore forming in the solvent casting process. Depending on the initial polymer concentration the polymer-rich phase or the polymer-lean phase separates initially from the solution. The mechanism disclosed here has a very generalized character and affords us a powerful method for controlling the average pore size and the pore size distibution of the polymeric membrane including cellulose and CA membrane (17).

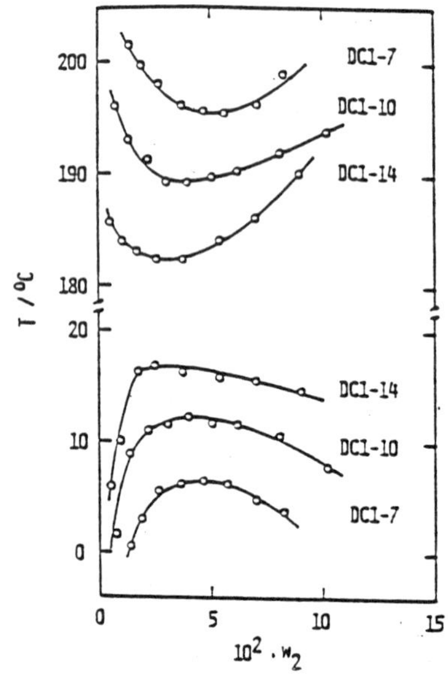

Fig. 5. Phase diagram for CDA fraction-2-butanone system, showing LCST and USCT for the fractions of indicated sample code.

THERMODYNAMICAL PROPERTIES

Linear expansion factor α_S ($=\langle S^2\rangle^{\frac{1}{2}}/\langle S^2\rangle_o^{\frac{1}{2}}$, $\langle S^2\rangle^{\frac{1}{2}}$ is the radius of gyration and o denotes the unperturbed state) of cellulose and CD solutions was determined by the penetration function

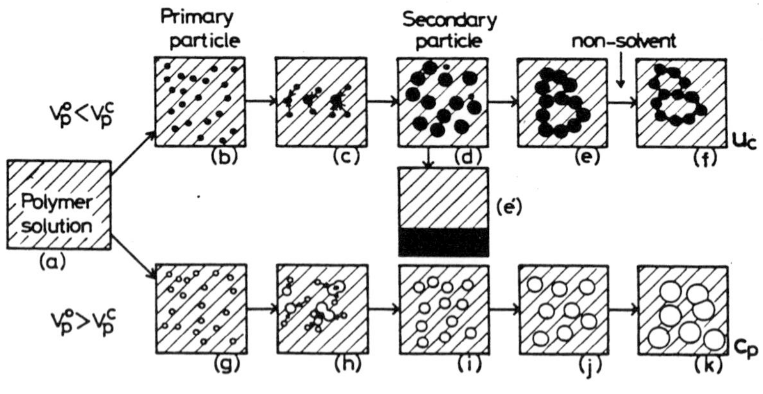

Fig. 6. Schematic representation of the formation of pores in the casting process.

ψ ($\equiv A_2M^2/4\ N_A\langle S^2\rangle^{3/2}$, M, molecular weight, N_A, Avogadro number). α_s values thus determined are not much larger than unity (Fig. 7), suggesting that the solvents of cellulose and CD are (configurationally) poor (19). The A_2 values are smaller than 1×10^{-3} (cm^3

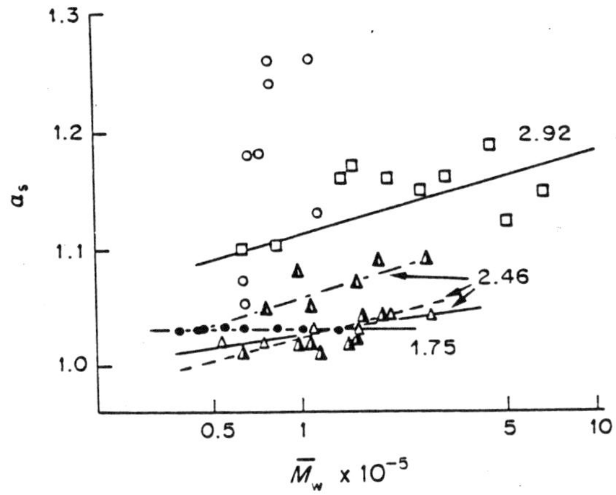

Fig. 7. Linear expansion factor α_s of cellulose acetate-solvent systems plotted as a function of M_w. ○, CA(0.49)-DMAc; ●, CA(1.75)-DMAc; △, CA(2.46)-DMAc; ▲, CA (2.46)-acetone; ▲, CA(2.46)-THF; □, CA(2.92)-DMAc.

$\cdot mol\cdot g^{-2}$). The specific interaction other than the van der Waals forces between the functional groups such as O-acetyl or hydroxyl groups in a CA molecule and the solvent molecules was proved to exist with NMR and adiabatic compressibility measurements, independently (20,21). Possible molecular model of interaction of CA(2.92) with various solvents was demonstrated in Fig. 8.

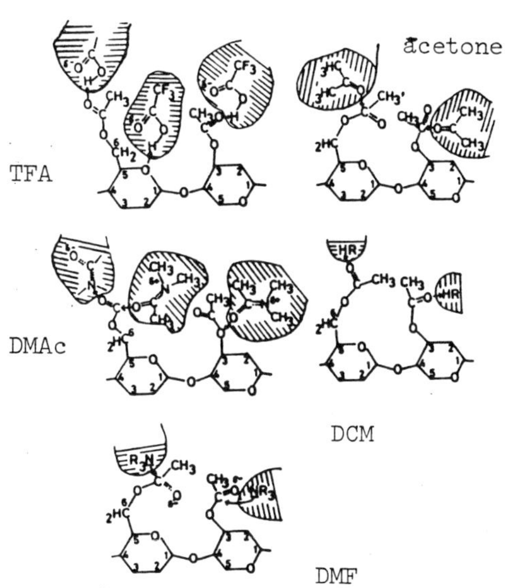

Fig. 8. Possible molecular model of interaction of CA (DS = 2.92) with various solvent employed in this study.

HYDRODYNAMICAL PROPERTIES

The slope of the log-log plot of limiting viscosity number $[\eta]$ against M, i.e.the exponent a in the Mark-Houwink-Sakurada (MHS) equation

$$[\eta] = K_m M^a \tag{1}$$

of cellulose and CD solutions was found to be 0.8 - 1.0. It should be noticed that the MHS parameters, reported in the literature for CA, were determined for polymer fractions, whose molecular weight distribution is wide, on the basis of M_n. The a values in MHS equation obtained for M_w of fractions with reasonably narrow molecular weight distribution are smaller than 0.8 (except for NaCS-0.5M NaCl aq.soln.) (8,9,12,22) and no peculia ity for the a values was observed for cellulose and CD solutions. The molecular weight dependence of the sedimentation constant s_o of CA(2.46) and CA(2.92) was also determined (23). The negative temperature coefficient of $[\eta]$ has long been discussed. Although all deductions were made from indirect measurements of their molecular dimensions. For CA solutions, negative temperature dependence of $[\eta]$ can be ascribed to the rapid decrease in $\langle S^2\rangle_o$ (24). The unperturbed chain dimension A and the number of the solvated solvent molecules per repeating unit of CA at infinite dilution s_o' of CA(2.46) in acetone decrease monotonically with rising temperature (Fig. 9). d $[\eta]$/d T can be finally interpreted in terms of the temperature dependence of solvation (21).

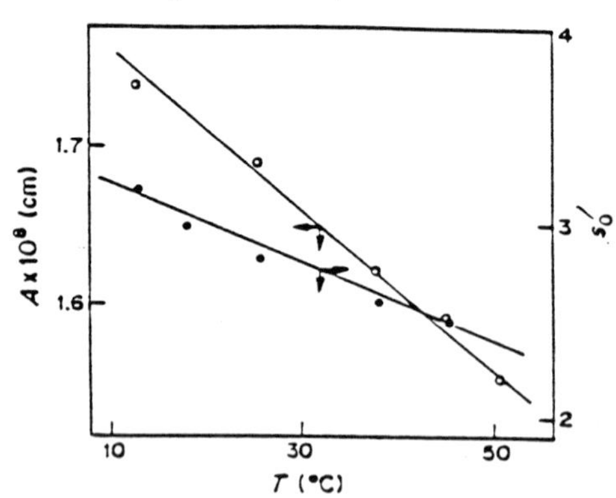

Fig. 9. Temperature dependence of unperturbed chain dimension A and number of the solvated solvent molecules at infinite dilution of CA(2.46) in acetone at 25 C. Open mark, A; closed mark, s_o'.

UNPERTURBED CHAIN DIMENSION

There were two conflicting theoretical views concerning the role of the partially free draining effect was demonstrated by the following facts: (a) the Flory viscosity parameter Φ significantly depends on the molecular weight (8-10,12,19,22),(b) the correlation of the values ($\gtrsim$ <5) for the draining parameter χ estimated by various methods was good, (c)for CA solution, the hydrodynamic diameter estimated using X agrees well with those by the first-order perturbation theory for the volume effect using θ and α_s(10). The free draining effect and non-Gaussian nature of cellulose and CD make us unable to determine A using so-called two parameter theory whose validity was widely confirmed for numerous vinyl type polymers(19). Many methods for estimating A were proposed (19,25-29). For the cellulose and CD, it was confirmed that A estimated by the methods considering the partial drainage and non-Gaussian nature is always larger than that estimated by two parameter theory (29). Judging from A values, cellulose and CD belong to the category of semi-flexible polymers and the flexibility of the cellulose chain varies to a large extent on the polarity of the solvents (19,21,29). In DMAc, A of CA molecule becomes maximum at <<F>>≅2.5. The A value of CA in hypothetical nonpolar solvent (dielectric const. ε=1) is relatively small, increasing monotonically with the increase of <<F>> (chain line in Fig.10), owing to the increase of steric hindrance of O-acetyl groups in CA molecules (21). Extrapolating the curve of A (ε=1) to <<F>> = 0, the A value of cellulose chain agrees with that in the free rotational state (closed rectangle in Fig.10). Therefore, it is concluded cellulose itself is a flexible chain.

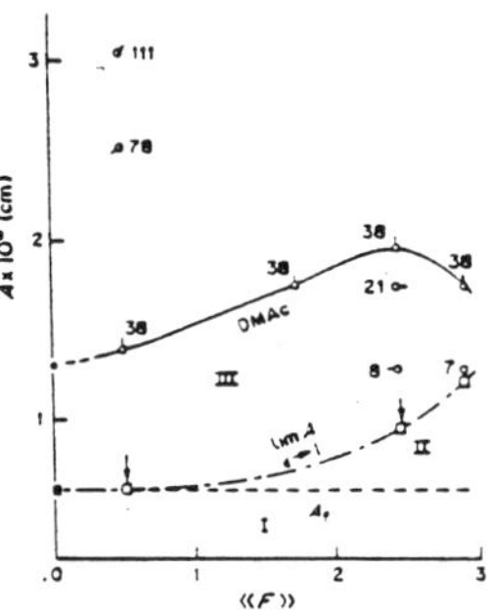

Fig. 10. Plot of A vs. <<F>> for CA-solvent systems. –, DMAc, -.-, A at ε = 1, ■, A at the limit of ε = 1 and <<F>> = 0. Number attached to the marks denotes ε.

REFERENCES

1)Polym.J.13,127('81),2)Cell.Chem.Tech.16,601('82), 3)Polym.J.15,309('83),4) ibid,17,909('85),5)Sen-i Gakkaishi,30,T-464('74), 6)ibid,31,T-410('75),7) Makromol.Chem.168,173('73),8)Polym.J.12,521('80),9)ibid,13,421('81),10) ibid,14,517('82),11)Sen-i Gakkaishi,34,T-144('81),12)Polym.J.11 285('79),13)Brit.Polym.J.15,91('83),14)Netsusokutei,8,67('81), 15)Eur.Polym.J.18,123('82),16)Brit.Polym.J.14,23('82),17)ACS Symp.Ser.269,1985,p197,18)Polym.J.16,825('84),19)ibid,10,409('78),20)ibid,13,115('81),21)Eur.Polym.J.20,903('84),22)Polym.J.15, 249('83),23)Makromol.Chem.183,3075('82),24)Eur.Polym.J.16,703('80),25)J.Polym.Sci.Polym.Lett.2,809('64),26)ibid,2,1029('64),27)Polym.J.10,539('78),28)Eur.Polym.J.17,1049('81),29)ibid,18,661 ('82),30)Polym.J.15,213('83),31)Makromol.Chem.Rapid,4,33('83).

Chapter 52

Membranes from cellulose and its derivatives

Wolfgang Pusch – Max-Planck-Institut für Biophysik, Frankfurt am Main.

ABSTRACT

The preparation, structure, and performance of typical membranes cast from cellulose and its derivatives such as cellulose nitrate (CN), cellulose acetate (CA), and blends of CA and CN are discussed.

INTRODUCTION

Cellulose and its derivatives (e.g. cellulose ethers and esters) are excellent film forming materials. For this reason, they have been used for membrane preparation since the beginning of membrane research. Ultrafiltration membranes, for instance, were cast from cellulose nitrate, and Collander [1] was the first who described collodion membranes which are cast from solutions of cellulose nitrate in mixtures of ether and ethanol.

TYPICAL CASTING SOLUTIONS

Table 1 lists typical casting solutions of hemodialysis, hemo-, micro-, ultra-, and hyperfiltration (RO) membranes cast from cellulose, CN, and CA, respectively. In addition, typical pore diameters and typical hydrodynamic permeabilities of the membranes are reported. Cellulose membranes can be prepared as Cuprophan and Cellophan membranes. Cuprophan membranes are obtained by dissolving celllulose as tetraamine copper complex in an ammoniac copper salt solution whereas Cellophan membranes result from solutions of cellulose in carbon disulfide (xanthogenate). In both cases, the cellulose is then regenerated from the corresponding casting solution by spinning the membrane in-

Table 1. Typical commercially available membranes from cellulose and its derivatives.

Membranes	Cellulose		Cellulose Nitrate	
	Cuprophan	Cellophan	Microfilter	Ultrafilter
Typical Casting Solutions	Cellulose dissolved in: Cuprammonium (Tetraammincopper complex)	Carbonsulfide (Xanthogenate)	Mixtures of Methylacetate or Acetone with Glycerine, Alcohols, Surfactants, and H_2O	Mixtures of Ether/Ethanol
Typical Pore Diameters	less than 50 Å		0.02 - 10 μm	0.003 - 0.03 μm
Typical Permeabilities (cm/s · atm)	hemodialysis: $1 - 10 \cdot 10^{-5}$ hemofiltration: $1 - 50 \cdot 10^{-4}$		$4 \cdot 10^{-3} - 15$	$1 - 100 \cdot 10^{-4}$

Membranes	Cellulose Acetate		
	Microfilter	Ultrafilter	RO Membranes
Typical Casting Solutions	Mixtures of Dichlormethane with Alcohols, Surfactants and H_2O	Acetone/Formamide Mixtures of Methylacetate with Acetic Acid and H_2O	Mixtures of Acetone, DMS, or Dioxan with Formamide
Typical Pore Diameters	0.02 - 10 μm	0.003 - 0.03 μm	less than 30 Å
Typical Permeabilities (cm/s · atm)	$3 \cdot 10^{-3} - 15$	$1 - 100 \cdot 10^{-3}$	less than $5 \cdot 10^{-5}$

to a sulfuric acid bath. Micro- and ultrafiltration membranes of CN are obtained from casting solutions of CN dissolved in mixtures of methylacetate or acetone with glycerine (microfilters) and mixtures of ether and ethanol (ultrafilters), respectively. CA can be cast into micro-, ultra-, and hyperfiltration membranes. Hyperfiltration membranes of CA possess a very thin layer (active layer) of about 1,000 up to 3,000 Å in thickness on top of a porous matrix possessing a gradient of porosity across the membrane.

TYPICAL ELECTRON MICROGRAPHS OF SYNTHETIC MEMBRANES

Figures 1a-d exhibit cross sections of cuprophan and cellophan membranes, respectively. The pictures at the left hand side show the same membrane as the ones at the right hand side. The only difference is the kind of water removal from the corresponding membrane samples. In one case (Fig. 1a+c), the water was removed by freeze drying, whereas in the second case (Fig. 1b+d), the water was removed by the so-called critical point method. With the critical point method, the water of the membrane is gradually exchanged for an organic solvent (e.g. acetone in case of cellulose) till the membrane contains 100% acetone. Then, the acetone is removed from the membrane by evaporation. As is obvious from Figures 1a-d, the membranes are not completely homogeneous but rather possess surface layers clearly differing in structure from the internal homogeneous membrane matrix. Electron micrographs of cellulose membranes taken at large magnifications (⩾ x 50,000) indicate pores in the inner membrane with diameters in the range of 40 to 60 Å whereas the surface layers do not show any pores even at very large magnification. Cellulose membranes apparently consist of a fine porous matrix located between two dense (non-porous) surface layers of about 1,000 to 3,000 Å in thickness.

Figure 2 exhibits scanning electron micrographs of the top and bottom surfaces of coarse porous membranes manufactured by Gelman Company from cellulose and CN, respectively. During the precipitation step of the preparation of coarse porous membranes, a phase separation takes place where a polymer-rich phase seperates from a polymer-poor phase. Pore formation occurs during this phase separation step. The average pore diameter, the pore size distribution, and the porosity gradient across the membranes depend on the casting solution composition, the casting procedure, and the kinetics of phase separation [2].

Figure 3 shows a transmission electron micrograph of the top and bottom surface of an asymmetric CA membrane. The top surface does not exhibit pores under the electron microscope meaning that the diameter of possibly existing pores is smaller than about 50 Å corresponding to the resolution of common electron microscopic pictures. The bottom surface possesses pores with diameters of the order of 0.1 μm. A cross section of an asymmetric CA membrane shows a porosity gradient across the porous matrix of the membrane. The transport properties of the membrane are essentially determined by the active layer.

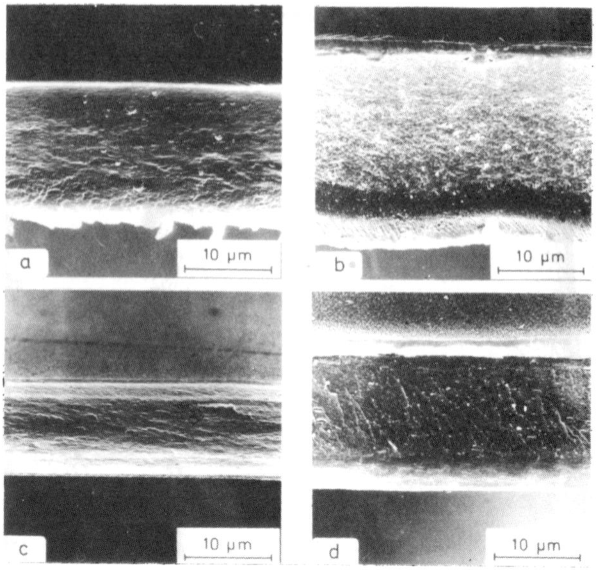

Fig. 1 a – d. SEMs of cross-sections of cuprophan (a, b) and cellophan (c, d) membranes. The water of the membrane samples was removed by freeze drying (a, c) and by the critical point method (b, d) [2].

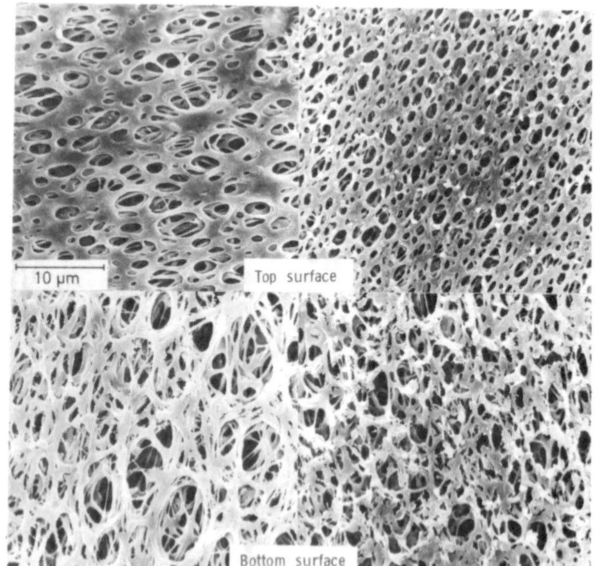

Fig. 2. SEMs of the top and bottom surfaces of microfilters manufactured by Gelman Company from cellulose and CTA [3].

In addition to flat sheet membranes, hollow fiber membranes are cast which are used to equip hollow fiber modules employed for artifical kidneys, gas separation, brackish and sea water desalination. Figure 4 exhibits scanning electron micrographs of cross-sections of hollow fibers cast from CA. The hollow fibers are made up of a finely porous matrix except at the bore side where larger cavities formed.

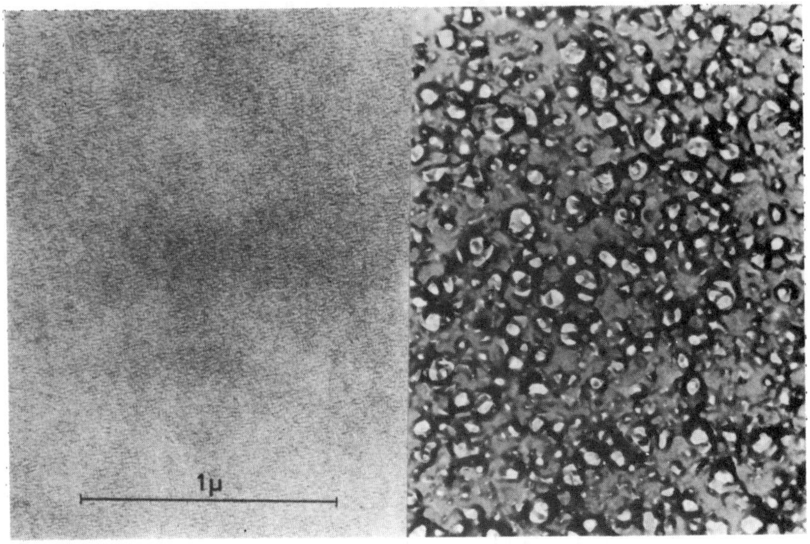

Fig. 3. TEM of top (left) and bottom (right) surface of an asymmetric CA membrane cast from a solution of BAYER Cellit K-700 in acetone/formamide [3].

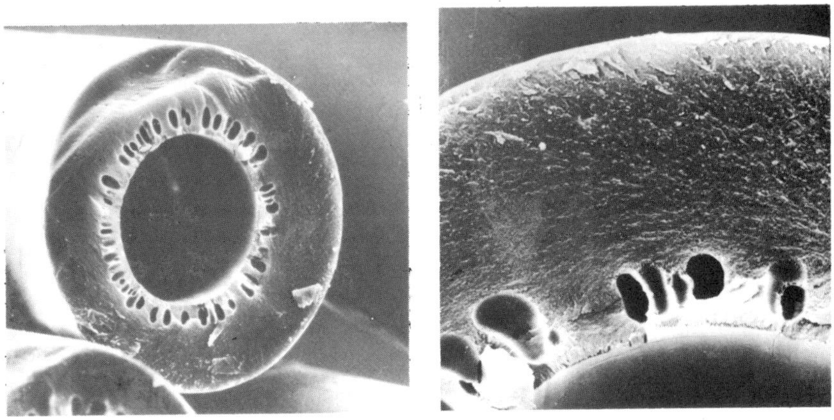

Fig. 4. SEMs of the cross-section of CA hollow fibers employed for gas separation; magnification × 1,000 (left) and ×10,000 (right).

TYPICAL TRANSPORT PROPERTIES

The transport properties of synthetic membranes can be characterized by the transport coefficients of the phenomenological relationships such as the hydrodynamic permeability, the osmotic or solute permeability, and the reflection coefficient of the membrane for solutes [4]. These transport coefficients can be obtained from so-called dialysis-osmosis and hyperfiltration experiments. Solute permeabilities, P_s, of a Cuprophan membrane PM-150, obtained from dialysis experiments with NaCl, $CaCl_2$, and glucose solutions, are reproduced as functions of the solute concentration in Figure 5. As is obvious from this figure, solute permeabilities are independent of solute concentration. The dependence of the normalized solute permeability, P_s/d, on the molecular weight of the solute is graphically shown in Figure 6 (d = effective membrane thickness). The logarithm of the solute permeability decreases nearly linearly with increasing molecular weight. Since there exists no analytical relation correlating the permeability with the molecular weight of the solute, no line is drawn in. Using solutes of different molecular weight which are generally employed for the characterization of artifical kidney membranes, the diffusive solute fluxes, Φ_s, are estimated as functions of the driving force Δc_s and are graphically compared in Figure 7. The diffusive solute flux decreases with increasing molecular weight of the solute as a consequence of the solute permeability decay with increasing molecular weight. For comparison, the convective solute transport (dashed line) is given as a function of the concentration $c_s^!$ where the convective flux is due to a pressure gradient of ΔP = 0.3 atm across the membrane. The convective solute flux, which is independent of the molecular weight of the solute, is of the same order of magnitude as the diffusive fluxes of high molecular weight solutes (e.g. inulin). In order to raise the convective flux to the diffusive flux of low molecular weight substances (e.g. NaCl, urea), the hydrodynamic permeability of the corresponding membranes must at least be ten times larger than that of common Cuprophan membranes. Today, such membranes are commercially available and are mainly employed for hemofiltration in order to extract from the plasma so-called middle molecular weight substances (molecular weight of 500 up to 5,000 Dalton).

APPLICATIONS OF SYNTHETIC MEMBRANES

Typical applications of synthetic membranes are summarized in Table 2. Todays main market is in the area of artificial kidney, micro- and ultrafiltration membranes. The market for gas separation is estimated to increase to $2 billion in 1995[5].

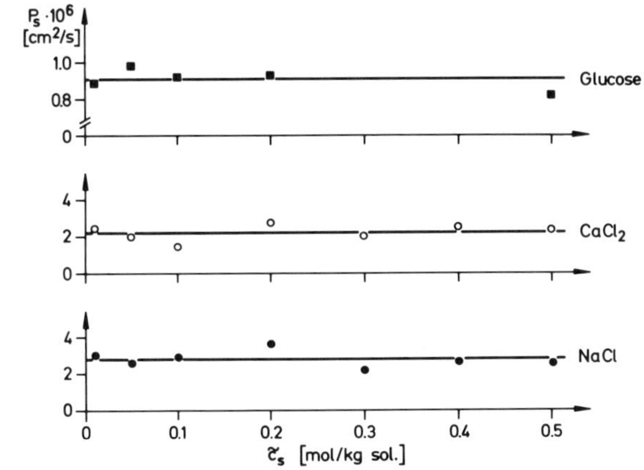

Fig. 5. Solute permeability, P_s, of a Cuprophan Membrane PM-150 as a function of solute concentration, c_s, at 25°C.

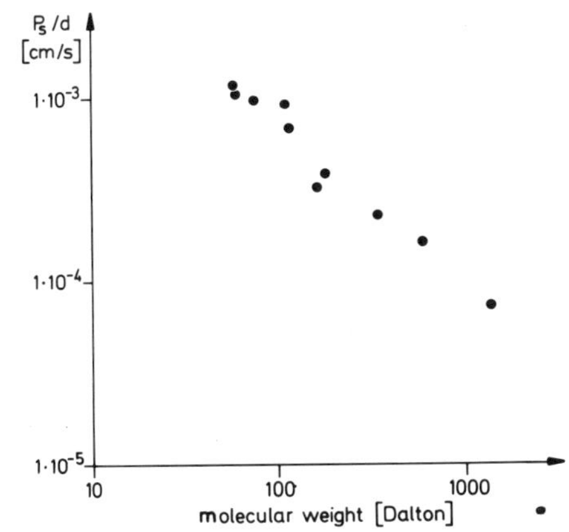

Fig. 6. Logarithm of normalized solute permeability, P_s/d, as a function of molecular weight using solutes of different molecular weight.

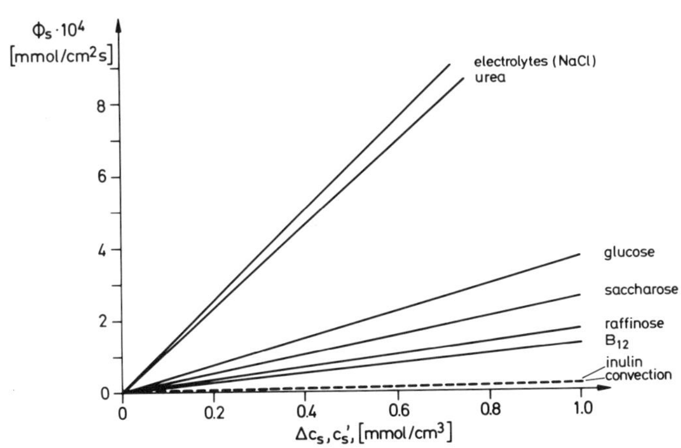

Fig. 7. Estimated solute fluxes across a cuprophan membrane as functions of $\triangle c_s$ or c'_s, respectively ($t = 25°C$, $\triangle P = 0.3$ atm).

Table 2.

Process	Membrane material	Membrane type	Field of application
Microfiltration	Cellulose CA, CTA	coarse porous FS, T, C	Isolation, cultivation of germs, sterile filtration, collection of cells (cytology) contamination control (aerosols), process filtration (beer)
Ultrafiltration	Cellulose CA, CTA	asym., comp. fine porous, dense FS, T, C, H-F	Food (dairy), car, chemical, pharma industry
Hyperfiltration (RO)	CA, CTA acetobutyrat	asym., comp. fine porous, dense FS, H-F	Production of boiler, ultrapure, and fresh water
Hemodialysis Hemofiltration	Cellulose CA	homogen., asym. fine porous FS, C, H-F	Artificial kidney
Gas separation	CA, CTA	asym. FS, H-F	Deacidification of natural gas
Controlled release	Diacetate	homogen. FS	Dosing of drugs
Xeroxcopy (Stencils)	CA, CTA	homogen. FS	Copying/reproduction
Packing	Cellulose CA	homogen. FS	Food industry

FS = Flat Sheet; T = Tubular; C = Capillary; H-F = Hollow Fiber

REFERENCES

1. R. Collander, Kolloid.Bei., 19 (1924) 72; 20 (1925) 273.
2. A. Walch und W. Pusch, Ang. Chem., Intl. Ed., 21 (1982) 660.
3. The author is grateful to Dr. Walch, Hoechst AG, Ffm., for the supply of the electron micrographs.
4. W. Pusch. Desalination, 59 (1986) 105.
5. Chemical Week/April 23, 1986.

Chapter 53

Biocompatibility of cellulose-containing composite materials

T. Miyamoto, H. Ito, S. Takahashi and H. Inagaki – Institute for Chemical Research, Kyoto University, Uji, Kyoto 611.

Y. Noishiki – Department of Surgery, Okayama University, Misasa, Tottori 682-02.

ABSTRACT

The characteristics of the polyion complexes(PICs) formed between anionic and cationic cellulose derivatives, cellulose graft copolymers with poly(γ-benzyl-L-glutamate) and poly(N^5-2-hydroxyethyl-L-glutamine)(Cell-g-PBLG and Cell-g-PHEG) as biomaterial was examined by in vivo as well as in vitro experiments. It was found that the cellulose-containing composite materials examined have an excellent blood compatibility. The results of in vivo tests on the tissue compatibility indicate that they are biodegradable and have good biocompatibility.

INTRODUCTION

Cellulose and its derivatives have been widely used in the biomedical and pharmaceutical fields: artificial kidney membranes, blood coagulants, coating materials for the controlled-release drugs, targeted drug carrier[Nakano et al., 1980] and stationary phases for optical resolution[Shibata, 1985]. Further attempts have been made to use cellulose derivatives as new biomedical materials such as blood anticoagulants[Teshirogi et al., 1980; Kamide et al., 1983] and drugs having antitumor activity[Matsuzaki et al., 1985]. However, few studies have been reported on the characteristics of cellulose-containing polymer alloys as biomedical materials, especially, as non-thrombogenic materials. Recently, we have found that the polyion complexes(PICs) formed between anionic and cationic cellulose derivatives[Ito et al., 1986a, b] and cellulose graft copolymers having polypeptide side chains[Miyamoto et al., 1986] exhibited excellent antithrombogenic properties. In this paper, the results on in vivo as well as in vitro experiments for thrombogenicity, absorbance by living body and the foreign body reaction of the cellulose-containing composite materials mentioned above are described.

EXPERIMENTAL

Cellulose derivatives

Two types of quaternary ammonium cellulose ether derivatives, Q-HEC and Q-Cell, were used as polycation derivatives. A quaternary ammonium derivative of hydroxyethyl cellulose, Q-HEC, was obtained from Union Carbide Co.(USA). A quaternary ammonium derivative of cellulose, Q-Cell, was prepared by homogeneous etherification of cellulose with glycidyl trimethylammonium chloride in a 10% LiCl/dimethylacetamide solvent. The details of the preparation of Q-Cell have been described elsewhere[Ito et al., 1986a]. Carboxymethyl cellulose(CMC) and cellulose sulfate(CS) were kindly supplied by Daicel Chem. Co.(Japan). The chemical structure and the degree of substitution(DS) of the sample derivatives employed so far are summarized in Table 1.

Table 1. Characteristics of cellulose derivatives employed.

Sample code	Substituent	Degree of substitution
Q-Cell	$-CH_2CH(OH)CH_2N^+(CH_3)_3{}^-Cl$	0.41
Q-HEC	$-(CH_2CH_2O)_nCH_2CH(OH)CH_2N^+(CH_3)_3{}^-Cl$	0.40
CMC	$-CH_2COOH$	0.85
CS	$-SO_3H$	0.54

Ethyl cellulose(EC) having a DS of 2.3 and a water-soluble methyl cellulose(MC, DS=1.8), which were used for the tissue compatibility experiments by living body, were purchased from Dow Chem. Co.(USA) and Shin-Etsu Chem. Co.(Japan), respectively. Unstretched rayon fibers were supplied by Unitika Rayon Co. (Japan).

Cellulose graft copolymers having polypeptide side chains

Cellulose graft copolymers with poly(γ-benzyl-L-glutamate) (Cell-g-PBLG) were prepared by polymerizing N-carboxy-γ-benzyl-L-glutamate using aminoethyl cellulose(AE-Cell) as macroinitiator. Graft copolymers with poly(N^5-2-hydroxyethyl-L-glutamine) (Cell-g-PHEG) were prepared by aminolysis of Cell-g-PBLG graft copolymers. AE-Cell having a DS of 0.05 was obtained from Serva Chem. Co.(USA). The DS value of AE-Cell used as substrate is relatively low(DS=0.05); every AE-Cell molecule possesses one possible grafting site per 20 AHG units. This is suitable for the preparation of the graft copolymers with a relatively low degree of grafting substitution. Table 2 shows the solubility of the Cell-g-PBLG graft copolymers. As can be seen from this ta-

Table 2. Solubility of cellulose, PBLG, and Cell-g-PBLG graft copolymers[a].

Solvent	Regenerated cellulose	PBLG	Cell-g-PBLG
Dimethylformamide	−	+	−
DMAc (b)	−	+	−
DMAc/LiCl	+	+	+
Dichloroacetic acid	−	+	+
Formic acid	+ (c)	−	+ (d)
Trifluoroacetic acid	+	+	+

a) +, soluble; −, insoluble.
b) Dimethylacetamide.
c) Dissolved as cellulose formate.
d) Graft copolymers with PBLG contents higher than 30 mol% are insoluble.

ble, the solubility behavior of the graft copolymers is largely determined by that of cellulose, reflecting that the degree of grafting substitution is relatively low. The details of the preparation and characterization of the graft copolymers have been reported in a previous paper[Miyamoto et al., 1986].

BLOOD COMPATIBILITY TESTS

In vivo test: The test, in vivo, for blood compatibility was performed according to a peripheral vein suture method[Noishiki, 1982]. In this test, a polyester suture was coated with a desired polymer. An 18-gauge needle was inserted into the peripheral vein(jugular and femoral vein) of a dog under general anesthesia. The suture was introduced through the needle into the lumen of the vessel. After the needle was withdrawn, the edge of the suture was ligated to the connective tissue near the puncture site of the vessel. After an adequate time(in this work, 1 day), the vein in which the suture had been inserted was opened gently and the suture was examined visually.

Lee-White test: Dog's whole blood was introduced into a test tube, the interior wall of which was coated with sample polymer, and the coagulation time of blood was measured by tilting of the test tube every minute after the first four minutes incubation at 37°C. The coagulation time is expressed in terms of a ratio to an untreated glass tube.

TISSUE COMPATIBILITY TESTS

Polyester fiber mesh cloth 1 x 3 cm in size was used as the supporter of polymer film and as a marker of polymer samples implanted in the tissue. The polymer samples were coated on the polyester mesh. The samples, implanted in the muscle tissue of the back of mongrel dogs for 4 weeks, were taken together with the tissue surrounding the polymer samples. The specimens obtained after fixation in 10% formalin were embedded in paraffin. The sliced specimens, stained with hematoxylin and eosin, were examined microscopically[Ito et al., 1982; Sato et al., 1985].

RESULTS AND DISCUSSION

Blood compatibility

As reported previously[Ito et al., 1986a], water-insoluble PICs can be obtained by mixing oppositely charged cellulose derivatives in aqueous solution, but the cellulosic PICs obtained are soluble in formic acid. In the present study, the coating of the cellulosic PICs on the polyester sutures was carried out by casting from formic acid solutions. The formic acid solutions were prepared by dissolving the water-insoluble PICs formed at pH 7.4.

Table 3. In vivo and in vitro results on blood compatibility of cellulosic PICs.

Sample code	In vivo Thrombus formation	In vitro Lee-White relative clotting time (glass = 1)a)
Q-HEC/CMC	−	5.8 ± 0.8
Q-HEC/CS	−	4.9 ± 0.3
Q-Cell/CMC	−	4.7 ± 0.3
Polyester suture	+++	
Glass		1.0

a) Expressed as a ratio with respect to untreated glass tube taken as 1.

The in vivo and in vitro test results on the cellulosic PICs are summarized in Table 3. The whole blood coagulation data are mean values of five experiments. To make the results reliable, in vivo experiments were carried out at least twice. The symbols evaluate the in vivo results as follows: The (+) symbols indicate the relative degree of the thrombus formation of the sample in 3 steps where (+++) denotes the thrombus formation along the entire length of the coated surface; (−) indicates the case where no thrombus was observed. It can be seen that by the criteria of the present in vivo test, the cellulosic PICs have a good blood compatibility.

The Lee-White test showed that these PICs have high relative coagulation times compared with glass as a control. In this connection, it should be mentioned that most of synthetic polymers studied so far fell into the range of 1.5 - 3.0. To be also noted is that the results on the coagulation time of whole blood

Table 4. In vivo results on blood compatibility of Cell-g-PBLG and Cell-g-PHEG graft copolymers.

Sample code	Peptide content (mol%)	Casting solvent	Thrombus formation
Regenerated cellulose	0	Formic acid	+
		DMAc/LiCl(a)	+
Cell-PBLG-10	7	Formic acid	−
		DMAc/LiCl(a)	−
Cell-PBLG-30	27	Formic acid	±
		DMAc/LiCl(a)	∓
Cell-PBLG-60	58	DMAc/LiCl(a)	−
PBLG	100	Chloroform	++ ~ +++
Cell-PHEG-10	7	Formic acid	−
Cell-PHEG-30	27	Formic acid	−

a) Water was used as regenerating media.

by Lee-White tests are in good agreement with those by in vivo experiments. For the purpose to clarify the correlation of the chemical structure of cellulosic PICs with the blood compatibility, further study on the effect of DS value of the component derivatives on the blood compatibility should be done. Such a study is in progress in our laboratory, and the results will be reported in a near future.

Results on the blood compatibility of Cell-g-PBLG and Cell-g-PHEG graft copolymers are shown in Table 4. It can be seen that the graft copolymers examined here have an excellent blood compatibility. However, the results also indicate that the blood compatibility depends not only on the polypeptide content but also on the coating conditions of the sample. The samples with polypeptide contents less than 20 mol% are not graft copolymers in a strict sence but cellulose derivatives with amino acids and /or oligopeptides as substituents, because such polypeptide sequences cannot assume a helical conformation characteristic of PBLG and PHEG. It should be noted that these graft copolymers with low polypeptide contents exhibited excellent antithrombogenic properties, independent of the casting conditions of the sample.

The implantation time in this work was relatively short, i,e., 24 h. To evaluate the biocompatibility of these graft copolymers, it is necessary to observe the blood compatibility at various time intervals after implantation. Such experiments are currently going on.

Tissue compatibility

Results on the tissue compatibility of the cellulosic PICs are listed in Table 5, together with those of regenerated cellu-

Table 5. Biocompatibility of regenerated cellulose, cellulose derivatives, and cellulosic PICs.

Sample code	Absorbance by living body	Foreign body reaction	Remarks
Cellulose A	−	± or +	Prepared by deacetylating CA cast film with aq. NH_3
Cellulose B	++	± or +	Unstretched rayon fibers
EC	−	± or +	Cast from CH_2Cl_2 solution
MC	++++	+	Crosslinked with epichlorohydrin
Q-HEC/CMC	+++	+	Cast from HCOOH solution
Q-Cell/CMC	++	+	Cast from HCOOH solution
Polyester	−	−	

Table 6. Biocompatibility of Cell-g-PBLG and Cell-g-PHEG graft copolymers.

Sample code	Absorbance by living body	Foreign body reaction	Remarks
Cellulose	−	± or +	Prepared by deacetylating CA cast film with aq. NH_3
Cell-PBLG-10	++	+	Cast from HCOOH solution
Cell-PBLG-30	++	±	Cast from HCOOH solution
Cell-PHEG-30	+++	±	Cast from HCOOH solution
PBLG	+	±	Cast from CH_2Cl_2 solution

lose and cellulose derivatives. Results on the Cell-g-PBLG and Cell-g-PHEG graft copolymers are shown in Table 6. The level for evaluating the tissue compatibility of the sample is as follows.

(−) cells do not recognize the entity as a foreign body;
(±) borderline recognition between (−) and (+);
(+) cells recognize it as a foreign body, giant cells, and fibroblasts gather close to the test materials, and 10 or fewer cell layers surround the foreign body, which corresponds to a light foreign body reaction;
(++) borderline recognition between (+) and (+++);
(+++) multiple layers of cells lump together close to the test material in a strong foreign body reaction;
(++++) the test material poisons the cells, causing necrosis.

The degree of absorbance by living body is evaluated according to the following indices:

(−) no low absorbance is observed;
(+) about 25% of test material is absorbed;
(++) about 50% of test material is absorbed;
(+++) about 75% of test material is absorbed;
(++++) more than ca. 95% of test material is absorbed.

The results show that the cellulosic PICs, Cell-g-PBLG and Cell-g-PHEG graft copolymers are biodegradable. It is of interest to note that MC is also highly biodegradable, but EC shows no absorbance by living body, indicating that the degree of absorbance depends on the chemical structure of the cellulose derivatives. Cellulose A, which was regenerated from CA cast film, has a high degree of crystallinity, while that of unstretched rayon fibers(Cellulose B) is very low. It should be also noted that the cellulose sample having a low degree of crystallinity is biodegradable.

On the other hand, the degree of the foreign body reaction was mild for all the sample polymers examined. The detailed results on the microscopic observation of these polymer samples and surrounding tissue will be reported elsewhere[Miyamoto et al., 1986b]. Thus, it may be concluded that the cellulosic PICs, Cell-g-PBLG and Cell-g-PHEG graft copolymers show considerable promise for biomedical applications.

REFERENCES

Ito, H., Miyamoto, T., Inagaki, H. and Noishiki, Y.(1982) Kobunshi Ronbunshu, 39, 249.

Ito, H., Shibata, T., Miyamoto, T., Inagaki, H. and Noishiki, Y. (1986a) J. Appl. Polym. Sci., 31, 0000.

Ito, H., Miyamoto, T., Inagaki, H., Noishiki, Y., Iwata, H.and Matsuda, T. (1986b) J. Appl. Polym. Sci., 31, 0000.

Kamide, K., Okajima, K., Matsui, T., Ohnishi, M. and Kobayashi, H. (1983) Polym. J., 15, 309.

Matsuzaki, K., Yamamoto, I., Sato, T. and Oshima, R. (1985) Makromol. Chem., 186, 449.

Miyamoto, T., Takahashi, S., Tsuji, S., Ito, H., Inagaki, H. and Noishiki, Y. (1986a) J. Appl. Polym. Sci., 31, 2303.

Miyamoto, T. Takahashi, S., Ito, H., Inagaki, H. and Noishiki,Y. (1986b) to appear.

Nakano, M., Ito, M., Juni, K., Sekikawa, H. and Arita, T.(1980) Int. J. Pharm., 4, 291.

Noishiki, Y. (1982) Jinko-Zoki(Jpn. J. Artif. Organs), 11, 794.

Sato, H., Nakajima, A., Hayashi, T., Chen. G.-W. and Noishiki, Y. (1985) J. Biomed. Mater. Res., 19, 1135.

Shibata, T. (1985) Proc. Daicel Cellulose Conf., Himeji, p. 77.

Teshirogi, T., Yamamoto, H., Sakamoto, M. and Tonami, H. (1980) Sen-i Gakkaishi(J. Soc. Fiber Sci. Technol., Jpn.) 36,560.

Chapter 54

Physicochemical properties of hydroxymethyl cellulose and other polysaccharide graft copolymers as super water absorbents

Namiko Miyata – Nakamura Gakuen College, Befu, Jonan-ku, Fukuoka, Japan.
Isao Sakata – Faculty of Agriculture, Kyushu University, Hakozaki, Higashi-ku, Fukuoka, Japan.

INTRODUCTION

In recent years super water-absorbents have come into great demand. Three kinds, namely starchy-, cellulosic-, and synthetic polymeric materials, are known. While most studies on them have dealt with preparations which are able to absorb a lot of water, physiological saline solution and other salt solutions , the problems in their properties have hardly been studied. Consequently, an attempt has been made to explain the physicochemical properties of the absorbents in this work.
In the previous work[1], hydroxyethyl cellulose (HEC) and the other polysaccharides (i.e., konjac mannan, alginic acid, agar, pectin, starch and chitin) were selected as the trunk polymers, These polysaccharides were prepared as graft copolymers with partially hydrolyzed polyacrylamide(PAM) or polymethyl acrylate(PMA), poly-2-(dimethylamino) ethylmethacrylate(PDM) or polyacrylic acid sodim salts(PAANa), to give super water-absorbents.

In this study the effects of the kinds of the trunk polymers and the branch polymers of the graft copolymers, the molecular weight(the viscosity) of the trunk polymers, the molar substitution(MS) of the hydroxyethyl groups of HEC, and the absorption conditions such as temperatures and aqueous absorbate compositions (i.e., H_2O:MeOH) on the absorbency were examined, and the gel strength of the absorbents was also evaluated in connection with the composition of the graft copolymers.

EXPERIMENTAL

1) The preparation of the super water absorbents: Grafting of acryl amide(AM) or the other monomers onto HEC and the other polysaccharides was carried out in the presence of N,N-methylenebisacryl amide as a bridging agent using Ce-salt as an initiator in a N_2 stream at 40°C. After grafting, homopolymers and unreacted trunk polymers were removed by extraction with water or other solvents for 1–2 days. PAM and the other branch polymers which could not be removed by extraction would supposedly be graft copolymerized. The PAM and PMA graft copolymers were hydrolyzed with 0.5N-NaOH solutions. The graft copolymers were dried in a vacuum for 24 hr. 2) The fluid absorbency tests were run as follows. About 0.1 g of each absorbent was added to 100–200 ml of deionized water or 50 ml of 0.85% NaCl solution or 100 ml of an aqueous MeOH solution. The mixtures were allowed to stand for 1–16 hr, and the swollen polymer particles were then separated from an unabsorbed solution by screening through a 80 mesh sieve. Then fluid absorbency was calculated as grams of solution per gram of absorbent. 3) The viscosity of the trunk polymers was determined by using a rotating viscometer(Tokyo keiki Co.,Ltd.). 4) The compressive strength of the gels of the absorbents was determined by using the Tom-200 type Universal Testing Machine at 20°C. The samples, in the form of a cylinder of 29 mm diameter, were compressed between flat parallel plates at a crosshead speed of 50 mm/min.

RESULTS AND DISCUSSION

The effects of the kinds of the trunk and the branch polymers on the absorbencies were HEC > konjac mannan, alginic acid > starch > agar > pectin, and partially hydrolyzed PAM > partially hydrolyzed PMA > PAA-Na > PDM > PAA > PAM, that is to say, partially hydrolyzed > ionic and hydrophilic > hydrophilic > hydrophobic, as shown in Figure 1. It was found that the absorbency of the HEC-PAM-Hyd graft copolymers was superior to that of the other polysaccharide graft copolymers, and that they could absorb water and 0.85% NaCl solution up

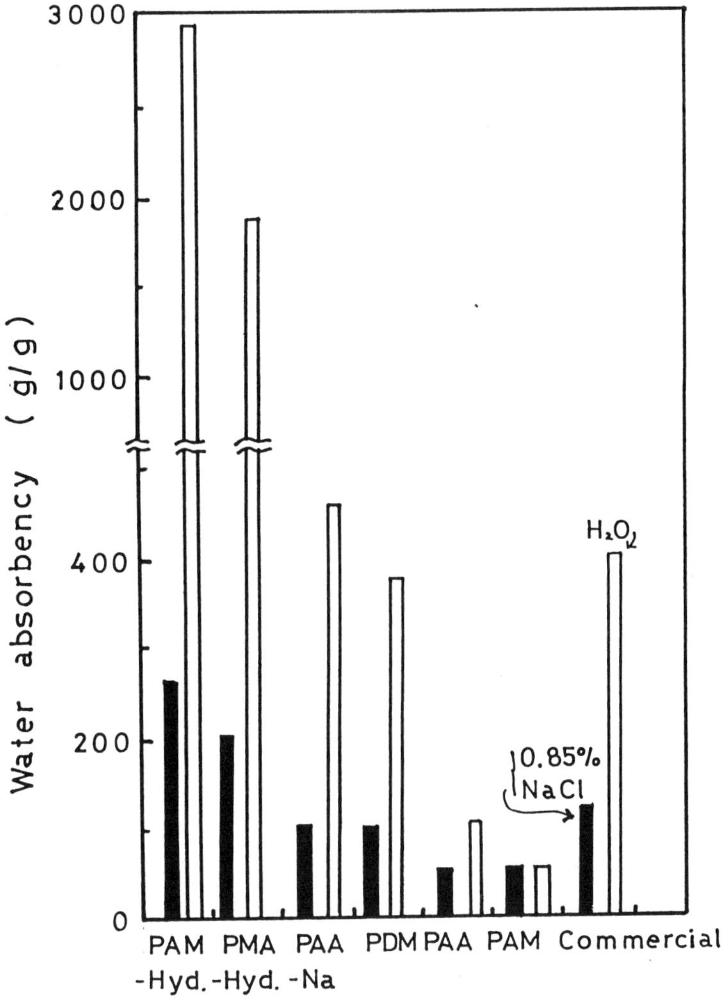

Fig. 1. Effects of the kind of the branch polymers of the HEC graft copolymers.

to 3000 g/g and 270 g/g of absorbent, respectively. Each value was 6 times and 2.5 times that of the commercial absorbents, respectively.

As the degree of grafting increased, the water absorbency reached the maximum value at 200–400% grafting and then gradually decreased. The graft copolymers containing lower molecular HEC gave greater absorbency compared with the graft copolymers containing higher molecular HEC. However, the compressive strength of the gels of the HEC graft copolymers increased with increaing the degree of grafting and that of the graft copolymers of higher molecular HEC was larger than that of the graft copolymers of lower molecular HEC. It can clearly be seen that the water absorbencies of the HEC graft copolymers were larger than those of the commercial absorbents.

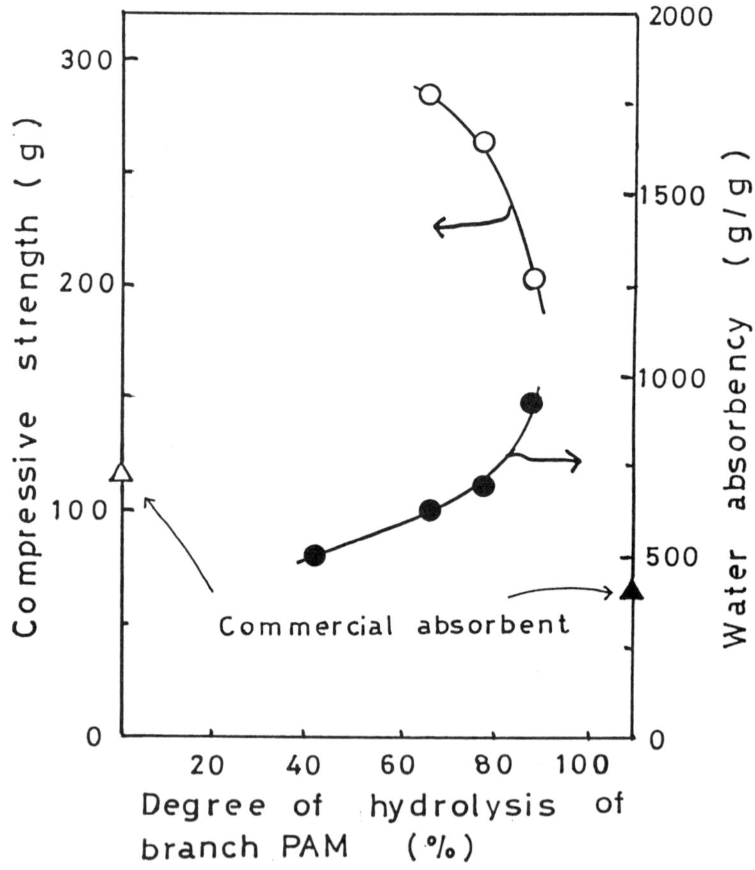

Fig. 2. Effects of the degree of the hydrolysis of the HEC-PAM graft copolymers.

As the degree of the hydrolysis of PAM of the branch polymers increased, the water absorbency increased; however, the compressive strength decreased as shown in Figure 2.

In Figure 3 it can clearly be seen that the water absorbency decreased and the compressive strength increased with increasing the concentration of the bridging agents. These data were in good agreement with Flory's equation(1) and equation(2) described below;

$$Q^{5/3} \cong [(i/2 \cdot V_u S^{*1/2})^2 + (1/2 - x_1)/V_1] / (\nu_e/V_o) \qquad (1)$$ [2]

$$G = RT \cdot \nu_e/V_o \qquad (2)$$ (see next page)

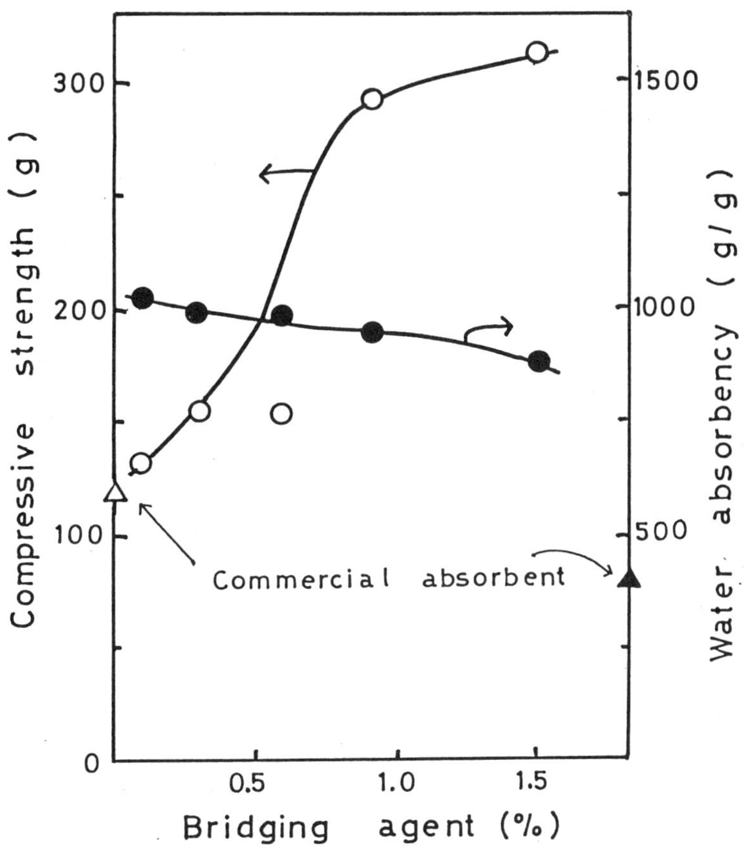

Fig. 3. Effects of bridging of the HEC-PAM-Hyd graft copolymers.

{Q:Water absorbency, i/Vu:Charge density of polymer, S*:Ionic strength of solution, $(1/2-x_1)/V_1$:Polymer-solvent affinity, ν_e/V_0 :Bridging density, G:Rigidity}

There were small effects of the MS between 1.2 and 4.0 on the water absorbency.

There were small effects of the temperature from about −15° C to 70° C on the fluid absorbency of HEC-PAM-Hyd graft copolymers. However, beyond that temperature the water absorbency decreased, and the swollen HEC graft copolymers shrank.
The water absorbency of the graft copolymers and the commercial absorbents depended largely on pH and their maximum absorbency was observed at pH 4-9, and both the graft copolymers and the commercial absorbents could hardly absorb water at the acidic and alkaline pH.

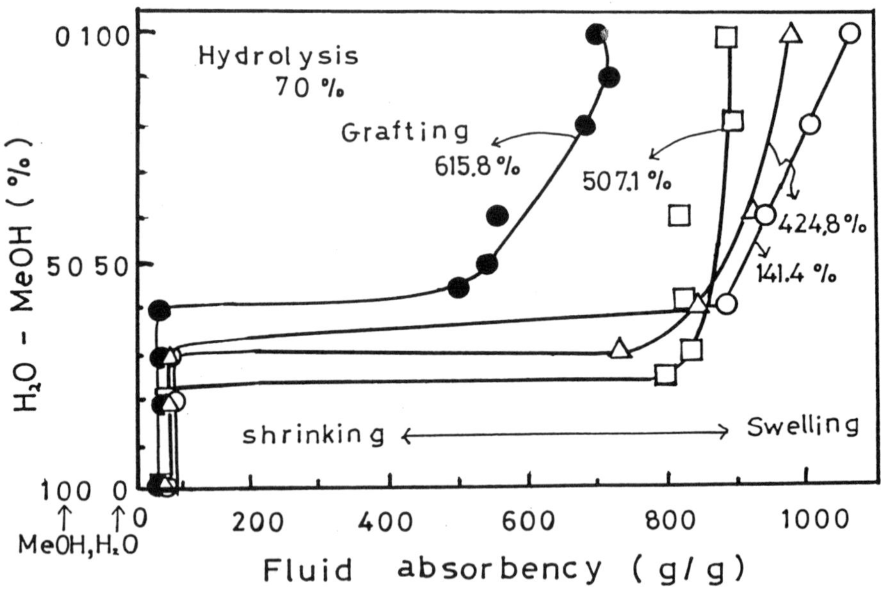

Fig. 4. Effects of the degree of grafting and the composition of H_2O-MeOH solution on the discrete volume change of the gels of the HEC-PAM-Hyd.

The gels of the HEC graft copolymers showed a discrete volume transition upon changing the composition of an aqueous MeOH solution as shown in Figure 4. The gels of the HEC graft copolymers swelled in the solutions of MeOH concentration lower than 60–80% volume, and shrank in the solution above 60–80%. The gel volume changed discontinuously about 50-fold at this MeOH concentration, and the degree of volume change varied with the degree of grafting and the hydrolysis of PAM of the branch polymers. These discrete volume changes are due to three forces on the gel; i.e., the positive osmotic pressure of the counterions, on account of charge density of polymers, polymer–solvent affinity and rubber elasticity of the polymer network as described by Flory's equation(1). The balance of these forces varies with changes in solution properties.
The HEC and the other polysaccharides graft copolymers as super water-absorbents would be applicable to sanitary materials, agricultural and gardening materials(water-keeping agents), industrial dehydration agents, sensors and other materials(drying preventatives, toys, and man-made muscles).

REFERENCES

[1] N. Miyata and I. Sakata; Abstracts of papers of 1984 Annual Meeting of the Society of Fiber Science and Technology, Japan, p.149(1984). Proceeding of the International Symposium on Fiber and Technology Organized of the Society of Fiber and Technology, Japan, p.292(1985)

[2] Paul J. Flory; Principles of Polymer Chemistry, copyright 1953 by Cornell University, p. 533.

Chapter 55

Recent progress in wood chemistry -fundamentals and applications

J. Nakano – Department of Forest Products, Faculty of Agriculture, The University of Tokyo, Japan.

INTRODUCTION

Major components of wood, namely cellulose, hemicellulose and lignin are discussed from the fundamental and applied viewpoints. The application of the major components to the production of wood chemicals, is discussed.

CELLULOSE

About 40-50% of the dry substances in most woods are cellulose, located mainly in the secondary cell wall. The supermolecular structure and the utilization of cellulose are discussed here.

Supermolecular structure

Although the chemical structure is well understood, its supermolecular structure is still debatable. It has now been accepted that all cellulose chains in native microfibrils (cellulose I) are oriented in the same direction, and they are parallel. Cellulose II is said to be an antiparallel chain. There still remains to be solved whether or not parallel chains could be changed to antiparallel chains. This problem was discussed by Prof. A. Sarko, State Univ. of New York, Syracuse at this conference.

Utilization of cellulose

(1) Degradation of cellulose

As shown in Figure 1, glucose is obtained from cellulose, and is a useful raw material for many end uses [1]. Although most ethanol is produced from ethylene at present, we can expect its production from cellulose in the future. Moreover, ethylene might be produced from ethanol. A small amount of ethylene is now produced from ethanol in Brazil, although an economic problem remains to be solved there. In order to convert cellulose into glucose,

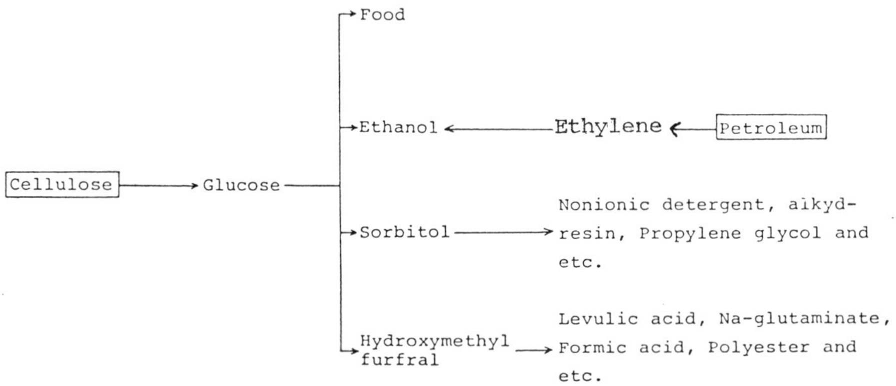

Fig. 1. Glucose chemicals from cellulose.

a dilute acid hydrolysis process has been practiced, but the yield of glucose is not more than 50% based on cellulose. Therefore, it is really important for the economical feasibility of wood hydrolysis to improve the reaction conditions and the accessibility of cellulose.

Biochemical degradation of cellulose might be an alternative method for the conversion of cellulose into glucose. This field has recently been most actively investigated all over the world.

(2) Cellulose derivatives

A number of industries are now using cellulose derivatives. The utilization of cellulose derivatives will be vastly developed in the near future not only as simple derivatives, but also as complementary functional polymers.

From the economical point of view, it may also be important to produce cellulose derivatives from high yield pulp. Here will be introduced our recent results on the carboxymethyl polysaccharides from refiner mechanical pulp [2]. As shown in Table 1, the carboxymethyl group introduced onto RMP was 4-5 meq/g. The water soluble portion of carboxymethylated RMP was only 30%. However, this value can be increased to more than 90% by hot water, sodium chlorite and ozone treatments. The utilization of carboxymethylated RMP is now under way to prepare liquid absorbents, chelating materials and so on [3].

Table 1. Characteristics of carboxymethylated RMP[a].

Time[b] hr	Wt. increase %	Carboxyl meq/g	H_2O sol[c] %
1	39.1	4.2	23.3
3	42.6	4.6	31.9
6	43.1	4.8	36.2
7	47.8	4.8	—

a: According to the standard method using isopropanol as a solvent
b: Reaction temp., 55 °C
c Extracted for 48hr at room temp.

(3) Contribution of nonaqueous solvent in utilization of cellulose

Concerning the production of regenerated cellulose, a new process which uses nonaqueous solvent is worthy of remark. This is a much more simplified process than the present viscose process. More than 80 different nonaqueous solvents have been proposed hitherto. These include dimethylsulfoxide-nitrogen dioxide, dimethylformamide-nitrogen dioxide, dimethylformamide -paraformaldehyde, liquid sulphur dioxide-amine etc [4]. However, in order to visualize a new process using nonaqueous solvent, there are several problems to be solved. Namely, we must find an excellent solvent in which the solubility of cellulose is more than 10%, and the recovery of the solvent must be easy.

The discovery of an excellent nonaqueous solvent would not only accelerate the development of a new process of cellulosic fiber and film production, but also contribute to the development of a new blend polymer of cellulose and the synthetic polymer, a new processing of cellulosic fiber, the sythesis of new cellulose derivatives and so on.

Tri-O-benzylcellulose can be prepared quantitatively, when an SO_2-dimethylamine-dimethylsulfoxide system is used as a cellulose solvent. Following such a method,

Table 2. Etherifications of Avicel by the use of SO_2-DEA-DMSO system as a reaction medium.

Reagents	DS	Reagents	DS
$(CH_3)_2SO_4$	< 2.2	p-$CH_3OC_6H_4CH_2Cl$	3
CH_3I	< 2.3	p-$ClC_6H_4CH_2Cl$	3
$ClCH_2COONa$	< 1.8	p-$BrC_6H_4CH_2Cl$	3
$(CH_3)CCl$	< 1.0	p-$(CH_3)_2CHC_6H_4CH_2Cl$	3
$CH_3(CH_2)_2CH_2Cl$	< 1.0	α-naphthylmethyl Cl	3
H_2NCOCH_2Cl	< 1.0	$C_6H_5CH(CH_3)Cl$	-
$(Et)_2NCH_2CH_2Cl\cdot HCl$	< 0.5	p- or m-$O_2NC_6H_4CH_2Cl$	-
CH_3OCH_2Cl	< 2.0	p-CH_2=$CHC_6H_4CH_2Cl$	-
$(CH_3O)_2CHCH_2Cl$	< 0.5	CH_2=$CHCH_2Cl$	3
$CH_3COOCH_2CH_2Cl$	< 1.0	CH_2=$C(CH_3)CH_2Cl$	3
CH_3OCOCH_2Cl	< 1.0	CH_3CH=$CHCH_2Cl$	3
$C_6H_5CH_2CH_2Cl$	< 1.0	C_6H_5CH=$CHCH_2Cl$	3
$C_6H_5CH_2Cl$	3	CH_2=$CHCH_2Br$	-
p-$CH_3C_6H_4CH_2Cl$	3	$C_6H_5COCH_2Br$	-

many tri-O-substituted cellulose ethers were recently synthesized as shown in Table 2 [5]. These new cellulose derivatives, which are soluble in many organic solvents, have chemically reactive groups such as an aromatic ring or a double bond. Accordingly, cellulose can be modified to some functionalized derivatives such as ion exchangers, polymer complexes, polymer catalysts, immobilized supports and so on.

Furthermore, some of these derivatives in Table 2 show the thermotropic liquid crystalline properties, smectic or nematic [5]. As shown in Table 3 and 4 [6], various tri-O-n-alkylcelluloses can be prepared with various n-alkyl iodides and bromides.

These cellulose derivatives have also thermotropic or lyotropic liquid crystalline properties. As shown in Table 3, the yields of cellulose ethers with n-C_{12} - C_{16} are more than 500% on the theoretical basis. These products should contain large amounts of by-products (di-n-alkyl ethers). However, the by-products could not be separated from cellulose derivatives by GPC. This suggests that large amounts of di-n-alkylethers must be intimately entangled with flexible long chains linked to the cellulose chain. This property may open a new usage of cellulose.

Nonaqueous solvent is also useful to prepare blend polymers from cellulose and synthetic polymers. Such

Table 3. Preparation of cellulose ethers with bromine compounds

Reagents	DS	Form	Comments
n-C6 Br (hexyl Br)	3	G	L.LC
n-C7 Br	3	G	L.LC
n-C8 Br	3	G	L.LC
n-C10 Br	3	G	L.LC
n-C12 Br	3	W.P.	Yield > 500%
n-C14 Br	3	W.P.	Yield > 500%
n-C16 Br	3	W.P.	Yield > 500%
iso-Amyl Br	3	W.P.	
$BrCH_2COOH$	0.8	W.P.	
$BrCH_2CH_2COOH$	0.4	W.P.	
$BrCH_2CH_2SO_3Na$	1.8	W.P.	
Br 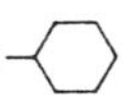	0.5		
$BrCH_2CH(OEt)_2$	0.5		
$BrCH_2CH_2Ph$	0.5		
$BrCH_2CH_2CH_2OPh$	3	W.P.	

G : Gummy solid W.P. : White powder

L.LC : Lyotropic liquid crystal

a method may promise the development of a new application of cellulose. For example, each polymer solution of polyvinyl acetate, cellulose acetate, polyacrylonitrile etc. was mixed with a nonaqueous solution of cellulose. Then, the mixed solution was regenerated mainly with water-alcohols. Thus, regenerated blend polymer is certainly very homogeneous and, as expected, shows the excellent properties of both constituent polymers [7].

HEMICELLULOSE

The amounts of hemicellulose in wood are usually between 20 to 30% on a dry basis. This paper now discusses the chemical structure of wood hemicellulose and utilization of sugar components in sulfite spent liquor.

Table 4. Preparation of cellulose ethers with iodine compounds.

Reagents	DS	Form	Comments	
Me I	3	W.P.	190°-220°C;T.LC	Td=320°C
Et I	3	W.P.	Tm=200°C	Td=330°C
n-Pr I	2.9	W.P.	180°-230°C;T.LC	Td=310°C
iso-Pr I	0.5			
n-Bu I	3	W.P.	Tm=195°C	Td=320°C
iso-Bu I	0.5			
n-C6 I	3	G	L.LC	
n-C7 I	3	G	L.LC	
n-C8 I	2.9	G	L.LC	
ICH_2COONa	0.6	W.P.		

W.P. : White powder G : Gummy solid
T.LC : Thermotropic liquid crystal
L.LC : Lyotropic liquid crystal

Chemical structure

The chemical structures of hemicellulose in softwoods are different in characteristics to those in the hardwoods. Figure 2 shows that the former is rich in glucomannan and the latter in glucuronoxylan.

Although the chemical structures of the major hemicelluloses have mostly been elucidated, some corrections to the fine structure can be anticipated. For example, as shown in Figure 3, it has been clarified that birch glucuronoxylan contains a galacturonic acid and a rhamnose residue at the reducing end [8]. Such a structure may be found in other hardwood glucuronoxylans. Moreover, a number of oak wood species may contain a small amount of xyloglucan which has never been found in wood hemicellulose [9].

Softwood hemicelluloses

(1) Galactoglucomannan (ca. 20%)

Gal:Glc:Man = 0.1:1:4
Gal:Glc:Man = 1:1:3

(2) Arabinoglucuronoxylan (5~10%)

Araf:GlcA:Xyl = 1.3:1:10

(3) Arabinogalactan

(4) Other polysaccharides

Hardwood hemicelluloses

(1) Glucuronoxylan (15~30%)

GlcA:Xyl = 1:10

(2) Glucomannan (2~5%)

Glc:Man = 1:1~2

(3) Other polysaccharides

Fig. 2. Chemical structure of wood hemicelluloses.

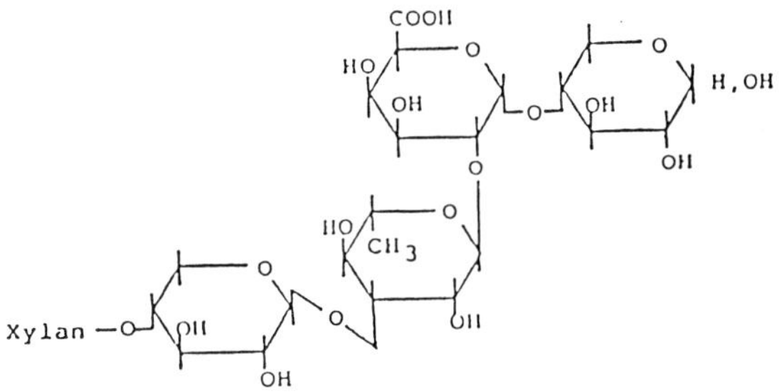

Fig. 3. Structure associated with the reducing end group of birch xylan.

Hydrolysis of hemicellulose

Various monosaccharides are obtained from wood hemicellulose by acid hydrolysis. In the production of dissolving kraft pulp, the prehydrolysis of wood chips is necessary to remove hemicellulose. Figure 4 shows a number of hydrolyzates from hemicellulose and their related products. Much of hemicellulose, however, comes from agricultural residues at present. In the case of glucose production from hardwood, it is basically important to utilize xylose which is produced at the prehydrolysis stage.

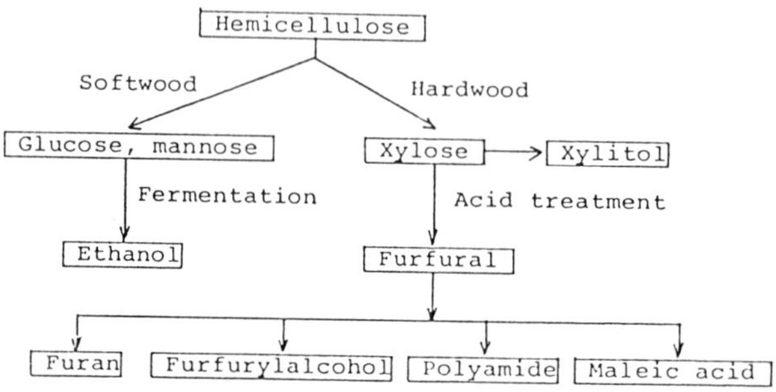

Fig. 4. Hydrolysis of hemicellulose and its related products.

Utilization of sugar components in sulfite spent liquor

Most of sulfite mills which produce the dissolving pulp for rayon, staple fiber and cellophane, have been utilizing sugar components in spent liquor for the production of ethanol in the case of softwood and yeast in the case of hardwood. The production of pulp yeast and RNA in Japan were 24,000 t/year and 1,500 t/year, respectively, a few years ago [1]. The production of pulp yeast, however, has now decreased to one-third due to the depression of dissolving sulfite pulp production.

LIGNIN

Normal softwood contains 26-32% lignin and hardwood 20-28%. The chemical structure of lignin and the utilization of industrial lignins are now discussed.

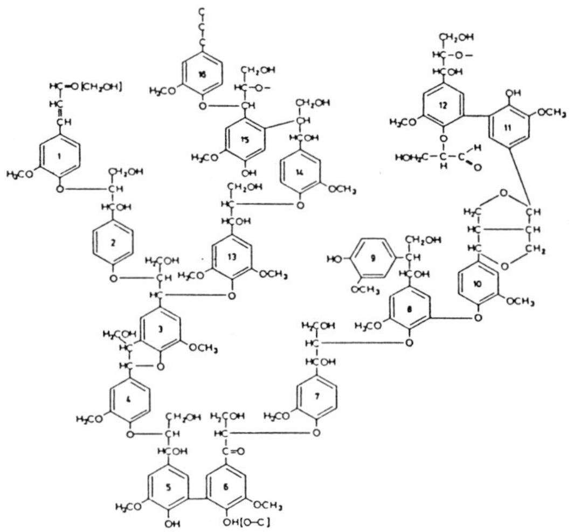

Fig. 5. Chemical structure of softwood lignin comprising 16 phenylpropane units.

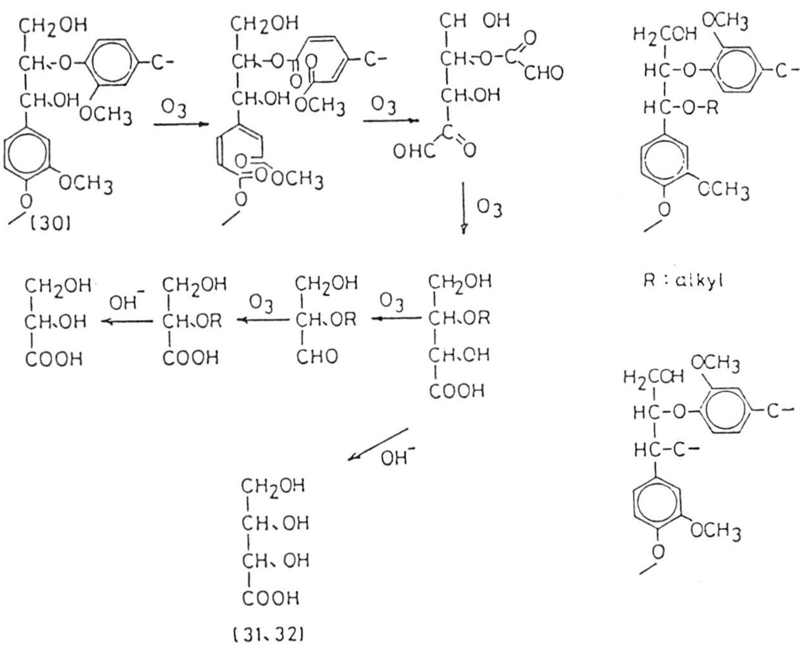

Fig. 6. Formation of monocarboxylic acids from guaiacylglycerol-β-guaiacyl ether structure by ozonation followed by saponification.

Chemical structure of lignin

The major structural elements have been elucidated gradually. These are such elements as the guaiacyl glycerol-β-aryl ether, phenylcoumarane and biphenyl type structures. More than two thirds of the phenylpropane units are linked by ether bonds and the rest by carbon-carbon bonds. The chemical structure of softwood lignin depicted by E. Adler, Sweden is shown in Figure 5 [10].

The ozonation method may be useful for determining various types of phenyl propane linkages. We have succeeded in determining the guaiacylglycerol-β-guaiacyl ether and the glyceraldehyde-2-aryl ether type structures by ozonation [11]. As shown in Figure 6 the former structure is determined on the basis of tetronic acids obtained by ozonation and then saponification. Figure 7 shows the determination scheme of the latter structure including sodium boronhydride and ozonation. Further determination of another type of structure is now under way.

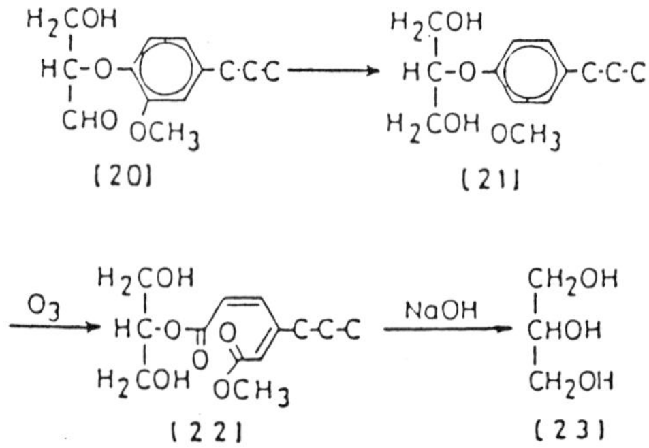

Fig. 7. Scheme for the determination of the structure [20] by the use of ozonolysis.

Utilization of industrial lignins

More than 200 kinds of commercial lignins which are lignosulfonate, thiolignin and hydrolysis lignin, are produced in the world. In future, exploded lignin will be also utilized. However, the utilization of lignin is mainly limited to lignosulfonate isolated from sulfite spent liquor at present. Lignin properties to be modified will be discussed from the viewpoint of dispersibility, solubility, color and degradation products. As shown in Table 5, the

Table 5. Demand of lignin products in Japan, 1982.

Main use	Demand $x10^3T/Y$	Ratio %
Cement and concrete admixture	54	47
Fine-ore binder	29	25
Fertilizer pelletizing agent	7	6
Cement dispersing and binder	6	5
Pesticide pelletizer and dispersant	3	3
Carbon black dispersant and binder for graphite, active carbon, etc.	2	2
Dyestuff, pigment dispersant and pigmentgrinding aid	2	2
Drilling mud conditioner	2	2
Plaster dispersant and gypsum slurry thinner	1	1
Molding sand binder	1	1
Others	7	6
Total	114	100

present demand of lignosulfonate in Japan is about 110,000 t/year [1]. Of these, concrete and cement admixtures are the most prominent.

(1) Dispersibility

Dispersion number was measured by the decrease in the viscosity of an aqueous suspension of titanium dioxide, when a known weight of lignosulfonate was dissolved in the medium [12]. At molecular weight below 10,000, the dispersion number decreased, and maximum dispersion was obtained by lignosulfonate of molecular weight 10,000 - 40,000. At present we are able to degrade the high molecular portion and to condense the lower molecular weight portion. However, it is expensive to do so.

Although the total portion with widely dispersed molecular weight is now used, the ultra-filtration of lignosulfonate will be adopted soon to increase the dispersion efficiency.

(2) Color

Commercial lignin will be accepted for very limited end use, because the color of lignin is rather dark. Decoloration of lignin is expensive at present, although lignin chromophores have been almost clarified by us. Table 6 shows the main chromophores to lignin color [13]. Protolignin is colorless or light yellow. The color of milled wood lignin is mostly due to the quinonoid structures. The color of thiolignin is darker than lignosulfonate. The distinct color difference between two industrial lignins is

Table 6. Main chromophores relating to lignin color.

Lignin	Fe(III) chelate	Quinoids	Conj. carbonyl
MWL	— %	100 %	— %
LSA	40-50	40	5-8
TL	10	35	5-8

MWL: milled wood lignin
LSA: lignosulfonate
TL: thiolignin

due to the extent of Fe(III)-chelate formation. It is worthy of note that the complete mechanism of thiolignin coloration is not clear at present. Thiolignin may have very unstable quinonoid structures.

(3) Solubility

Poor solubility of modified lignin causes another problem of lignin utilization. As shown in Table 7, thiolignin and modified thiolignin were treated with epichlorohydrin, and then cured with phthalic anhydride as a hardner. An aluminum test piece was used instead of wood, because lignin epoxide showed very high shearing strength. Among the results, phenolated lignin epoxide showed the excellent shearing strength, higher than 80 kg/cm^2. However, this lignin resin was not practically acceptable, because of the lack of good and cheap solvent, in spite of its attractive adhesive properties [14].

Table 7. Shearing strength of lignin epoxide cured with phthalic anhydride.

Resin	Epoxy value eq/100g	Hardner %	Resin mg/cm^2	Shearing strength, kg/cm^2 Min.	Max.	Av.	Failure[a] %
L[b]	0	10.0	5.5	14.8	21.9	17.3	0
LE[c]	0.10	11.8	6.7	38.9	51.3	45.6	0
	0.14	16.6	5.9	54.5	80.0	62.8	20
	0.16	19.2	7.4	65.9	80.0	74.5	60
	0.19	22.6	6.2	44.9	69.9	57.1	0
PLE[d]	0.28	33.3	4.9			80.0	100
ALE[e]	0.17	20.4	6.6	24.7	39.9	29.6	0
HLE[f]	0.16	19.2	7.0	26.2	40.3	35.1	0

a Alminium (90x25mm) failure : 80kg/cm^2
L: Thiolignin; LE: Lignin epoxide; PLE: Phenolated lignin epoxide; ALE: Epoxide of bisguaiacylated lignin with acetone; HLE: Epoxide of bisguaiacylated lignin with methyl hexyl ketone

(4) Low molecular weight phenols
Low molecular weight phenols by the degradation of lignin might have a large potential possibility in the utilization of lignin. Here will be referred to the phenols in acidic oil produced by the hydrogenation of lignosulfonate. Unfortunately, the hydrogenation process is not economically feasible at present, because many kinds of common phenols are obtained after the degradation as shown in Table 8 [14]. In other words, it is really important to develop a new economical process with controlled degradation of lignin.

Table 8. Phenols in acidic oil produced by hydrogenation of lignosulfonate.

(1) Guaiacol(1.6-1.9)	(2) Phenol(4.1-6.3)
(3) o-Cresol(4.4-5.0)	(4) m-, p-Cresol(12.9-20.0)
(5) o-Ethyl phenol + 2,4-Xylenol(6.3-9.3)	
(6) Propyl guaiacol(0.9-1.5)	
(7) m-,p-Ethyl phenol(3.2-5.8)	
(8) 3,4-Xylenol(3.2-5.8)	
(9) p-Propyl phenol(5.6-5.8)	(10) Catechols(5.6-13.6)

Yield of acidic oil: 10-12% lignin basis
Data in ( ): acidic oil basis
Catechols: methyl, ethyl and propyl catechols

SUMMARY

The age of wood-based chemicals has already arrived. Cellulochemicals will have greater practical feasibility in the near future. A new process for cellulochemicals will be developed in the production of cellulose derivatives, not only from highly purified pulp, but also from high yield pulp. Alcohol production from wood is also important. In this case, a route to the utilization of residual lignin after the hydrolysis must be found.

Concerning the utilization of lignin, we are now in a position to reconsider the direction of development on the background of lignin chemistry. It is worthy of note that the modification or the activation of industrial lignin is fundamentally important. Efforts are being continued to arrive at a better understanding of the relationship between the lignin structure and performance properties of lignochemicals.
Plants have the ability to prepare many kinds of

substances. Therefore, according to the improvements of cultivated plants or the use of tissue culture methods, we shall be able to expect the possibility finding various plants for medicines, perfumes, dyestuffs and so on.

REFERENCES

1) J. Nakano: Proceedings of IV Japan-Brazil Symposia on Science and Technology, vol.1 (Polymers), p.13 (Aug. 1984), Rio de Janeiro, Brazil.
2) L.T.B. Tuyet, A. Ishizu and J. Nakano: Japanese Tappi, 35, 798 (1981); 36, 1080 (1982).
3) L.T.B. Tuyet, K. Iiyama and J. Nakano: Mokuzai Gakkaishi, 31, 752 (1985); 32, 44 (1986).
4) A.F. Turback, R.B. Hammer and H.L. Hergert: Chemtech., 51 (1980).
5) A. Isogai, A. Ishizu and J. Nakano: J. Appld. Polym. Sci., 29, 2097 (1984).
6) A. Isogai, A. Ishizu and J. Nakano: J. Appld. Polym. Sci., 29, 3873 (1984).
7) R.G. Schweiger: Tappi, 57, 86 (1974).
8) K. Shimizu et al: Mokuzai Gakkaishi, 22, 616 (1976).
9) T.J. Eom: Master Thesis, Kyoung Buk Univ., Korea (March, 1985).
10) E. Adler: Wood Sci. Technol., 11, 169 (1977).
11) Y. Matsumoto, A. Ishizu and J. Nakano: Proceedings of 1983 ISWPC, vol.4, p.68 (May, 1983), Tsukuba, Japan.
12) A. Rezanowich, J.E. Jaworzyn and D.A.I. Goring: Pulp and Paper Mag. Can., 62, T172 (1961).
13) K. Iiyama and J. Nakano: Japanese Tappi, 27, 182, 530 (1973).
14) J. Nakano: R and D Report, No.40, 1 (1983), CMC, Japan.

Chapter 56

Adhesive bonding to wood

D. M. Brewis, J. Comyn and C. Phanopoulos – Senior Lecturer, Reader, Research Student, School of Chemistry, Leicester Polytechnic.

ABSTRACT

The effects of accelerated aging on adhesively bonded wood joints are examined using extracted, partially extracted and non-extracted wood. Surface analysis by XPS and SIMS are used to examine the various wood surfaces. The level of surface extractives is seen to be a major influence on the quality of the joints, both in terms of durability and strength.

INTRODUCTION

This paper summarises some of our investigations on the interactions between formaldehyde-containing adhesives and softwoods, the work being sponsored by The Building Research Establishment.

Two areas of the work will be considered here. Firstly, durability studies of joints constructed with woods extracted to different levels and secondly some surface analysis, by X-ray photoelectron spectroscopy (XPS) and secondary ion mass spectroscopy (SIMS).

The adhesives used were urea-formaldehyde (UF), and phenol-resorcinol-formaldehyde (PRF) which were kindly supplied by Ciba-Geigy Ltd; they were free from any fillers or extenders. It has long been known that PRF adhesives are superior to UF's in both durability. Our principal interest, however, has been to discern the mechanism of adhesion to wood and to determine if possible which groups in wood may be involved in adhesive bonding.

DURABILITY

Scots pine (Pinus sylvestris) was cut and planed to the dimensions required by British Standard 1204 (BSI, 1979). Materials for single lap joints were subsequently prepared in one of the following ways.

(a) Non-extracted;

(b) Water extracted. This was achieved by immersing the substrates in boiling water for 7 hours;

(c) Solvent extracted. This was achieved by extraction in a Soxhlet apparatus with each of the following solvents in sequence for more than 12 hours each (Rodwell, 1983); dichloromethane, benzene/ethanol (2:1 by vol.), acetone, methanol. This procedure is responsible for the removal of extractives. Only a small amount of lignin is removed.

Single lap joints (25 mm x 25 mm overlap) of Scots pine wood were constructed with UF or PRF and left to cure. After two weeks, the bonded joints were placed in a glove box at >95% relative humidity and 30°C, and left for various lengths of time up to 1800 hours. The aged joints were then tested to destruction on a Monsanto Tensometer with a crosshead speed of 5 mm min^{-1}. Strengths of aged joints are shown in Figures 1 and 2.

X-RAY PHOTOELECTRON SPECTROSCOPY

This work was carried out at Loughborough University of Technology on a VG ESCALAB Instrument with X-ray energy (AlK$_{\alpha}$) = 1487 eV the analysis area was approximately 5 mm x 5 mm and the beam diameter 6 mm. The samples were 100 μm sections of spruce (Picea abies) cut with a microtome. Some samples were aged at >95% RH at 30°C for six weeks. The results of these analyses are shown in Table 1. XPS has a sampling depth of 5 - 10 nm and is highly surface sensitive (Briggs and Seah, 1983).

SECONDARY ION MASS SPECTROSCOPY

This work was carried out at ICI Plc. (Petrochemicals and Plastics Division) on a VG MM 12-12 instrument with a quadrupole mass spectrometer. The analysis area was 6mm x 6mm, primary beam current 0.4 x 10^{-3} A and the primary ion was Xe^{+} (2 keV). Secondary ions are ejected from the surface and analysed in a mass spectrometer. SIMS Spectra of wood contain a large number of fragments and Figures 3 and 4 are given as examples.

Table 2 lists the samples that were studied by SIMS. In order

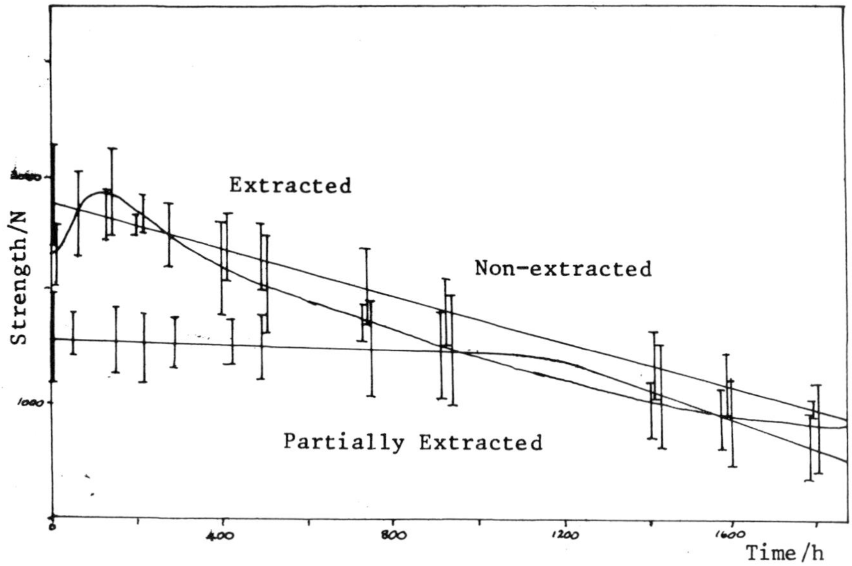

Fig. 1. Joint strengths of UF bonded Scots pine during aging (95% rh, 30°C) using extracted, non-extracted and partially extracted wood.

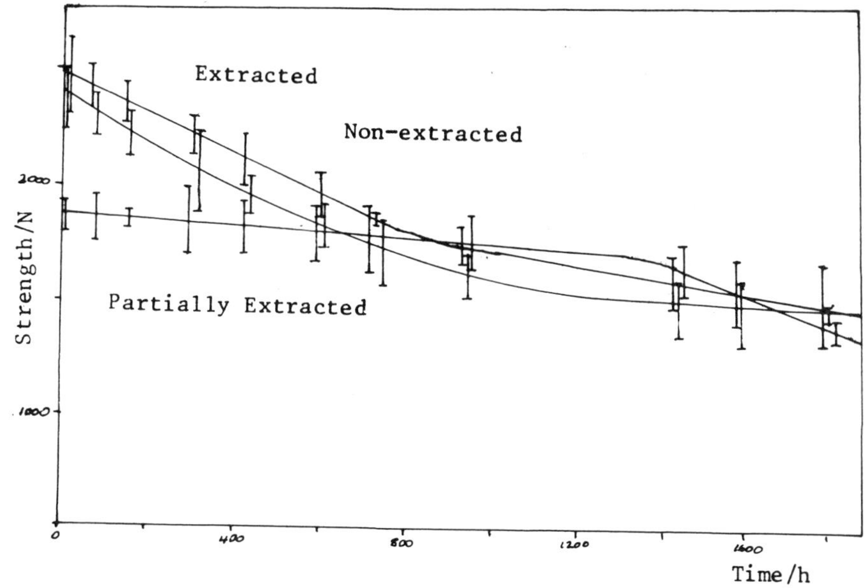

Fig. 2. Joint strengths of PRF bonded Scots pine during aging (95% rh, 30°C) using extracted, extracted, non-extracted and partially extracted wood.

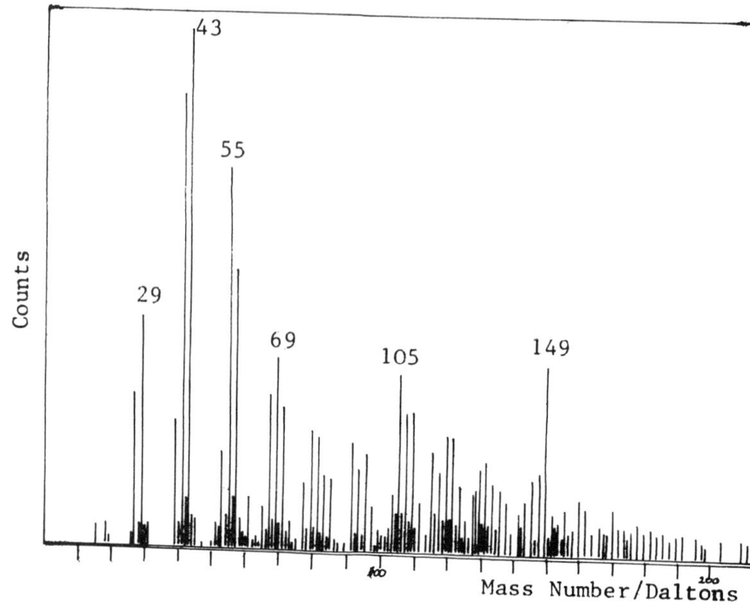

Fig. 3. Positive ion SIMS spectrum of urea cloped non-extracted spruce.

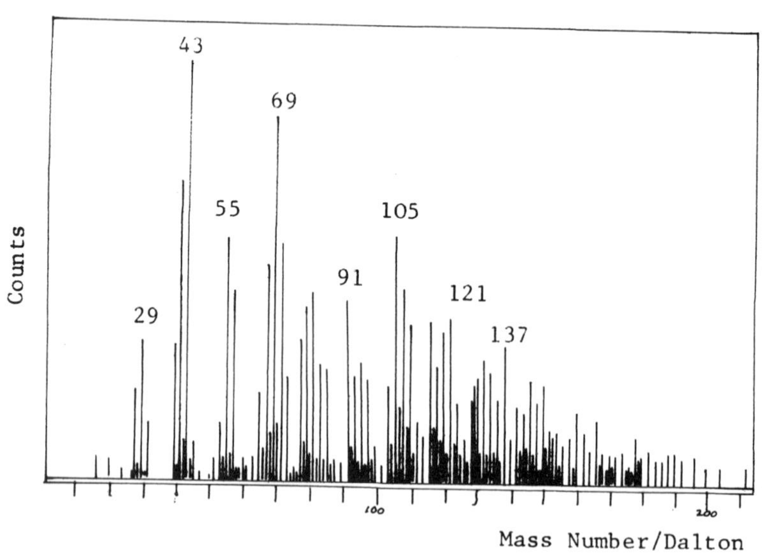

Fig. 4. Positive ion SIMS spectrum of non-extracted spruce.

Table 1. XPS analysis of extracted, non-extracted and partially extracted sprucewood – before and after aging.

	% ATOM COMPOSITION				
Sample	C	N	O	Contaminants	O/C Ratio
Non-Extracted	80.9	0.0	18.3	0.9 (Si)	0.23
Part-Extracted	84.3	0.0	14.6	1.1 (Si)	0.17
Extracted	70.9	0.0	25.6	2.6 (Si)	0.36
Aged non-Extracted	83.3	0.0	16.7	0.0	0.20
Aged part-Extracted	69.5	1.7	27.1	1.7 (Fe)	0.39
Aged Extracted	63.5	0.0	36.5	0.0	0.57

to investigate the mechanism by which adhesives bond to wood, some samples were treated with the adhesive components, intermediates or model compounds which are indicated in Table 2. By comparing spectra of treated and untreated woods, we were able to attribute changes to interactions between the adhesives and wood.

DISCUSSION

The results of tensile strength tests of aged joints (Figures 1 and 2) show a number of features common to both adhesives. Firstly, for the first 800 hours of aging, extracted wood provides stronger joints than non-extracted wood, and these in turn are stronger than partially extracted wood joints. Secondly, for approximately the first 1100 hours of aging, the water extracted joints show no weakening. Finally, after approximately 1000 hours aging, the joint strengths of the three types of wood converge to approximately the same strengths.

How can these changes be explained? Traditionally, loss of strength of adhesive joints is usually said to be due to hydrolytic degradation of the resin and of any resin-substrate bonds. This may still be true; but what of the differences due to the different treatments of wood? It seems that the extractives content at the surface influences bond strengths. This is indi-

Table 2. Sample analysed by SIMS.

1. Extracted and non-extracted spruce.
2. Extracted and non-extracted Scots pine.
3. Softwood lignin.
4. Urea non-extracted spruce and urea.
5. Dimethylol urea treated non-extracted spruce and dimethylol urea.
6. Methylene diurea treated non-extracted spruce and methylene diurea.
7. Phenol treated non-extracted spruce.
8. 2-Hydroxybenzyl alcohol (2.HBA) treated non-extracted spruce and 2.HBA.
9. Bis(2-hydroxyphenyl)methane (B2-HBA) treated non-extracted spruce and B2-HPM.
10. Paraformaldehyde treated non-extracted spruce and paraformaldehyde.
11. Cured detached films of VF, PRF and melamine-formaldehyde (MF) adhesives.

cated by the XPS data which shows the oxygen to carbon ratio variation with level of extraction. However, the amounts of extractives are also related to the availability of cellulose, high levels of extractives may restrict adhesives entering the wood pores.

For water extracted wood, resin penetration is reduced and some of the resin-extractive bonds will form to the detriment of the resin-cellulose bonds. Indeed, strengths of the partially extracted wood joints are initially stable and do not lose strength. This suggests a certain amount of hydrolytic stability of these interactions.

The extracted wood allows increased adhesive penetration and has more cellulose available for bonding compared to partially extracted wood. These two factors could account for the superior strengths of joints with these types of wood.

SIMS may provide information on the manner in which molecules are attached to wood, and as an example spectra of non-extracted spruce both before and after treatment with urea are shown in Figures 3 and 4. These reveal the formation of new fragments (e.g. M = 64) and the loss of others (e.g. M = 148). Further, there are numerous changes in relative intensities (counts) of other fragments. By comparing non-extracted and extracted spruce spectra, it is possible to identify whether the fragments originated from cellulose plus lignin or from the extractives. Thus, spectral changes on doping can be related to dopant-wood component interactions. In this way, Table 3 shows the number of spectra changes observed for urea and other dopants which indicate interactions with either cellulose-lignin or extractive components. We have regarded a spectral change as worthy of counting only when its intensity changes by more than 20%. Thus, it seems that substantial interactions occur between dopants (and therefore also the adhesives) and wood. So, interactions between the adhesives and cellulose-lignin may account for the increase joint strengths noted in the aging of joints work.

XPS has shown that during aging the remaining extractives migrate to the wood surface. Hence, the surfaces of the three types of wood come to resemble each other. This would account for the convergence of the strengths of joints constructed with the three types of wood.

Table 3. Number of SIMS spectral changes on doping non-extracted spruce.

Dopant	Cellulose	Extractive
Urea	77	48
Phenol	9	20
Formaldehyde	41	47
Methylene Diurea	23	33
Dimethylol Urea	15	23
B2-HPM	14	32
2.HBA	47	51

REFERENCES

BS1204 : Part 2 : 1979 Synthetic Resin Adhesives (Phenolic and Aminoplastic) for Wood Specifications for Close Contact Adhesives.

Rodwell, D., (1983), Private Communication.

Briggs, D. and Seah, M.P., (1983), Practical Surface Analysis by Auger and X-ray Photoelectron Spectroscopy. John Wiley and Sons, New York.

Chapter 57

Economics of accelerated bonding

George B. Harpole and Robert L. Geimer – Research Forester and Research Forest Products Technologist USDA Forest Service, Forest Products Laboratory.

ABSTRACT

With consideration of potential investment and operating cost savings, isocyanate binders used with steam-injection techniques appear to be economically competitive with phenolic binders and conventional pressing systems.

INTRODUCTION

Research related to the development of structural flakeboard production has created interest in use of isocyanate binders in combination with, or as a substitute for, traditional phenolic binders [1,2,3,7,9,10,12]. The good technical characteristics of panels produced using isocyanate binders and potential for cost savings continue to encourage their consideration. Incentives for using isocyanate binders include attainment of higher internal bondings, tolerance for higher moisture contents in press furnish, shorter press times, lower press temperatures, and potential savings in quantity of wood raw material. The most serious obstacle to the use of isocyanate binders is currently recognized as the direct cost of substituting isocyanate binders for phenolic binders.

In this paper computerized simulations of structural flakeboard manufacture are used to assess manufacturing costs for a diversity of processing assumptions related to binders [6,8]. Costs are based on facilities of an efficient size producing 150 million square feet per year of 7/16-in-thick panel per year [11]. Costs are computed on an after-tax basis, allowing for investment tax credits equal to 10% of processing equipment costs; an effective tax rate of 46% with straight-line depreciation for a 10-yr facility life and 10% salvage value; and an after-tax return (internal rate of return) on investment of 10%.

PRODUCTION COSTS

We estimated production costs for three factory designs. The first is a standard design for use of phenolic binders with dryers and press capacities sized to allow a reduction of furnish moisture content to 6% and to permit button-to-button press cycles to average approximately 7 min [10]. A 24-opening 8- by 24-ft press was assumed. According to Hse's work, with minor modifications such a facility should allow serial application of isocyanate and phenolic binders, or conceivably use of isocyanate alone [7]. Estimated total production costs for phenolic-bonded panels and isocyanate-bonded panels are shown in table 1.

Production costs for combined use of phenolic and isocyanate binders can be assumed to fall somewhere between computed costs for exclusive use of either binder.

The second design is for a factory using only isocyanate binder. Considerable reduction of capital investment requirements might be achieved by designing manufacturing facilities specifically for use of isocyanate binders. Because of the significant costs related to the press section, reductions in the number of press openings will have the greatest effect toward reducing costs for producing panels. Other ways in which capital costs might be reduced by designing facilities specifically for use of isocyanate binders are:

- with tolerance for higher moisture content, dryer capacities might be reduced 50% to 75%;
- with reduced resin requirements, resin storage facilities might be reduced 20%;
- with shorter pressing times, press load capacities might be reduced 30% to 65%;

Table 1. Estimates of structural flakeboard production costs for three factory designs and two binders.

Cost	Phenolic design		Isocyanate design	
			Conventional	Steam-injection
	Phenolic binder	Isocyanate binder	Isocyanate binder	Isocyanate binder
	- - - - - - - US$ million - - - - - -			
Construction (1985)	35.25	35.25	33.00	30.73
Panel costs:	- - - [1]US$/mft[2], 7/16 in thick - -			
Variable costs:				
Wood ($25/cord)	22.85	22.85	22.85	22.85
Resin[2]	34.14	38.48	38.48	38.48
Wax (0.5%)	1.99	1.95	1.95	1.95
Electricity (4¢/kWh)	8.75	8.75	8.75	8.75
Dryer heat	.63	.38	.38	.63
Process steam	.18	.07	.07	.18
Subtotal	68.54	72.18	72.18	72.84
Process labor	10.53	10.53	10.53	10.53
Fixed costs:	27.56	27.56	26.15	24.78
Capital recovery	15.75	15.75	14.75	13.67
Taxes (46%)	9.48	9.51	9.06	8.55
After-tax profit (10%)	14.00	14.05	13.16	12.40
Total panel cost	145.86	149.88	146.13	142.77

[1]mft[2] = thousand square feet.

[2]Phenol formaldehyde binder at US39¢/lb, dry solids; isocyanate binder at US70¢/lb, dry solids.

- with lower press temperature requirements, low-pressure boilers might be used in place of high-pressure systems;
- with reduction in processing equipment, an associated reduction in maintenance and repair costs may be realized, and profit and taxes will be reduced proportionately while still yielding the same return on investment.

In the second design, moisture content of furnish from the dryer is allowed to be 12%, press cycles to average 5 min with pressing temperatures of 180°C [10,11]. Therefore, we assume design requirements and costs are reduced through reduction of dryer capacities, decrease in press openings to 17 openings, and use of a lower cost boiler system. We assumed reductions of US$2,250,000 in processing equipment costs with investment tax credit and depreciation allowances reduced proportionately. Repair and maintenance costs and requirements for taxes and profits are also reduced. The production cost for isocyanate-bonded panels was approximately equal to that for phenolic (Table 1).

The third design is for a factory using isocyanate binders and steam injection into the mat during pressing. The injection of steam into the flakeboard mat during pressing significantly reduces press times and permits production of thick panels (2 in thick or thicker) with little more than proportional changes in press cycle time and steam energy from producing 7/16-in-thick panels (4,5). For 7/16-in-thick panels, press cycles are assumed to average 2.5 min, and press openings thereby reduced to 9. Moisture content of furnish from the dryer is taken at 6%, with addition of moisture via steam injection at the press to bring the cured panel moisture content to 7%--the same as from the other two factory models. Production costs are US$3 per thousand square feet (mft^2) less than for phenolic-bonded panels.

We assumed binder costs of US39¢/lb (dry solids) for phenolic and US70¢/lb (dry solids) for isocyanates. Costs for isocyanate-bonded panels are somewhat reduced on the assumption that 3.5% (dry weight of panels) of isocyanate binder sufficed to give panel properties comparable to those obtained with 5.5% of phenolic binder. The variable costs (for wood, chemicals, and energy) are increased by the higher isocyanate resin costs. We assumed that, for isocyanate bonding, furnish would be dried only to 12% moisture content as opposed to 6% for phenolic and steam-injection isocyanate bonding, and that press temperatures could be 325°C for isocyanates as opposed to 400°C for phenolics. We allowed the same costs for wood, wax, and electricity in all simulations.

If reductions in wood requirements can be accomplished--possibly up to 15% [10]--further reductions in production costs might be achieved.

SENSITIVITIES OF PRODUCTION COSTS

Because of the variability of costs at diverse locations and over time, sensitivities of various cost elements are calculated (Table 2).

Table 2. Sensitivity of panel costs.

Factors	Change	Change in panel cost	
		Phenol	Isocyanate
		- [1]$/mft²-7/16 in -	
Direct Costs:			
Wood	$1/cord	0.91	0.91
Resin	1¢/lb	1.033	0.644
Wax	1¢/lb	0.094	0.092
Electricity	1¢/kwh	2.582	2.581
Natural gas	1¢/mft²	0.011	0.006
Percent of Panel Weight:			
Wood	1%	0.583	0.634
Resin	1%	7.32	12.97
Wax	1%	4.70	4.65

[1]mft² = thousand square feet.

Panel costs are reduced US63¢/mft² (7/16 in) for each percentage point reduction in wood requirements when using isocyanate (Table 2). This reduction reflects reduced requirements and costs for wood, resin, wax, and energy for drying and pressing, but assumes electrical costs will remain constant. The drop in panel costs associated with reducing wood requirements when using phenolic binders is US58¢/mft² (7/16 in) per percentage point of change. The reduction is less than when using isocyanate because resin cost per cubic foot of bonded wood is less. Production costs appear significantly sensitive also to the percentage of binder used and the amount of capital investment for processing equipment.

As indicated in Table 2, a change of 1% in dry resin weight effects a change of US$7.32/mft² (7/16 in) for phenolic binder and a change of US$12.97/mft² (7/16 in) for isocyanate binders when cost for these binders are respectively US29¢/lb and US70¢/lb of solids changes may be necessary to attain given board properties. These responses to change will be proportionate to any change in costs for binders.

Our analysis shows a reduction of US$1 million in initial investment costs can be expected to reduce panel costs by US$1.45/mft² (7/16 in). Thus, the potential for reducing production costs by reducing processing equipment requirements offers a way in which isocyanate binders may become economically competitive with phenolic binders.

REFERENCES

1. Deppe, J. J. 1977. Technical progress in using isocyanate as an adhesive in flakeboard manufacture. Proceedings of 11th Washington State University Symposium on Particleboard, p. 12-21.

2. Ellingson, G. P. 1977. ELCOBOARD: A structural composite panel. Proceedings of 11th Washington State University Symposium on Particleboard, p. 269-282.

3. Frink, William J., and H. Daryle Layton. 1985. Polymeric MDI in board production--it really works (an overview of mill experiences). Proceedings of 19th Washington State University Symposium on Particleboard, p. 323-347.

4. Geimer, Robert L. 1982. Steam injection pressing. Proceedings of 16th Washington State University Symposium on Particleboard, p. 115-134.

5. Geimer, Robert L. 1985. Steam-injection pressing of isocyanate-bonded aspen flakeboards--latitudes and limitations. Res. Pap. FPL 456. Madison, WI: U.S. Department of Agriculture, Forest Service.

6. Harpole, George B. 1978. Cash flow analysis computer program for analyzing manufacturing investment. Res. Pap. FPL 205. Madison, WI: U.S. Department of Agriculture, Forest Service.

7. Hse, C. Y. 1978. Development of a resin system for gluing southern hardwood flakeboards. Gen. Tech. Rep. WO-5. Washington, DC: U.S. Department of Agriculture, Forest Service, p. 81-92.

8. Ince, Peter J., and George B. Harpole. 1977. PARVCOST: A flakeboard variable cost program. Gen. Tech. Rep. FPL-14. Madison, WI: U.S. Department of Agriculture, Forest Service.

9. Johns, William E. 1980. Is there an isocyanate in your future? Chemical aspects. Proceedings of 14th Washington State University Symposium on Flakeboard, p. 177-184.

10. Mobay Chemical Corporation. 1978. Evaluation of Mondur E-441 and Mondure MR polymeric isocyanates as binders for aspen waferboard. Unpublished report: A joint project of Mobay Chemical Corporation, Pittsburgh, PA, and the Canadian Eastern Forest Prod. Lab., Ottawa, Ontario.

11. Vajda, Peter. 1978. Plant facility considerations for structural flakeboard manufacture. Gen. Tech. Rep. WO-5. Washington, DC: U.S. Department of Agriculture, Forest Service, p. 122-129.

12. Wilson, James B. 1980. Is there an isocyanate in your future? Property and cost comparisons. Proceedings of 14th Washington State University Symposium on Flakeboard, p. 185-192.

Chapter 58

Improving the performance of paint on timber

E. M. Pearce A.I.W.Sc. – Timber Treatments Division, Fosroc Ltd., Marlow, Bucks, U.K.

CONCEPT

Over the years there have been many opinions, proposals and suggestions as to how paint and decorative finishes can be made to last longer on wooden joinery. It is the intention of this paper to describe one particular approach to the problem of improving the paint/decorative finish performance in service.

However before the research and development that was undertaken is discussed, it is useful to review the factors which contribute to the breakdown of a paint or decorative finish in service. These can be listed as :

a) chemical
b) UV light
c) heat
d) water
e) biological e.g. staining fungi, bacteria

These factors have not been listed in any order of importance. In fact it is well documented that it is usually a combination of these factors which is involved in the breakdown of surface finishes, and so ideally a finish should provide protection against all of these factors.

However, since protection against each of these factors requires specific properties, it was considered that formulating a product which could provide protection against breakdown by all of them would be extremely difficult and prohibitively costly.

Therefore, effort was concentrated on providing protection from the effects of biological degradation, water, and to a limited extent, UV degradation. The other factors listed were not considered because it is the combination of UV light, water and biological agents which produce the most dramatic adverse effects on timber in the shortest time.

The concept that was developed by Fosroc, (Timber Treatments Division), based at Marlow, was for an organic solvent wood preservative that could be impregnated by a double vacuum process into timber using a large scale industrial apparatus such as Protim® Prevac. The chemistry of this preservative should be such that, when timber treated with it is over coated with a decorative finish, the preservative and the finish should lock together, thus providing an intergrated barrier to the factors listed above.

FORMULATION DEVELOPMENT

Protection against water

Protection of timber against water may be achieved both by treating the timber to make it water repellent, and by ensuring that it is covered by a sound decorative finish.

The usefulness of water repellent additives in wood preservatives has been recognised for a long time. A report prepared by Browne and Schwebs 1944 (1) discussed methods for measuring water repellency and also of the effectiveness of various water repellents, indicating the advantages of using these materials on timber. Banks (2) further confirmed the benefit of water repellent in wood preservatives. In the light of this, it was decided to include in the formulation a hydrophobic ingredient.

To achieve the goal of providing a treatment which would bond the timber with a paint or stain and so enhance water resistance, a good deal of research was required. Stalker (3) had indicated that the inclusion of a resin assisted in the overpainting of preservative treated timber. To determine which resin would have the most beneficial effect numerous resins of the types shown in Table 1 were screened.

The initial screening involved assessment for water uptake, splitting, cupping and film flaking of slotted panels which had been treated by double vacuum impregnation with simple solutions of the various resins and then dipped primed with an ICI fast drying alkyd primer. After three days drying the panels were subjected to natural exposure on a horizontal rack for 6 months.

Table 1. Range and the type of resins subject to initial screening.

Oil Type	Esterfying alcohol
Linseed	Pentaerythritol
Linoleic rich	Pentaerythritol
Dehydrated caster	Glycerol
Tung/linseed	Pentaerythritol
Linseed	Glycerol
Segregated marine	Pentaerythritol
Tung/linseed phenol modified	Pentaerythritol
Semi drying vinyl toluene modified	-
Soya	Glycerol

PROVING THE SYSTEM

Having established the basic ingredients that could achieve improved decorative finish durability, a series of tests was carried out to quantify the improvement that the product would provide. These included evaluation of paint adhesion, natural weathering exposure and water absorption.

Table 2 shows the results obtained from paint adhesion tests using an Elcometer Tensiometer. These tests were carried out on matched *Pinus sylvestris* specimens variously treated with Protim® R* and a conventional organic solvent preservative containing a non reactive resin. Before testing, the panels were subjected to 1, 6 and 24 hours immersion in water at 22°C.

* Name selected for the resin based preservative developed by Fosroc Ltd.

Table 2. Paint adhesion (after immersion in water). Comparison of Protim R® with a conventional preservative based on a non-reactive resin system.

Immersion period	Preservative	Ave. failing load kg/cm^2	Type of failure
1 hour	Non reactive conventional	26.9	Paint film
	Protim R	32.7	Paint/ wood
6 hours	Non reactive conventional	18.1	Paint film
	Protim R	37.9	Paint film
24 hours	Non reactive conventional	0.9	Paint film
	Protim R	24.9	Paint film

These results endorse the results published by Kyte (8) who also showed that the use of a reactive resin in a wood preservative improved paint adhesion.

The results from the initial screening showed that the oil length was important. Based on these initial results, a secondary screening was carried out and this yielded results which enabled the resin which is now used to be selected.

Fungicidal protection

The task of finding a suitable fungicide which would give protection from wood rotting fungi was more straight forward than identifying the most suitable resin. This was because the number of fungicides which were suitable and acceptable to the British Wood Preserving Association (4) and British Standards (5) for use in wood preservatives for the treatment of external joinery was limited to 7. Out of the 7 only 2 (namely Pentachlorophenol and Tributyl tin oxide) were found to be suitable from a compatability point of view with the resin system selected and these were used in candidate formulations for assessment by mycological testing according to BS 838 (6). From these tests tributyl tin oxide was selected for use in all further candidate formulations and tests.

UV PROTECTION

It is well documented that UV light can cause changes to the surface of timber which have an adverse effect on the performance of applied paints etc. It is also well documented that pigments are very efficient UV absorbers (7). It was therefore decided to incorporate a pigment into the formulation to improve the overall performance of the treated decorated timber, but there were several factors which had to be taken into account before this was achieved successfully:-

1) pigment type
2) pigment size
3) product stability

Iron oxide based pigments were chosen for study because they are the most cost effective means of providing UV protection. Pigment particle size was investigated bearing in mind that the product was to be supplied in bulk and applied to timber in a large industrial process, and adequate penetration into *Pinus sylvestris* (European pine) was essential, this being the predominant timber species used in the manufacture of joinery in the U.K. After exhaustive testing a pigment was selected which fulfilled the requirements of type, particle size and, when correctly formulated, produced a stable suspension.

NATURAL WEATHERING

Having established by adhesion tests that improved bond strength was achieved between Protim R® treated timber and paint systems compared with timber treated with a conventional organic solvent preservative, the question remained "was this improvement of paint performance on Protim R® treated timber maintained when the timber was exposed to the weather"? To answer this question a programme of natural weathering tests was carried out.

The performance of treated and painted panels during natural weathering exposed at 45° facing south was monitored for 4½ years.

Panels treated with Protim R® and Protim R® Coloured and painted with a conventional fast drying oil based primer/undercoat/gloss top coat provided a sound finish for approximately twice as long as panels treated with a conventional preservative and finished with a similar paint system.

Also on panels where a conventional primer coat had been omitted and the finish consisted of only undercoat and gloss top coat, the durability of this system was as good as a 3 coat paint finish on panels treated with a conventional preservative.

Weathering tests of panels with decorative stains were also carried out. Although these were not as extensive as those with primer/undercoat/gloss coat, the tests gave similar results to those obtained with painted panels.

WATER INGRESS

So far this paper has discussed data which shows that the bond between an oil based decorative system and Protim R® treated timber is significantly improved when treatment was carried out with Protim R® compared with a conventional organic solvent preservative. The benefits of this are improved paint performance both in decorative terms and also in enabling the paint to maintain another important function in providing a barrier to water ingress. The presence of a sound paint film obviously limits water penetration through lateral surfaces.

However the most likely source of entry of water into a window frame is via an end grain, either in the sill or in the mortice and tenon joints. Miller (9) strongly advocates the application of a good quality primer paint to these areas as a means of preventing water ingress into the joinery.

Unfortunately modern methods of primer paint application such as electrostatic spray to assembled units do not achieve high paint loadings on end grain areas and as a consequence the level of protection against water ingress is very low.

We were interested to know whether Protim R® provided any benefit in limiting water entry through an end grain. This was examined by testing specimens of timber *Pinus sylvestris* (European pine) by the technique described by Miller.(9) This involved taking matched 150 x 38 x 38 mm pieces of sapwood and applying the candidate treatment to one piece. The treated and untreated pieces were then placed on cotton wool saturated with water, and the water rise into the test pieces was monitored by measuring the moisture content along the length of the piece at 10mm intervals using an electrical resistance moisture meter. The following graph shows the effectiveness of Protim ® treated timber in minimising the water ingress.

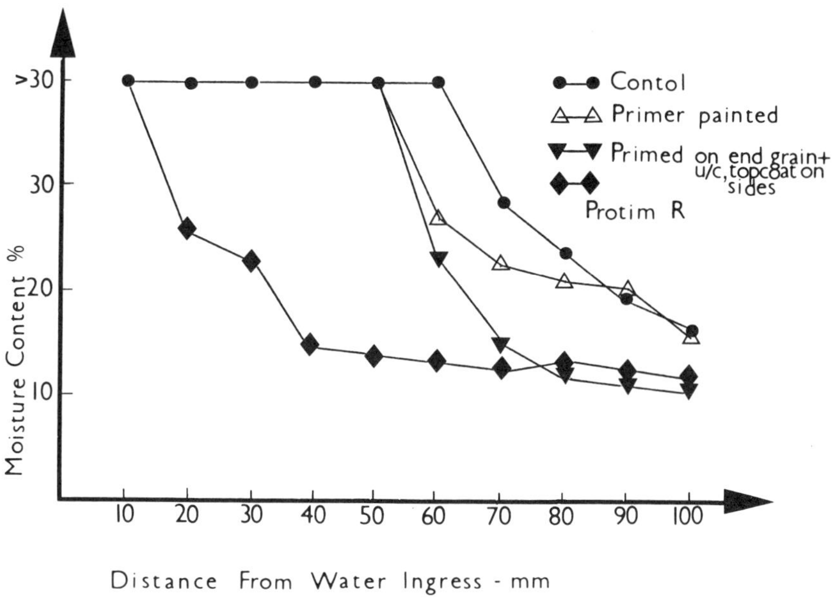

PRESERVATIVE LOADING

A vital aspect of the evaluation of Protim R® was a study of treatment characteristics during double vacuum impregnation. The British Wood Preserving Association (10) have indicated in their preservation standards that a minimum loading of 8% m/m in the outer 3 mm of sapwood should be achieved when using an conventional organic solvent based preservative. When such a preservative is applied by the double vacuum cycle suitable for the treatment joinery a preservative loading averaging around 15% w/w in the outer 3mm of sapwood of *Pinus sylvestris* is normally found. As Protim R® and Protim R Coloured® were formulated using a reactive resin system, and the resin level in the product was significantly higher than the quantity of non reactive resin normally used in a conventional preservative, the effect of resin content on preservative absorption was examined using three progressively more severe treating schedules.

Table 3. Effect of resin concentration on preservative absorption.

Resin Content	Treatment Schedule (1,2,3) and absorption of preservative (l/m^3)		
	1	2	3
6.5%	7.2	10.8	90.8
10.0%	6.0	10.6	77.6
15.0%	5.2	8.8	63.6

These results showed quite clearly that increased concentrations of resin significantly reduced the preservative loading, but this reduction could be counteracted by using a more severe schedule. In timber treatment it is important that the requisite quantity of preservative is deposited in the wood and so a series of treatment tests were carried out to ascertain which treatment schedule should be used to achieve a loading of Protim R® which would be similar to that of timber treated with a conventional preservative applied by the accepted cycle for the treatment of joinery.

As a result of these tests a schedule was developed, where the level of the initial vacuum was increased and the immersion period extended, and this resulted in average preservative loading in the outer 3 mm of sapwood of 14.9% w/w.

CONCLUSIONS

By the use of a correct combination of alkyd resin and iron based pigments and formulating this combination into a wood preservtive and impregnating it into *Pinus sylvestris* (European pine), it has been demonstrated that the effective life span of an applied oil based paint or decorative stain system is improved. Furthermore it has been shown that treatment of *Pinus sylvestris* (European pine) with Protim R® is a very efficient method of preventing water penetration into the vulnerable end grain areas of wooden joinery components providing further benefit to the joinery.

These benefits have considerable practical advantages for the joinery producer and the users of wooden joinery.

The advantages in joinery production are that in one industrial process which does not involve handling the timber, the joinery is protected against biological degradation and is also impregnated with resin and pigments which fulfil certain aspects of the function of a primer paint coat.

Subsequently applied paint coats may be either undercoat followed by a gloss coat which gives a performance equivalent to a primer/undercoat/gloss coat system applied to timber treated with a conventional preservative,or, for a superior performance the normal 3 coat paint system may be applied. Timber so treated will have advantages in service because of reduced costs of maintenance. In addition the restriction in water penetration through the end grain will also help substantially in extending the service life of joinery.

Modification of timber and the added value benefits as achieved when using a treatment such as Protim R®, are important in ensuring that timber products maintain their rightful place as the premier material used in building construction.

REFERENCES

1) Browne F.L. and Schwebs A.C. (1944) A study of methods of measuring the water repellency of water repellents and water repellent preservatives for wood (Ref. for FDR, Prod. Lab, Madison)

2) Banks W.B., (1971) BWPA Convention Record 1971 129 - 147

3) Stalker I (1972) BWPA Convention Record 1972 161 -181

4) British Wood Preserving Association Standards Commodity Specification C5 : 1975

5) British Standard BS 838 : 1961 : Methods of test for toxicity of wood preservatives to fungi

6) British Standard BS 5589 : 1978: Code of Practice for Preservation of Timber

7) Marvuglio P., Sharrock P.F., Kennedy R.J., (1978) Journal of Oil Colour Chemists Association 61, 79 - 85

8) Kyte C.T. (1975) Holz als Roh und Werkstoff (33) S. 94-96

9) Miller E.R. (1983) JOCCA (10) 308 - 315

10) British Wood Preserving Association Specificiation : Standards 112 - 118 : 1975

Chapter 59

Effects of weathering on wood-coating systems

E. R. Miller – Head of Finishes Section, BRE Princes Risborough Laboratory.

ABSTRACT

The function of exterior coatings is to protect wood against the action of weathering agencies, and ultimate performance should seemingly depend on the weather resistance of the coating. However experience shows that it can often be the wood which is the determinant of performance, for if the integrity of the protective system is breached a rapid and possibly irretrievable degeneration in substrate condition can occur. The development of high moisture contents in permeable timbers imposes physical stresses on coatings and can lead to chemical and biological deterioration of the substrate and coating. Photodegradation leads to a loss in the cohesive strength and paint holding capabilities of wood at the interface.

This paper will present results of recent PRL research into the chemical and physical aspects of weathering processes occurring in wood-coating systems.

INTRODUCTION

Considerable difficulties are often encountered in providing effective surface protection for exterior wood components. Individual instances of unsatisfactory coating performance may have a variety of causes, but there have been two general factors of particular importance. The first has been the use of surface coatings which are inadequate for the challenging conditions often presented by exterior timber. The second has been the failure to ensure, in the design, construction and maintenance of wooden components, that acceptable substrate condition will be retained in service.

The surface coatings industry has made considerable efforts to introduce improved wood coatings and there is now available a growing range of natural and opaque finishes designed for use on exterior wood. Some are developments of well-tried alkyd resin technology, others, such as the water-borne acrylics, break new ground. However much higher their intrinsic durability, these coatings will only demonstrate their full potential if the degeneration in substrate condition which commonly occurs in practice can be prevented.

The effects of the two most important factors involved in the weathering of wood, namely water and photodegradation are considered in this paper.

Effects of water

The traditional philosophy for the protection of exterior timber components against wetting has involved the application of a surface coating intended to provide a barrier to entry of water and stabilise moisture contents within the range of 16–18 per cent. Experience and testing have shown that jointed components of permeable timber can achieve moisture levels very much higher than this after a few weeks of service. The main route for water entry is via end-grain in the joints, since moisture permeability in the longitudinal direction can be several hundred times higher than through lateral wood surfaces (Miller and Boxall 1984).

High moisture contents have profound consequences for the wood-coating system. They provide the conditions necessary for the sequential colonisation of the wood by a range of micro-organisms, which first cause increased wood porosity, and hence susceptibility to wetting, and eventually decay (Carey and Bravery 1985). Wetting also causes dimensional wood movement which imposes mechanical stresses on the coating, and results in impairment of adhesion which is particularly serious for aged coatings. Observations by Banks and Evans (1984) that long-term exposure of wood to water alone at 55–65°C can cause losses in tensile strength and toughness suggest that wetting may adversely affect the wood-coating interface.

The solution to the problem of protecting fabricated components which has tended to gain acceptance has essentially been the general introduction of adequate levels of fungicide throughout the timber to prevent decay. The use of coatings which possess greater moisture permeability has also been favoured, though there is now some evidence that the long-term overall wood moisture content under such paints does not differ significantly from that under a traditional paint system. A degree of protection against water ingress into joints can be achieved by the use of water repellent preservatives containing combinations of hydrophobic resins and waxes. Such treatments can perform well initially but water repellent effectiveness generally declines during the first few years. This loss of performance could stem from the disruption of wood tissue on wetting (Banks and Voulgaridis 1980) or uneven distribution of hydrophobes within the wood (Banks and Carragher 1984).

Recent work in BRE has focused on the issue of water entry into end-grain and has included a reappraisal of the once traditional practice of sealing end-grain by the application of primer to the joints, and assembling while the paint is wet. Results of an 80 week natural weathering trial on L-joints have been reported (Miller and Boxall 1984). Subsequent work confirms that end-grain sealing can produce a dramatic improvement in moisture content control, and substantial enhancement of paint performance. It also reduces the increase in wood porosity which commonly occurs during exposure and which is so detrimental to long-term performance.

Effects of photodegradation

The profound effects on the properties of wood substrates which result from exposure to sunlight and weathering are well recognised. Boxall (1977) has demonstrated that weathering adversely affects the paint-holding properties of Baltic redwood, and is thus an important factor in the high maintenance costs of exterior woodwork. It is also accepted that the general limitation on the service life of clear and semi transparent wood coatings out of doors is degradation of the substrate by light transmitted by the coating.

Consideration of the energy requirements for the photolysis of wood constituents indicates that the most destructive wavelengths should lie in the ultra-violet region of the spectrum, and accordingly, numerous studies have been made of the effect of ultra-violet light on wood and cellulosic materials, usually employing artificial sources. The photochemical degradation of ligno-cellulosic materials has been the subject of a comprehensive review by Hon (1981).

It has been established that all of the structural wood constituents, cellulose, lignin and hemicelluloses are degraded by ultra-violet light, and that free radical mechanisms are involved. Pure cellulose is not a good light absorber, but chromophoric groups capable of absorbing light are often present and impurities, eg metal ions, may also act as photosensitisers. Lignin absorbs light readily to generate free radicals, and is believed to be the key component in initiating oxidative chain reactions in cellulose in wood. Oxygen and moisture influence the rates of degradation.

The effects of light on the microstructure of the wood are well established. They include degradation of the intercellular middle lamella region, destruction of pits and ray cells, and checking and delamination within the wood cell walls. Work at Princes Risborough on the photodegradation of wood has been reported (Derbyshire and Miller 1981). This involved the determination of the losses in tensile strength occurring in wood strips about 130 μm in thickness exposed to natural weathering directly or behind transmitting filters. The work showed that exposure of wood to sunlight results in a rapid reduction in its surface integrity. Table 1 shows typical results obtained for pine veneers exposed for five months during June-October behind a filter (G1) which transmits virtually the full solar spectrum, and a P4 filter which transmits only visible wavelengths.

Table 1. Chemical and tensile strength changes in weathered scots pine veneers.

	Unexposed	G1 filter cut-off 290 nm	P4 filter cut-off 400 nm	Opaque filter
Tensile strength Mnm^{-2}	59.94	9.80	17.0	55.2
Solubility in cold water (%)	1.6	26.6		
Solubility in hot water (%)	2.3	38.0		
% methoxyl	4.7	2.05		
Degree of polymerisation of holo-cellulose	1300	470		

The results indicate that strips exposed behind a filter transmitting the full range of solar wavelengths suffer a marked loss in strength, and that this is attributable to a breakdown of the lignin, as reflected in the fall of methoxyl content, and to a pronounced degradation of the holocellulose to low molecular weight water-soluble products. From the standpoint of integrity and solubility characteristics the weathered surface cannot be considered to be wood any longer, and it is totally unsuitable as a coating substrate.

Although, as can be seen, photodegradation is most severe in wood exposed to the full range of solar wavelengths, strips exposed only to visible light are also degraded. Subsequent work has indicated that visible light may account for half of the degradation occurring in Scots pine. This finding is of great practical importance because it imposes a limitation in principle on the performance of clear and semi transparent finishes. It implies that the route to durable natural finishes lies in the use of pretreatments which stabilise the wood against photodegradation.

CONCLUSIONS

The two main weathering agencies, water and sunlight have been shown to adversely affect the properties of wood as a substrate to a marked degree. End-grain sealing can prevent water entry into permeable timbers and considerably enhance paint performance. The only effective means presently available for protecting wood against photodegradation is through the application of an opaque finish. There is a need for an increased understanding of the mechanism of wood photodegradation and for improved types of photostabilisers.

REFERENCES

Banks, W B and Voulgaridis, E 1980. Rec Ann Conv Brit Wood Pres Assn 43-53.

Banks, W B and Carragher, J G 1984. Rec Ann Conv Brit Wood Pres Assn.

Banks, W B and Evans, P D. The degradation of wood surfaces by water. 15th Ann Meeting of International Research Group on Wood Pres, Sweden 1984.

Boxall, J. Painting weathered timber. Building Research Establishment Information Sheet IS 20/77.

Carey, J K and Bravery, A F (1985). Developments in the assessment of joinery. preservatives. Rec Ann Conv Brit Wood Preserv Assoc 3-11 1985.

Derbyshire, H and Miller, E R 1981: The photodegradation of wood during solar irradiation Part 1: Effects on the structural integrity of thin wood strips. Holz Roh-Werkstoff, 39 341-350.

Hon, D N-S. Photochemical degradation of ligno cellulosic materials. Developments in polymer degradation - 3 - 1981 Chapter 8 Applied Science Publishers.

Miller, E R and Boxall, J. The effectiveness of end-grain sealers in improving paint performance on softwood joinery. Holz Roh-Werkstoff 42 27-34 1984.

Chapter 60

Protecting wood against the weather

E. A. Hilditch – Technical Director, Cuprinol Ltd., Adderwell, Frome, Somerset, BA11 1NL.

INTRODUCTION

When wood is exposed to the weather it is subject to:-

Radiation from the sun
Wetting by rain or dew
Changing atmospheric humidity
Changing temperature
Wind and wind borne particles
Fungal, bacterial and algal infection
Atmospheric chemicals

The result of this exposure is initially a change in colour which is followed by surface degradation, progressively getting severe and deeper. This paper discusses these aspects together with ways in which the wood can be protected.

SOLAR RADIATION

The first effect of weathering on strongly coloured woods is likely to be a fading or lightening of colours and on light coloured woods a darkening. In most coloured woods, the colour is due to extractives (i.e. materials other than cellulose, hemicellulose or lignin). These may be washed out or be degraded by photochemical action or by both. In all woods, the wood structure suffers photodegradation. Breakdown is greatest in the lignin, although cellulose is also degraded. This leads to a darkening of the surface, obvious on light coloured woods e.g. pine, but in darkly coloured woods masked by lightening due to loss of extractives.

Lignin is the main component of the middle lamella. Its breakdown leads to separation between the cells leaving a grey unstructured mat of loose fibres on the surface.

WETTING

Water through its effects is very damaging to wood. It causes dimensional changes and provides conditions under which micro-organisms can attack. When rain falls, or dew condenses on the surface, wood cells close to the surface quickly become saturated. The hydrogen bonded capacity of the cell wall is fully satisfied and the lumen fills with water. Penetration of water deeper into the wood only progresses very slowly.

Wood movement occurs across the range of moisture contents corresponding to equilibrium with atmospheric humidity changes. At the higher moisture contents, reached when in contact with liquid water, maximum swelling has occurred and no further swelling results from even more wetting.

Wood is in dynamic equilibrium with the moisture vapour in the air. In dry weather with atmospheric humidity at say 60% (relative), at equilibrium there will be no free water in the lumen and about only one third of the hydrogen bonding capacity of the walls will be satisfied. Wetting and drying of the cell wall results in a change in wall thickness and hence in the size of the cell and of the whole wood. The extent to which an individual cell shrinks or swells depends on the size of the cell and on the wall thickness. These differ in different parts of the wood especially between early wood and late wood. Stresses resulting from these differential changes lead to mechanical separation between cells and splitting of the wood. Splits provide a route by which water, dirt and spores can penetrate deeply into the wood. Control of moisture movement is one of the most important requirements of protective coating.

TEMPERATURE

Temperature changes at the wood surface can be large. In hot sun a dark surface can exceed 60°C. On a cold winter's night it could reach minus 20°C. On a sunny showery day especially where a drying wind causes extra cooling, changes of 50°C in surface temperature can be expected to occur over a short period and be repeated frequently. Wood is a poor transmitter of heat, surface changes are not therefore transmitted throughout the wood. Temperature changes of this magnitude cause pressure changes, driving air exchange between the wood and the atmosphere. Humidity relationships depend on relative humidity, which is temperature dependent.

This combination of physical factors leads from an initial colour change to a break up of the wood substance at the surface. Effects only progress slowly into the wood so that even after several years, break up may be no more than a millimetre deep. The wood will look unattractive.

BIOLOGICAL DEGRADATION

Amongst the general dirt deposited on anything out doors are a whole host of fungal, bacterial and algal spores. If the wood surface is hospitable - and a degraded damp surface is - they will grow.

Bacteria will infect new wood in a few days, moulds and mildews not much later and decay fungi within a month.

PROTECTION

The prime requirement for any protective treatment or coating is that it protects the wood from solar radiation, that it controls movement of moisture into and out of the wood and that it protects against biological attack. Almost invariably protective coatings are also required to be decorative.

Protection from solar radiation is usually achieved by the blocking effect of solid pigments or by the use of UV absorbers. Usually these are iron oxide pigments so that good protection is obtained even with transparent wood stains.

Uptake of liquid water must be prevented. Movement of moisture vapour across the protective coating should be optimised, neither maximised nor minimised. If the rate at which water vapour can move across the protective coating is too low the wood will not dry. In any situation where water can enter through unprotected ends or sides, the wood will eventually get wet enough for decay to occur. If the rate of vapour transmission is too high, dimensional changes will be frequent with maximum damaging effect.

Biological degradation of wood in outdoor situations is mostly fungal. Fungi that attack wood exposed to the weather are conveniently divided into decay fungi (mainly wet rots), blue stain (fungi that cause stains deep into the wood) and surface moulds. Protection against each has different requirements. A product that protects against blue stain may not protect against decay, perhaps because the spectrum of the fungicide selected is wrong, perhaps the amount used is too low or it has not penetrated deep enough into the wood. Equally, a treatment protecting against decay may not provide good protection against staining fungi or moulds.

Materials used to protect and decorate wood exposed to the weather are:

Paint

Varnish

Exterior Wood Stains

Coloured Organic Solvent Wood Preservatives

Creosote

PAINT

A traditional paint system of primer, undercoat and top coat is the most common finish for wood on houses and other "superior" uses. Asthetically, painted wood looks like paint not wood, no different to painted metal, stone etc.

While intact, paint very adequately protects the surface of the wood by providing a total, opaque barrier between it and the weather. An intact conventional paint film is more or less impermeable to the passage of liquid or vapour. This is its strength in preventing water entering through the exposed painted surface, but this very impermeability also prevents water getting out.

During the last decade a new type of paint has been developed with improved permeability and flexibility. These are known as moisture vapour permeable or breathable or microporous paints and give much better performance on wood.

Vapour permeable coatings can be produced with both solvent and water-based technology.

Water-based systems have a higher specific permeability which allows application of a thicker film while maintaining overall permeability.

A second benefit of these aqueous systems is their higher initial flexibility and retention of flexibility on ageing. Any water absorbed into the film acts as a temporary plasticiser, so increasing the film's extensibility at a time when the wood itself is expanding due to moisture absorption, thus they are more able to follow dimensional changes in the timber.

The main disadvantage of aqueous coatings is their dependence on evaporation of water as the first stage in the drying process. If applied in adverse weather conditions i.e. low temperature, condensation or high humidity, drying will be protracted.

Aqueous systems are particularly well suited to full factory finishing where drying conditions can be controlled, and many examples exist of full factory finished components performing well in practice after ten years service with no subsequent maintenance repainting.

Solvent-based systems penetrate better into timber. This is especially important in low build stains or base coats containing wood preservatives or blue-stain inhibitors.

Solvent-based paints dry much better under adverse weather conditions.

Vapour Permeable Paints and Exterior Wood Stains, share the property of giving a surface finish that allows water vapour to pass through. They are linked in some companies sales promotions.

Currently certain aspects of their performance are subject to debate. In future years, this debate will be settled by practical performance, but until then we must attempt to predict total performance from basic properties, from accelerated tests and from such service data as is available. Some aspects of this illustrate well the problem of determining and predicting the performance of protection products. Accordingly I will dwell on it.

The first consideration must be "How long do they last?" Here there is ample evidence for both paints and stains from both controlled comparative tests and from actual useage. Microporous paints, particularly those based on polymer emulsions can have a life exceeding ten years, which is well in excess of that expected from a conventional paint system. Stains have a shorter life but generally failing in a way that makes maintenance easier.

The water vapour permeability is high compared with a conventional paint system. This is considered to play a major part in determining the performance of these products. In particular, through allowing water vapour to escape rather than putting pressure on the film, blistering and the like is avoided.

The second consideration is to how far these finishes protect the underlying wood from decay. Some contain fungicides but it is not expected that they perform adequately in the manner of a true wood preservative in that they will prevent the growth of decay fungi under conditions under which these fungi would otherwise grow.

Decay fungi will only attack wet wood. The area of current contention is to how far the use of a permeable coating, allowing more drying out of water, may help to keep the wood too dry for decay.

Constantly or cyclically water is always entering wood and drying out. The moisture content of the wood reflects the balance.

Because of their known permeability it is safe to predict that more water will escape from wood with a permeable coating than with an impermeable one. But how does this change the balance, after all, more may get in and unless the moisture content is reduced below the level at which decay can occur, the reduction will not be useful.

Assuming that in actual situations there is a wide spread in the rates at which water enter wood, we need to know the relative number of

1. Situations in which the rate of water entry is such that the greater drying through a permeable MVP coating will keep the timber moisture below the level at which decay can occur.

2. Situations at which the rate of water entry is too great and the wood will reach a moisture content at which it is liable to decay.

Clearly, this cannot be determined by tests so designed that the rate of water entry is high, for it can be predicted that this will show the coatings not to perform well in this respect. Equally, isolated tests under less severe conditions, while perhaps proving the contention that in some situations MVP coatings are capable of keeping wood dry, will not show whether they are of general value or are only valuable in isolated instances.

A true answer will only be found through regular monitoring, over a period of wood in actual use - with measurements of moisture content under both Vapour Permeable and conventional coatings.

While this uncertainty exists, there will be debate, but I believe that there is sufficient grounds for recommending the general use of these products, after all, there is evidence that they are long life decorative coatings. The debate centres around the extent of possible additional benefits.

Even if they are found to be effective in this way, even beyond reasonable expectation, there will still be a need to provide protection against decay by use of a full wood preservative treatment and it will still be worthwhile adopting all practical measures to minimise the amount of water entering the wood.

VARNISH

Varnish has many of the attributes of an ideal wood finish. It retains fairly closely the full original wood appearance. It provides a barrier that prevents the weather getting at the wood and on which dirt does not readily accumulate. Unfortunately, varnishes do not last long (maximum 2 years under full exposure conditions). When varnish fails, failure is uneven and patchy with rapid discolouration of the wood in these areas. Maintenance is not easy. The main reason for the short life of varnish is UV embrittlement of the film accompanied by cracking when it is no longer sufficiently flexible to move with the wood. Water getting through cracks accelerates deterioration.

Varnish incorporating UV inhibitors could be expected to give longer life. Many so called inhibitors have been found ineffective and only a very few materials exist which give good results without producing unacceptable colouration.

EXTERIOR WOOD STAINS

Of all the finishes for outdoor wood, which allow the wood features to show through and leave it apparent that it is wood, this group of products offer the greatest choice of decorative finishes which also meet most requirements for the protection and maintenance of outdoor timbers.

These products consist of a combination of light fast pigments to provide colour, resins and drying oils to fix the pigments and to provide physical protection to the wood surface, a solvent to make the product easy to apply and usually (but not universally) a fungicide capable of protecting against staining fungi and/or decay fungi.

Either through the type of pigment chosen or through other additants, both the film itself and the underlying wood is usually protected against light degradation.

A wide variety is available. Products differ in colour, opacity (or transparency), which governs the extent to which wood features are visible and in surface texture and lustre. The different products differ substantially in the extent to which they penetrate into the wood and in the thickness of the surface film.

In this respect, there is a dilemma facing the formulator. Maximum penetration is desirable for fungicide containing products to achieve maximum protection of the wood against either staining or decay fungi. Maximum film thickness is desirable for physical protection and durability.

With stains, as with paints and varnishes, moisture movement in and out of the wood is controlled by the rate at which it can move through the surface film, which depends, of course, on the porosity of the film. The film porosity depends partly on the intrinsic (or specific) permeability of the film forming material and partly on the film thickness, thus the control achieved in practice depends on the application as well as the product. Too many coats, giving too thick a film of even a well formulated product, may have a permeability below that required for optimum performance. Some stains, particularly those produced for the use on rough sawn timber, are intended to give only a very thin film and the rate at which water vapour can be transmitted will be relatively high. Use of such products should be restricted to those situations where this is acceptable.

CONCLUSION

Weather is most aggressive and damages wood. Nonetheless, wood has been used by man outdoors since man began. Today we are more demanding than our ancestors and have the techniques to satisfy, to a greater or lesser extent, this demand. A wide choice of materials is now available to protect wood in the outdoor environment. These have a performance profile to match any situation and offer a wide choice of decorative effects.

REFERENCES

1. Hilditch E.A. Crookes J.V.
Exterior Wood Stains, Variety, Performance and Appearance

Proceeding Annual Convention 1981
British Wood Preserving Association

2. Hilditch E.A. Woodbridge R.J.
Progress in Timber Finishing in Great Britain

Journal of Oil and Colour Chemists Association
68 No. 9 217-228 September 1985.

Chapter 61

Time dependent strain development of wood loaded at right angles to the grain

A. J. Bolten and M. C. Breese – Department of Forestry and Wood Science, University College of North Wales, Bangor.

INTRODUCTION

A long standing interest in the physics of the manufacture and behaviour of composites made from densified wood (e.g. particleboards) has stimulated us to study the physics of strain development during sustained loading, of stress relaxation, and of strain reversal after the load has been removed, in wood stressed normal to the grain.

Wood is generally considered to be a viscoelastic material, and attempts to interpret its behaviour have been fairly typical of the approaches used to model viscoelasticity. These have, of course, been largely empirical; amongst the most popular have been the spring and dashpot mechanistic models following the ideas of Eyring and Halsey.

We have attempted to interpret our observations in terms of four element spring and dashpot models (Figure 1) with a marked lack of success. Some of the difficulties we have experienced are:

1. difficulty in defining the instantaneous elastic component i.e. in separating it from the delayed elastic component (Figures 1 and 2).
2. a lack of agreement between instantaneous elastic and delayed elastic components determined from the loading curve and those determined from the unloading curve.
3. the apparent need for a viscous component which is non-linear in time. This arises because the theory

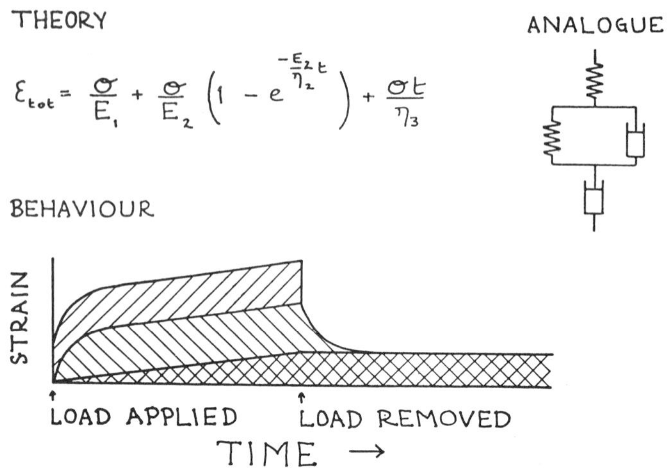

$$\varepsilon_{tot} = \frac{\sigma}{E_1} + \frac{\sigma}{E_2}\left(1 - e^{-\frac{E_2 t}{\eta_2}}\right) + \frac{\sigma t}{\eta_3}$$

Fig. 1. A four element spring and dashpot model often used in attempts to explain viscoelastic behaviour in wood.

suggests that the viscous component should be linear in time and that the slope of the total strain curve should tend to that of the viscous component. In practice the slope of total strain curve is less than the slope of the viscous component calculated on the assumption of linearity. If a non-linear viscous component must be used, it is difficult to know what type of equation to select and how this might be justified.

4. difficulty in fitting a curve with only one exponential term to experimental data which appear as a curve tending to linearity irrespective of the time scale selected. The parts of the curve which appear linear on one time scale turn out to be curvilinear if replotted on a smaller time scale (Figure 2).
5. the so-called constants in the theory E_1, E_2, η_2, η_3 are in reality dependent on the moisture content of the wood in the range oven dry to fibre saturation point, and on temperature. They may also be dependent on stress level. The theory does not allow for this.
6. the viscous component left behind after the load has been removed for a period greater than the loading period is supposed to be irreversible. However, much or all of this strain will be recovered if the wood moisture content is subsequently increased (providing that the moisture content during loading was less than the fibre saturation point) or if the temperature is subsequently increased.

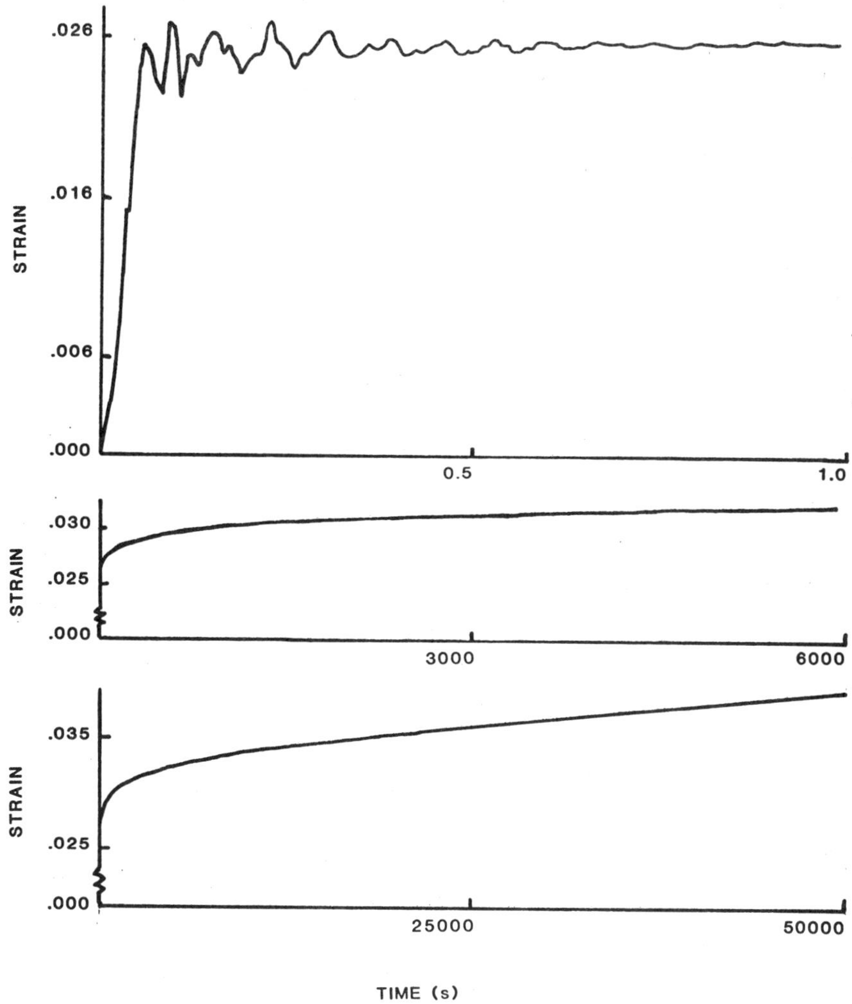

Fig. 2. Observed strain in air dry wood loaded to 760 kN/m^2 in the radial plane at room temperature. The data are presented on three different time scales; note that an apparent tendency to linearity on one scale is seen to be misleading when the data are replotted on a smaller scale.

These difficulties, together with a distrust of the use of such empirical constants in extrapolation and simulation exercises have led us to attempt a more rigorous approach. This is based on consideration of the compounds in the cell wall, how they are bonded together, and how we might expect them to behave under stress.

SOME BACKGROUND CONSIDERATIONS

A proportion of the cellulose in wood is known to be present in a paracrystalline form sufficiently ordered for it to show a unit cell in X-ray crystallography. There is no evidence that the dimensions of the unit cell or the degree of crystallinity change when the material is stressed. We may therefore conclude that the majority of strain development involves bond breakage and reformation in the less ordered cellulose, the hemicellulose and the lignin.

Hermans (1946) has considered the relative densities of ordered and disordered cellulose at great length, and concludes that there is a density difference of about 6%. Densification and reorganisation of disordered cellulose can therefore, at a maximum, lead to a fractional strain of 0.06 in that material, and even less in a mixture of ordered and disordered cellulose. While the densities of hemicellulose and lignin in native wood are uncertain, they are unlikely to be very different from that of disordered cellulose. This makes it difficult to see how densification of the cell wall could lead to a fractional strain in excess of, say, 0.1. In practice, external strains of 0.50 are quite common in wood composite manufacture. So, much of the external strain observed cannot be <u>directly</u> attributed to cell wall strain. The majority of the external strain must be produced by obliteration of the voids in the middle of cells, the cell lumina. This is then a structural deformation, not a cell wall deformation. But ultimately the structural deformation is only possible because the cell wall material is deforming at least in certain localised areas. In a complex structure such as wood there can be no guarantee that the plane of the localised material deformation is the same as that of the plane of the applied stress.

Finally, it is known that water adsorbed in the cell wall is hydrogen bonded onto hydroxyl groups. It is generally presumed that the majority of the hydroxyl groups in question are on polysaccharide chains, and not on lignin polymers. If this presumption is correct, granted that the rate of strain development is so sensitive to moisture content, it is not unreasonable to conclude that the breaking and reforming of hydrogen bonds between polysaccharide chains is crucial to strain development. (For the time being, and with considerable reservations, we have ignored lignin in our model). So while we recognise that in other polymer systems strain development and stress relaxation have been interpreted in terms of entropy effects, leading to concepts such as that of the Drunkard's Walk in rubber elasticity (Rodriguez,1970), the model presented below explores the possibility that strain in wood could be dominated by the behaviour of hydrogen bonds.

THE MODEL

It is useful to consider a population of hydrogen bonds between polysaccharide monomers which will respond to stress in some way. If we accept that a range of bond strengths is likely to be present (Nissan, 1959) it is nonetheless difficult to know what the proportion (in statistical terms, the frequency) of different strengths present will be. We have assumed, somewhat arbitrarily, that in unstressed air dry wood there is a normal distribution of bond strengths with mean 14.65 kJ/mol and standard deviation 4.45 kJ/mol, such that 99.9% of all values fall between 0 and 29.3 kJ/mol. We have then ignored the 0.1% statistically possible bonds of higher or lower strength. The upper limit is one suggested by Marrinan and Mann (1954).

If we now consider a hexose sugar monomer present in a wood polysaccharide, hydrogen bonding is most likely to occur through the three hydroxyl groups. From statistical sampling theory, the average strength of these three hydrogen bonds will again vary between the limits 0 and 29.3 kJ/mol, with mean 14.65 kJ/mol, but now with standard deviation 2.57 kJ/mol. From this the total strength of bonds which must be broken to allow a particular monomer to move any substantial distance can readily be calculated. We assume that strain development involves a large number of such monomer movements.

The concept of an energy barrier is quite commonly applied to materials under stress. This originates from the work of Eyring and colleagues (see Tobolsky _et al_ (1943), for example). More recently, Caulfield (1985) has used it to analyse duration and rate of loading effects on wood, and we have followed his approach to some extent. If we consider two possible adjacent monomer positions, which can be termed left and right hand, then in the unstressed material, movements from left to right, V_+, will be given by (Caulfield, 1985):

$$V_+ = \frac{k\ T}{h} \exp \frac{-\Delta F}{R\ T} \qquad [1]$$

where k, h, and R are Bolzmann's, Plank's and the Gas Constants respectively. ΔF is the energy barrier height or activation energy. A number of workers (including Caulfield, 1985) have reported reasonable agreement between ΔF and the strength of hydrogen bonds at room temperature. We have assumed this equivalence at room temperature. (At higher temperatures, depending on the magnitude of entropy effects, this assumption may not be acceptable). Thus the monomer will only move if ΔF is made available. In the unstressed material, there will be an exactly similar number of movements from right to left, V_-; so, on average, no net movement occurs. In the presence of an external stress, the energy barrier is distorted so that there is a greater tendency towards movements in the direction of the applied stress, than in the opposite direction. As a result there is a net number of movements per unit time, N, given by:

$$N = \frac{k\,T}{h} \exp\left(\frac{-\Delta F}{R\,T}\right)\left[\exp\left(\frac{fA\lambda}{2\,kT}\right) - \exp\left(-\frac{fA\lambda}{2\,kT}\right)\right] \qquad [2]$$

where f is the applied stress applied to area A, and where λ is an average distance moved (Caulfield, 1985).

The main problem with this approach in the context of wood rheology is that it suggests that there is no decrease in the rate of movement with time (i.e. that the behaviour might conventionally be termed purely viscous). Yet the rate of strain development in wood is clearly seen to decrease with time. We presume that this is because at each new position a monomer will require more energy for the next move than it did for the previous one. So, if we start with a normal distribution of ΔF values in the unstressed material, at any time after stressing the distribution will be skewed to the right. The longer the stressing period, the greater the degree of skewing. At time infinity we presume that the total ΔF value would reach the limit value possible for three hydrogen bonds, i.e. 87.9 kJ/mol. At this stage we make no attempt to explain quantitatively why ΔF increases as strain develops. Presumably the stress causes a greater proximity of the groups participating in bonds, and this should allow the formation of some stronger bonds. In any event, if the development of strain by wood is determined by an equation such as [2], time must affect one of the variables, and ΔF seems to be the most likely candidate.

If we accept this, the question then becomes, by how much does ΔF increase for each monomer which moves? Here we have presumed that probabilities are determined by the normal distribution curve described above. If we consider a monomer with an energy barrier for movement ΔF_1, it must, from our assumptions above, have an energy for further movement greater than ΔF_1, after it has moved. If the new value is randomly related from those possible and greater than ΔF_1, some new ΔF values will clearly be more likely to arise than others. A simulation programme has been written which evaluates the outcome of this process. The programme takes a given time, t, and

1. takes the first (lowest) value of ΔF, and calculates from Equation [2] how many of the monomers which have this energy will move during the next time increment, Δt. The distribution of resultant new total energies per monomer is calculated from probability theory, based on the area under the curve and to the right of the starting value of ΔF.
2. similarly considers values of ΔF to the right of the first one, in turn.
3. when all values of ΔF have been considered in this way, the number of monomers of each total energy at time ($t_1 + \Delta t$) is recalculated.
4. The entire procedure is then repeated for a later time, defined by $t_2 = t_1 + \Delta t$, and so on.

It will be noted that monomers with a higher value of ΔF in Equation [2] will move at a slower rate. That is, the energy barrier for these is higher. Because with time the frequency distribution of ΔF becomes more and more skewed to the right, monomers move at a progressively slower rate on average. We have thus introduced a time dependency into Equation [2].

It also seems reasonable to postulate a second source of time dependence. Equation [2] as stated presumes a limitless reservoir of monomers about to move, and a limitless sink for monomers which have moved. This is reasonable in a situation where the monomer has the same energy barrier for the first and all subsequent movements. In our model, where we are postulating an increase in the height of the average energy barrier it is more questionable. If we consider first order reaction kinetics, the theory tells us that the rate of formation of species B from species A is dependent on the <u>concentration</u> of A. It seems reasonable to consider the breaking and reforming of H-bonds necessary for a monomer movement as a first order chemical reaction. We have therefore introduced into the above simulation procedure a rate modification factor based on the relative concentration of monomers at each energy barrier. The introduction of this factor has the effect of increasing the time dependency of the model.

Selection of values for the effective stress, f, the area over which this is applied, A, and the typical distance moved, λ, in Equation [2] is not easy. Because we believe that most strain is the result of monomers slipping past one another, we have considered that the stress is applied over the area of a glucose residue as seen from the end of a cellulose chain. Thus A can be taken from the dimensions of the unit cell. See Figure 3. We have used a value of ac/2 = 0.33 nm^2. The stress applied to this area we have calculated from the stress applied to the specimen, and the proportion of the specimen surface area that could be considered to be hemicellulose or disordered cellulose (on the assumption that these materials are distributed isotropically through the material). The typical distance involved in a monomer movement is harder to quantify. For want of any evidence suggesting otherwise we have assumed that this is of the order of one monomer length. Se we have used the value for glucose of 0.515 nm. The number of monomers in series that contribute to strain is hardest of all to quantify. This is particularly true since much observed strain must arise from structural deformation, rather than from deformation of the cell walls. For convenience we have chosen a number which will give us a strain similar to that observed experimentally after long loading periods. This is legitimate because we are more interested in matching the shape of the simulated curves with those obtained experimentally than with quantitative comparisons. Change in the number of monomers in series simply scales the curves up or down, and does not change their shape.

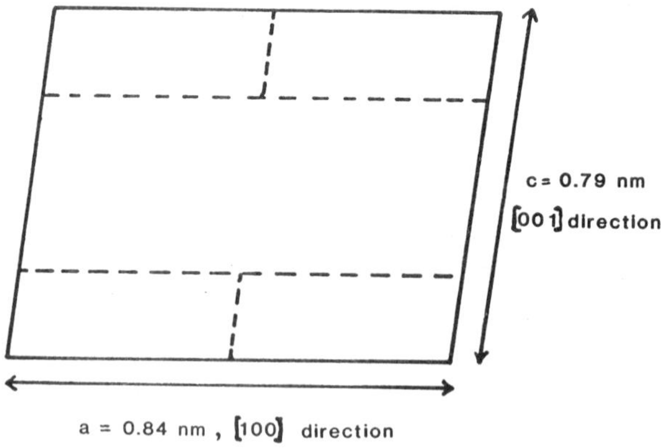

Fig. 3. Basal plane of unit cell of Cellulose I. Since the unit cell consists of a central chain and a quarter of four other chains, the "effective area" of the central chain may be taken as ac/2.

THE PREDICTIONS OF THE MODEL

Using the above assumptions and procedures, we have simulated the rate of strain development we would expect from the postulated monomer movements following H-bond breaking and reforming. The results of one such simulation are shown at three different time scales as Figure 4.

The shape of the predicted strain curves is basically similar to those obtained experimentally. The predicted strain curves shown here have a more rounded shoulder, but a much sharper shoulder can be obtained by increasing the assumed standard deviation of H-bond strengths. Having carried out no experimental determination of the distribution of H-bond strengths at the same moisture contents at which the experimental data shown here were collected, we cannot be certain that the standard deviation we assumed for the curves shown here is correct. This apart, the predicted curves always show the characteristic, noted earlier in experimental curves, of appearing to tend to linearity at whatever scale they are drawn. In general we find the shape of the predicted curves encouraging at this stage.

The simulation model also produces curves showing the predicted distribution of ΔF values for the moving monomer's H-bonds after various times. See Figure 5. It can be seen that the distribution gradually becomes more skewed to the right with time. However the time interval between the curves shown increases by a power function and very long times indeed would be required for all bonds on all monomers to achieve the assumed maximum value of ΔF.

POTENTIAL DEVELOPMENTS OF THE MODEL TO COVER OTHER EXPERIMENTAL VARIABLES

The long term behaviour of wood under load varies with stress level, moisture content, temperature, and the plane of stress application. Since the applied stress, f, appears in Equation [2], it should be possible in future work to compare the effect of stress level on both experimental and predicted results. As for moisture content, it is recognised that the first water molecules bound onto the cell wall have a much higher heat of wetting associated with them than those bound on just below the fibre saturation point (Skaar, 1972). This implies that the frequency distribution of H-bond strengths should vary with moisture content. It may prove possible to quantify this spectroscopically. Temperature variation is allowed for in Equation [2], but a second temperature effect must also be considered: we have assumed that entropy effects are negligible at room temperature. It will be necessary to test this at higher temperatures. Once this has been done, it should be possible to predict the effect of high temperatures on strain, and to test these predictions experimentally.

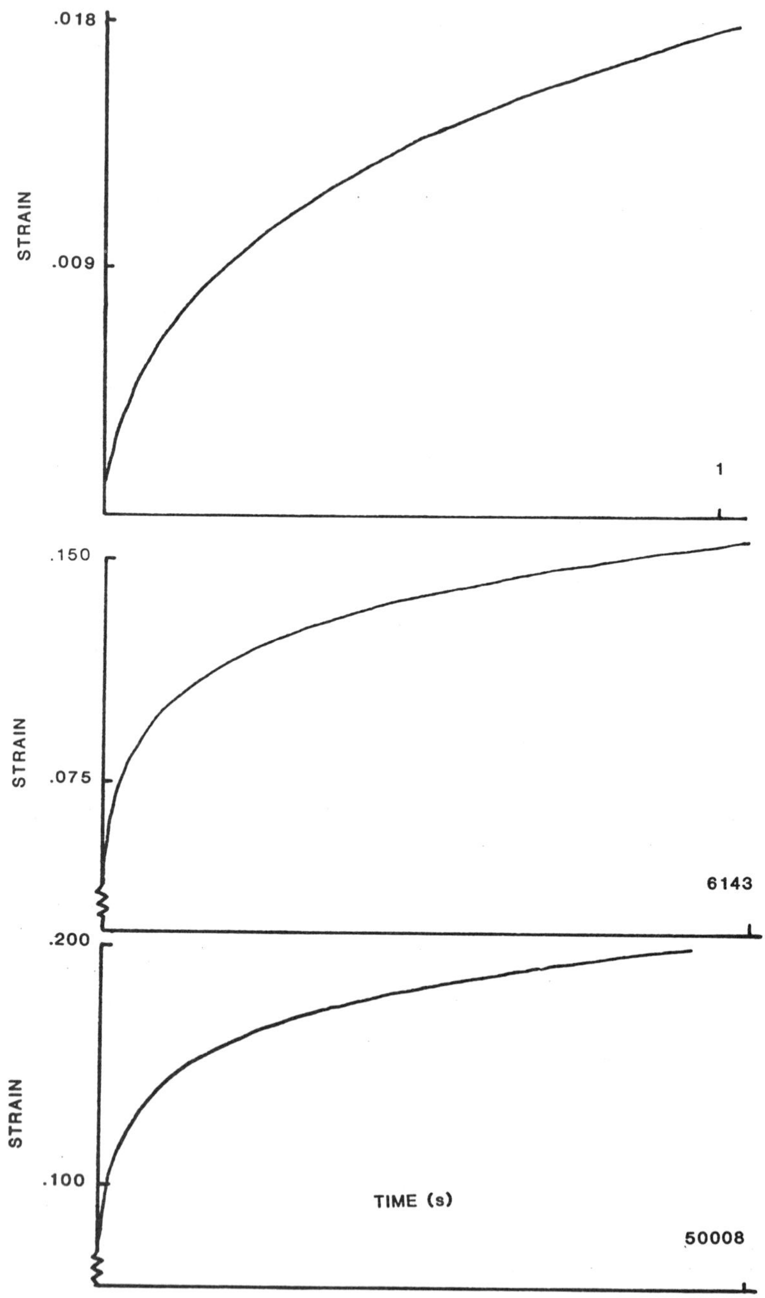

Fig. 4. Simulated strain development at three different time scales.

The present model does not obviously allow for the anisotropy of wood with respect to strain development. While the stress applied to each moving monomer, and the typical distance moved, may well vary slightly with the plane of the applied stress, the principle effect is likely to be through the interrelationship between structural deformation and cell wall deformation. That is, cell wall deformations may be of similar order of magnitude when stress is applied in any plane, but the structural deformations that these cell wall deformations make possible may vary much more, because of anisotropy of structure and consequent bending moment effects, etc. Quantification of the relationship between structural and cell wall deformations is likely to be very difficult. The relationship may well turn out to be dependent on variables such as moisture content and temperature. Until the relationship is fully defined quantitative understanding of the anisotropy of wood rheology is likely to remain elusive.

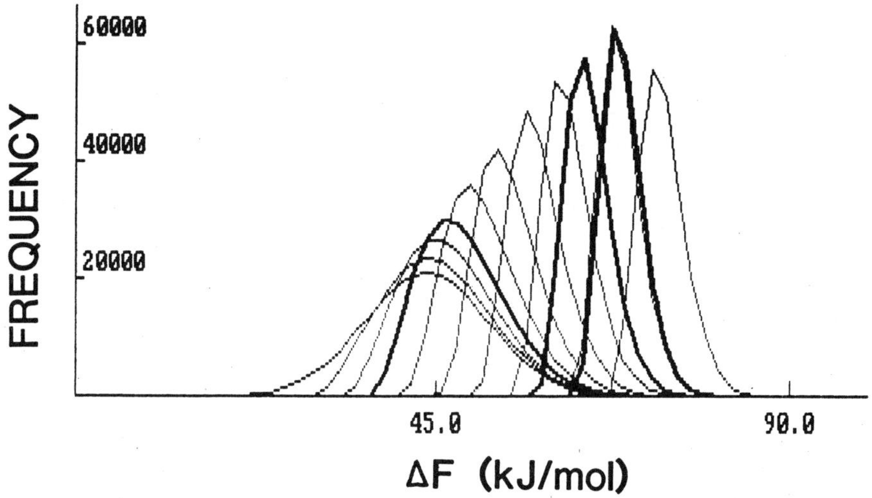

Fig. 5. Frequency distribution of ΔF values for moving monomers for a variety of different times. Curves for the maximum times shown on Fig. 4 are shown as heavy lines. With increasing time the curves become more skewed and displaced to the right from the original normal distribution.

Most conventional theory suggests that instantaneous and delayed elastic strain should be recovered on unloading in the same way as it is developed on loading. It was noted earlier that this is not always the case with wood. The present model offers a possible explanation for this. Wood which has been loaded will have a totally different distribution of H-bond strengths from wood which is unstressed (see Figure 5). As a result its behaviour on unloading can be expected to be very different. This area is under investigation at present.

CONCLUSIONS

The present model may well turn out to be at least as good as any of the empirical ones for the prediction of strain development in wood, and have the added benefit of some rigour in its derivation. If in future work it can be shown to predict the effects of variation in stress levels, moisture content, and temperature on strain development, it will be more useful than conventional models.

BIBLIOGRAPHY

Caulfield, D.F. 1985. A chemical kinetics approach to the duration-of-load problem in wood. Wood and Fiber Science 17 (4) 504-521.

Hermans, P.H. 1946. Contribution to the physics of cellulose fibers. Ch.IV. Elsevier, Amsterdam.

Marrinan, H.J. and Mann, J. 1954. A study by infra-red spectroscopy of hydrogen bonding in cellulose. Journal of Applied Chemistry 4, 204.

Nissan, A.H. 1959. Fundamentals of adhesion from molecular forces in cellulose. Tappi 42 (12) 929-933.

Rodriguez, F. 1970. Principles of polymer science. McGraw-Hill, New York.

Skaar, C. 1972. Water in wood. Syracuse University Press.

Tobolsky, A.,Powell, R.E. and Eyring, H. 1943. Elastic-viscous properties of matter. In R.E. Burk and O. Grummitt, eds. The chemistry of large molecules. Interscience Publishers, New York.

Chapter 62

Coatings for exterior wood

Peter S. Collins – Development Chemist, Imperial Chemical Industries PLC, Paints Division, Wexham Road, Slough, SL2 5DS.

SUMMARY

Exterior woodwork poses what is arguably the greatest challenge to the coatings formulator. Firstly, because of the tremendous variety of types and forms in common usage, secondly, because of its hygroscopicity, and thirdly, because of its susceptibility to natural degradation.

This paper identifies the problems posed by the substrate and examines a variety of formulating options available to overcome them.

PROBLEMS POSED BY WOOD AS A SUBSTRATE

Before considering the choices of technologies available today, it is necessary to take note of how substrate differences affect the choice, and the requirements placed on the coating by the substrate.

How do the substrate differences affect the choice?

If wood was an inert, immobile and consistent material, then matters would be far more straightforward, but it is not, and so it is essential to consider the requirements of the substrate. To do this, it is necessary to make a simple sub-division of timber substrates.

SUBSTRATES REQUIRING DIMENSIONAL STABILITY	SUBSTRATES NOT REQUIRING DIMENSIONAL STABILITY
↓	↓
SMOOTH PLANED WOOD (WINDOWS, DOOR FRAMES)	ROUGH SAWN TIMBER (SHEDS, FENCES)

Wood not requiring dimensional stability

For substrates of this type (eg fences) a highly permeable zero or low build woodstain may be most suitable. Such products are designed to soak into the wood taking fungicides into the surface layers of the wood, but they leave little or no film on the surface to act as a barrier against moisture. They may, however, have some water-repellant properties. Zero build stains tend to perform best on rough sawn timber rather than on smooth planed wood; this is because they are more able to penetrate and be retained in the fibrous surface layers.

Wood requiring dimensional stability

Where dimensional stability is required (eg window frames), then a moderately permeable, medium build woodstain, or an exterior paint system, will be more appropriate. Zero build stains are totally unsuitable for reasons outlined above. Low build stains may seem like a good idea because they will wear away failing by erosion rather than flaking. However on smooth planed wood, the rate of erosion is often fairly rapid making it necessary to redecorate on a frequent basis.

Also their highly permeable nature results in poor dimensional stability, and can lead to splitting and cracking of the substrate, which is particularly unsightly on smooth planed wood. Medium build stains are better suited than low build stains because they will provide better dimensional stability, and they are less likely to allow wood to crack or split. Their maintenance cycle (although far less than a quality exterior paint system) will be far longer than a low build stain, and if correctly formulated they will still tend to fail by erosion rather than flaking making for easier maintenance.

Exterior paint systems offer the greatest protection for exterior wood in this category, providing the longest period between maintenance cycles, however redecoration takes longer because of the different mode of failure.

Properties required of the coating

From here on, most of my remarks will relate to smooth planed wood which is required to be dimensionally stable.

For good durability, there are five key properties required of a coating for exterior wood:-

1. Moisture Control
2. Flexibility
3. Adhesion
4. Resistance to Attack by Micro-Organisms
5. Protection from Sunlight

Moisture control

Unless the rate of change of moisture content within the wood is adequately controlled, then dimensional changes will be large and rapid, also if the level of moisture is allowed to remain above 22% for prolonged periods then wood decay fungi can develop and attack the substrate. The crucial question then is how best to control the moisture content of exterior wood. Given that wood takes up water far more rapidly through end grain than through its lateral surfaces, there is an obvious need to slow down the uptake of water by sealing its end grain. This can pose problems where joinery has already been assembled, installed and, most usually, previously decorated. However, low viscosity penetrating resin solutions can be extremely beneficial, even in redecoration situations. What about the use of so called "microporous" paints? Well, any paint which allows water out must also let water in, and no good evidence has been shown to date which demonstrates that highly permeable paints maintain exterior wood at significantly lower levels of moisture content than moderately permeable paints.

Flexibility

Because all coatings are permeable to some extent, and because film defects will always occur, moisture movement will always take place even if it has been minimised. Therefore it is essential that any coating is sufficiently flexible to cope with the resulting wood movement. From data obtained from commonly used species of timber, a value of about 8% extensibility is required for a coating to be able to withstand dimensional changes at joints.

Adhesion

For good durability, coatings must have good adhesion, not only to bare wood but also to previously applied coatings, since the majority of decoration involves redecoration.

Resistance to attack by micro-organisms

Although preservative treatment of the wood can only be completely effective when carried out under carefully controlled factory conditions, it is very important to ensure that the coating is well protected against surface disfigurement moulds. These organisms are not only airborne, attacking the coating from above, but are found in the wood and can erupt through, damaging the coating as well as impairing its appearance.

Protection from sunlight

Sunlight consists of infra-red (IR) visible and ultra-violet (UV) light, the latter two of which initiate attack upon the cell walls of the wood, unless it is protected. Left exposed to sunlight, wood becomes denatured, looking grey to the naked eye and appearing very fibrous when examined under a microscope. Ideally both visible and UV light need to be screened out for full protection. If it is required that the coating remains transparent or semi-transparent, then only UV light can be screened out, however, even this offers significant protection to exterior wood.

In formulating coatings, it is paramount to optimise the composition so that the best overall balance of the properties listed above is achieved.

THE CHOICE OF APPROPRIATE TECHNOLOGIES

To obtain this balance, coatings formulators have a wide range of options open to them. Even to the professional user, it may seem that coatings manufacturers exercise selfish choices rather than ones to aid the customer. It is therefore appropriate at this point to review the various options in detail and consider the attributes that they confer upon the coating.

Pigment

Little needs to be said on this subject, except as mentioned above that fully opaque pigmentation affords more protection from sunlight than transparent or semi-transparent materials. In semi-transparent woodstains, very fine iron oxides are used to colour and provide protection against UV light. This accounts for at least some of the differences in performance seen across the range of varnishes, semi-transparent woodstains and fully opaque paints.

Thinner

Two materials account for almost all commonly-found decorative products; these are white spirit and water. It is instructive to consider in turn the properties conferred by these widely-differing liquids.

White spirit	Water
Flammable	Non-flammable
Able to use unlacquered metal cans	Can corrosion - needs lacquered metal or plastic containers
Solvent smell	Odour-free
Controlled evaporation rate independent of humidity	Evaporation rate varies widely dependent on humidity
Well suited to UK climate	Rain is a disaster during application and drying
Naturally resistant to microbial attack in can	Susceptible to microbial attack in can
Grain raising less prevalent	Grain raising problematic
Brushes wash out in white spirit	Brushes wash out in water

It can be seen that the majority of these properties relate to the material in the can or during application. It should be noted that the type of thinner used has no direct bearing on durability either for good or bad. However, given the vagaries of the UK climate, there is much to be said for the use of white spirit.

Binder

The binder or polymer matrix needs to be considered from two points of view - a) Physical form, and b) Chemical form.

Physical form

The physical form of a binder may be either a solution or a dispersion and the following properties stem from whichever is selected.

Solution	Dispersion
Good penetration of wood	Poor penetration
Slower evaporation of thinner	Fast evaporation of thinner
Good flow	Poor flow
Good lapping	Poor lapping
High gloss appearance	Lower gloss levels
Higher solids	Lower solids
Viscosity limits molecular weight	Viscosity and molecular weight largely independent

These properties mostly relate to application characteristics and initial film appearance. However, it is clear that solution polymers are of greatest benefit in formulating penetrating preservative primers and high gloss topcoats. Dispersion polymers may well play a useful role in undercoats, where penetration and high gloss are unimportant, but the ability to produce brushable high molecular weight paints not requiring further crosslinking is a definite asset.

Chemical form

Three main types of polymer are commonly found in decorative coatings:-

i) PVA copolymer
ii) Acrylic copolymer
iii) Alkyd resin

The former two form films by means of a physical process only, whereas the latter reacts chemically after the thinner has evaporated.

PVA/acrylic copolymer	**Alkyd resin**
Non-reactive dry	Autoxidative crosslinking
Odour-less	Secondary odours during drying
Fast drying	Long dry times
Poor block resistance	Good long term block resistance
High flexibility	Tend to embrittle with age
Poor dirt pick-up	Easier to keep clean
Non-yellowing	Yellows with age

Acrylic copolymers offer maintained long term flexibility but have a number of drawbacks, particularly when used in topcoats. Alkyds do not suffer from problems associated with thermoplasticity, but because of this they require very skilful formulation to maintain their long term flexibility. Flexibility decreases markedly as pigment levels are increased, so enhanced polymer flexibility is of paramount importance in products such as undercoats which are highly pigmented to give good obliteration.

It should be realised that in considering these formulating options, no account has been taken of the current practical limitations on our ability to exploit them. The table below displays the options that are currently available.

	WATERBORNE	SOLVENTBORNE
Dispersion Coalescing	Emulsion Paints	
Solution autoxidising		Oil or Alkyd Paints

CONCLUSIONS

In selecting appropriate coatings for exterior wood, careful attention should be paid to the substrate type.

In formulating products for exterior wood, the coatings technologist must have a strategy which addresses all of the problems posed by the substrate in a hostile environment.

The technology selected needs to take account of the above points and be well suited to the prevailing conditions. Furthermore, the technology used should be that which is most appropriate to the role required of it within the overall system.

Chapter 63

Dissolution and spinning of exploded wood in amine oxide systems

H. Chanzy, M. Paillet, A. Peguy and R. Vuong – Centre de Recherches sur les Macromolécules Végétales, C.N.R.S., B.P. 68, 38402 St Martin d'Hères Cedex, France.

ABSTRACT

The dissolution of a series of steam exploded aspen wood in N Methyl Morpholine N Oxide (MMNO) was investigated as function of the wood steaming conditions. When steaming times of 40-90 seconds at temperature of 250°C were selected, the resulting exploded wood became entirely and rapidly soluble in MMNO, yielding wood-based solutions possessing a good spinnability. Continuous spinning of these solutions was achieved, leading to fibers of exploded wood, which were analyzed for mechanical, structural and ultrastructural characteristics. In the best cases, these fibers had strengths and moduli comparable to those of conventional rayon but presented a poor value for their elongation at break.

INTRODUCTION

The rayon production, following either the cuprammonium or the viscose process, is slowly regressing throughout most of the industrialized countries. New cellulose fibers, spun from cellulose solutions in organic solvents are likely to come on the market in the near future as several important industrial laboratories are actively pursuing research and development with such systems (1). Among the organic solvents for cellulose, which have been recently investigated, the cyclic amine oxides and in particular N-methylmorpholine N-oxide (MMNO) appear to be the most promising (2), especially since the recycling and recovery of the solvent has been mastered (3). MMNO is not only extremely powerful for the dissolution of fairly large quantities of cellulose but also for the solubilization of a wealth of other synthetic and bio-polymers. In particular, MMNO will dissolve easily hemicelluloses (4) and even small amounts of lignin (5) in combination with cellulose. For this reason, the dissolution of cellulose in MMNO is not only restricted to high quality dissolving grade cellulose but also to a series of pulps of lower purity such as ordinary kraft or sulfite pulps.

In the present work, which follows a preliminary communication (6), we describe how still coarser wood-based material, namely samples from exploded wood, can also be dissolved in MMNO in order to prepare spinnable wood solutions from inexpensive and readily available raw materials. For the purpose of this study, exploded wood samples from aspen were prepared under specific conditions where the degree of polymerization (DP) of cellulose was kept high and where the lignin was depolymerized to the point where it became soluble in many organic solvents and in MMNO in particular. This paper which describes the preparation and spinning of such exploded wood solutions, gives also some details on the properties of the fibers which were obtained in a dry jet wet spinning continuous operation.

EXPERIMENTAL

Materials

The exploded aspen wood samples were prepared at the Centre Technique des Industries des Papiers, Cartons et Cellulose de Grenoble. In the process, chips of aspen wood were pressurized for 20 sec to 165 sec in saturated steam at 40 bars (250°C) and were submitted to an explosive decompression. After washing with hot water (in order to remove most of the soluble sugars) the exploded wood was dried and disintegrated with a Thomas apparatus, fitted with a 1 mm mesh screen.

The product obtained was analyzed by analytical gel permeation chromatography after tricarbanilation. Depending on the reaction times used in the process, the exploded material had compositions and cellulose DP described in Table I.

Bleached softwood kraft pulp was used in some experiments, in combination with the exploded wood. The pulp which had a DP_w of 1200 was shredded before mixing with the exploded wood.
The N-methylmorpholine N-oxide (MMNO) monohydrate was prepared as described elsewhere (7) from an aqueous solution purchased from Texaco.

Table 1. Characteristics of the exploded wood samples as function of the steaming time at 40 bars.

Time	20 sec.	60 sec.	165 sec.
Composition* after water extraction			
- lignin	35.5	38.9	44.3
- hemicellulose	11.1	3.4	0.6
- cellulose	53.4	57.7	55.1
Cellulose DP_W	1500	420	210

*Initially, the samples had the following composition : cellulose 48 %, lignin 21 %, xylan 24 %, glucomannan 3 % and pectin 4 %.

Dissolution and spinning

The dried exploded wood was mixed with N-methylmorpholine N-oxide (MMNO) monohydrate (Mp 73°C) in concentrations ranging from 15 to 20 % (W/W). Total dissolution of the product took place within a few minutes by stirring the mixture at 120°C in a glass vessel immersed in an oil bath.

The solutions were then extruded with a Davenport Melt Extruder apparatus, equipped with a single hole spinneret, having a diameter of 0.5 mm. The temperature of extrusion was comprised between 75 and 110°C and the dry jet/wet spinning technique was used, with an air gap of 10 cm. Coagulation took place in a cold water bath along a 1 m path and the resulting fibers were wound on bobbins, at linear speeds ranging from 50 to 70 m/minute.

Properties of the fibers

The as spun fibers were extensively soaked in tap water to leach out the occluded solvent. They were air-dried and mounted for scanning electron microscopy, optical microscopy and X-ray analysis.

The mechanical properties of the filaments were determined at 23°C and 65 % relative humidity with an Instron Tensile Tester. The samples had a gauge length of 100 mm and the testing speed was 16 mm/min or less.

Ultrathin sections of the fibers were also achieved. For this purpose, the fibers were embedded in epoxy, after staining with permanganate. With some fibers, the lignin was removed by soxhlet extraction in water-dioxane. These fibers which displayed only the porous cellulose skeleton were also stained and embedded.

RESULTS AND DISCUSSION

Spinning of exploded wood

The spinning characteristics of the various exploded wood samples are summarized in Table II.

Table 2. Spinning characteristics of the various exploded wood samples.

Steaming time at 40 bars of the various samples time in seconds	20	40	60	90	90*	165
DP of the cellulose	1500	875	420	465	-	210
Solution concentration m % (w/w)	no solution	17.5	15	15	15	15
Spinning t°		75	110	73	90	spinning impossible
Spinning speed m/min		50	70	60	70	

*The steaming in this case was achieved at 35 bars.

It has to be noted that for a hydrolysis time of 20 seconds, the solution was too viscous to obtain a spinnable solution and the exploded wood was not entirely soluble.
On the contrary a long hydrolysis time gives a fluid solution which is no longer spinnable.

Morphological aspect

When observed in polarized light between crossed nicols, the fibers display a moderate birefringence indicating that the alignment of the cellulose molecules during spinning is rather poor. This result was confirmed by X-ray analysis which gives a fibre diagram corresponding to the expected cellulose II lattice (6). Nevertheless the orientation of the fibre diagram is rather poor, well in line with the observed weak birefringence of the corresponding fibre.

Structural and ultrastructural studies were undertaken. A cross section of a fiber as viewed with the transmission electron microscope displays a nearly perfect cylindrical shape and indicates that the inner structure of the fiber is homogeneous and compact. The microdomains of lignin and cellulose are about 10 to 20 nm in size as seen in Figure I. In Figure 1a are displayed the lignin domains stained with permanganate. In Figure 1b, only the porous cellulose skeleton is left after removal of the lignin by dioxane-water extraction.

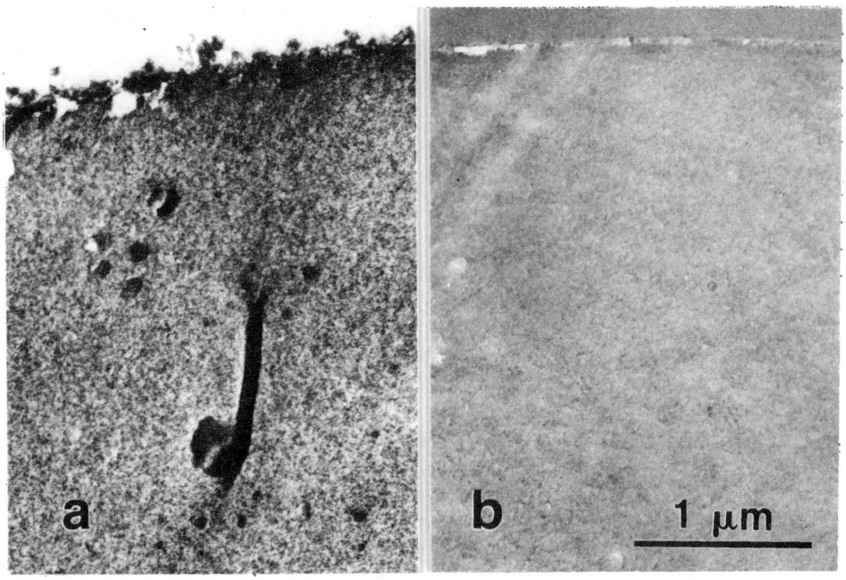

Fig. 1. (a) Transmission electron micrograph of a cross section of a wood fiber stained with permanganate. (b) same as in 1a but after extraction of lignin by water-dioxane treatment.

Mechanical properties

When tested for their mechanical properties it was found (Table III) that such fibers from exploded wood had tenacities and moduli corresponding to those of regular viscose rayon. On the other hand, their elongation at break was lower, as only values of 3 to 4 % could be achieved in the best cases. This low value is certainly due to the presence of the lignin matrix in the fibers.

Table 3. Mechanical properties of the various exploded wood samples.

Steaming time at 40 bars of the various samples time in seconds	20	40	60	90	90*	165
Solution concentration m % (W/W)	no solution	17.5	15	15	15	15
Fiber properties						
- diameter (μm)		20	30	20	20	
- modulus (GPA)		9	14	9	14	
- tenacity (GPA)		0.3	0.3	0.2'	0.3	
- elongation at break		3 %	4 %	3 %	4 %	

*The steaming in this case was achieved at 35 bars.

An attempt was made to improve the mechanical properties of fibers from exploded wood. When a percentage of Kraft pulp was added to the exploded wood (which in fact decreased the lignin content and increased the overall DP of the cellulose fraction), the spinning conditions were markedly improved and a spinning speed of more than 100 m/minute could be reached.

Under these conditions, fibers of smaller diameter could be obtained which displayed improved mechanical properties as listed in Table IV.

With a mixture consisting in 80 % of exploded wood, hydrolysed for 90 seconds and 20 % of kraft pulp, the Young's modulus and the tenacity of fibers went up to 27 GPa and 0.5 GPa respectively. In that case the elongation at break was about 4 %.

This improvement of mechanical properties corresponds to an improvement of the orientation of the material as confirmed by X-ray analysis. Not surprisingly the orientation of the cellulose chains increased with the relative decrease of lignin content and relative increase of DP of cellulose.

Table 4. Spinning characteristics and mechanical properties of the fibers resulting from mixtures of exploded wood and kraft pulp.

Ratio Kraft pulp/ exploded wood	0.1	0.21	0.33
Solution concentration % (w/w)	18	16.5	15
Spinning temperature	100	100	100
Spinning speed m/min	100	100	90
Fiber properties - diameter (μm)	20	18	16
- modulus (GPa)	16.5	27	27
- tenacity (GPa)	0.35	0.5	0.45
- elongation at break	4	4	4

CONCLUSION

This study describes a fast process to convert wood into continuous filaments. We have shown that under specific conditions the cellulose and the lignin were soluble in tertiary amine oxides. The solutions obtained were spinnable between 70 and 110°C at linear speeds comprised between 50 and 100 m/minute. The fibers obtained had mechanical properties comparable to those of normal rayon, except for their elongation at break.

A final aspect of this work is that it gives access to interesting fibrous structures where lignin and cellulose are blended in microdomains having around 10-20 nm in diameter. We have shown that the structure of these fibers is homogeneous and compact.

REFERENCES

(1) HUDSON, S.M., and CUCULO, J.A., J. Macromol. Sci. Rev., Macromol. Chem. C18 (1) 1-82 (1980).

(2) FRANKS, N.E., and VARGA, J.K., US Patent 4,145,532 (1979) McCORSLEY, C.C., and VARGA, J.K., B.F. 2,398,758 (1978).

(3) LAITY, P.R., PCT Int. Appl. WO 83 04,415 (1983).

(4) CHANZY, H., CHUMPITAZI, B., and PEGUY, A., Carbohydr. Polym. 2, 35-42 (1982).

(5) JOHNSON, D.L., U.S. Patent 3,447,956 (1969).

(6) CHANZY, H., PAILLET M. and PEGUY A., Polymer Comm., 27, 171-172 (1986).

(7) CHANZY, H., NAWROT, S., PEGUY, A., SMITH, P. and CHEVALIER, S., J. Polym. Sci., Polym. Phys. Edn, 1982, 20, 1909.

Chapter 64

Internally reinforced fibres

J. R. Collier, B. J. Collier, R. L. Rabe and D. L. Wright – Ohio University, Athens, Ohio 45701 USA.

ABSTRACT

A new fiber has been demonstrated that has a synergistic combination of synthetic fiber mechanical properties and natural fiber surface properties, as contrasted to the normal averaging of properties obtained from blends. An oriented nylon fiber was used as the core, fed through a die and thereby coated with a thin surface layer of unoriented rayon.

INTRODUCTION

The advent of synthetic fibers has led to fabrics for which it is possible to control many of the bulk dominated physical properties by altering the chemical structure, processing conditions and cross-sectional shape. As a result fabrics with high wrinkle resistance and strength are available. However, since many consumers still prefer fabrics of natural or regenerated fibers much research and development effort has been concentrated on conferring the perceived comfort of these preferred materials upon synthetic fibers. It has been shown that the perception of comfort is strongly related to the transport of water through the fabric (1-3), and therefore fiber, yarn, and fabric construction characteristics that affect this transport also influence comfort (4-8). Moisture is transmitted through fabric both in the form of a vapor and also as a liquid. The vapor transport is essentially controlled by fabric construction since the mode of transport is primarily through the fabric interstices (6,9). On the other hand liquid moisture transport is apparently dominated by capillary action and is therefore influenced by fiber surface characteristics as well as fiber packing within the yarn (1,4,5,10).

Untreated natural and regenerated fibers apparently have surface characteristics which contribute to liquid moisture transport (6). Researchers have emphasized that the surface properties and not the bulk moisture regain are most influential in this respect (5,9). Recently it has been reported that water transport through fabrics of synthetic fibers can be altered by processing conditions to have similar comfort characteristics (6,7,11). The primary variables being altered are fiber cross-section, surfactant coatings, and yarn and fabric construction. Changes in these properties that enhance capillary action of moisture through the fabric are desirable. For example fabrics made from fuzzy yarns are not as comfortable as fabrics made from the same fibers but with smooth yarns.

The researchers altering the characteristics of synthetic materials are not claiming that these fibers in the same construction will be equivalent to natural and regenerated fibers. Instead they are asserting that by using different fabric construction characteristics, altered processing conditions, and surface treatments, it is possible to achieve comfort properties similar to cotton and other natural fibers. Whether or not truly equivalent states are achieved, or the construction changes are acceptable, they are effectively demonstrating that water transport through fabric is dependent upon more than just bulk moisture regain.

Many consumers have been willing to accept some loss of water transport ability to gain desirable mechanical properties, such as wrinkle resistance and strength, that can be achieved by blending synthetic and natural fibers. This blending leads to an averaging of properties based upon the blend ratios of the fibers, if all other parameters are maintained the same. However since mechanical properties are essentially dominated by the bulk properties of the fiber and the water transport behavior by the surface characteristics, a synergistic combination of properties can result from a skin/core fiber. In this paper a unique process for producing an internally reinforced rayon fiber is described, in which an already oriented synthetic fiber core is coated with an unoriented rayon skin (12). Since this rayon skin can be coated onto any cross-sectional shape fiber, control over resultant fabric properties should be possible. A patent application is pending on this fiber and the process for producing it. This continuing research effort is being supported by the National Science Foundation.

FIBER PREPARATION

In the fiber coating process a previously oriented 20 micron diameter nylon 66 monofilament fiber supplied by duPont was coated with a commercial viscose rayon provided by Avtex Inc. Subsequently a 25 micron diameter nylon 6 monofilament supplied by Allied was also coated. The Ohio University (OU) Intergraph Computer Aided Design / Computer Aided Manufacturing (CAD/CAM) system was used to design a die which is similar to wire coating dies but with smaller openings and altered flow channels. In this die the core nylon fiber was contacted with the viscose solution and then exited through an 800 micron diameter opening into a commercial strength rayon spinning bath. The bath, containing 9 w% sulfuric acid and 13 w% sodium sulfate, was followed by a water rinse, drying, and fiber take-up. A take-up mechanism designed for use in a Brabender blown film system was altered to increase the maximum peripheral velocity of the driven roll from 70 to 380 ft/min. For line speeds higher than 170 ft/min, a dryer had to be used. This consisted of a "cooling" air ring from the same blown film system, altered so that warm air from a rheostat controlled industrial air gun was discharged axially, countercurrent to the line direction. The viscose pressure was varied from 0.5 psig up to 20 psig by pressurizing the reservoir.

RESULTS AND DISCUSSION

At the minimum viscose reservoir pressure of 0.5 psig, obtained with only the gravity head of the fluid, a one micron thick rayon skin was coated on the 20 micron diameter nylon 66 fiber. Since the coated core fiber exited the die through an 800 micron diameter opening, this thickness corresponded to drawdown and coagulation ratio of approximately 400 to 1. Figure 1 and 2 are scanning electron micrographs of the fiber produced at these conditions and a line speed of 70 ft/min. The relative thicknesses of the skin and core are evident in Figure 1 in which a fiber cut by a razor blade is shown. The skin thickness is consistent in both the fiber axis direction and around the fiber. Apparently

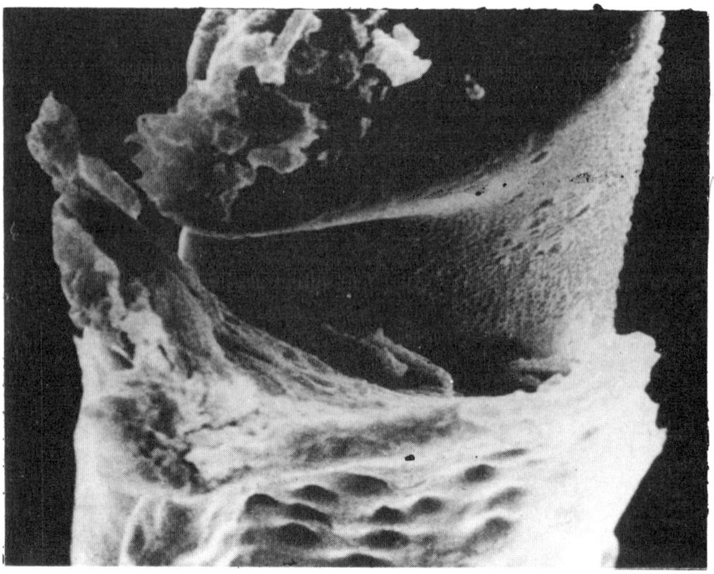

Fig. 1. SEM of an internally reinforced rayon fiber cut by a razor blade, 850 ×

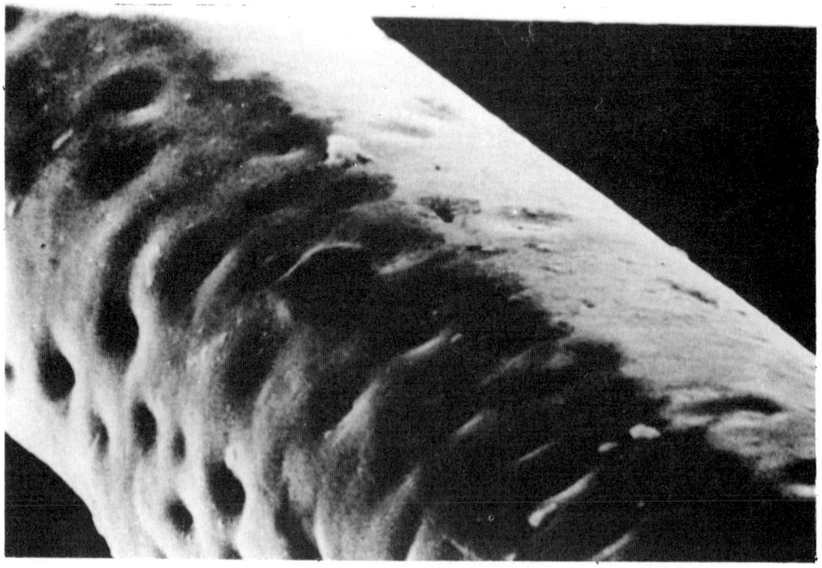

Fig. 2. SEM of an internally reinforced rayon fiber, 1,020 ×.

the core remained centered in the die as a result of rheological forces, and the drawdown and coagulation were consistent in both directions. As shown in Figure 2, the rayon skin is dimpled indicating an enhanced surface area which would be useful for water and dye sorption. Since the dimples do not appear to be elongated in either the axial or hoop directions, the skin has very low, if any, orientation, which should also contribute to its sorptive capacity.

Coating quality is dependent upon system conditions with a very good fiber being produced using the following: fresh viscose, a line speed of 70 to 380 ft/min (the highest line speed achievable with the available equipment), a commercial concentration spinning bath consisting of 9 w% sulfuric acid and 13 w% sodium sulfate, and a simple running tap water rinse. By maintaining the acid concentration in the spinning bath, sufficient adhesion between the skin and core layers is achieved making it difficult to mechanically remove the skin layer. Adequate drying of the fiber prior to takeup is necessary to prevent adhesion between adjacent segments, whereas commercial rayon fibers can be taken up in the wet state and subsequently dried. This difference in wet fiber characteristics could be due to the lack of orientation in the internally reinforced fibers as contrasted to the more highly oriented surface of commercial rayon fibers. This adhesive property could be useful in some applications such as nonwoven materials.

Processibility results indicate that periodic skin thickness variation noted at low speeds gradually disappears as the line speed approaches 70 ft/min, and does not reappear even at thicker skin coatings. The only apparent process limitations to be avoided at higher line speeds are insufficient residence time in the acid bath for complete coagulation and water removal prior to reaching the take-up mechanism. If the unoriented rayon skinned fibers contact each other prior to sufficient drying they tend to adhere at the contact points without the application of adhesives.

Data for a line speed in the 100 to 170 ft/min range indicate that drag flow dominates the total flow rate by contributing about 75% of the total when the pressure is between 10 and 20 psig. In the same operating region the coating thickness is three times more sensitive to line speed than it is to pressure.

Mechanical property determinations using an Instron Universal Testing Unit indicate that the core layer does indeed dominate the modulus of these internally reinforced fibers. A slight decrease in elongation to break was noted probably due to the effect of the spinning bath on the nylon core fiber.

CONCLUSIONS

A new fiber has been demonstrated at OU, that is an internally reinforced rayon fiber. This composite fiber apparently maintains the desirable mechanical properties of the synthetic core fiber yet retains the water sorption and dye affinity of the rayon skin. A skin/core structure was used so that bulk dominated properties could be decoupled from the normally dependent surface controlled properties. Attempts to increase the strength and wrinkle recovery of rayon, or other regenerated textile fibers generally lower the surface related properties of water sorption and dye affinity. However by using a continuous oriented nylon core fiber with an unoriented skin, this desired decoupling of normally related properties was achieved and a synergistic combination of properties resulted.

REFERENCES

1. Pontrelli, G.J., Clothing Comfort, ed. by N.R.S. Hollies and R.T. Goldman, Ann Arbor Science, Ann Arbor, MI, 1977, 71.
2. Chandler, J. and Zeronian, S.H., Text. Chemist and Colorist, 11, 56 (1979).
3. Wallenberger, F.T., et al, Journ. of Eng. for Ind., 102, 1 (1980).
4. Hollies, N.R.S., et al, Text. Res. Journ., 26, 829 (1956).
5. Yoon, H.N. and Buckley, A., Text. Res. Journ., 54, 289 (1984).
6. DeMartino, R.N., et al, Text. Res. Journ., 54, 447 (1984).
7. Hollies, N.R.S., et al, Text. Res. Journ., 54, 544 (1984).
8. Scheurell, D.M., et al, Text. Res. Journ., 55, 394 (1985).
9. Wallenberger, F.T., Text. Chemist and Colorist, 14, 154 (1982).
10. Harnett, P.R. and Mehta, P.N., Text. Res. Journ., 54, 471 (1984).
11. Latta, B.M., Clothing Comfort, ed. by N.R.S. Hollies and R.F. Goldman, Ann Arbor Science, Ann Arbor, MI, 1977, 33.
12. Wright, D.L., M.S. Thesis in Ch.E., OU, June 1985, Advisor J.R. Collier.

Chapter 65

Aspects of new generation fibre technology in Japan

Taksuya Hongu – Nisshinbo Industries, Inc. 3 Yokoyamacho Nihonbashi Chuo-ku, Tokyo 103, Japan.

FOREWORD

The textile industry in Japan has greatly changed structurally due to the keen domestic as well as foreign competition over the past thirty years and has now entered a new stage of development. In this talk I wish to present an overview of the present and future textile industry with a special emphasis on the so-called new-generation cellulose technology. Two specific examples will be given which illustrate the newly emerging technology.

FUTURE DEVELOPMENT OF THE TEXTILE INDUSTRY

The future developments of fiber science and technology will not be made solely in their traditional application to textile, but more often than not in close relationship with other areas of science and technology, such as biotechnology, optoelectronics, communication and data processing, and energy and space science and technology. Particularly important for the materials to be used in the future developments of these areas will be their higher-order fine structures: molecular orientation, degree of crystallinity, etc. The scientists and engineers of fiber science and technology, through many years of experience are in an excellent position to pursue this goal. This has already been manifested in the successful development of materials for hollow fiber type artificial kidneys and lungs by several textile companies.

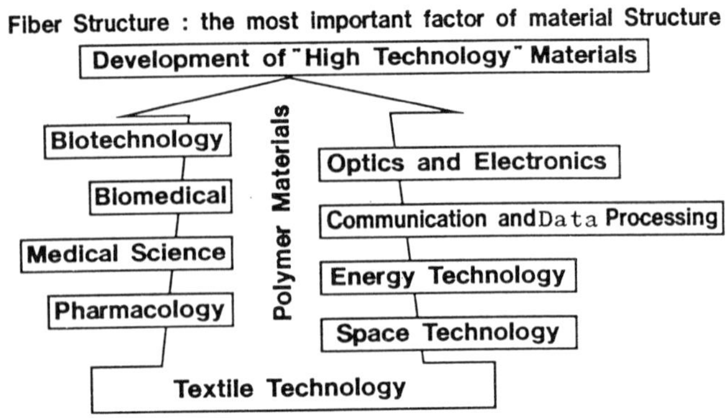

Fig. 1. Achieving added value by controlling fine structure.

NEW-GENERATION CELLULOSE TECHNOLOGY

The new-generation cellulose technology includes functionalization, immobilization of functional materials, surface modificaiton, and plasticization.

What is meant by functionalization here is to introduce substituent groups which themselves do not possess any particular functions, but when incorporated will give new functions or properties to cellulose fibers. Such functions and properties include optical resolution, anticancer action, action against blood coagulation, liquid crystal formation, and controlled solubility.

Immmobilization of functional groups or materials is accomplished by making use of the hydrophilicity, biodegradability, high chemical reactivity, and castability of cellulose. Cellulose can thus be used as an excellent support material for various functional materials. Functional materials immobilized so far are enzymes, deodorizing agents, agricultural chemicals, insecticides, and fertilizers.

The third area of new cellulose technology will be surface modification by exploiting cellulose's excellent fiber and membrane forming property as well as its high chemical reactivity. A representative example will be modification of cellulose fiber and membrane surface against blood coagulation. Another example will be carboxymethylation which gives cellulose fibers a high absorption capability.

Plasticization here does not simply mean plasticization of cellulose or wood for use as convential plastics, but it is aimed at making components for composite materials and polymer blends.

Other promising new aspects of cellulose technology will include microfibrillation and gelation of cellulose for biomedical uses. Cellulose's biocompatibility and its ease for hydrophilic/hydrophobic adjustment are promising aspects for biomedical applications.

DEODORIZING FIBERS

Among many examples of next-generation cellulose technology mentioned so far, the deodorizing function of metal phthalocyanine incorporated into rayon fibers, I believe, is one of the most promising functions investigated so far(1-4).

Iron octacarboxyphthalocyanine has a trivalent high-spin electronic state and has a function quite similar to that of iron(III) hematoporphyrin IX in the in vivo oxidase and thus catalyzes a series of reactions depicted below in a pseudoenzymatic manner.

Fig. 2.

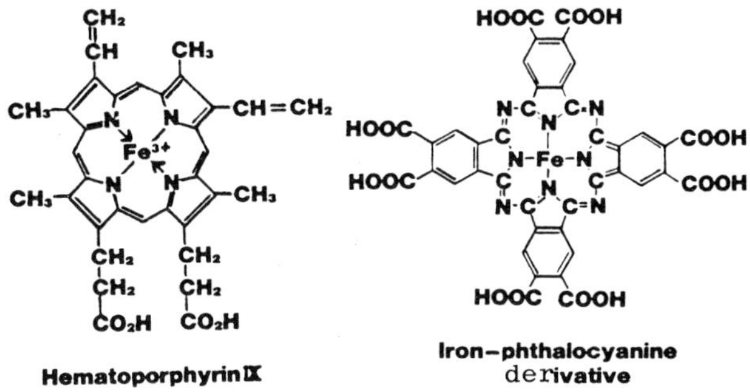

SH_2 oxidation (AH_2: L-ascorbic acid)

(1) $SH_2 + O_2 \longrightarrow S + H_2O_2$

(2) $S + O_2 \longrightarrow SO_2$

(3) $S + O_2 + AH_2 \longrightarrow SO + H_2O + A$

(4) $2H_2O_2 \longrightarrow O_2 + 2H_2O$

(5) $SH_2 + H_2O_2 \longrightarrow S + 2H_2O$

Mercaptoethanol oxidation

(6) $2HOCH_2CH_2SH + O_2 \longrightarrow (HOCH_2CH_2S)_2 + H_2O_2$

(7) $H_2O_2 + 2HOCH_2CH_2SH \longrightarrow (HOCH_2CH_2S)_2 + 2H_2O$

It is water-soluble and strongly adsorbs onto cellulose and other naturally-occurring polymers. Thus, it is adsorbed onto rayon of approximately 16% in crystallinity. Adsorption follows Freundlich type adsorption behavior and the maximum adsorption is approximately 4 wt% with respect to the weight of rayon. X-ray microanalysis of the fiber cross-section reveals a uniform distribution of iron throughout the fiber. As in the free state iron phthalocyanine when adsorbed onto rayon fiber catalyzes air oxidation of mercaptoethanol. The rayon fiber also promotes air oxidation of mercaptan, hydrogen sulfide, various aldehydes, phenols, indols, scatols and other substances of unpleasant smells, thereby converting them to odorless substances.

Figure 3 shows how hydrogen sulfide, mercaptan, and ammonia are removed when 1 g of a 1-to-1 mixture of polyester and rayon fibers incorporated with 3 wt% iron phthalocyanine is placed in a 1.5-l bag into which is introduced 100 to 400 ppm of a bad smelling gas. As is shown here, these gaseous substances are completely oxidized and made odorless within 1 hour.

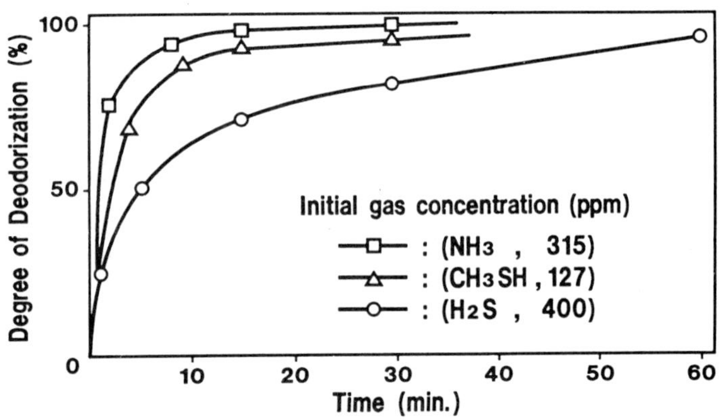

Fig. 3. Rate of deodorizing (room temperature) 1.5-l bag with 1 g mixed fibers (Pc:Cu:PET = 1:1:2) and 0.5 g water.

Figure 4 shows the duration of deodorizing abilities against various gases. Since the deodorizing mechanism is catalytic, duration of deodorizing power is excellent. This biomimetic fiber can thus be used to remove body odor as well as odors from animal wastes. In fact, it has been put into use for care of incontinent patients. Its use for underwear, masks, various filters, bedding materials and interior decoration materials is also being investigated.

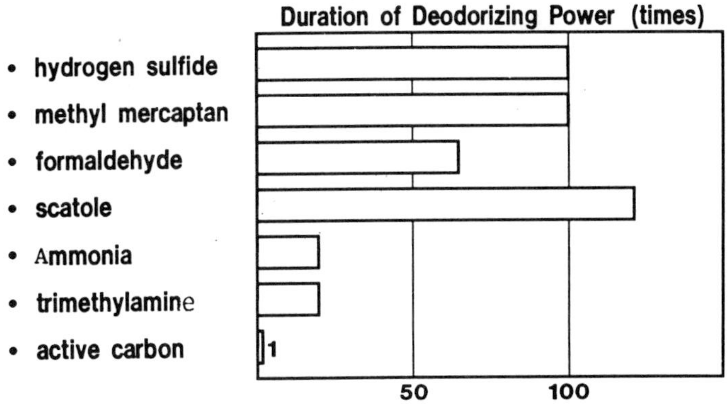

Fig. 4. Comparison of duration of deodorizing power between rayon-bound iron phthalocyanine and active carbon.

COATING AGENTS FOR INTESTINALLY-DISSOLVING DRUGS

Another promising application of new cellulose technology is preparation of coating materials for intestinally-dissolving drugs. The requirements of the coating agents for intestinally-dissolving drugs are: resistance against the high acidity of gastric juice which is normally in the pH range of 1.5 to 2.5, and dissolution or disintegration in the intestine the pH of which is 3.6 to 7.9. For this purpose, cellulose derivatives having carboxyl groups, soluble in organic solvents, are most suited. Presently being used for this purpose are cellulose acetate and hydroxypropylmethylcellulose derivatized with phthalic acid. These, however, have ester linkages and thus are susceptible to hydrolysis when drugs are stored for a long period of time. Recently, a new coating material has been developed from carboxymethylcellulose of degree of substitution in the range of 0.4 to 0.8(5).

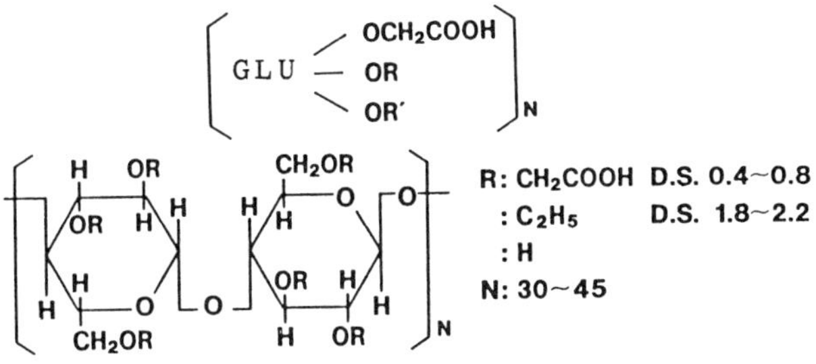

Fig. 5. Chemical structure of CMEC.

Thus, the sodium salt of carboxymethyl cellulose is ethylated with ethyl chloride, the degree of ethylation being 1.8 to 2.2. It is then converted to acid form, and then hydrolyzed with acid to lower its molecular weight. The obtained carboxymethylethylcellulose is soluble in ethanol /water mixtures with ethanol contents larger than 55%.

As shown in Figure 6, the material shows excellent resistance against dissolution under acidic conditions, but dissolves fairly rapidly under slightly acidic or neutral conditions. It is thus quite promising as a coating agent for intestinally dissolving drugs.

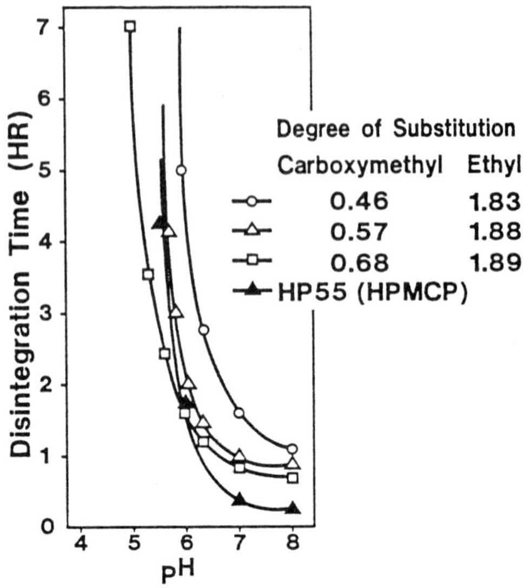

CONCLUSION

1) The naturally-occurring fibers including cellulose are biological in nature so that their functions are best realized when they are used under the conditions close to those of a living body. Under other conditions, synthetic polymeric materials will be better suited.

2) The biocompatibility of cellulose as well as its hydrophilicity adjustability are most advantageous in developing new and high-performance biomedical materials. The fiber industry has long been engaged in research in that direction and, no doubt, will be a front-runner in the above-mentioned and other biomaterial fields in the 21st century.

REFERENCES

(1) H. Shirai et al., Makromol. Chem., 181, 565 (1980)
(2) H. Shirai, et al., Makromol. Chem., 181, 1003 (1980)
(3) H. Shirai, Sen-i Gakkaishi, 41, 226(1985)
(4) Chikujou, Hanabusa, and Shirai, a paper presented at the annual meeting of Society of Fiber Science and Technology, Japan, 1986
(5) K. Yamagishi, a paper presented at the 16th Res. Conf., Foundation for Research Conferences in Chemistry, Osakc, 1983

POSTER PRESENTATIONS

Chapter 66

^{13}C-NMR studies ofsome cellulose derivatives

T. Miyamoto, T. Fujimoto, S. Takahashi, T. Yamagishi, O. Hasegawa and H. Inagaki – Institute for Chemical Research, Kyoto University, Uji, Kyoto 611, Japan.

ABSTRACT

^{13}C-NMR spectra of trityl cellulose, tosyl cellulose, cellulose formate, cellulose S-methyl xanthate, O-methyl cellulose, cyanoethyl cellulose, and 6-azido-6-deoxy cellulose in dimethylsulfoxide-d_6 were analyzed at 50.4 or 100.8 MHz. It was found that the distribution of substituents in the anhydroglucose units of these cellulose derivatives can be estimated from their ring carbon spectra. The calculation results on the distribution of substituents were discussed in terms of the reactivity of hydroxyl groups at C-2, C-3 and C-6 positions.

INTRODUCTION

^{13}C-NMR has shown its feasibility and applicability for the structural characterization of different types of cellulose derivatives. The advantage of this method is that the degree of substitution(DS) of a cellulose derivative and the distribution of substituents in the anhydroglucose(AHG) units, i.e., the relative DS at three different types of hydroxyl groups, can be determined. Few ^{13}C-NMR studies have yet been reported on the important cellulose derivatives such as triphenylmethyl(trityl) cellulose (Tr-Cell), p-toluenesulfonyl(tosyl) cellulose(Ts-Cell), cellulose formate(CF), cellulose sodium xanthate(Cell-Xan), cyanoethyl cellulose(CE-Cell), which are classical but synthesized according to the so-called fundamental methods in cellulose chemistry.

In our previous papers[Miyamoto et al., 1984,1985], we reported that the distribution of O-acetyl groups in the AHG units of cellulose acetate(CA) can be estimated quantitatively from the ^{13}C-NMR spectra of ring carbons. This allowed us to expect that ^{13}C-NMR spectra of ring carbons of other cellulose derivatives can be also analyzed in a similar way as applied to CA. This paper deals with the 50.4 or 100.8-MHz ^{13}C-NMR spectra of ring carbons of cellulose derivatives mentioned above.

Concerning O-methyl cellulose(MC), the ^{13}C-NMR spectra of 22.6 MHz have been already reported by Parfondry and Perlin[1977]. In the present study, the 100.8 MHz spectra of MC were also examined. Furthermore, the ^{13}C-NMR spectrum of 6-azido-6-deoxycellulose(Cell-6-N), which is considered to be one of the cellulose derivatives selectively substituted at C-6, was investigated to elucidate the degree of regioselectivity.

EXPERIMENTAL

Cellulose samples

The preparation of CA sample has been reported in a previous paper[Miyamoto et al., 1985]. A commercial MC sample(DS=1.8) was purchased from Nakarai Chem. Co., Ltd. Tr-Cell and Ts-Cell samples were prepared according to the method of Honeyman[1947]. The details of the preparation of CF and cellulose S-methyl xanthate (Cell-M-Xan) samples have been reported elsewhere[Fujimoto et al. 1986; Takahashi et al., 1986a]. A water-soluble CE-Cell was obtained by reaction of alkali cellulose with acrylonitrile under homogeneous conditions during part of the reaction time[Lukanoff et al., 1979]. Cell-6-N was prepared from 6-chloro-6-deoxycellulose(Cell-6-Cl) according to the method of Horton et al.[1973].

NMR measurements

^{13}C-NMR spectra of CA, Tr-Cell, Ts-Cell, CF, and Cell-M-Xan samples were obtained from a Varian Model XL-200 NMR spectrometer operating at 50.4 MHz in the proton noise-decoupled mode by using a 10-mm probe and DMSO-d_6 as solvent at 80°C. The spectra were recorded by using a spectral width of 12.5 KHz, a repetition time of 3 sec, and a flip angle of 45°. Scans(10,000 - 20,000) were accumulated.

The spectra of MC, CE-Cell, and Cell-6-N samples were taken on a JEOL Model GX-400 NMR spectrometer operating at 100.8 MHz. The measurement conditions were almost the same as those at 50.4 MHz.

RESULTS AND DISCUSSION

Trityl and tosyl celluloses

Tr-Cell and Ts-Cell derivatives have been used as intermediates for preparing new, functional cellulose derivatives. However, the relative DS in these derivatives is believed to be very different among the three hydroxyl groups. At first, we examined the ^{13}C-NMR spectra of these derivatives. The peak assignments for the derivatives were made in a similar manner as applied to CA. In advance of discussing ^{13}C-NMR analyses of Tr-Cell and Ts-Cell, the results on CA are briefly described below.

Figure 1a shows the ring carbon spectrum of CA having a DS value of 1.23. In the case of CA, it is confirmed that the peaks C6 and C6s are due to C-6 carbons bearing an unsubstituted and a substituted hydroxyl group, respectively. The relative DS at C-6 carbons can be estimated from these two peaks. C-1 and C-4 carbons in the AHG units have no hydroxyl group. However, it is known that the esterification or etherification of a hydroxyl group of glucopyranosic compounds causes an upfield shift of the resonance peak of the adjacent carbons. By taking this fact into account, a peak designated by C1s may be assigned to C-1 carbons adjacent to C-2 carbons bearing a substituted hydroxyl group and a peak designated by C1 to C-1 carbons adjacent to C-2 carbons with a free hydroxyl group, respectively. The same argument may be also applicable to the peaks C-4 carbons designated by C4 and C4s. These assignments have been supported by the following results: (i) the relative intensities of C1, C4 and C6 peaks decreased as DS value increased, while those of C1s, C4s and C6s peaks increased, (ii) the combined area of peaks C1 and C1s was almost identical to that of peaks C6 and C6s, and (iii) the relative DS values at individual hydroxyl groups, which were estimated according to the assignments mentioned above, were in good agreement with those by ^{1}H-NMR method[Miyamoto et al., 1984].

Figure 1b shows the ring carbon spectrum of Tr-Cell having a DS of 1.0. To be noted here is that the signal of C-6 carbons shows a single peak and that the chemical shift of this C-6 carbon signal is much lower than that expected for C-6 carbons with a free hydroxyl group. By taking into account the reactivity of primary hydroxyl groups at C-6 and the total DS of the sample, the peak at 62.5 ppm may be assigned to C-6 carbons bearing a substituted hydroxyl group. A detailed discussion concerning the peak assignments has been given elsewhere[Takahashi et al. 1986a]. It may be concluded that the Tr-Cell sample obtained was very close to 6-O-trityl cellulose, showing that trityl chloride reacts mainly with primary hydroxyl groups.

Figure 2a shows the ring carbon spectrum of Ts-Cell having a DS of 1.0. It can be seen that the downfield shift of C-6 carbon signal and the upfield shift of C-1 carbon signal by tosylation are both much larger than those by acetylation and tritylation. The assignments of the peaks C1, C1s, C6, and C6s were simple. On the other hand, the relative DS at C-3 position can be estimated from the DS values at C-6 and C-2 positions, because the total DS value of Ts-Cell can be easily obtained by elemental analysis. From the spectrum, it is obvious that the difference in the relative DS value among the three hydroxyl groups is not so high, although the relative reactivities of hydroxyl groups toward tosylation decrease in the order C-6 > C-2 > C-3.

Cellulose formate

CF is one of the well-known ester derivatives, but little attention has been paid to this derivative because of its instabil-

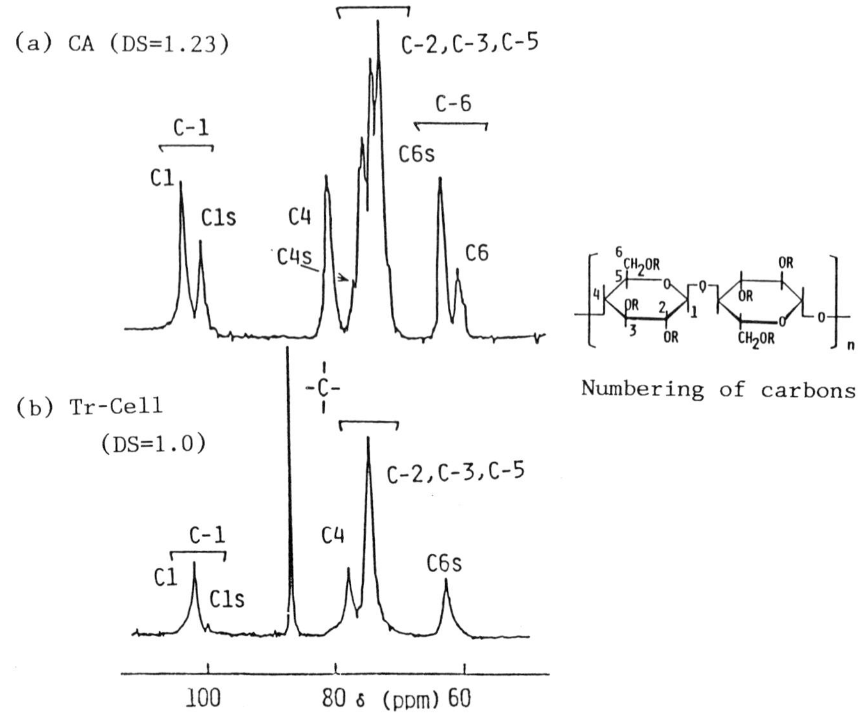

Fig. 1. 50.4-MHz ^{13}C-NMR spectra of the ring carbon region of CA and Tr-Cell samples in DMSO-d_6 at 80°C.

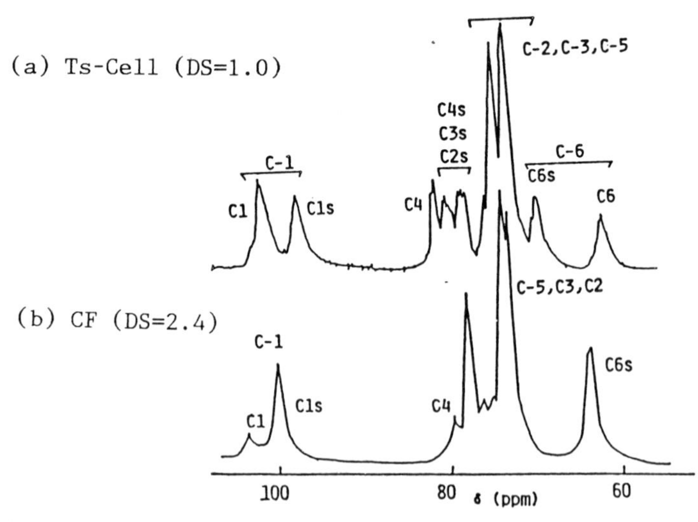

Fig. 2. 50.4MHz ^{13}C-NMR spectra of the ring carbon region of Ts-Cell and CF samples in DMSO-d_6 at 80°C.

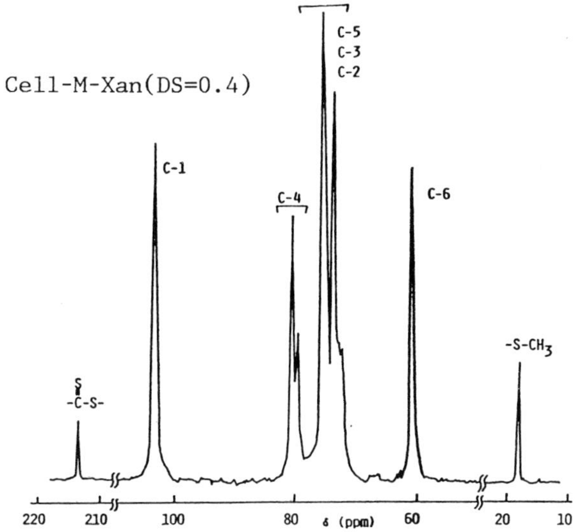

Fig. 3. 50.4-MHz ^{13}C-NMR spectrum of Cell-M-Xan in DMSO-d_6.

ity. However, stable forms can be obtained by immersing regenerated cellulose in formic acid without catalyst and adding hot water[Fujimoto et al., 1986]. These characteristics should be interesting from scientific as well as practical points of view. This section deals with the ^{13}C-NMR spectrum of CF.

Figure 2b shows the ring carbon spectrum of CF having a DS of 2.4. The partial assignments of the signals were made on the basis of the comparison with spectra of CA, by taking into account the total DS value of the sample. The details of the peak assignments have been reported elsewhere[Fujimoto et al., 1986]. It can be seen that the ease of formylation is C-6 > C-2 > C-3.

Cellulose s-methyl xanthate

Several attempts have been made to elucidate the distribution of substituents in Cell-Xan, but no convincing experimental evidence has been obtained so far. We attempted to obtain the ^{13}C-NMR spectrum of this derivative. However, its DMSO solution was so unstable that gelation occurred during the NMR measurements at 80°C. Thus Cell-Xan was converted to a much more stable and more readily refinable derivatives by reacting with methyl iodide.

Figure 3 shows the ^{13}C-NMR spectrum of Cell-M-Xan obtained. The signals corresponding to C-1 and C-6 carbons give only a single peak, while the signal of C-4 carbons splits into two peaks, suggesting that the hydroxyl groups at C-3 position are mainly substituted. In fact, the relative DS at C-3 position, which was calculated from the peaks designated by C4 and C4s, was ca. 0.35, being in good agreement with the total DS value of the sample. It is of interest to note that the ease of xanthation is different

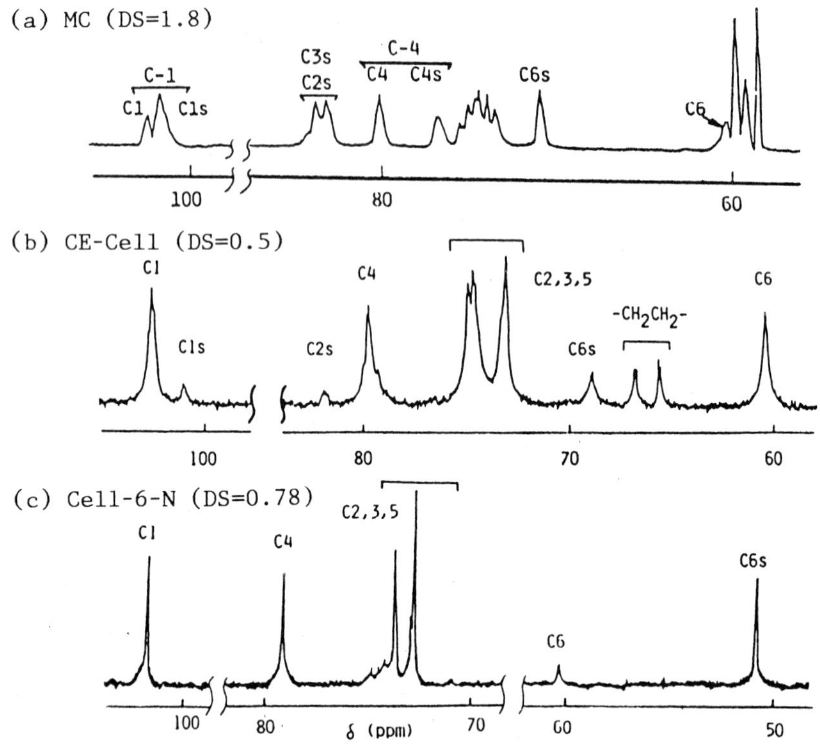

Fig. 4. 100.8-MHz ^{13}C-NMR spectra of the ring carbon region of MC, CE-Cell and Cell-6-N samples in DMSO-d_6.

from the reactivity order in other cellulose derivatives prepared from alkali cellulose. The reason why the reactivity of the hydroxyl groups at C-3 position is higher toward xanthation is still open for discussion.

O-methyl cellulose

As to MC, the ^{13}C-NMR spectra at 22.6 MHz have already been reported by Parfondry and Perlin[1977]. The signal of C-1 carbons adjacent to C-2 carbons with a free hydroxyl group(C1) was indistinguishable from that of C-1 carbons adjacent to C-2 carbons bearing a substituted hydroxyl group(C1s) in the spectra at 22.6 MHz. During the course of study on the relationship between the distribution of substituents in MC and its water solubility by ^{13}C-NMR at 100.8 MHz, we found that the peak for C1s is separated from that for C1[Takahashi et al., 1986b]. This splitting enables us to estimate the relative DS at C-1 carbons.

Figure 4a shows the ring carbon spectrum of a commercial MC having a DS of 1.8, which was prepared by the alkali cellulose process. The notation of each peak is the same as already mentioned.

The relative DS at individual hydroxyl groups can be estimated from the relative intensities of peaks C1, C1s, C4, C4s, and C6s, because the sum of individual carbon peaks should be equal to each other. The relative DS value in the sample examined was found to be 0.6, 0.6, and 0.4 for hydroxyl groups at C-6, C-2, and C-3 positions, respectively, showing that the ease of methylation is C-6 ≃ C-2 > C-3.

Cyanoethyl cellulose

CE-Cell, which is prepared via alkali cellulose, is one of the reactive cellulose derivatives[Miyamoto et al., 1986]. Figure 4b shows the ^{13}C-NMR spectrum of CE-Cell having a DS of 0.5. The notation of each peak is the same as mentioned above. The relative DS value was found to be 0.3, 0.15 and 0.05 for hydroxyl groups at C-6, C-2, and C-3 positions, respectively, showing that the ease of cyanoethylation is C-6 > C-2 > C-3.

6-Azido-6-Deoxycellulose

It has been reported that Cell-6-Cl can be prepared by treating cellulose with methanesulfonyl chloride in N,N'-dimethylformamide-chloral[Ishii et al., 1977]. The cellulose derivative selectively substituted at C-6 position is very interesting for the preparation of functional cellulose derivatives. Unfortunately, Cell-6-Cl was insoluble in common organic solvents, and converted into the corresponding Cell-6-N, which is soluble in DMSO, according to the method of Horton et al.[1973].

Figure 4c shows the spectrum of Cell-6-N obtained. It can be seen that the chlorination of cellulose by methanesulfonyl chloride is regioselective and that its conversion to the corresponding Cell-6-N quantitatively proceeds.

REFERENCES

Fujimoto, T., Takahashi, S., Tsuji, M., Miyamoto, T. and Inagaki H.(1986) J. Polym. Sci. Polym. Lett. Ed., in press.

Honeyman J. (1947) J. Chem. Soc., 168.

Hortman, D., Luetzow, A.E. and Theander, O.(1973) Carbohyd. Res. 26, 1.

Ishii, T., Ishizu, A. and Nakano, J.(1977) Carbohyd. Res.,59, 155.

Lukanoff, B., Philipp, B. and Schleicher, H.(1979) Cellulose Chem. Technol., 13, 417.

Miyamoto, T., Sato, Y., Shibata, T., Inagaki, H. and Tanahahsi, M.(1984) J. Polym. Sci. Polym. Chem. Ed., 22, 2363.

Miyamoto, T., Sato, Y., Shibata, T., Tanahashi, M. and Inagaki, H.(1985) J. Polym. Sci. Polym. Chem. Ed., 23, 1373.

Miyamoto, T., Takahaski, S., Tsuji,S., Ito, H., Inagaki, H. and Noishiki, Y.(1986) J. Appl. Polym. Sci., 31, 2303.

Parfondry, A. and Perlin, A.S.(1977) Carbohyd. Res., 57, 39.

Takahashi, S., Fujimoto, T., Barua, B.M., Miyamoto, T. and Inagaki, H.(1986a) J. Polym. Sci. Polym. Chem. Ed., in press.

Takahashi, S., Fujimoto, T., Miyamoto, T. and Inagaki, H.(1986b) J. Polym. Sci. Polym. Chem. Ed., in press.

Chapter 67

Enzymatic degradation and fermentation of cellulosic materials following hydrothermal pretreatment with and without organic solvents as additive

G. Bonn, H. F. Hörmeyer, M. Berreck and O. Bobleter – Institute of Radiochemistry, University of Innsbruck, Austria.

ABSTRACT

Various lignocellulosic materials (poplar wood, birch wood, wheat straw) were subjected to pretreatment processes that increase accessibility of the cellulose to cellulolytic enzymes. The substrates so obtained are shown to yield, upon hydrolysis catalyzed by a **Trichoderma viride**-derived cellulase, glucose solutions which are readily fermentable to ethanol by the brewer's yeast strain of **Saccharomyces carlsbergensis** W 34. Nearly quantitative glucose-to-ethanol conversion could be achieved within 1.5-3 hours' fermentation time.

INTRODUCTION

In order to improve the accessibility of lignocellulosic materials to the action of cellulases, effective pretreatment methods are necessary /1/. Some of these processes consist in percolating pressurized solvents through the biomass, whereby the products of hemicellulose and, in part, of lignin degradation can rapidly be removed from the reaction zone. In the case of hydrothermolysis, pure water is employed for this purpose /2,3/; alternatively, organic solvents, e.g. methanol, may be used as additive /4/. The glucose liberated in the enzymatic hydrolysis stage can be fermented to yield ethanol /5-7/.

EXPERIMENTAL

Substrates

The lignocellulosic materials (poplar wood, birch wood, wheat straw) were percolated with either deionized water or aqueous methanol (1:1 v/v) at approx. 200°C and 100 bar pressure. This was done in a hydrothermolysis apparatus /2,3/. In other experiments, native cellulose (Merck) was used as a substrate.

Enzymic hydrolysis

Hydrolysis of the cellulose present in the residues was brought about by incubation with a cellulase preparation derived from **Trichoderma viride** ("Cellulase 12" of A.B.M. Chemicals Ltd., Stockport/England) under conditions given in Ref. /8/. For each gram of cellulosic residue, 0.25 g of the lyophilized cellulase powder were applied, which corresponds to an activity of 1.75 units releasing 1 mg glucose/min in the filter paper disintegration assay procedure specified by the manufacturer /9/. Experiments with pretreated wheat straw and poplar wood were performed in Erlenmeyer flasks under reciprocal shaking, whereas for all other hydrolyses a laboratory fermenter was employed (F 2000 KLF, Bioengineering AG, Wald/Switzerland; total volume 2.5 l). In this case, about 1 l of the 3% substrate suspension could be maintained at 50 °C and stirred at 100 rpm.

Ethanol fermentation

All fermentations were performed using brewer's yeast **Saccharomyces carlsbergensis** strain W 34, supplied in wort by Adambräu/Innsbruck, and which has been described in /7/. The yeast was centrifuged in a Beckman centrifuge J21C at 4°C and 1950 g for 10 min; under a nitrogen atmosphere, a weighed amount was resuspended in sterile bidist. water. Yeast mass and water volume were chosen to yield a 10 % yeast suspension after mixing with the original hydrolysis solution. Degassing with nitrogen was continued for a few minutes; thereupon, the fermentation was started at 25°C and 0.2-0.4 bar overpressure.

In the flask experiments 120 strokes/min of reciprocal shaking were maintained, whereas in the laboratory fermenter the suspension was stirred at 85 rpm. The samples withdrawn were analyzed by HPLC on a Ca^{2+}-loaded ion exchanger column as in Refs. /10,11/ for quantification of glucose a:d ethanol.

RESULTS AND DISCUSSION

Determination of yeast activity

Most preliminary experiments designed to check yeast activity were performed in the laboratory fermenter. Standard glucose solutions (10% w/v in sterile deionized water) gave ethanol yields of 90-95% of the theoretical value calculated from the formula,

$$\frac{g(\text{ethanol})}{g(\text{glucose})} = \frac{2 \times \text{m.wt.}(\text{ethanol})}{\text{m.wt.}(\text{glucose})} \times 0.92 = 0.471 .$$

The correction factor 0.92 accounts for yeast growth and glycerol formation /12/.

In further experiments, native cellulose was suspended in citrate buffer and hydrolyzed enzymatically. Saccharification proceeded until approx. 70%, whereupon all glucose present in the hydrolysate could be fermented to ethanol within 2 hours. On the other hand, 3 hours´ fermentation time was necessary for achieving quantitative conversion to ethanol in smaller-scale flask experiments if solutions of glucose in citrate buffer, and of similar concentration, were fermented. This clearly demonstrates the advantage of stirring over reciprocal shaking.

Enzymatic hydrolysis of pretreated cellulosic substrates with subsequent fermentation

In the following, the pretreatment processes specified in the Experimental section are shortly termed hydrothermolysis (HT; eluent: dist. water) and organosolv treatment (OS; eluent: methanol/water, 1:1 v/v), respectively. The data obtained for various samples of pretreated wheat straw are listed in Table 1.

Table 1. Effect of pretreatment conditions on the degradability of wheat straw and on the fermentability of the hydrolysates.

Temp. (°C)	Treatment	Residue (% dry weight)	% Saccharification yield after 70 h	% Ethanol yield after 3 h
192-195	HT	55	54	82
211-215	HT	47	68	77
224-229	HT	45	72	100
212-216	OS	50	56	98
223-229	OS	48	72	91

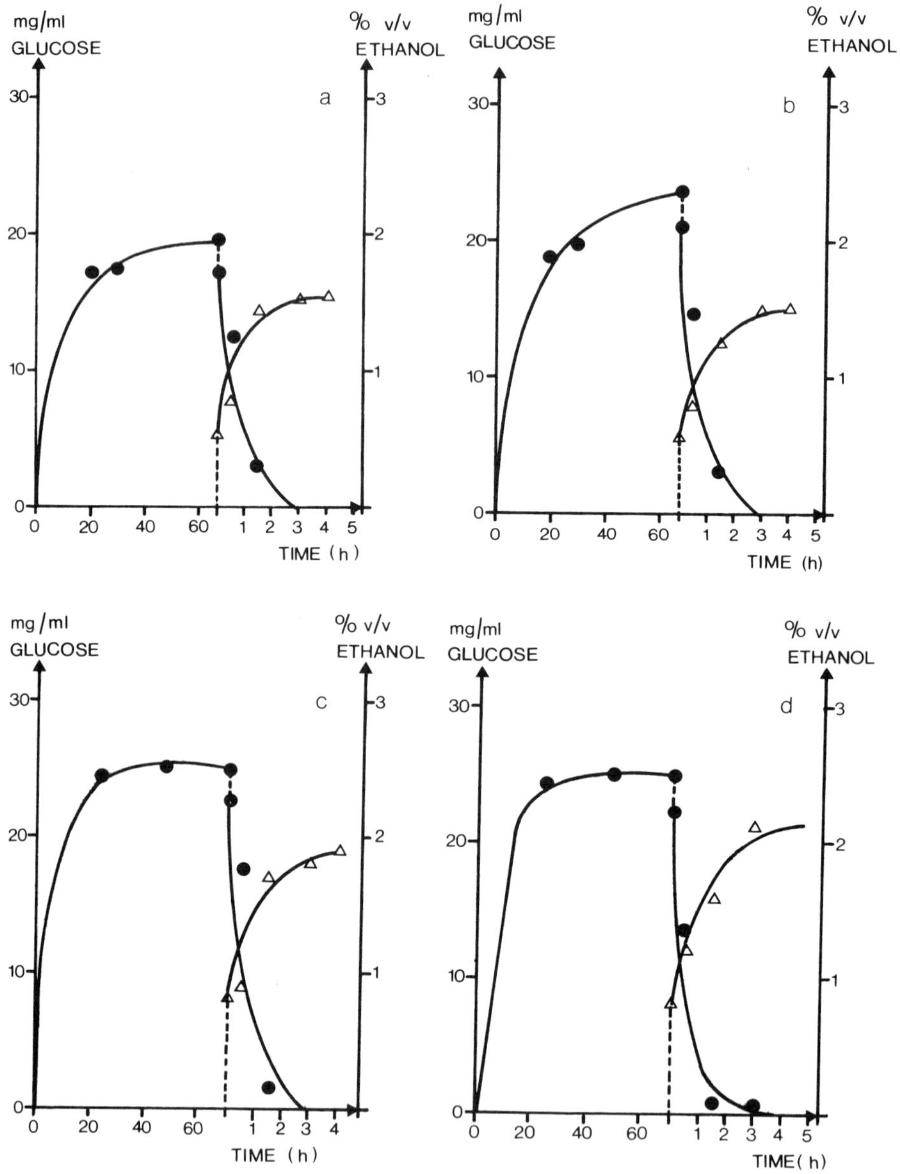

Fig. 1. Wheat straw cellulolysis and fermentation; ●, glucose, △, ethanol; a) OS, 212–216°C, b) HT, 211–215°C, c) OS, 223–229°C, d) HT, 224–229°C; flow: 10 ml/min.

The hydrolysis yields given in the preceding table were calculated as

$$\text{mg/ml glucose x ml } V_{final} \times \frac{0.9}{\text{mg substrate}} \times 100\% ,$$

where the factor 0.9 represents the molecular weight ratio of glucan monomers and glucose. Substrate weights are based on absolutely-dry matter.

Several profiles of the course of enzymatic hydrolysis and subsequent fermentative conversion are shown in Figs. 1a-1d. These results are in line with the generalizations made in /8/, namely that hydrothermal pretreatment of wheat straw is superior to the organosolv process in the lower temperature range (Figs. 1a and 1b), whereas above 220°C, both methods give rise to similar saccharification yields (Figs. 1c and 1d). From Table 1, it is also evident that hydrothermolysis tends to leave less solid residue than the organosolv pretreatment.

For comparison, birch wood chips were also subjected to hydrothermolysis, followed by cellulolysis and fermentation. In the laboratory fermenter, saccharification by **Trichoderma** cellulase was followed by an ethanol fermentation which after only $1^1/_2$ hours gave 93% yield /13/.

As with wheat straw hydrolysates, high fermentation yields (96%) could also be observed with glucose solutions obtained from poplar wood. However, in contrast to wheat straw, this substrate has been shown to be more efficiently pretreated by the organosolv process in the entire range of 200-250°C /8/.

ACKNOWLEDGEMENTS

One of the authors (H.H.) would like to thank the Austrian Research Centre Seibersdorf for granting a bursary. Furthermore, we are grateful to Mr. Gabriel of Adambräu Brewing Company, Innsbruck.

REFERENCES

/1/ Philipp, B. (1983). Acta Chim. Hung. 112:445-459

/2/ Bobleter, O. and Pape, G. (1968). Aust. Pat. 263661

/3/ Bonn, G., Concin, R. and Bobleter, O. (1983). Wood Sci. Technol. 17:195-202

/4/ Edel, E. and Feckl, J. (1985). Wood as a source of chemicals, presentation at CEC Contractors' Meeting in Friedrichshafen, FRG, 1985/05/08-09

/5/ Ghose, T.K. and Tyagi, R.D. (1979). Biotechnol. Bioeng. 21:1401-1420

/6/ Becker, D.K., Blotkamp, P.J. and Emert, G.H. (1981). In: Fuels from Biomass and Wastes, D.L. Klass and G.H. Emert (Eds.), Ann Arbor Science Publ., Ann Arbor/Michigan, 375-391

/7/ Pfeifer, P.A., Bonn, G. and Bobleter, O. (1984). Biotechnol. Lett. 6:541-546

/8/ Bonn, G., Hörmeyer, H.F. and Bobleter, O. (to be published)

/9/ A.B.M. Chemicals Ltd., Technical Bulletin on Cellulase

/10/ Bonn, G., Pecina, R., Burtscher, E. and Bobleter, O. (1984). J. Chromatogr. 287:215-221

/11/ Bonn, G. (1985). J. Chromatogr. 322:411-424

/12/ Harrison, J.S. and Graham, J.C.J. (1970). In: The Yeasts, A.H. Rose and J.S. Harrison (Eds.), Academic Press London, New York; 3:329

/13/ Bobleter, O., Schwald, W., Bonn, G. and Hörmeyer, H.F. (1986). Presentation at: Symposium on the Pretreatment of Lignocellulosic Materials (IEA/FE, Project CPD-2), Graz, Austria, 1986/06/23-27

Chromatographic optical resolution on cellulose and other polysaccharide derivatives

H. Namikoshi, T. Shibata, H. Nakamura, I. Okamoto, K. Shimizu, Y. Toga – Daicel Chemical Industries.

ABSTRACT

Derivatives of cellulose and some polysaccharides were synthesized to examine their ability of chiral resolution as packing material for HPLC. Cellulose derivatives synthesized under homogeneous conditions - cellulose triacetate (CTA-II), cellulose tribenzoate, cellulose triphenylcarbamate, cellulose tricinnamate and so on - showed very high abilities of chiral resolution for many racemic compounds when they were supported on macroporous Silica-Gel. Polysaccharide derivatives having regular and rigid structure such as chitosan, curdlan and α-1,3-glucan have also good ability for chiral recognition. Some mechanistic aspects of chiral resolution by means of HPLC using polysaccharide derivatives as packing material were also investigated.

INTRODUCTION

Cellulose and other polysaccharides have been known to have an ability of chiral recognition.

Cellulose has been known to perform an optical resolution for amino acids and some other compounds since about 1950 (1). The remarkable success in chiral resolution with cellulose derivatives was achieved by Hesse and Hagel using cellulose triacetate (2). They used microcrystalline cellulose triacetate (MCCTA) produced by heterogeneous acetylation of microcrystalline cellulose. In Fig.1-a is shown the reported chromatogram of Tröger's base on MCCTA. It is very interesting that dissolution

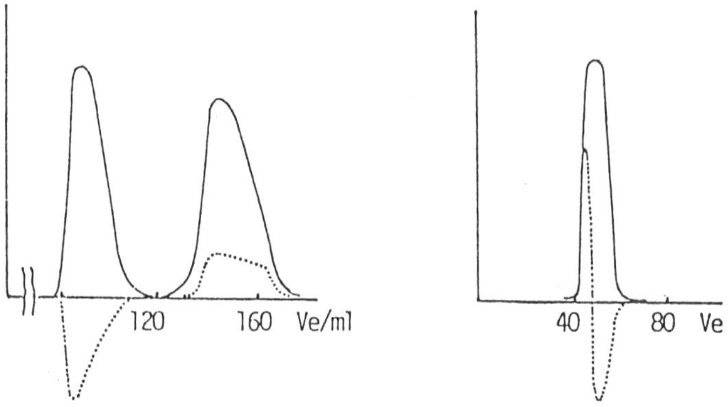

Fig. 1. Chromatograms of Trogers base (a) on "MCCTA" and (b) on CTA from acetone. Two 16 cm columns, eluent, ethanol. – uv, ... optical rotation.

and reprecipitation of MCCTA resulted in drastic loss of ability of chiral recognition while small reversed enantioselectivity was recognized (Fig.1-b). Though these observations did not necessarily mean the immediate participation of the crystalline lattice, it seems to be widely accepted that crystallinity is an essential requirement for chiral recognition. In spite of its excellent ability of chiral recognition, MCCTA has some limitations. The most important one is that it looses ability by dissolution. That makes it impossible to use solvents for the purpose of preparing a packing material with an improved efficiency. Therefore we started to investigate chiral resolution using CTA II, not MCCTA, postulating that CTA II should also show chiral recognition ability under the high crystalline state.

RESULTS AND DISCUSSION

Optical resolution with cellulose triacetate II

Crystalline fine powder of CTA II was prepared and Tröger's base was resolved on it (Fig.2). Though comparative efficiency of separation was attained for Tröger's base both on MCCTA and on CTA II, two striking differences in the behavior of adsorption were observed between those two kinds of cellulose triacetates. One is inversion of elution order of enantiomers and the other is that MCCTA (CTA I) showed much stronger adsorption for substrate. These observed differences are considered to be attributed to the differences of supermolecular structure between CTA I and CTA II.

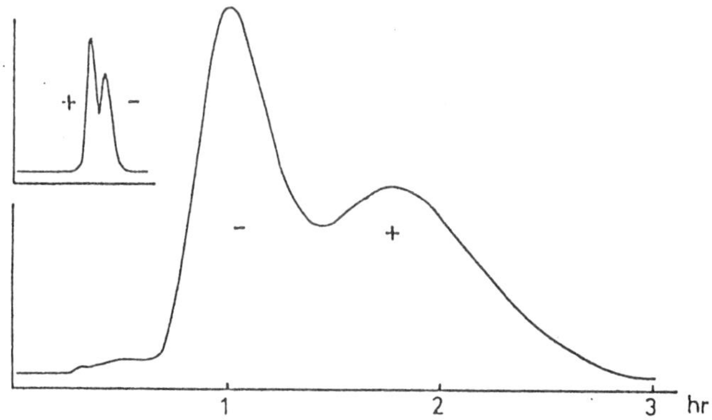

Fig. 2. Chromatogram of Troger's base on crystalline CTA II (upper) and "MCCTA" (lower) eluent, ethanol, 0.2 ml/min; column 25 cm.

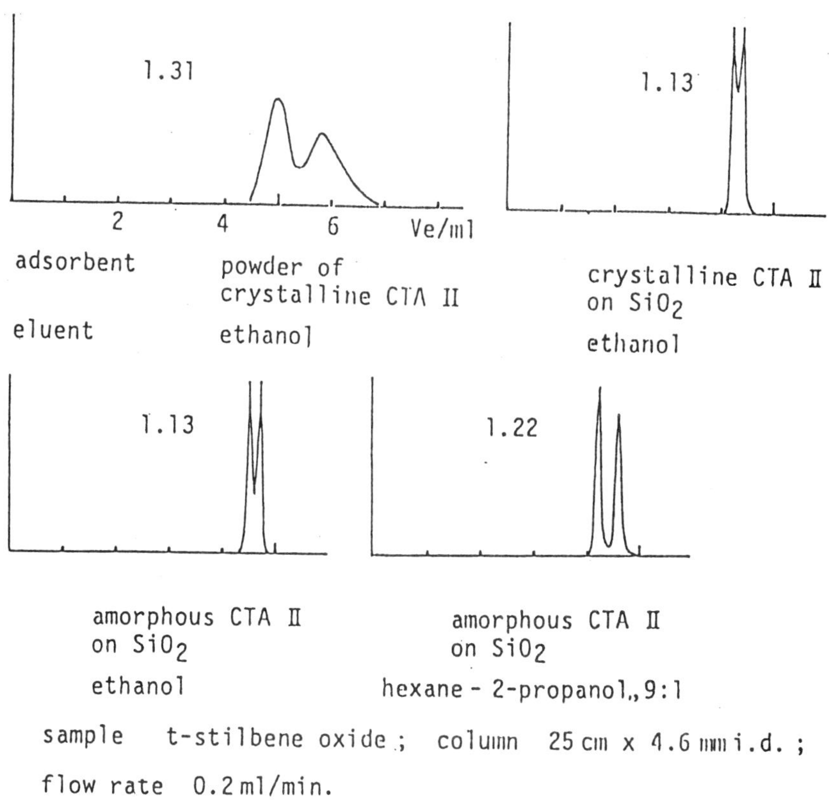

Fig. 3. Improvement in column efficiency and in choice of eluent.

In order to develop a more efficient packing material, macroporous Silica Gel was examined as a supported material. Two important results were obtained. (1) Amorphous CTA on Silica Gel gives an even better separation of enantiomers than crystalline CTA does. (2) Use of non-polar eluent (a mixture of hexane and 2-propanol, 9 : 1) resulted in much better resolution (Fig.3).

Studies on various derivatives of cellulose **Results are summarized as follows**

It became apparent that the occurrence of a particular natural supermolecular structure is not essential for optical recognition. So various cellulose derivatives supported on Silica Gel (20 wt% on Silica Gel) were examined.

Cellulose ethers showed only poor ability, and no apparent peak splitting was observed except for cellulose tribenzyl ether. Cellulose triesters e.g., propionate, nitrate, phenylcarbamate, benzoate, cinnamate and the related derivatives showed good abilities of chiral resolution (Table 1 and 2).

Table 1. α-Values given by cellulose esters containing no aromatic ring.

COMPOUND \ R		$COCH_3$	$CO(CH_2)CH_3$	$CO(CH_2)_2CH_3$	CO– 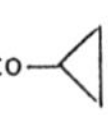	CO– 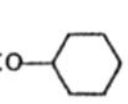	NO_2
[η]		0.50	0.70	0.62	0.68*1	0.62*1	1.90*2
1	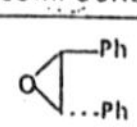	1.22(–)	1.08(–)	1.0	1.21(–)	1.0	1.61(–)
2	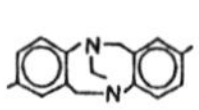	1.31(+)	1.0	1.0	1.23	1.0	1.33(+)
3	Ph–CH(OH)–C(=O)–Ph	1.05(–)	1.07(+)	1.0	1.10	1.21	1.0(–)
4	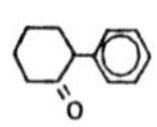	1.07(+)	2.03(+)	1.0	1.0	1.0	1.14(+)
5	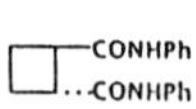	1.13(–)	1.18(–)	1.29	1.27	1.0	1.22(–)
6	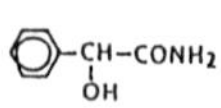	1.08	1.56	1.0	1.0	1.0	1.10

eluent : hexane–2–propanol(9:1) flow rate 0.5 ml/min

[η]: in CH_2Cl_2-CH_3OH (9 : 1) *1 in CH_2Cl_2 *2 in acetone 25°C

Separation factors , α and optical rotatory sign (in the eluent) of the first eluted enantiomer, if known, will be given along with some comments.

Table 2. α-Values given by cellulose derivatives containing aromatic rings.

R.
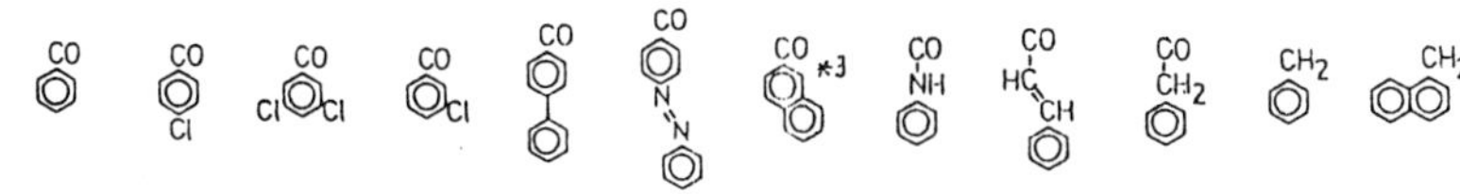

[η] / compd.	0.29		0.20		0.27	0.24 *1	0.22	0.61 *2	0.33 *1		1.17 *1	1.04 *1
1	1.47(+)	1.27	1.21(+)	1.33	1.31	1.28	1.58	1.32(+)	1.15(+)	1.0	1.0(+)	1.0
2	1.0 (-)	1.25	1.0	1.0	1.84	1.42		1.32(+)	2.82(+)	1.09	1.40(+)	1.0
3	1.12(+)	1.0	1.23(−)	1.26	1.14	1.15	1.0	1.0 (+)	1.08(−)	1.0	1.0 (-)	1.0
4	1.47(−)	1.0	1.34(−)	1.10	1.16	1.12	1.34	1.14(−)	1.26(−)	1.0	1.0 (+)	1.0
5	2.06(+)	1.0	1.22(+)	1.0	1.0	1.0	1.0	1.25(+)	1.52(−)	1.0	1.0	1.0
6	1.0	1.0	1.39(+)	1.07	1.15		1.0	1.0	1.0	1.0	1.0	1.0

R = CO 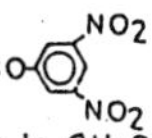 partial resolution

[η] : in CH_2Cl_2 − Ch_3OH (9:1) *1 in CH_2Cl_2 *2 in acetone 25°C *3 a lorge band width

eluent : hexane-2-propanol(9:1) flow rate 0.5 ml/min

Results are summarized as follows

(1) Every derivative has its own characteristic selectivity for substrates.

(2) Aliphatic cellulose triesters and cellulose trinitrate showed similar enantioselectivities. Large substituents reduced efficiency of chiral recognition.

(3) Aromatic derivatives showed the tendency of different enantioselectivity from that of aliphatic ones.

(4) Free rotating bonds in substituent reduced efficiency.

(5) In some cases, the difference in selectivity of substrate is attributed to length of substituent on cellulose. For example Tröger's base was resolved with high efficiency only when benzoyl groups have large substituents at para position.

OPTICAL RESOLUTION WITH DERIVATIVES OF OTHER POLYSACCHARIDES

To make comparison with cellulose, some derivatives of a variety of polysaccharides were prepared and examined (Table 3 and 4). Though some degree of ability of chiral recognition was observed in most polysaccharide triesters examined, no general ruling on the adsorption behaviours have been found.

Results are summerized as follows.

(1) Of the 1,4-glycans, chitosan triacetate showed the best ability for chiral recognition.

(2) 1,3 α-D-Glucan triacetate has a very high ability which may be due to the space made by three glucose units.

(3) 1,4 β-D-Glucan triacetate showed an excellent ability due to its helical chain conformation.

Table 3. α-Values given by polysaccharide acetates.

Polysaccharide	Cellulose	Amylose	Xylan	Mannan A	Chitin	Cyclosophoraose	Curdlan	α-1,3-Glucan
configuration	β-1.4	α-1.4	β-1.4	β-1.4	β-1.4	β-1.2	β-1.3	α-1.3
[η]	0.25						1.00	1.30
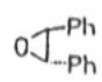	1.22(−)	1.14(+)	1.0(+)	1.0(−)	1.0	1.0(+)	1.60(+)	3.42(+)
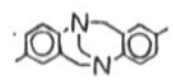	1.31(+)	1.0	1.0	1.0(−)		1.0	1.0	(+)
PhCH(OH)–C(=O)Ph	1.05(−)	1.0(−)	1.0	1.0	1.07(+)	1.0	1.65(+)	1.09(+)
	1.07(+)	1.0(−)	1.0	1.0(−)	1.0(−)	1.0(−)	1.43(+)	1.0(−)
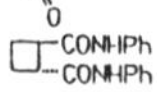	1.13(−)	1.0	1.0	1.0	1.22(+)	1.0	1.41(+)	1.0
PhCH(OH)–$CONH_2$	1.08	1.0	1.0	1.0	1.15(−)	1.0	1.42(+)	1.0(−)
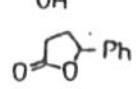	1.39(+)	1.0	1.0	1.0		1.0		1.0(−)

eluent: hexane-2-propanol (9:1) flow rate 0.5ml/min [η]: in CH_2Cl_2-CH_3OH (9:1)

Table 4. α-Values given by polysaccharide benzoates.

Polysaccharide configuration	Cellulose β-1.4	Amylose α-1.4	Xylan β-1.4	MannanA β-1.4	Cyclosophoraose β-1.2	Curdlan β-1.3	α-1.3-Glucan α-1.3	Dextran α-1.6	Pullulan α-1.4 α-1.6
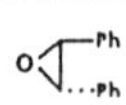	1.42(+)	1.0	1.17(+)	1.13(+)	1.0(−)	1.0(+)	1.0	1.18(−)	1.0
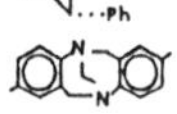	1.0(−)	1.0	1.0	1.0	1.0(+)	1.0(−)	1.0	1.0(+)	1.0
Ph–CH(OH)–C(=O)–Ph	1.14(+)	1.0	1.0	1.0	1.0	1.0	1.0	1.0(−)	1.0
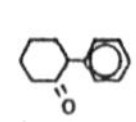	1.19(−)	1.0	1.17(−)	1.09(−)	1.0(+)	1.0	1.0	1.0(+)	1.0
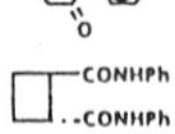	1.33(+)	1.0	1.0	1.0	1.0(−)	1.0	1.0	1.0	1.0
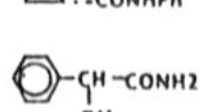	1.0	1.0	1.0	1.0	1.0(−)	1.0	1.0	1.0	1.0
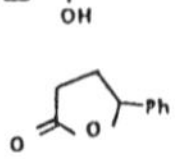	1.17(−)	1.07(−)	1.11(−)	1.24(−)	1.0	1.18(+)	1.0	1.0	1.0

eluent hexane—2—propanol(9:1) flow rate 0.5 ml/min.

(4) the macrocyclic 1,2 β-D-glucan derivative has poor ability.
(5) 1,4-Glycan (cellulose, amylose, xylan and mannan) benzoates tend to show similar enantioselectivities for the same substrate
(6) Pullulan and dextran showed poor abilities. In general, glucans having regular and rigid structure are suitable for chiral resolution.
(7) 1,3-α and β-D-glucan benzoates have only poor abilities of chiral recognition while acetates have excellent ability. This may be attributed to destruction of a suitable conformation of the polysaccharide chain by large substituents.

CONCLUSION

New chiral adsorbents derived from cellulose and some polysaccharides were realized. They have very unique properties compared with conventional MCCTA. Chiral recognition seems to be a property of the macroporous structure of adsorbent, not by the crystallinity of polysaccharide. Some cellulose esters deposited on macroporous Silica Gel were found to be practically useful.

ACKNOWLEDGEMENT

We thank Professor Y. Okamoto for the mutural exchange of information and Dr. K. Ogawa for giving us 1,3-α-D-glucan triacetate.

REFERENCE

(1) C.E. Dent, Biochem. J., 43, 169 (1948)
(2) G. Hesse and R. Hagel, Chromatographia, 6, 277 (1948)

Adsorption of n-butanol and i-propanol from dilute aqueous solutions

Torvald Vestberg and Johan-Fredrik Selin – Neste Oy, Technology Speciality Polymers and Biotechnology, SF-06850 Kulloo, Finland.

ABSTRACT

Batch experiments with adsorption of n-butanol and i-propanol from dilute aqueous solutions at room temperature are described. Three different adsorbents, activated carbon, a commercial adsorption polymer, and a silicalite were tested. Activated carbon proved the best adsorbent. In competitive adsorption with both alcohols n-butanol is preferentially adsorbed, especially on silicalite.

INTRODUCTION

Downstream processing is an ever more important part of biotechnology. The dilute aqueous solutions should first be filtered to remove cell mass and other solid material, and then the product isolated and purified. The traditional methods of distillation and crystallization are currently often replaced or complemented by chromatography and other methods based on adsorption.

The novel organosolv pulping processes fractionate the main components of wood and other lignocelluloses. These methods are generally best suited for hardwoods and grasses. The hemicellulose component consists mainly of pentoses. The impetus for the current work is fermentation of the pentoses to n-butanol and i-propanol by strains of **Clostridium beijerinckii.** The fermentation produces these two alcohols in concentrations of less than 10 g/l each. In order to assess the recovery of the components, adsorption of alcohols from aqueous solutions on a few adsorbents was studied. A few similar studies are known from the literature [1, 2].

MATERIALS AND METHODS

Three adsorbents were tested in this study. They were a) a silicalite synthesized at Neste according to U.S. Pat. 3,702,886, but without aluminium. To get mechanically stable particles, 25 % (w/w) SiO_2 (Ludox HS-40) was added. The surface area was 370 m^2/g, and the particle size 0.5 - 2mm. b) Activated carbon, Norit ROW 0.8, with a surface area of 900 m^2/g and a particle diameter of 0.8 mm, and c) an adsorption polymer XAD-4 made by Rohm and Haas. This is a crosslinked polystyrene type polymer with a surface area of approx. 800 m^2/g and a particle size of 0.5 mm. Before use the silicalite was heated for 2 h at 600 °C, and the XAD-4 polymer was rinsed with pure water.

The adsorption of n-butanol and i-propanol was studied by addition of different amounts of the adsorbent to 20 ml of the alcohol in dilute aqueous solutions, and by analyzing the alcohol content in solution initially and at equilibrium when left over night at 23 °C. The analyses were performed on a Hewlett Packard 5890 GC with a 26 cm long WCOT Fused silica column with CP Wax 52 CB as the liquid phase.

RESULTS AND DISCUSSION

Figures 1 and 2 show the adsorption of n-butanol and i-propanol on the three studied adsorbents. Activated carbon is the best adsorbent. The adsorption of n-butanol from water on silicalite initially increases rapidly, but levels off at the concentration 80 mg/g adsorbent. The channel system in the silicalite is then close to being saturated with the alcohol.

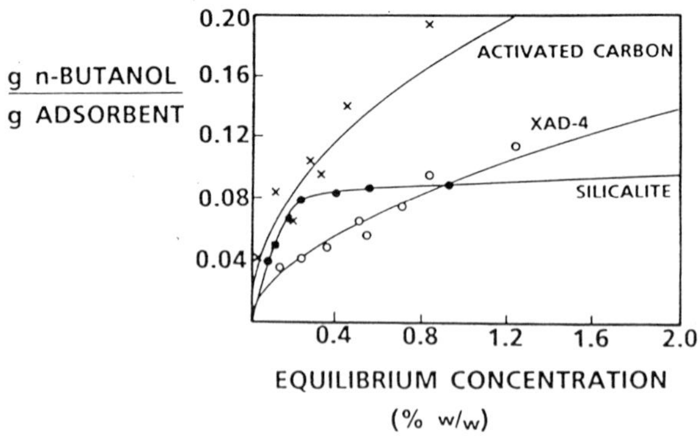

Fig. 1. Adsorption isotherms of n-butanol from aqueous solutions on some adsorbents (at 23°C).

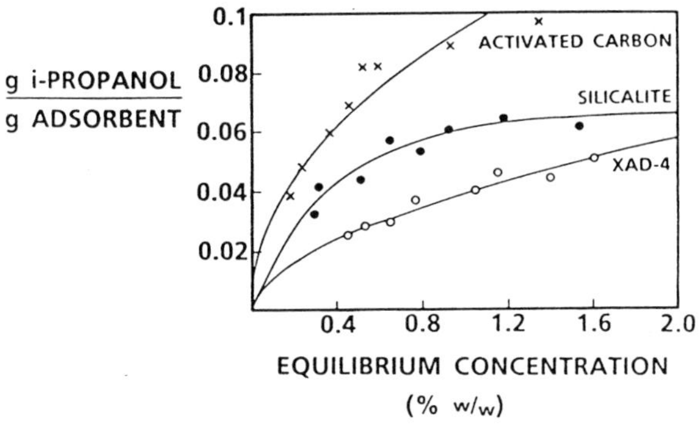

Fig. 2. Adsorption isotherms of i-propanol from aqueous solutions (at 23°C).

The final value around 85 mg/g is in good accordance with other published values [3]. In the case of isopropanol this same saturation is not seen. However, the adsorbed amounts are substantially lower than for butanol. This decreased adsorptivity is ascribed to the reduced organophilic properties of i-propanol. Steric limitations are probably not significant as the values conform with adsorption values for n-propanol [3], and also because adsorption of i-propanol is lower on the other adsorbents which lack the channel structure of silicalite [4].

The Freundlich isotherms have been calculated for the activated carbon and the polymer XAD-4. Silicalite does not adsorb in the same way, which is clearly seen with butanol (Figs. 1 and 2). Table I shows the factors a and b in the expression $y = ax^b$.

Table 1. Experimental constants a and b in the Freundlich isotherm ($y = ax^b$). r^2 is the correlation factor.

Alcohol	Adsorbent	a	b	r^2
n-Butanol	Activated carbon	0.180	0.474	0.84
n-Butanol	XAD-4	0.094	0.468	0.92
i-Propanol	Activated carbon	0.094	0.573	0.94
i-Propanol	XAD-4	0.039	0.535	0.95

In Fig. 3 the competitive adsorption of n-butanol and i-propanol on silicalite is shown. When small amounts of adsorbent are added, butanol is preferentially adsorbed, and its concentration in solution rapidly declines. Only when the butanol concentration is lower than 0.2 % is i-propanol adsorbed. This same concentration for butanol is seen in Fig. 1 as the saturation equilibrium concentration for n-butanol.

The data are also given in Fig. 4 as the adsorption isotherms. The lines connect the equilibrium concentrations of each respective alcohol, and they correspond to 1, 2, 3 and 4 g of added silicalite, respectively, from right to left. Amounts of adsorbed butanol are identical with those in Fig. 1. Initial concentrations of both alcohols were 1 % (w/w).

Figure 5 shows the adsorption on activated carbon. In this case, too, n-butanol is more readily adsorbed. The similar preference for butanol at low adsorbent quantities in not seen, however.

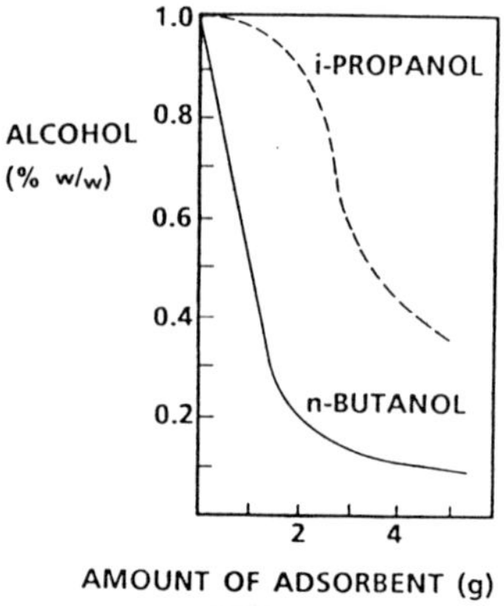

Fig. 3. Competitive adsorption of n-butanol and i-propanol on silicalite.

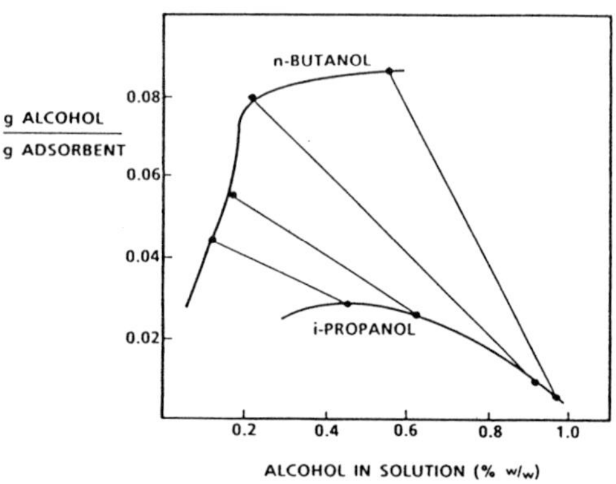

Fig. 4. Adsorption isotherms for competitive adsorption of n-butanol and i-propanol on silicalite. Initial concentration of respective alcohol was 1% (w/w) and the lines represent equilibrium solution concentrations after adsorbent addition.

CONCLUSIONS

No attempts were made in this study to recover the alcohols, i.e. to desorb them. However, it can be seen that for all adsorbents the adsorption of n-butanol is significantly higher than that of i-propanol, and especially with silicalite the adsorption is highly selective. In actual fermentations n-butanol is the major constituent, and also more harmful for bacterial growth. A partial removal of n-butanol from a fermentation liquid thus seems plausible.

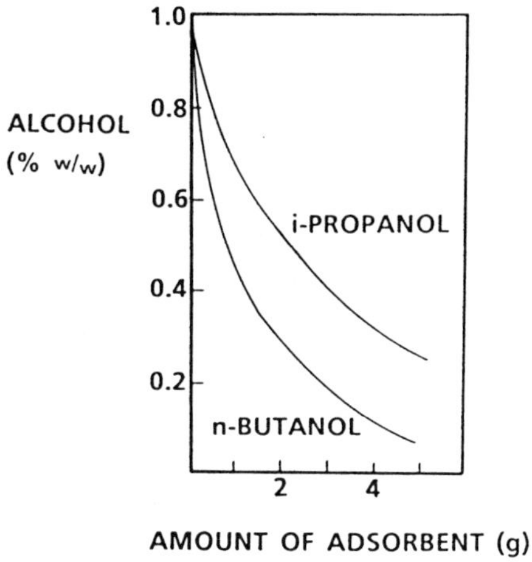

Fig. 5. Competitive adsorption of n-butanol and i-propanol on activated carbon.

REFERENCES

1) Maddox, I., Biotechnol. Lett. **4** (1982) 759-760
2) Larsson, M., Holst, O. and Mattiasson, B., VII Int. Biotechnol. Symp. New Delhi, 1984, Abstr. 1 pp. 216-217.
3) Milestone, N.B. and Bibby, D.M., J. Chem. Tech. Biotechnol. **34 A** (1983) 73-79.
4) Kokotailo, G.T., Lawton, S.L., Olsen, D.H. and Meier, W.M., Nature (London) **272** (1978) 437.

Penetrability of boracol solutions in Scots pine sapwood (Pinus Sylvestris) of varying moisture contents and diffusion periods

J. Bright Amofa – Department of Wood Science and Technology, Moi University, Eldoret, Kenya.

ABSTRACT

Boracol is a novel preservative developed as an alternative method of preserving timber with intermediate technology which can be applied to unseasoned timber. The preservative system consists of boron (disodium octaborate tetrahydrate), the biological toxicant in a water miscible carrier solvent, monoethylene glycol. The system provided wood preservative solutions for diffusion impregnation of wood for on-site application.

The prime advantage of the system over alternative formulations was found to be the potential for the Boracol to establish a dynamic diffusion process as a result of its hydroscopicity, total miscibility with water and a relatively low vapour pressure.

Boracol appeared to penetrate fast into wet wood and to be effective within a wide spectrum of moisture contents.

Moisture contents at or just above the fibre saturation point appeared to be the lower limit for satisfactory treatment.

INTRODUCTION

The literature on diffusion indicates that a truly aqueous solution of chemicals will diffuse into wood the degree of penetration being influenced by wood moisture content, solution concentration, length of post-treatment storage (diffusion period), wood substrate, orientation of wood surface and temperature of the treatment solution.

Preliminary experiments on Boracol suggest a likely validity of these observations for ethylene glycol (EG) based formulations. This study investigates the effect of some of the diffusion variables on Boracol penetration and the performance of Boracol compared with that of aqueous boron solution (Timbor) complying to BWPA Standard 105.

MATERIALS AND METHODS

Conditioning of test blocks

The wood used in this study was kiln-dried Scots pine sapwood of average growth rate (15-19 rings per 25 mm), radially and tangentially orientated and finished to the size of 30 x 30 x 30 mm. The blocks were conditioned to various moisture contents.

At all moisture contents except at the fibre saturation point (FSP), the blocks were vacuum impregnated with water to saturation, coated with 'ABS' co-polymer (Acrylonitrile - butadiene - styrene) in acetone sealant leaving the desired tangential or radial face for a unilateral penetration test, then air dried in a conditioned room of 15% RH and 23°C. At suitable intervals, samples were randomly taken and stored in polythene bags for 24 hours. Sub-samples were taken to determine the moisture content by the oven drying method.

For moisture contents at the FSP, blocks were conditioned in a humidity chamber of 100% RH for constant weight, coated and then returned to the chamber overnight.

All blocks were frozen-stored in polythene bags and later thawed for use.

Preservative – Boracol 40 (B40), patented

The potential formulation consists of the biological toxicant boron (40% polybor, $Na_2B_8O_{13}4H_2O$) in carrier solvents ethylene glycol (51%, EG) and water (9%).

Drop and spread method

Drops of test solution(s) were evenly placed on the exposed surface of each block while resting on a weighing machine and spread over uniformly with a small paint brush until a specified weight of the solution had been registered. The blocks were stored in airtight containers to prevent loss of water due to evaporation from the system.

Penetration assessment

Test blocks were split into four equal parts from the opposite untreated face. Three different cleaved faces were air dried for up to 30 minutes, sprayed generously with a Boron reagent and left for 5 minutes for a bright colour due to the presence of Boron to develop. The colour zone was marked to record the penetration. Two depths (10 mm from the opposite edges) were measured to the nearest 0.5 mm for each of the three cleaved faces per block.

Modified Boron spot test

The procedure according to BWPA (Standard 105), Williams (1968) was modified for spot testing of Boron. 0.12g of curcumin and 6 g of salicylic acid were dissolved in 200 ml of 95% ethanol with 20 ml of concentrated hydrochloric acid by boiling under a reflux condenser for one hour. The solution was cooled and stored in a polythene bottle.

During spot testing Boron changes the yellow curcumin (diferylmethane) to bright red resocyanine. The bright red colour indicates a concentration of not less than 0.2% boric acid.

EXPERIMENTATION

Moisture content x diffusion period x Boracol solutions

Sets of 8 replicated Scots pine blocks of various moisture contents (8%, 15%, 26%, 37% and 55%) were treated with 0.5 g of B10, B20 and B40 solutions of Boracol, (B10, Boracol 40 formulation diluted to 10% polybor with EG; B20, 20% Polybor concentration) by the Drop and Spread method. Each set was stored for the desired periods between 3 to 28 days for unilateral radial penetration. After the desired period had elapsed, test blocks were assessed for

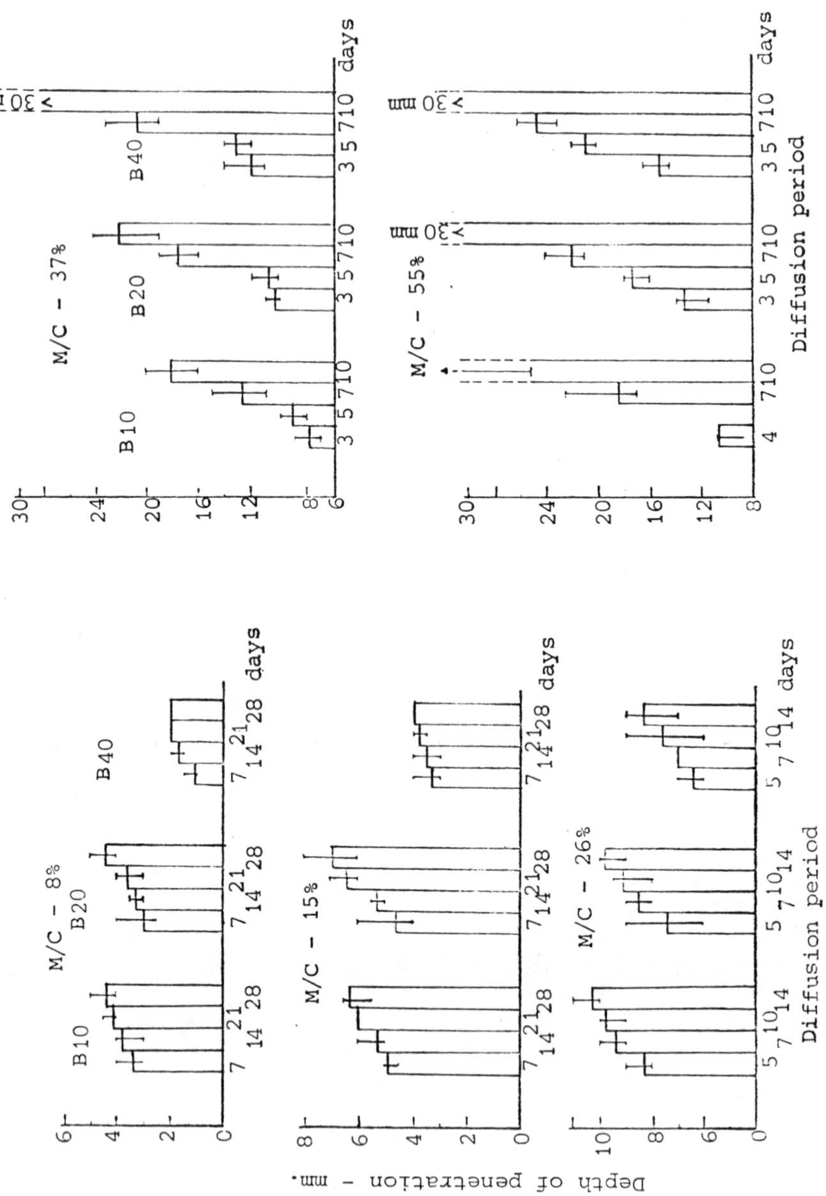

Fig. 1. Illustration of comparative depth of penetration of Boracol solutions.

depth of penetration. There were 48 measurements (2 depths x 3 faces x 8 blocks) per set. The mean and range are presented in Figure 1.

RESULTS AND DISCUSSION

It is evident from Figures 1 and 2 and reported by other workers such as Fuse and Nishimoto (1961), Torres (1969) that dilute solutions penetrate lower moisture content materials below FSP faster than concentrated (more viscous) solutions. The effect of the dilute solutions (B10 and B20) was greater up to 26% moisture content, aided by capillarity and was not increased when the moisture content was above the FSP. The concentration ratio 1:2:4 of B10, B20, B40 made B20 a favourable solution at all moisture contents. The relatively low viscosity and high concentration of B20 made it suitable for flow and diffusion into wood.

The FSP, which occurs within the limits of 25 - 30% in most woods, is illustrated in Figure 1 to coincide with the zone where the effect of concentration on penetration was reversed from B10 > B20 > B40 to B40 > B20 > B10 with increasing moisture content.

Fig. 2. The effect of moisture content on the radial penetration of Boracol solutions into Scots pine in 7 days diffusion period.

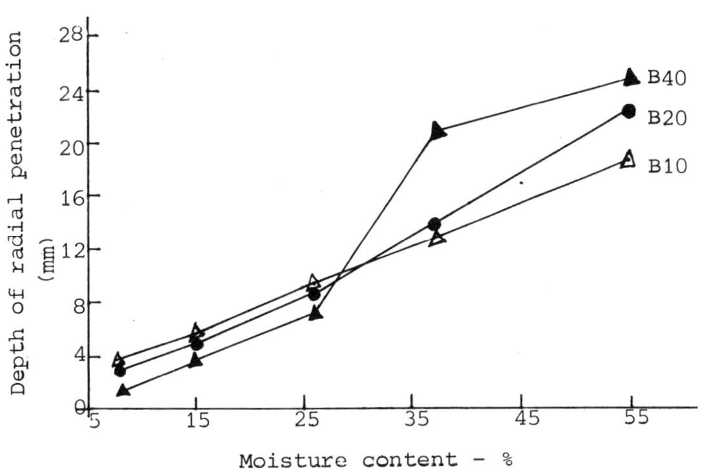

Above the FSP, at 37% and 55% moisture contents, penetration increased so much for concentrated solutions that blocks treated with B40 were completely penetrated well before the 10th day. From moisture contents of 26% to 37%, the average depth of penetration increased about four fold to more than 30 mm and that for B10 and B20 about two fold. Moisture content at 37% or just above the FSP would appear to be a lower limit for satisfactory and economical treatment with Boracol and B40 the most favourable concentration. In contrast, the minimum diffusion storage period for Timborising 25 mm of softwood timber is 4 weeks under the BWPA Standard 105 and that recommended for aqueous Boric acid treatment of 25 mm radiata pine and tawa (Beitschmedia tawa) are 3 weeks and 4 to 6 weeks respectively (Harrow, 1955) in New Zealand.

It was observed that not all of the B40 solution applied to 8% and 15% moisture content blocks had penetrated even at the 28th day. At these moisture contents, the depth of penetration was even obvious and highly marked (before splitting and spraying with the Boron reagent) by the undesirable localised swelling and/or bulking. This observation (Figure 3) indicates the swelling and bulking capability of B40 and the need for complete or deeper penetration with Boracol solutions to prevent this localised effect and may imply unsuitability of Boracol solutions for in situ remedial treatment of wood with moisture content below FSP.

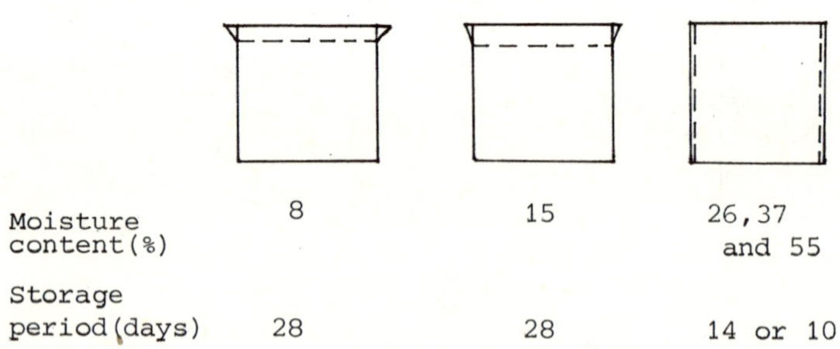

Fig. 3. Diagrammatic representation of Scots pine sapwood applied with Boracol 40 for unilateral penetration indicating localised swelling and bulking of penetrated depth.

It is worth noting that although as much as 5.6 kg/m2 (i.e. 0.5 g per 30 x 30 mm2) of Boracol was applied during the unilateral penetration test, greater quantities could be applied to yield even higher loadings of Boron at ambient temperature. In addition the penetration was determined for a minimum loading of 0.2% or more Boric acid equivalent. The treatments satisfy the requirement of the BWPA Standard 105.

Comparative test: T40 versus B40 at various moisture contents

The rate of penetration of Boracol solutions over a wide range of moisture contents at ambient temperature appears rapid, it was then compared with treatments with Timbor.

Timbor is polybor ($Na_2B_8O_{13} - 4H_2O$) in water and Boracol is polybor in ethylene glycol (EG). Any observable comparable difference is justifiably attributed to the role of the solvents.

Solvents of low molecular weight and low volatility (vapour pressure) at ambient temperature and capable of hydrogen bonding diffuse more readily into wood (and the cell wall matrix) with the toxicants they carry so long as there is no breakdown of the solution phase.

High molecular weight solvents diffuse into wood with difficulty due to molecular size effects. (Tarkow et al., 1966; Stamn, 1964; Bauch, 1964).

Theoretically aqueous Borate solution should diffuse faster than Boracol solution in wood because of fewer interactions.

For general cases of a polar solvent (EG) and ions (Borate) diffusing in a polar medium (water in wood), three interactions may be considered, probably in the following order of importance.

1. Hydrogen bonding between water molecules and the solvent.
2. Competition for water molecules between solvent and surrounding ions.
3. Special dipole field interaction between polar solvent, water and ions.

Their interactions and other gross properties of water and EG in diffusion impregnation were assayed by evaluating the effect of water and EG on penetration and absorption of Polybor in Scots pine sapwood from solutions of T40 (40% Timbor solution) and B40 respectively. Both solutions were maintained at 55-60°C to bring all the Timbor into aqueous solution during treatment.

Sampling and treatment

Four sticks (30 x 30 x 500 mm) were obtained from two selected boards of Scots pine sapwood designated 'A' and 'B' of the same stock. Each stick was cross-cut into 30 mm sizes to give 12 end-matched blocks and numbered sequentially. A T40 treatment block was selected adjacent to that for a B40 to give odd-numbered T40 blocks and even-numbered B40 blocks.

Blocks from the same stick were conditioned together to each of the test moisture contents (7.5%, 20%, 39%, 53%). Boards A and B were allocated for samples of radial and tangential penetrations respectively.

Blocks were immersed with the unsealed faces face down in 500 g of T40 or B40 solutions for 3 minutes and weighed immediately to find the uptake. The treatment was repeated after 24 hours to increase the uptake.

Sets of 6 replicated blocks of the same moisture content group treated with B40 or T40 were stored together to effect penetration for 4 to 14 days. Each block was assessed for depth of penetration. There were 36 measurements (2 depths x 3 faces x 6 blocks) per set.

Statistical analysis

The penetration data were subjected to statistical analysis to assess the reality of the difference between two sets of comparable mean values by the student 't' test.

The significance was judged at a probability level of 99% or p = 0.01. Thus declaring significant any difference having $p^{+} < 0.01$ when the null hypothesis is discredited i.e. the difference is real. p^{+} is the calculated level of significance (Table 3). Such a high significance level was necessary to give sufficient assurance that the observation will be realised in practice.

RESULTS AND DISCUSSION

With the probable exception of the tangential absorption at 53% m/c, there was no significant difference between the uptakes of B40 and T40 at moisture contents above the 20%, as shown in Table 1. At 7.5% m/c more T40 was absorbed than B40 but the latter penetrated further. Water-borne preservative treatments result in relatively high absorption but slight penetration at low moisture content below the FSP (Bryan and Purslow, 1961).

Table 1. Immersion treatments of timbor (T40) and boracol (B40) at 55–60°C.

AVERAGE UPTAKES IN GRAMMES (g)

M/C %		T40		B40	
AV.	RANGE	TAN	RAD	TAN	RAD
53	47-57	0.22(0.07)	0.15(0.06)	0.16(0.00)	0.16(0.02)
39	35-44	0.16(0.04)	0.29(0.06)	0.17(0.00)	0.30(0.12)
20	19-23	0.16(0.02)	0.16(0.02)	0.15(0.02)	0.17(0.03)
7.5	7.3-7.7	0.34(0.05)	0.20(0.06)	0.20(0.06)	0.12(0.01)

Table 2. Average penetrations in millimeters (mm).

	T40		B40		Storage period
M/C %	TAN	RAD	TAN	RAD	(DAYS)
53	10.7(0.7)	10.6(0.7)	11.0(0.4)	11.9(0.6)	4
39	10.4(1.1)	9.8(0.6)	10.2(0.7)	11.0(0.7)	4
20	3.0(0.7)	2.2(0.3)	5.2(0.6)	4.1(0.9)	7
7.5	1.4(0.4)	1.2(0.4)	2.6(0.4)	1.8(0.4)	14

(Standard deviation in brackets. Density of Scots pine at 7.5% m/c = 0.48 ± 0.001g/cc. Uptakes are averages of 6 replicates.)

Below FSP and especially at 7.5% m/c, pure diffusion is minimum as flow by capillarity is superimposed on diffusion. Both solvents being polar, inter-cell wall diffusion motivated by concentration gradient as a result of absorbed solution is eventually set up. At this stage, migration favoured EG of lower vapour pressure. The water carrier of T40 breaks from the Timbor mainly due to the relatively high vapour pressure of water to evaporate. EG is, therefore, a better penetrant than water (Hockman, 1971). The difference in penetration at the 14th day was significant.

Above 20% m/c, it is apparent from Table 2 and confirmed by the significant test that B40 (or EG) penetrated deeper than T40 (or water). At 20% m/c, B40 penetration was about one and a half times to twice as much as T40 (i.e. 2.2 vrs 4.1 for radical penetration and 3.0 vrs 5.2 for tangential penetration).

Table 3. Significance test of the difference between the average penetrations, significant limit, P = 0.01.

	RAD			TAN		
M/C %	T40	B40	p+	T40	B40	p+
53	10.6	11.9	<0.001	10.7	11.0	>0.1
39	9.8	11.0	<0.001	10.4	10.2	0.5
20	2.2	4.1	<0.001	3.0	5.2	<0.001
7.5	1.2	1.8	<0.001	1.4	2.6	<0.001

p+ $\leqslant$ 0.01, significant, p $\geqslant$ 0.01 not significant
p+ = probability of the difference between two sample means

Deeper penetration was observed at higher moisture contents. The free water provided additional continuous medium for diffusion. For radial treatments and in tangential treatments below the 20% moisture content, the differences in penetration was statistically significant, though in practice such differences (1 to 2 mm) are negligible.

Thus at high moisture contents (i.e. above the 39% m/c) it appeared that EG did not provide any significant practical advantage over water under comparable diffusion impregnations. The free water at this level provided equally an effective diffusion medium for the two formulations.

The advantage of the Boracol (B40) over Timbor (T40) was realised at low moisture contents due to the relatively low vapour pressure of ethylene glycol which provides long contact time against evaporation to favour prolonged movement for deeper penetration. In addition, the high concentration at ambient temperature (40% g/g of Polybor in Boracol 40 solution, cf 9.5% g/g of Polybor soluble in water at 25°C) and the reservoir on the wood surface needed to effect rapid diffusion of Boron is obtainable with Boracol.

ACKNOWLEDGEMENTS

This work was carried out at Imperial College of Science and Technology, London, whilst studying for a PhD. My gratitude goes to Professor J.F. Levy, Mr. Carl Beckgaard and Dr. D.J. Dickinson for their invaluable advice during the course of this work.

REFERENCES

1. Bauch, J. (1964) Axial permeability of pine sapwood to aqueous solutions 1, Non-electrolytes. Planta 61(3) 196-208.

2. Bryan J. and Purslow, D.F. "Non-pressure methods of applying wood preservatives". FPL Record No. 31.

3. BWPA, Standard 105. The preservation of softwood timbers by means of 'Timbor' (disodium octaborate tetrahydrate). British Wood Preserving Assoc.

4. Harrow, K.M. (1955). Diffusion impregnation of building timbers with Boron compounds. N.Z. Timber J. For. Res. 2 (3 & 4).

5. Hockman, H. (1971). Effect of solvent on the penetration of wood preservatives by non-pressure process. Proc. Am. Wood Preservers Assoc. 67, 170-173.

6. Fuse, G. and Nishimoto, K. (1961). Studies in the wood preservation by the diffusion process. J. Jap. Wood Res. Soc. 7(5). 183-189 (Eng.Summ).

7. Stamn, A.J. (1964). Swelling in aqueous solution and non-aqueous liquid. Wood and Cellulose Science. Ronald Press. N.Y.

8. Torres, J.J. (1969). The preservation of dry timber by immersion. N.Z. For. Serv. Wellington.

9. Tarkow et al. (1966). Interaction of wood with polymeric materials. Penetration versus molecular size. For. Prod. J. 16 (10) 61-65.

10. Williams, J.A. (1968). Extraction and determination of disodium octabrate in sitka spruce. Analyst 93, 111-115.

Surface analysis of cellulose treated by melamine resin

Li Hua and Per Flodin – Department of Polymer Technology, Chalmers University of Technology 412-96 Göteborg, Sweden.

ABSTRACT

The water absorption of cellulose fibres has been reduced after surface treatment with mono- or di-methylolmelamine (DMM) resins. Surface analysis using transmission electron microscopy (TEM) offered information about the melamine resin retained on the cell wall.

INTRODUCTION

In our previous work, the surface of cellulose fibres was modified by mono- or di-methylolmelamine resins. The water sensitivity of cellulose was thereby reduced (1).

TEM gives an insight into the fine structure. Ultra-microtomy is an important method for providing thin slices of specimens for internal microstructural elucidation. Staining of the ultrathin sections of the polymer can be achieved by using the staining methods for biological materials. (2)
In this study, photos obtained from TEM analysis for melamine resin absorbed on cellulose fibres are presented.

EXPERIMENTAL

Bleached softwood kraft paper was modified by dispersing in 5% DMM, then curing in an oven. The paper increased in weight by 10-12% due to absorption of the resin.

The following cellulose fibres were tested:-

Sample 1. untreated.
Sample 2. cellulose fibres treated by 5% DMM dispersion (1).
Sample 3. cellulose fibres treated by 5% DMM dispersion then immersed in water for one hour, and dried at ambient temperature.

The pretreatment of specimens followed the method for plant materials (3) to some extent, but improvements were made. The test specimens were kept over 1.5% OsO_4 solution for 3.5 hr. at room temperature in order to fix the shape of cellulose fibres. The water in the specimens was extracted progressively by 30% to 99.5% alcohol solutions. The alcohol was displaced progressively by small amounts of a low viscosity epoxy resin according to Spurr (1969). The treated specimens were individually embedded in Spurr epoxy resin, then cured in an oven.

The ultrathin sections of cellulose cross sections were prepared using the ultramicrotome LKB III and had a thickness of 1000-2000 Å. They were stained using 4% uranyl acetate and lead citrate (Reynolds 1963) solution, which made the melamine resin visible in TEM.

The treated slice specimens were observed using the Philips TEM model 301, with a filament voltage of 60-100 KV.

RESULTS AND DISCUSSION

The observations by TEM show the cross sections of various cellulose fibres.

Untreated cellulose fibre possesses thin inner and outer boundary layers in the cell wall (Fig. 1).

The cell wall of sample 2 retains a layer of melamine resin about 0.2 µm thick (Fig. 2). The thickness of melamine resin on cellulose varies with the concentration of the DMM dispersion. The heterogeneity of the process results in the outer surface becoming more modified than the inner surface.

The surface of melamine resin layer of sample 3 was found to be rough (Fig. 3). Some melamine resin on the surface was washed away possibly due to swelling or mechanical factors.

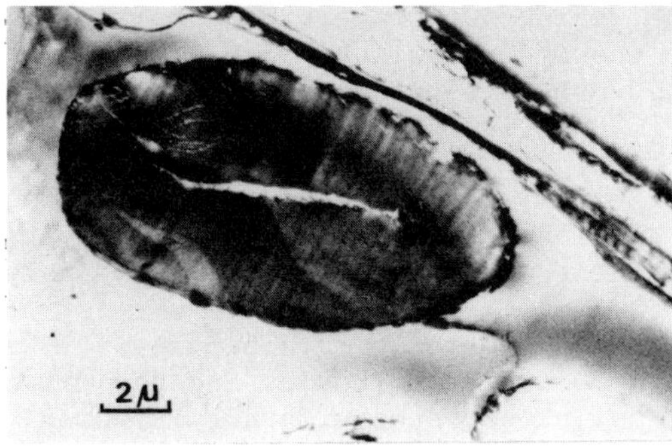

Fig. 1. Cross section of sample 1.

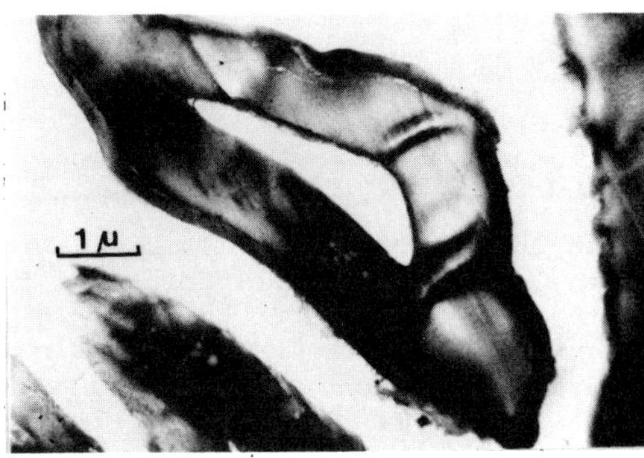

Fig. 2. Cross section of sample 2.

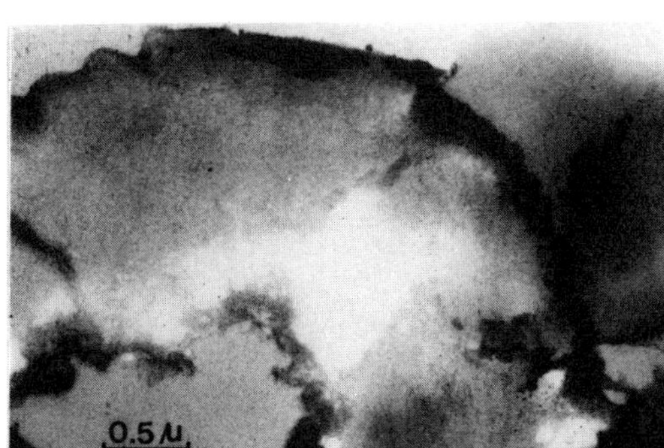

Fig. 3. Cross section of sample 3.

Melamine resin that penetrates into the cell wall is retained because the DMM crosslinks with cellulose molecules through the OH groups hence reducing the water sensivity.

CONCLUSION

(1) The thickness of melamine resin on the surface of cellulose can be measured by TEM analysis.

(2) The melamine resin retained in cell wall crosslinks with cellulose molecules and the water sensitivity of cellulose is considerably reduced.

ACKNOWLEDGEMENTS

The authors wish to express their thanks to Mr. Z.H. Lai, Dept. of Physics, Chalmers Univ. of Technology for TEM analysis, and Ms. Lena Töyrä, Dept. of Botany, Univ. of Göteborg, for making ultrathin section specimens.

REFERENCES

(1) Li Hua and Per Flodin, Cellulose Fibres Treated by Dimethylolmelamine as Reinforcement in Cellulose-Unsaturated Polyester Composites, in press.

(2) Willard C. Hamilton, Applied Polymer Symposium, No. 19, 105 (1972)

(3) Christer Sundqvist and Hans Rybert, Physiol. Plant. 47: 124-128, (1979).

List of participants

J B Amofa	Wood Science and Technology Department, Moi University, Box 3900, Eldoret, Kenya.
D M W Anderson	Chemistry Dept, University of Edinburgh, West Mains Road, Edinburgh.
B A K Andrews	Southern Regional Research Centre, New Orleans, LA 70179, USA.
F S Baker	Royal Ordnance, R + D Centre, Stewardstone Road, Waltham Abbey, Essex, EN9 1AY.
M. Bailey	Biotechnical Lab, Technical Research Centre of Finland, Tietotie 2, SF-02150 Espoo, Finland.
P. Bale	Hercules Ltd, 20 Red Lion Street, London, WC1R 4PB.
W B Banks	Dept of Forestry and Wood Science, University College of North Wales, Bangor, Gwynedd, LL57 2UN.
R Bates	Hercules Ltd, 20 Red Lion Street, London WC1R 4PB.
R. Beatson	Pulp and Paper Research Institute of Canada, 570 St. John's Blvd, Pointe Claire, Quebec, Canada, H9R 3J9.
H Beckers	Hercules BV, PO Box 71, 3330 AB Zwijndrect, The Nerthelands.
C Birkinshaw	National Institute for Higher Education, Limerick, Ireland.
A Blazej	Slovak Technical University, 812 43 Bratislava, Czechoslovakia.
A J Bolton	Dept of Forestry and Wood Science, University College of North Wales, Bangor, Gwynedd, LL57 2UN.
A D J Broome	Laporte Industries Ltd, Moorfield Road, Widnes, Cheshire, WA8 OJU.

C Bucke	School of Biotechnology, Polytechnic of Central London, 115 New Cavendish Street, London, W1M 8JS.
R Burley	Courtaulds Research, PO Box 5, Spondon, Derby, DE2 7BP.
R Burton	Dept of Chemical Engineering, Loughborough University, Loughborough, Leics, LE12 8QJ.
C M Cahalan	Dept of Forestry and Wood Science, University College of North Wales, Bangor, Gwynedd, LL57 2UN.
D G Calver	Berol Kemi (UK) Ltd, 23 Grosvenor Road, St. Albans, Herts, AL1 3AW.
A Chanse	Rm 262, Old Medical School, Newcastle University, Newcastle, NE1 7JU.
P Chipperfield	ICI Paints Division, Wexham Road, Slough, SL2 5DS.
S Coleman-Kammula	Shell Research, Badhuisweg 3, 0131 CM Amsterdam, The Netherlands.
C Colin	Spontex, 74 Rue St Just des Marais, BP 309, 60026 Beauvais, Cedex, France.
B J Collier	Tupper Hall 208, Ohio University, Athens, Ohio 45701, USA.
J R Collier	Stocker Engineering Centre 176, Ohio University, Athens, Ohio 45701, USA.
P Collins	ICI Paints Division, Wexham Road, Slough, SL2 5DS.
J Comyn	Chemistry Dept, Leicester Polytechnic, PO Box 143, Leicester, LE1 9BH.
A Conner	US Forest Products, One Gifford Pinchot Drive, Madison, WI 53705-2398, USA.
J J W Coppen	Tropical Development and Research Institute, 56-62 Grays Inn Road, London, WC1X 8LO.
L Dahlgren	Berol Kemi (UK) Ltd, Box 851, 44401 Stenungsund, Sweden.

B Delval	Spontex, 74 Rue St Just des Marais, BP 309, 60026 Beauvais Cedex, France.
A Detemmerman	BP Chemicals Belgium, Scheldedijk 50, B-2730 Zwijndrecht, Belgium.
S Dixon	D + A Research K80, Nobels Explosives Ltd, ICI, Ardeer, Stevenson, Ayrshire, KA20 3LN.
M Diamantoglu	ENKA, Kolpingstr. 4, D-8765 Erlenback/M, West Germany.
S E Dobson	Hercules Ltd, 20 Red Lion Street, London, WC1R 4PB.
R Donges	Kalle AG, Rheingaustr. 190, D-6200 Wiesbaden, West Germany.
K Edelmann	R + D Dept, Metsaliiton Teolisuus OY, SF 44100 Aanekoski, Finland.
M El Kashouti	National Research Centre, Dokki, Cairo, Egypt.
M Ellerby	Courtaulds plc, 101 Lockhurst Lane, Coventry, CV6 5RS.
A S El Nawawy	Kuwait Institute for Scientific Research, PO Box 24885, Safat 13109, Kuwait.
K Engelskirchen	Henkel KGaA, TFC/Makromolekulare Chemie, Postfach 1100, D-4000 Dusseldorf 1, West Germany.
W C Feist	US Dept of Agriculture, Forest Products Lab, One Gifford Pinchot Drive, Madison, WI 53705, USA.
R Fernandes	Pira, Randalls Road, Leatherhead, Surrey, KT22 7RU
A H K Fowler	Courtaulds plc, PO Box 111, Lockhurst Lane, Coventry, CV6 5RS.
G J Francis	Forestry Commission, 231 Corstorphine Road, Edinburgh, EH12 7AT.
U P Gandhi	Applied Chemistry Dept, UWIST, PO Box 13, Cardiff, CF1 3XF.

K Geddes	Crown Decorative Products, Crown House, Hollins Road, Darwen, Lancs. BB3 OBG.
A M Gibbons	Dept of Pure and Applied Chemistry, University of Strathclyde, 295 Cathedral Street, Glasgow, G1 1XL.
E J Gibson	Forestry and Wood Science Dept, University College of North Wales, 9 Amersham Hill Gardens, High Wycombe, Bucks, HP13 6QP.
J A Gibson	Nobels Explosives plc, PSO Ardeer, Stevenston, Ayrshire.
D N Gore	Carri-Med, Poynings, Pilgrims Close, Westhumble, Surrey, RH5 6AR.
J J Gray	Hercules Ltd, 20 Red Lion Street, London WC1R 4PB.
A Grayson	Forestry Commission, Alice Holt Lodge, Wrecclesham, Farnham, Surrey, GU10 4LH.
M Graziosi	Fratelli Lamberti SpA, Via Piave 18, 21014, Albizzate Va, Italy.
T Greenway	Berol Kemi (UK) Ltd, 23 Grosvenor Road, St. Albans, Herts, AL1 3AW
P J F Griffiths	Applied Chemistry Dept, UWIST, PO Box 13, Cardiff, CF1 3XF.
F Guttfield	Finnforest (Chemicals) Ltd, Old Court House, Hughenden Road, High Wycombe, Bucks.
J Harun	University Pertanian Malaysia, Serdang, Malaysia.
H Hatakeyama	Industrial Products Research Institute, 1-1-4 Yatabe Higashi, Tsukuba, Ibaraki 305, Japan.
T Hatakeyama	Research Institute for Polymers and Textiles, 1-1-4 Yatabe Higashi, Tsukuba, Ibaraki, 305, Japan.
T Hattula	The Finish Pulp and Paper Research Institute, PO Box 136, SF-00101 Helsinki, Finland.

J Hayashi	Faculty of Engineering, Hokkaido University, Kita-ku N1 3W8, Sapporo 060, Japan.
A Hilditch	Cuprinol Ltd, Adderwell, Frome, Somerset, BA11 1NL
J R Holker	Shirley Institute, Didsbury, Manchester, M20 8RX.
T Hongu	Nisshinbo Industries Inc, No 3-10 Nihonbashi, Yokoyama-cho, Chou-ku, Tokyo, Japan 103.
H Hornmeyer	Institut fur Radiochemie der Universtitat Innsbruck, Innrain 52a, A-6020 Innsbruck, Austria.
P Howard	Chemistry Dept, University of Surrey, Guildford, Surrey, GU2 5XH.
Li Hua	Dept of Polymer Technology, Chalmers University of Technology, S-41296, Gottenburg, Sweden.
H Hughes	Research Division, The North East Wales Institute, Connah's Quay, Deeside, Clwyd, CH5 4BR.
M Huglin	Dept of Chemistry and Applied Chemistry, University of Salford, Salford, M5 4WT.
S Huybrechts	BP Chemicals Belgium, Scheldedijk 50, B-2730 Zwijndrecht, Belgium.
R Hwang	2466 Webb Avenue, Bronx, NY 10468, USA.
H Inagaki	Institute for Chemical Research, Kyoto University, Uji, Kyoto 611, Japan
T W Jeffries	Institute for Microbial and Biochemical Technology, Forest Products Lab, One Gifford Pinchot Drive, Madison, Wisconsin, USA.
J D Jenkins	Dow Chemicals Europe, Badenstr. 1701, 8036 Zurich, Switzerland.
E B G Jones	Dept of Biological Sciences, Portsmouth Polytechnic, King Henry Building, King Henry I Street, Portsmouth, PO1 2DY.
M A Kalnins	US Forest Products Lab, One Gifford Pinchot Drive, Madison, Wisconsin 53705, USA.

K Kamide	Fundamental Research Lab of Fibres, Asahi Chemical Industries Co, 11-7 Hacchonawate, Takatsuki, Osaka 569, Japan.
M Karasawa	Dept of Polymer Chemistry, Yamagata University, Jonan 4-3-16, Yonezawa, 992 Japan.
J F Kennedy	Department of Chemistry, University of Birmingham, Birmingham, B15 2TT.
S W Kim	Dept of Chemical Engineering, University of Birmingham, Birmingham, B15 2TT.
H Klusik	Wolff Waltstrode AG, Postfach 3030, Walstrode 1, West Germany.
P Laity	Courtaulds Research Laboratories, 101 Lockhurst Lane, Coventry, CV6 5RS.
T J Lewis	Ministry of Defence, RARDE, Waltham Abbey, Essex.
M Linko	Biotechnology Lab, Technical Research Centre of Finland, Tietotie 2, SF-02150 Espoo, Finland.
P E Linnett	Shell Research Ltd, Sittingbourne Research Centre, Sittingbourne, Kent, ME9 8AG.
R Loveland	Director of Wintech, Pearl House, Grayfriars Road, Cardiff, CF1 2YX.
G Marshall	Courtaulds Acetate, PO Box 5, Spondon, Derby DE2 7BP.
G Meister	Lenzing AG, Abtlg, Forschung und Entwicklung, A-4860 Lenzing, Austria.
E R Miller	Princes Risborough Lab, Princes Risborough, Aylesbury, Bucks, HP17 9PX.
K K Mistry	Allied Colloids Ltd, Low Moor, Bradford, W. Yorks.
T Miyamoto	Institute for Chemical Research, Kyoto University, Uji, Kyoto 611, Japan.
N Miyata	Nakamura Gakuen College, 5-7-1 Befu, Jonan-ku, Fukuoka, 814 Japan.

R Mouzouras	School of Biological Sciences, Portsmouth Polytechnic, King Henry I Street, Portsmouth, PO1 2DY.
H S Munro	Dept of Chemistry, University of Durham, South Road, Durham, DH1 3LE.
M Musallam	Kuwait Institute for Scientific Research, PO Box 24885, Safat 13109, Kuwait.
K Nakamura	Industrial Research Institute of Kanagawa Perfecture, 3173 Showa-machi, Kanuzawa-ku, Yokohama, 236 Japan.
Y Nakamura	Gunma University, Kiryu, Gunma, Japan.
J Nakano	Dept of Forest Products, University of Tokyo, 1-1-1 Yayoi, Bunkyo-ku, Tokyo, Japan 113.
H Namikoshi	Daicel Chemical Industries, 1239 Shinzaike, Alioshi-ku, Kimeji, Hyoga, Japan.
S J O'Donohue	Dept of Chemistry and Applied Chemistry, University of Salford, Salford, M5 4WT.
W B Painting	Surfactants Dept, Hoechst UK Ltd, Hoechst House, Salisbury Road, Hounslow, TW4 6JH.
G S Park	Applied Chemistry Dept, UWIST, PO Box 13, Cardiff, CF1 3XF.
E M Pearce	Fosroc Ltd, Fieldhouse Lane, Marlow, Buckinghamshire, SL7 1LS.
A Peguy	CERMAV-CNRS, BP 68, 38042, St. Martin d'Heres, France.
C Persson	Berol Kemi AB, Box 851, 44401 Stenungsund, Sweden.
R A Pethrick	Dept of Pure and Applied Chemistry, University of Strathclyde, 295 Cathedral Street, Glasgow, G1 1XL.
C Phanopoulus	School of Chemistry, Leicester Polytechnic, PO Box 143, Leicester, LE1 9BH.
G O Phillips	The North East Wales Institute, Connah's Quay, Deeside, Clwyd, CH5 4BR.

G Privett	Royal Ordnance, R + D Centre, Stewardstone Road, Waltham Abbey, Essex, EN9 1AY.
W Pusch	Max Planck Institut fur Biophysik, Kennedy-Allee 70, D-6000 Frankfurt am Main, West Germany.
F Redman	Oil and Colour Chemists Association, Crown Decorative Products, Crown House, Hollins Road, Darwen, Lancs.
A Reveley	Courtaulds Chemicals and Plastics, PO Box 5, Spondon, Derby, DE2 7BP.
O Rogues	Spontex, 74 Rue St Just des Marais, BP 309, 60026 Beauvais Cedex, France.
L Rose	Courtaulds Chemicals and Plastics, PO Box 5, Spondon, Derby, DE2 7BP.
M Saito	Fundamental Research Lab of Fibres, 11-7 Hacchonawate, Tatatsuki, Osaka 569, Japan.
A Sarko	Dept of Chemistry, State University of New York, Syracuse, NY 13210, USA.
M Scandola	University of Bologna, Via Selmi 3, 40126, Bologna, Italy.
H Schneider	Division of Biological Sciences, National Research Council of Canada, Ottawa, Canada, K1A OR6.
D Schweizer	Henkel KGaA, D-4000 Dusseldorf, Postfach 1100, West Germany.
J-F Selin	Neste OY, Technology Centre, SF-06850, Kullo, Finland.
M Sellar	Hercules Ltd, 20 Red Lion Street, London, WC1.
A Service	Berol Kemi (UK) Ltd, 23 Grosvenor Road, St. Albans, Herts, AL1 3AW.
S Shackleford	US Dept of the Air Force, European Office of Aerospace Research and Development, 223/231 Old Marylebone Road, London NW1 5TH.
R D Short	Dept of Chemistry, University of Durham, South Road, Durham DH1 3AW.

V R Srinivasan	Dept of Microbiology, Louisianna State University, Baton Rouge, LA 70803, USA.
R W Stephens	Chelsea College, 49 West Hill Road, Wandsworth, London, SW18 1LE.
P Stephenson	3M UK plc, European Technical Centre, Gorseinon Road, Swansea, SA4 1GD.
S Suto	Dept of Polymer Chemistry, Yamagata University, Jonan 4-3-16, Yonezawa, 992 Japan.
M Takai	Hokkaido University, Sapporo, Hokkaido, 060 Japan.
R Tanaka	Oji Paper Co Ltd, 1-10-6 Shinonome, Kotu-ku, Tokyo, 135 Japan.
T C Theobald	Courtaulds Research Ltd, 101 Lockhurst Lane, Coventry, CV6 5RS.
D A Thompson	Forestry Commission, Alice Holt Lodge, Wrecclesham, Farnham, Surrey, GU10 4LH.
B Townsend	Courtaulds Research Laboratories, 101 Lockhurst Lane, Coventry, CV6 5RS.
H T'Sas	Hercules BV, PO Box 71, 3330 AB Zwijndrecht, The Netherlands.
C Viney	Dept of Metallurgy and Materials Science, Cambridge University, Pembroke Street, Cambridge, CB2 3QZ.
L Wallbacks	University of Umea, 90187 Umea, Sweden.
A F A Wallis	Division of Chemical and Wood Technology, CSIRO, Private Bag 10, Clayton, Vic 3168, Australia.
R A J Warren	Institut fur Biologie III, Albert Ludwigs Universitat, Schenzlestr. 1, D-7800 Freiburg i. Br, Federal Republic of Germany.
J O Warwicker	2 Stuart Avenue, Marple, Cheshire, SK6 6JX.
D A J Wase	Dept of Chemical Engineering, University of Birmingham, PO Box 363, Edgbaston, Birmingham, B15 2TT.

H West	Dept of Forestry and Wood Science, University College of North Wales, Bangor, Gwynedd
C A White	Biotechnics, 1 Imperial Avenue, Kidderminster, Worcs, DY10 2RA.
R N Wilkins	Dept of Agricultural Biology, University of Newcastle-upon-Tyne, Newcastle-upon-Tyne, NE1 7RU.
P A Williams	Research Division, The North East Wales Institute, Connah's Quay, Deeside, Clwyd, CH5 4BR.
R A Young	Dept of Forestry, University of Wisconsin, 1630 Linden Drive, Madison, WI 53705, USA.
P. Zugenmaier	Institut fur Physikallsche Chemie, Universitat Clausthal, Adolf Rome Strasse 2A, D-3392 Clausthal-Zellerfeld, W. Germany.

Subject index

A

D

E

W

X

Y